HIGHWAY ENGINEERING

SIXTH EDITION

HIGHWAY ENGINEERING

Paul H. Wright

With Contributions By
James S. Lai
Peter S. Parsonson
Michael Meyer
Georgia Institute of Technology

JOHN WILEY & SONS, INC.
New York • Chichester • Brisbane • Toronto • Singapore

ACQUISITIONS EDITOR Cliff Robichaud
ASSISTANT EDITOR Catherine Beckham
MARKETING MANAGER Debra Riegert
PRODUCTION EDITOR Tony VenGraitis
DESIGN SUPERVISOR Ann Marie Renzi
MANUFACTURING MANAGER Dorothy Sinclair
ILLUSTRATION COORDINATOR Eugene Aiello
PRODUCTION SERVICE Ruttle, Shaw & Wetherill, Inc.

This book was set in Times Ten Roman by Ruttle, Shaw & Wetherill, Inc., and printed and bound by Hamilton Printing Company. The cover was printed by Lehigh Press.

Recognizing the importance of preserving what has been written, it is a policy of John Wiley & Sons, Inc. to have books of enduring value published in the United States printed on acid-free paper, and we exert our best efforts to that end.

The paper on this book was manufactured by a mill whose forest management programs include sustained yield harvesting of its timberlands. Sustained yield harvesting principles ensure that the number of trees cut each year does not exceed the amount of new growth.

Library of Congress Cataloging-in-Publication Data
Wright, Paul H.
    Highway Engineering / by Paul H. Wright, with contributions by
James S. Lai . . . [et al.]. — 6th ed.
        p.   cm.
    Includes index.
    ISBN 0-471-00315-8 (cloth : alk. paper)
    1. Highway engineering.   I. Title.
TE145.W74   1996
625.7—dc20   95-20035
                                                                            CIP

Printed in the United States of America

10 9 8 7 6 5 4 3

PREFACE

This book is successor to the book by the late L. J. Ritter and Professor R. J. Paquette, which was first published in 1951. The three editions of that publication served as a valuable textbook for thousands of students over a period of 28 years. The fourth and fifth editions, which I prepared, appeared in 1979 and 1987, respectively.

My goal in this, the sixth edition, is to picture the broad field of highway engineering as it is currently practiced and as it is likely to be for several years to come. While giving appropriate emphasis to fundamentals of highway engineering design, my colleagues and I have attempted to include materials in areas of emerging significance.

Readers familiar with the previous edition will find a better organized and more interesting chapter on transportation planning and additional material on social and environmental impacts. Faculty members who wish to treat the latter subject in greater depth will be provided (in the Instructor's Manual) a case study of Atlanta's controversial Freedom Parkway Project, which, because of public opposition, extended over 30 years.

There is a more substantial treatment of traffic engineering and new material on emerging technologies in traffic control. The material on computer applications in highway engineering has been made a part of the chapter on geometric design of highways. A new chapter on roadside design has been added. The chapters on highway materials, flexible pavement design, and concrete pavements have been thoroughly edited and updated.

In recognition of the changes brought about by the Omnibus Trade and Competitiveness Act of 1988, dual units are used, but metric units appear as the primary units.

I am indebted to three of my colleagues at Georgia Tech who made important contributions to this edition. Dr. Michael Meyer revised Chapter 3, Transportation Planning, and contributed new material on social and environmental impacts. Dr. Peter Parsonson revised and updated Chapter 12, Traffic Engineering, and suggested some changes on geometric design. Dr. James Lai did a thorough revision of Chapters 15, 16, and 20, which deal with materials and pavements. Without the help of these friends, I would never have completed the book on time.

Many others made significant contributions to this edition of the book. Mr. Jeffrey Davis revised Chapter 7, Geometric Design, improving it in many ways. Several Georgia Department of Transportation professionals were most helpful. Mr. James Kennerly, Mr. Mike Gannaway, and Mr. Tony Duke helped with the transition to metric units and arranged for the preparation of the typical plan and profile and cross-section sheets. Mr. Jeff Holt prepared those sheets. Mr. Bascombe Hughes and Mr. Warren Beverly reviewed Chapter 13 and made recommendations to modernize those sections dealing with surveying, photogrammetry, and mapping. Mr. Johnny Morris, Hydraulic Engineer, Federal Highway Administration, made helpful suggestions and provided reference materials for the revision and updating of Chapter 11. For all of these contributions, I am most grateful.

In addition, many organizations and individuals supplied information, photographs, and sketches for this book. Direct credit is given in the text for these sources.

Finally, I wish to thank Cliff Robichaud and Catherine Beckham, John Wiley & Sons, Inc., for their encouragement and support.

Paul H. Wright

CONTENTS

CHAPTER 1

INTRODUCTION

In the United States and other developed nations, the influence of the motor vehicle is both profound and pervasive. More than 13 million U.S. workers (1 in every 7) are employed in the manufacture, distribution, service, and commercial use of motor vehicles. Highway passenger and freight transportation expenditures in 1992 exceeded $840 billion or about 14 percent of the total gross domestic product (1). Twenty-five percent of all U.S. retail sales are automotive.

By 1992, there were over 144 million automobiles and 45 million trucks and buses registered in the United States, which comprise 32 percent and 33 percent of the world totals, respectively (2). Motor carriers operating on the nation's highways transport 43.4 percent of the intercity tonnage of freight and account for 27.2 percent of the total intercity ton-miles (3). Highway transportation overwhelmingly dominates the transportation of people, accounting for 94.3 percent of all person trips. In 1993, highway travel in the United States exceeded 2.2 trillion vehicle-miles. (1)

Good highways are so interwoven with every phrase of our daily activities that it is almost impossible to imagine what life would be like without them. We depend on highways for the movement of goods, for travel to and from work, for services, for social and recreational purposes, and for many other activities necessary to the functioning of our complex society. The planning, design, construction, operation, and maintenance of highways depend largely on the efforts of the highway engineer, who must translate the desires of the people for better highway transportation into physical being.

THE DEVELOPMENT OF HIGHWAYS

Before beginning the study of highway engineering, it is interesting and informative to briefly trace the development of highways and highway systems from early historical periods to the modern era.

1-1 Ancient Roads

The great highway systems of our modern civilization have their origin in the period before the dawn of recorded history. Even before the invention of the wheel, which is popularly supposed to have occurred some 10,000 years ago, individual and mass movements of people undoubtedly took place. The earliest travel was on foot; later, pack animals were utilized, crude sleds were developed, and simple wheeled vehicles came into being. Many of the migrations of the

1

early historical period involved large numbers of people and covered relatively great distances. More or less regularly traveled routes developed, extending to the limits of the then-known world.

As various civilizations reached a higher level, many of the ancient peoples came to a realization of the importance of improved roads. The streets of the city of Babylon were paved as early as 2000 B.C. History also records the construction of a magnificent road to aid in the building of the Great Pyramid in Egypt nearly 3,000 years before the birth of Christ. Traces of early roads have been found on the island of Crete, and it is known that the early civilizations of the Chinese, Carthaginians, and Incas also led to extensive road building.

By far the most advanced highway system of the ancient world was that of the Romans. When Roman civilization was at its peak, a great system of military roads reached to the limits of the empire. Many of these roads were built of stone and were 3 ft or more in thickness. Traces of this magnificent system are still in existence on the European continent; in fact, some of these roads still serve as bases for sections of modern highways.

After the decline and fall of the Roman Empire, road building, along with virtually all other forms of scientific activity, practically ceased for a period of 1,000 years. Even as late as the early part of the eighteenth century, the only convenient means of travel between cities was on foot or on horseback. Stage-coaches were introduced in 1659, but travel in them proved exceedingly difficult in most instances because of the extremely poor condition of rural roads.

1-2 Later European Development

Interest in the art of road building was revived in Europe in the late eighteenth century. During this period, Trésaguet, a noted French engineer, advocated a method of road construction utilizing a broken-stone base covered with smaller stones. The regime of Napoleon in France (1800–1814) gave a great impetus to road construction, chiefly for military purposes, and led to the establishment of a national system of highways in that country.

At about the same time in England, two Scottish engineers, Thomas Telford and John L. McAdam, developed similar types of construction. Telford urged the use of large pieces of ledge stone to form a base with smaller stones for the wearing surface. McAdam advocated the use of smaller broken stone throughout. This latter type of construction is still in extensive use, being the forerunner of various types of modern macadam bases and pavements.

DEVELOPMENT OF HIGHWAYS IN THE UNITED STATES

During the early history of the American colonies, travel was primarily local in character, and rural roads were generally little more than trails or cleared paths through the forests. (See Figure 1-1.) Toward the end of the eighteenth century, public demand led to the improvement of various roads by private enterprise. These improvements generally took the form of toll roads or "turnpikes" and were located principally in areas adjacent to the larger cities. The first American turnpike made use of existing roads in Virginia to connect Alexandria with settlements in the Blue Ridge Mountains. Another important toll road of this period was the Philadelphia and Lancaster Turnpike Road, from Philadelphia to Lancaster, Pennsylvania.

FIGURE 1-1 During the pre-Revolutionary War period, the journey from Philadelphia to New York required 2 days; today, it requires only 2 hours. (Courtesy Federal Highway Administration.)

Another famous early road in American history was the Wilderness Road, blazed by Daniel Boone in 1775, which led from the Shenandoah Valley in Virginia through the Cumberland Gap into Kentucky.

In 1806 the federal government entered the field of highway construction for the first time with the authorization of construction of the National Pike, or Cumberland Road. The first contract on this route was let in 1811, and by 1816 the road extended from Cumberland, Maryland, the terminus of the Chesapeake and Ohio Canal, to Wheeling, West Virginia. When construction was completed (1841), it had reached Vandalia, Illinois. This route, which was surfaced largely with macadam, was nearly 800 miles in length and was built at a cost of approximately $7 million.

During the middle of the nineteenth century, the great western expansion of the country began, and many roads or "trails" figured prominently in this development. Probably the most famous of the great trails was the Oregon Trail. This route began at Independence, Missouri, went overland to the Platte River, near Grand Island, Nebraska, and thence along the Platte River and its North Fork to Fort Laramie, Wyoming. Continuing westward, the trail followed the Sweetwater and crossed the Continental Divide through South Pass, in Wyoming. West of this point travelers followed the path of the Snake River into Oregon. Other famous western roads, each of which has a fascinating history, included the Santa Fe Trail, the Mormon Trail, the California Trail, and the Overland Trail.

In 1830, however, as the steam locomotive demonstrated its superiority over horse-drawn vehicles, the interest in road building began to wane. By the time the first transcontinental railroad (the Union Pacific) was completed in 1869, road-building activities outside the cities virtually ceased until near the beginning of the twentieth century.

By 1900 a strong popular demand for highways again existed. The principal demand came from farmers, who clamored for farm-to-market roads so that they might more readily move their agricultural products to the nearest railhead. Some demand was also being felt for improved routes connecting the larger centers of population. During this period certain states began to recognize the need for state financial aid for road construction. The first state-aid law was enacted by New Jersey in 1891, and by 1900 six other states had enacted similar legislation.

1-3 Advent of the Motor Vehicle

The year 1904 marked the beginning of a new era in highway transportation in America with the advent of motor vehicles in considerable numbers. Almost overnight an enormous demand was created for improved highways, not only for farm-to-market roads but also for through routes connecting the metropolitan areas. Additional state-aid laws were enacted, and by 1917 every state participated in highway construction in some fashion. By this time, also, most states had established some sort of highway agency and had delegated to these bodies the responsibility for the construction and maintenance of the principal state routes. Figure 1-1 shows the evolution of one U.S. road from those early days to the present time.

This period was also marked by radical changes in road construction methods, particularly with regard to wearing surfaces. The early roads of American history were largely natural earth. Since timber was readily available, plank and corduroy roads were numerous, while later, wood blocks were used. Some gravel was used on early surfaces, while many city streets were paved with cobblestones. The invention of the power stone crusher and the steam roller led to the construction of a considerable mileage of broken-stone surfaces. Development of this type of surface generally paralleled that in Europe.

In the cities relatively large concentrations of wheeled vehicles and the need for abatement of noise and dust brought various improved surfaces into being. The first brick pavement in this country is supposed to have been built in Charleston, West Virginia, in 1871, and asphalt was used for paving Pennsylvania Avenue, in Washington, D.C., in 1867. Concrete pavements were introduced about 1893, and the first rural road of concrete was built in 1909 in Wayne County, Michigan.

1-4 Development of the Federal-Aid Program

World War I intensified highway problems in the United States and led the federal government to again actively enter the field of highway construction. Actually federal participation in highway affairs on a continuing basis began in 1893, when Congress established the Office of Road Inquiry, an agency whose work was primarily educational in nature. This agency later became the Bureau of Public Roads of the Department of Agriculture, and in 1939 it became the Public Roads Administration under the Federal Works Agency. In 1949 this agency was transferred to the Department of Commerce, and its name again changed to the Bureau of Public Roads. In 1967 a National Department of Transportation (DOT) was established, and the Bureau of Public Roads was made a part of the Federal Highway Administration within that Department. The name Bureau of Public Roads was dropped in 1970, and the organization was merged with the Federal Highway Administration (FHWA).

The modern era of federal aid for highways began with the passage of the Federal Road Act of 1916, which authorized the expenditure of $75 million over a period of five years for improvements of rural highways. The funds appropriated under this act were apportioned to the individual states on the basis of area, population, and mileage of rural roads, the apportioning ratios being based on the ratio of the amount of each of these items within the individual state to similar totals for the country as a whole. States were required to match the federal funds on a 50-50 basis, with sliding scale in the public land states of the West.

The Federal-Aid Highway Act of 1921 extended the principle of federal aid in highway construction and strengthened it in two important respects. First, it required that each state designate a connected system of interstate and intrastate routes, not to exceed 7 percent of the total rural mileage then existing within the state, and it further directed that federal funds be expended on this designated system. Second, the act placed the responsibility for maintenance of these routes on the individual states.

During this period many states, experiencing difficulty in securing the necessary funds to match the federal appropriations, looked for new sources of revenue. This fact led to the enactment of gasoline taxes by several states in 1919, and other states rapidly followed suit.

During the ensuing years appropriations for federal aid were steadily increased, while the basis of participation remained practically the same. Funds were provided (as now) for the improvement of roads in national parks, national forests, Indian reservations, and other public lands. The principle of federal aid was broadened to allow the use of federal funds for the improvement of extensions of the federal-aid system into and through urban areas and for the construction of secondary roads.

Another important step taken during this period was the authorization by Congress, in 1934, of the expenditure of not more than 1.5 percent of the annual federal funds by the states in making highway planning surveys and other important investigations. The first planning survey was inaugurated in 1935, and work of this nature was being conducted by all the states in 1940.

World War II focused attention on the role of highways in national defense, and funds were provided for the construction of access roads to military establishments and for the performance of various other activities geared to the war effort. Normal highway development ceased during the war years.

The Federal-Aid Highway Act of 1944 provided funds for highway improvements in postwar years. Basically the act provided $500 million for each of the three years. Funds were earmarked as follows: $225 million for projects on the federal-aid system; $150 million for projects on the principal secondary and farm-to-market roads; and $125 million for projects on the federal-aid system in urban areas. The law also required the designation of two new highway systems. One is the National System of Interstate Highways, which will be discussed later in this chapter, and the other is one composed of the principal secondary routes (Federal-Aid Secondary System).

The 1952 act, for the first time, contained specific authorization of funds for the National System of Interstate Highways. In 1954 funds for the Interstate System were greatly increased, with apportionment based one-half on the traditional formula for the primary system and one-half on population. The Secretary of Commerce was directed to make studies of highway financing, with special attention to toll roads and to problems of relocating public utilities services. In 1955 the President's Advisory Committee on a National Highway Program (Clay Committee) reported the need for a greatly increased program for modernization of the Interstate System. Intensive efforts by all interested groups in 1955 and 1956 led to the passage of the Federal-Aid Highway Act of 1956.

1-5 Federal-Aid Highway Act of 1956

The Federal-Aid Highway Act of 1956 was a milestone in the development of highway transportation in the United States. It marked the beginning of the largest peacetime public works program in the history of the world. The act authorized completion, within a period of 13 to 15 years, of the 41,000-mile National System of Interstate and Defense Highways. Federal funds to be expended on this system were set at $24.825 billion, beginning July 1, 1956, and ending June 30, 1969.

Federal funds were made available on a 90-10 matching basis; state expenditures were expected to be approximately $2.6 billion, making a total of more than $27 billion available for the program. Subsequent increases in cost have led to upward revision of estimates of the cost of completing the system. By 1992 the Interstate System, which had been extended to approximately 45,500 miles, was virtually complete. It is expected that the cost of the completed system will exceed $100 billion.

Several new provisions that were included in the 1956 act are applicable only to the Interstate System. One of these provisions denies federal funds to any state that permits vehicles with excessive weight and length to use the Interstate

System. Approved design standards, including control of access, must be used, and the design was to be adequate to accommodate traffic expected in 1975. (In 1963 Congress amended the highway act to require designs to be based on 20-year traffic forecasts, dating from the time when the plans for each project are approved.) Toll roads could be included in the Interstate System, if suitably located and adequately designed. Congress indicated its intent to consider at a future time whether or not, and how, the states should be reimbursed for previously constructed toll and free portions of the Interstate System.

Title II of the 1956 act, cited as the Highway Revenue Act of 1956, required revenues derived from taxes levied by the federal government to be earmarked and put into a special trust fund (the Highway Trust Fund).

Increases in the estimated cost of completing the Interstate System and a desire to complete the system within the time period originally established led Congress to increase highway user taxes in subsequent years.

1-6 Highway Legislation of the 1960s

The highway legislation of the 1960s recognized the growing transportation needs of urban areas. The 1962 act required that after July 1, 1965, all federal-aid highway projects in urban areas with a population of more than 50,000 be based on a continuing, comprehensive, and cooperative transportation planning process. The act permitted more extensive use of federal-aid secondary funds in urban areas and required that urban highway improvements be "an integral part of a soundly based, balanced transportation system" for the involved area.

In 1965 Congress authorized a $1.2 billion, 5-year program to build 2350 miles of "developmental highways" in the 11-state Appalachian area, plus another $50 million to build access roads. The purpose of the program, in which the federal government paid 70 percent of the highway costs, was to spur economic development of the region. Congress also inaugurated a highway beautification program in 1965 under which funds are provided for the control of billboards and junkyards adjacent to highway rights of way and for roadside landscaping.

The Federal-Aid Highway Act of 1968 reflected the continuing concern of Congress for urban transportation problems. It authorized the use of federal-aid urban funds for parking facilities in urban fringe areas and established Traffic Operation Programs to Increase Capacity and Safety of urban facilities (TOPICS). The act also increased the compensation to families and businesses displaced by highway projects.

1-7 Highway Legislation of the 1970s

The highway legislation of the 1970s reflected the desire of Congress to cope with the trends of increasing urbanization, traffic congestion, environmental degradation, and shortages of fuel.

The Federal-Aid Highway Act of 1970 recognized the growing importance of urban highways by providing for the establishment of a federal-aid *urban* highway system.

The 1970 legislation provided that under certain circumstances federal highway funds could be used for exclusive or preferential bus lanes, traffic control devices, and fringe and corridor parking facilities that serve public mass transportation.

The 1970 act also required the Secretary of Transportation to promulgate guidelines designed to ensure that possible adverse economic, social, and environmental effects are properly considered during the planning and development of federal-aid highway projects. The intent of this requirement is to ensure that decisions on federally funded projects be made in the best overall public interest, taking into consideration any adverse effects due to air, noise, and water pollution; the displacement of citizens, farms, and businesses; and the disruption or destruction of natural resources, public facilities, community cohesion, and aesthetic values.

The Federal-Aid Highway Act of 1973 greatly expanded the federal-aid urban system and provided for a major realignment of the federal road systems based on anticipated functional usage in 1980. (See Section 1-10.) Meanwhile, efforts were being made to revitalize the economy of rural areas and smaller communities and to reverse the pattern of migration from rural to urban areas by the judicious use of highway construction funds. To further these goals the 1970 act authorized demonstration grants to states for the construction and improvement of *economic growth center development highways* on the federal-aid primary system. The development of economic growth center development highways was upgraded to a standing program by the 1973 act and expanded to include all federal-aid highway systems except the Interstate.

The 1973 act made it possible for local officials in urban areas to decide not to use revenues from the Highway Trust Fund for road improvements but instead to receive an equal amount of public mass transit aid from the general fund. The law gave conditions under which trust funds could be used for mass transit and placed a limit on the amount of such transfers.

Public Law 93-643, entitled Federal-Aid Highway Amendments of 1974, included the following landmark provisions:

1. It required the highway projects receiving federal aid to be planned, designed, constructed, and operated to allow effective utilization by elderly and handicapped persons.

2. It established a rural highway mass transportation demonstration program to encourage the development, improvement, and use of public mass transportation systems utilizing highways that serve rural and small urban areas.

3. It authorized the Secretary of Transportation to make grants for demonstration projects designed to encourge the use of carpools in urban areas. Under the act, funds may be provided for establishing carpooling systems and for marking, signing, and signalizing preferential carpool or carpool/bus highway lanes.

4. It authorized similar grants to states for demonstration projects for the construction of bikeways. The federal share for such projects is 80 percent of the total costs.

5. It increased the maximum permissible weight restrictions that states may impose on vehicles that use the Interstate System. Single-axle maximum weights were increased from 18,000 to 20,000 lb. The overall gross weight limit was changed from 73,280 to 80,000 lb. The act also gave a formula for the calculation of the allowable overall gross weight of two or more consecutive axles.

6. It established a uniform national maximum speed limit of 55 mph on public highways.

1-8 Highway Legislation of the 1980s

The Surface Transportation Assistance Act of 1982 (STAA) was passed to address the problem of a deteriorating highway infrastructure. The act extended authorizations for the highway and transit programs from 1983 to 1986 and raised highway user fuel taxes from 4 to 9 cents per gallon. The act also instituted substantial increases in truck user fees. It specified that the additional funds were to be used for (1) accelerating the completion of the Interstate Highway System, (2) increasing the Interstate resurfacing, restoration, rehabilitation, and reconstruction (4R) program, (3) expanding the bridge replacement and rehabilitation program, and (4) providing greater funding for other primary and secondary projects.

The Surface Transportation Assistance Act allowed larger trucks to operate on the Interstate Highway System and certain other federal-aid highways. The law specified that states must allow twin-trailer combination trucks and may not establish overall length limits on tractor-semitrailer or tractor–twin-trailer combinations. The law established a width limitation of 102 in. and required that states allow

1. semitrailers in a tractor-semitrailer configuration 48 ft in length
2. trailers in a tractor-semitrailer-trailer configuration 28 ft in length

The law required that states allow the longer vehicles on interstate highways and on "those classes of qualifying Federal-aid Primary System highways as designated by the Secretary" (of Transportation). It provided that the Secretary designate as qualifying highways those Primary System highways that are capable of safely accommodating the longer vehicle lengths. The implementation of this part of the law has been accompanied by controversy, and by the mid-1980s research was being performed to determine the classes of highways that could safely accommodate the larger trucks.

The Surface Transportation and Uniform Relocation Assistance Act of 1987 authorized $68.8 billion for federal-aid highway programs over a five-year period, fiscal years 1987–1991. Title II of the act authorized an additional $795 million for the National Highway Traffic Safety Administration's safety programs.

A major provision of the law allowed the states to raise the 55 mph speed limit on rural portions of the Interstate System to 65 mph. The higher speed limit is optional with the states and is allowed only along areas of the Interstate System that lie outside the designated areas of 50,000 or more.

1-9 The Intermodal Surface Transportation Efficiency Act of 1991

Landmark highway legislation was passed in 1991 as the Intermodal Surface Transportation Efficiency Act of 1991 (ISTEA), making significant changes to the way highways are funded and administered.

With this legislation, the Congress authorized* federal-aid programs for a period of 6 years at a total funding level of $151 billion. This authorization

---

*It should be remembered that Congressional authorizations set the maximum funding levels that can be made available for programs. The actual annual appropriations may be less than these amounts.

amounted to an overall increase in funding of 63 percent for highways and 91 percent for transit. Under the law the federal matching share of transportation expenditures is generally 80 percent federal, but 90 percent for Interstate construction and maintenance and 100 percent for projects on federal lands. In addition, certain safety, commuter, and transportation operations projects are solely funded by the federal government.

The act dropped the primary, secondary, and urban highway classifications and provided a new 155,000-mile National Highway System. The new system will be comprised of the Interstate system and other major highways designated by the various states in collaboration with the U.S. Department of Transportation. (See Section 1-10.)

Under the ISTEA, additional responsibilities for transportation policy were given to state and local agencies, giving these jurisdictions more flexibility in how federal transportation monies may be spent. The law placed greater emphasis on funding for public transportation programs.

The law required that the states have a statewide transportation planning process and specified the various elements that must be included in such a process. It required that 2 percent of a state's federal-aid funds be devoted to highway planning and research programs.

In addition, the Intermodal Surface Transportation Efficiency Act of 1991 authorized funds for:

- the inspection, maintenance, and rehabilitation or replacement of bridges
- transportation projects that will help air quality nonattainment areas implement projects to facilitate compliance with the Clean Air Act
- an Intelligent Vehicle Highway Systems (IVHS) Program, under which advanced technologies for traffic control can be researched and developed for the nation's most heavily trafficked corridors
- highway safety construction and management programs as well as funding of rulemaking requirements designed to improve the safety of vehicles
- traffic surveillance and control equipment, computerized signal systems, motorist information systems, incident management systems, and other operational improvements designed to mitigate traffic congestion
- environmental transportation protection and enhancement measures
- improvements on toll roads

HIGHWAY SYSTEMS IN THE UNITED STATES

The highways of the United States now total more than 3.9 million miles. These highways are commonly classified by two systems: functional and administrative.

1-10 Functional Classification of Highways

To facilitate orderly highway development and efficient fiscal planning and to ensure logical assignment of jurisdictional responsibility, highways are classified according to function. Functional classification involves grouping streets and highways into classes or systems according to the character of service they are intended to provide. Highways generally serve a dual role in a highway system, providing both travel mobility and access to property. Thus highways may be

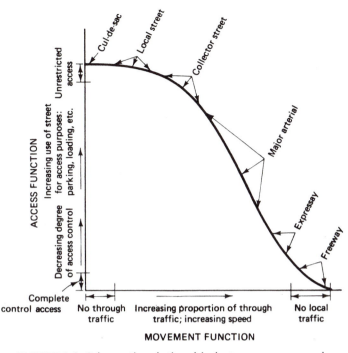

FIGURE 1-2 Schematic relationship between access and movement functions of streets and highways. (Source: Institute of Transportation Engineers, Transportation Planning Handbook, Prentice Hall, Englewood Cliffs, NJ, 1992.)

viewed as a hierarchical system consisting of highway classes with differing purposes and levels of importance. (See Figure 1-2.) Functional classes of highways commonly used include principal arterial, minor arterial, collector, and local highways. The dominant function of arterial highways is mobility, while the principal function of local roads is to provide access to abutting property. Collector roads provide more or less balanced service for both of these functions. A breakdown of the approximate mileages for various functional classes of highways in the United States is given in Table 1-1, along with percentages of mileage and vehicle-miles of travel for each class. Great variations in the performance demands for the various categories can be observed in the table. Interstate highways (a subclass of principal arterial), for example, comprise only 1.2 percent of the mileage but accommodate 22.6 percent of the vehicle-miles of travel. On the other hand, local streets and roads constitute 68.8 percent of the mileage but serve only 13.2 percent of the travel. These percentages vary significantly between rural and urban areas, as the table shows.

1-11 Administrative Classification of Highways

Administrative classification systems are used for the purpose of allocation of highway funds and to define the authority for planning, design, construction, maintenance, and operation of specific highways. Administratively, highways may be grouped into three broad systems: federal-aid systems, state systems, and local systems.

TABLE 1-1 Mileage and Percentage of Mileage and Travel of U.S. Highways
By Functional System

Functional Class	Mileage	Percentage by Mileage	Percentage by Vehicle-Miles
Rural			
Interstate	33,027	0.9	9.1
Other principal arterial	94,798	2.4	8.8
Minor arterial	137,637	3.5	6.6
Major collector	434,175	11.1	8.2
Minor collector	284,706	7.3	2.2
Local	2,132,212	54.7	4.4
Subtotal Rural	3,116,555	79.9	39.3
Urban			
Interstate	12,466	0.3	13.5
Other freeways/expressways	8,465	0.2	6.2
Other principal arterial	52,165	1.3	15.4
Minor arterial	80,368	2.1	11.6
Collector	82,657	2.1	5.2
Local	549,039	14.1	8.8
Subtotal Urban	785,160	20.1	60.7
Total Rural and Urban	3,901,715	100.0	100.0

Source: Federal Highway Administration, *Highway Statistics 1992,* U.S. Department of Transportation, Washington, DC (1993).

Federal-Aid Systems

Since the passage of the Federal-aid Highway Act of 1921, Congress has required that federal road funds be spent on specified systems of roads. The federal-aid highway systems are segments of state and local mileage that are eligible for federal funds. That is, the designation of a road or street as a federal-aid highway does not alter its status as a state or county road or city street. It simply means that, because of its service value and importance, it has been made eligible for federal funds (2).

The highway system designated in 1921 came to be known as the Federal-aid Primary system. Later, the Federal-aid Secondary System and the Federal-aid Urban System were designated and made eligible for federal funds. (See Table 1-2.)

The Intermodal Surface Transportation Act of 1991 eliminated the historical federal-aid systems and created the National Highway System and other federal-aid categories. The law required that the final system be designated and approved by Congress by September 30, 1995. In the interim, the principal arterial system, consisting of the Interstate, other freeways and expressways, and other principal arterial functional systems became the National Highway System. Maps of the Interstate System and the National Highway System are shown as Figures 1-3 and 1-4, respectively.

State Systems

Each state has its own state-designated system, made up principally of arterial and collector routes of statewide importance. In some states, however, county

TABLE 1-2 Historical Federal-Aid Highway Systems

Highway System	Year Established	Description
Federal-aid Primary System	1921	Heavily traveled, well-designed system. Originally consisted of about 7% of a state's rural mileage. Approximately 300,000 miles.
Federal-aid Secondary System	1944	Main secondary and feeder roads connecting farms and smaller communities. About 400,000 miles.
Federal-aid Urban System	1970	Served major centers of activity in urbanized areas and high-volume arterial routes. Approximately 140,000 miles.
National System of Interstate and Defense Highways	1956	Controlled access system connecting the nation's principal cities and industrial centers. Mostly multilane with high design standards. A part of the Federal-aid Primary System. Approximately 45,500 miles.

roads and municipal streets have been placed under state control. In 1992 state highway systems included 800,000 miles of streets and highways.

Local Road Systems

By far the greatest percentage of mileage of roads in the United States is under the control of local governmental units. Local governments include counties, townships, and municipalities.

All states have organized county governments except Connecticut and Rhode Island. Counties, however, have limited or no responsibility for roads in the New England states, or in Delaware, North Carolina, Virginia (with some exceptions), or West Virginia. In Alabama and Maryland, the State has assumed responsibility for roads in certain counties. Counties are called parishes in Louisiana and boroughs in Alaska, and townships are known as towns in New England, New York, and Wisconsin (2).

In 1992 there were 2,918,418 miles of local streets and roads categorized as follows:

County systems	1,727,075 miles
Town and township systems	483,567 miles
Municipalities and other local systems	707,776 miles

1-12 U.S. Numbered Highways

Every tourist is familiar with the comprehensive network of interstate highways designated as "U.S.-Numbered Highways." These routes are marked by charac-

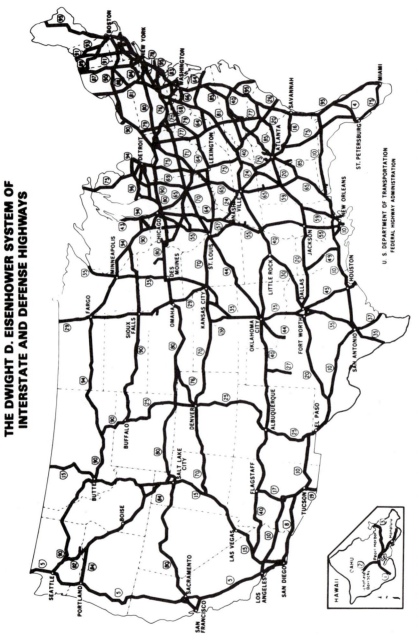

THE DWIGHT D. EISENHOWER SYSTEM OF INTERSTATE AND DEFENSE HIGHWAYS

FIGURE 1-3 The Dwight D. Eisenhower System of Interstate and Defense Highways. (Courtesy Federal Highway Administration.)

14

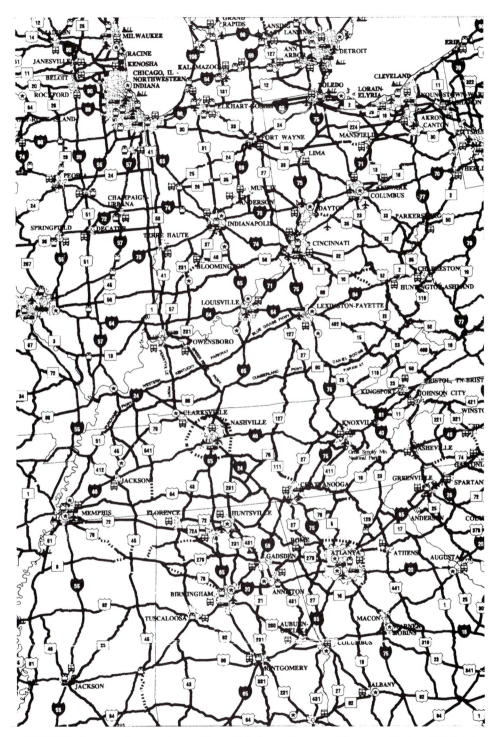

FIGURE 1-4 A portion of the National Highway System. (Courtesy Federal Highway Administration.)

teristic shield-shaped emblems bearing the initials "U.S.," the name of the state, and a route number. They have been designated by joint action of the various state highway departments, acting through the American Association of State Highway and Transportation Officials. Highways included in this system are normally a part of the federal-aid and state systems. Some have been incorporated in or replaced by elements of the Interstate System. Routes traversing the country from east to west bear even numbers, while those with characteristic north and south directions have been given odd numbers. Among the more famous of the routes are U.S. 1, "from Maine to Miami," and U.S. 40, the "Lincoln Highway."

1-13 Toll Roads

During the period 1950 to 1958 many toll roads or turnpikes were built and placed in service in various sections of the country. The second toll road era of American history was brought about by the need to provide high-type facilities capable of relieving pressing traffic problems and the corresponding inability to construct such facilities from current revenue. Toll roads were, in general, built by special authorities created by the various state governments. Revenue bonds were sold and the income from tolls collected used to retire the bonds and interest charges.

The initial toll road of this era was the original Pennsylvania Turnpike, completed in 1940. Toll roads were built and placed in operation in Colorado, Connecticut, Florida, Illinois, Indiana, Kansas, Kentucky, Maine, Massachusetts, New Hampshire, New Jersey, New York, Ohio, Oklahoma, Pennsylvania, Texas, Virginia, and West Virginia. Total mileage of toll roads in operation was 3210 miles, built at an estimated cost of more than $5 billion.

With passage of the Federal-Aid Highway Act of 1956, interest in the construction of new toll roads waned in most states. However, toll road mileage has been built in the intervening years in several states, and many of the original toll roads have been incorporated into the Interstate System. By 1992, approximately 2,750 miles of the Interstate System were toll facilities (2).

FUTURE HIGHWAY DEVELOPMENTS

The highway system of the United States has reached a level of development without parallel in the modern world. In terms of highway mileage, the system is approaching a state of completion. Yet the nation's highway needs are great, and the challenges of future highway engineers surpass those of their predecessors.

As the Interstate System nears completion there is a special need to improve other road systems. A 1983 Report of the Secretary of Transportation to the U.S. Congress (4) indicated that a total of $456 billion (1980 dollars) would be required during the period 1981 to 2000 to bring all U.S. roads and streets to an acceptable level of design, based on an estimated 2.8 percent average annual growth in traffic. The report indicated that $162 billion would be needed for improvement of the urban road system and $294 billion would be required for the rural system.

As we look to the last years of the century, shortages of energy will likely be accompanied by greater demands for separate access-controlled "busways" and express lanes for public mass transit vehicles. Selected improvements in highway gradient and curvature will be required. In order to lessen the impact of urban congestion, more extensive use will be made of sophisticated traffic control measures, and some physical improvements will be needed along urban streets. Heavier permissible wheel loads for commercial vehicles will likely bring a need for stronger pavements, improved pavement design techniques, and better maintenance. Deficient bridges will need to be repaired or replaced. Special efforts will need to be undertaken to decrease the number of highway accidents and to decrease the number of injuries and fatalities by controlling the energy dissipated during crashes.

As these improvements are made, extraordinary efforts will be needed to provide highways that are aesthetically pleasing, socially acceptable, and environmentally compatible. Highway engineers will continue to make a vital contribution to providing the nation with safe, efficient, and convenient transportation.

REFERENCES

1. American Automobile Manufacturers Association, *AAMA Motor Vehicle Facts and Figures 94*, Detroit, MI (1994).

2. Federal Highway Administration, *Highway Statistics 1992*, U.S. Department of Transportation, Washington, DC (1993).

3. Eno Transportation Foundation, Inc., *Transportation in America*, 11th ed., Lansdowne, VA (1993).

4. U.S. Department of Transportation, *The Status of the Nation's Highways: Conditions and Performance*, Report of the Secretary of Transportation to the U.S. Congress, 98th Congress, 1st Session (July 1983).

CHAPTER 2

HIGHWAY ADMINISTRATION

Administration of public highways in the United States is a governmental function, responsibility for which is delegated, in whole or in part, to appropriate agencies of the federal government, to the various state governments, and to numerous local governmental units. Although there is and must be a separation of function and responsibility among the governmental units mentioned, close contact and cooperation among all governmental units concerned with road and street construction are necessary if the development of the overall highway system is to proceed in an intelligent, adequate, and economical fashion.

Highway transportation is influenced to some extent by all of the three major divisions of government: legislative, judicial, and executive. Three standing Congressional committees have jurisdiction over highway-related legislation: the Commerce, Science, and Transportation Committee and the Environment and Public Works Committee in the Senate and the Public Works and Transportation Committee of the House of Representatives.

The economic regulation of interstate carriers (bus companies, trucking firms, brokers, and freight forwarders) is the responsibility of the Interstate Commerce Commission, a quasi-judicial arm of the Congress. Similar state commissions regulate intrastate carriers. Decisions of these regulatory bodies may, of course, be appealed in the courts.

THE FEDERAL HIGHWAY ADMINISTRATION

In the executive branch, the Federal Highway Administration (FHWA) acts as the representative of the federal government in all matters relating to public highways. FHWA is one of ten* principal organizational units that comprise the U.S. Department of Transportation (DOT). Prime responsibility within the Department of Transportation rests with the Secretary of Transportation, a member of the President's cabinet.

> The Federal Highway Administration encompasses highway administration in its broadest scope, seeking to coordinate highways with other modes of transportation to achieve the most effective balance of transportation systems and facilities under cohesive federal transportation policies pursuant to the act.

---

*The other nine units are the Coast Guard, Federal Aviation Administration, Federal Railroad Administration, Federal Transit Administration, St. Lawrence Seaway Development Corporation, National Highway Traffic Safety Administration, Maritime Administration, Research and Special Programs Administration, and the Bureau of Transportation Statistics.

The Administration is concerned with the total operation and environment of highway systems, including highway and motor carrier safety. In administering its highway transportation programs, it gives full consideration to the impacts of highway development and travel; transportation needs; engineering and safety aspects; social, economic, and environmental effects; and project costs. It ensures balanced treatment of these factors by utilizing a systematic, interdisciplinary approach in providing for safe and efficient highway transportation (*1*).

2-1 Organization

As Figure 2-1 indicates, the Federal Highway Administration is headed by an administrator under whom serve the executive director and six associate administrators. The Washington headquarters staff is responsible for policy formulation and general direction of public roads operations in engineering, finance, management, and legal fields.

The Federal Highway Administration has nine regional offices located across the country, each supervising the federal-aid program in from four to eight states. The regions have division offices in every state, Puerto Rico, and the District of Columbia. It is through this field organization that most relations with the state highway departments are carried on.

FEDERAL HIGHWAY ADMINISTRATION

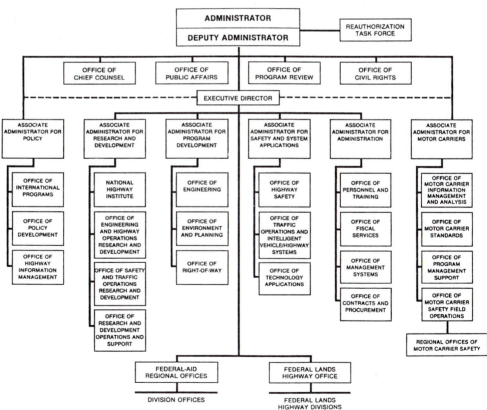

FIGURE 2-1 Organization chart for the Federal Highway Administration. (Courtesy Federal Highway Administration.)

2-2 The Federal-Aid Highway Program

FHWA administers the federal-aid highway program providing financial assistance to the states for highway construction and improvement of the efficiency of highway and traffic operations. This program provides funds for the improvement of approximately 155,000 miles of the National Highway System, including the 45,530-mile System of Interstate and Defense Highways. Construction and preservation of the Interstate System is financed generally on a 90 percent federal, 10 percent state basis.

The Administration also administers the Surface Transportation Program, providing federal aid for state, county, and city routes not functionally classified as local or rural minor collectors. Financing for these roads and streets is generally on an 80 percent federal, 20 percent state basis and is available for highway and bridge improvement projects as well as capital costs for transit projects and bicycle and pedestrian facilities.

The federal-aid highway program also provides funds for acquiring real property for rights-of-way, providing relocation assistance for those displaced by highway construction, improving access for the handicapped, and preserving the natural beauty of lands along highways. A special program of FHWA directs funds to transportation improvement designed to contribute to attainment of national ambient air quality standards.

2-3 Administration of the Federal-Aid Program

Some of the principal functions of the Federal Highway Administration in its administration of federal aid to the states are described in the following paragraphs.

Authorization and Apportionment

The history of a federal-aid project begins with the authorization by Congress of federal funds for highway construction. The federal funds for each year are apportioned among the states, usually from 6 to 12 months before the beginning of the indicated fiscal year, and are available for that fiscal year and for two years thereafter. Within this time limit the funds apportioned to a state must be committed to specific construction work by formal agreement between the state and the FHWA; once such commitments are made, the actual expenditures may continue beyond the three-year limit. Federal-aid authorizations are proportionally divided among the states by formulas that take into account the area, population, and postal route mileage in each state, as provided by law. Amounts available to the states also are limited by mandatory operation of the program on a pay-as-you-go basis within the limits of revenue available to the Highway Trust Fund.

Programming and Subsequent Steps

Knowing the amount of federal aid available to it in each fiscal year, each state draws up programs of projects to be built with those funds, based on preliminary surveys and cost estimates. Since there is close contact between the Federal Highway Administration and the states, questions as to the selection of projects and their location, design, and cost are settled in the early stages of the program planning.

After acceptance of the program by the Federal Highway Administration, the state makes detailed surveys for each project and prepares plans, specifications, and final cost estimates. These must be approved by the FHWA division engineer. The state then advertises for bids and awards a contract for construction subject to concurrence by the Federal Highway Administration. The day-to-day supervision of the work is performed by state engineers resident on the job, but an FHWA engineer or investigating team makes periodic inspections on the site.

FHWA does not prescribe detailed design standards and construction specifications for federal-aid work. The general policies and design standards adopted by the American Association of State Highway and Transportation Officials (AASHTO) are used by most of the states and are endorsed by the Federal Highway Administration. Details of design and construction specifications are prepared by each state to fit its individual needs and are subject to review and approval by the federal agency.

Payments

As the work progresses on a project, the state periodically pays the contractor and claims reimbursement from FHWA for the allowable federal share of the cost. When the project has been completed, inspected, and accepted, a final detailed accounting of costs is made by the state and reviewed by federal auditors, after which final payment of the federal share is made. Thus the federal-aid funds are needed over a period of time rather than all in advance, and the federal aid for each fiscal year is initially made available only as an authorization. After completion of a federal-aid project, FHWA engineers continue to inspect it at intervals to see that it is being properly maintained.

Certification Acceptance

FHWA provides an alternative procedure for administering certain federal-aid highway projects in lieu of the detailed procedures that have traditionally been employed. The process is known as "certification acceptance." A state certification is a written statement prepared by a state highway agency that sets forth the state's organization and capability and the laws, regulations, and standards it will use in the administration of certain federal-aid projects. Once the process is approved, a state highway agency may use state inspectors and auditors to ensure and certify that established procedures are followed on applicable federal-aid projects.

2-4 Federal Lands Highway Program

FHWA, through cooperative agreements with other federal agencies, administers a coordinated federal lands program relating to roads in national parks and forests, Indian reservations, and other federally controlled lands. This program provides for the funding of more than 80,000 miles of federally owned or public authority-owned roads that are open to public travel and serve federal lands. The administration provides the program coordination and directs transportation planning, engineering studies, design, and construction engineering assistance (1).

Many routes of outstanding engineering accomplishment and scenic beauty have resulted from this work of the federal agency. Included among these are the Blue Ridge Parkway in Virginia and North Carolina; the Trail Ridge Road in Rocky Mountain National Park, Colorado; and the Natchez Trace Parkway, which extends from Nashville, Tennessee, to Natchez, Mississippi.

2-5 Research Activities of the Federal Highway Administration

Research was one of the purposes for which the Office of Roads Inquiry was created in 1893 and is one of the important functions of the Federal Highway Administration today.

> The Administration coordinates varied research, development, and technology transfer activities consisting of six principal programs: Intelligent Vehicle/Highway Systems, Highway Research and Development, Long-Term Pavement Performance, Technology Applications, Local Technical Assistance, and the National Highway Institute. The Intelligent Highway/Vehicle Systems program involves developing and deploying state-of-the-art vehicle and information systems to improve safety and alleviate congestion on highways. Other programs focus on highway safety, structures, pavements, materials, environment, policy, and planning to discover ways to improve the quality and durability of highways to reduce construction and maintenance costs, and to offset the negative impact of highway transportation. The Long-Term Pavement Performance Program is a massive evaluation of existing street and highway surfaces. A major effort of the agency is to transfer technology developed through research and other means to state, county, city, and other local highway jurisdictions.

> Through its National Highway Institute (NHI), the Administration develops and administers, in cooperation with State highway agencies, instructional training programs designed for public sector employees, private citizens, and foreign nationals engaged in highway work of interest to the United States (*1*).

2-6 Other Activities of the Federal Highway Administration

The Federal Highway Administration also gathers and disseminates highway statistics, research reports, and planning and design manuals. Its quarterly publication *Public Roads* contains articles describing the results of recent research and research in progress and lists highway documents available from the FHWA and the National Technical Information Service.

FHWA's Bureau of Motor Carrier Safety has jurisdiction over the safety performance of approximately 275,000 motor carriers engaged in interstate commerce. Its field personnel make vehicle inspections, control the transportation of hazardous cargoes, analyze truck and bus accidents, and enforce federal regulations regarding driver qualifications and their hours of service. The administration provides grants to qualified states for the development or implementation of programs to enforce federal or compatible state regulations applicable to commercial vehicle safety and hazardous materials transportation by highway. In fiscal year 1992 the states performed more than 1.6 million roadside inspections and decommissioned over 497,000 vehicles and 126,000 drivers for safety regulation violations (*1*).

STATE HIGHWAY AND TRANSPORTATION DEPARTMENTS

State governments, through the various state highway and transportation departments, occupy the key position in the development of highway systems in the United States at the present time. The relation of the Federal Highway

Administration to highway development has already been discussed, but it must be emphasized that federal-aid programs are undertaken at the option of the individual states and that the states are responsible for the planning, design, construction, operation, and maintenance of routes constructed with federal participation, subject only to review and approval by the FHWA.

In the majority of states, the state highway agency has the responsibility for the development of roads included within the state-designated system, which is generally made up of primary and secondary routes of statewide importance, including urban facilities. As previously mentioned, certain states have assumed responsibility for all local roads.

2-7 Functions of State Highway and Transportation Departments

Functions of the various state highway organizations vary widely because of differences in the laws, directives, and precedents under which they are organized and operated. However, the principal function of any state highway agency is the construction, operation, and maintenance of highways contained within the state-designated system, including those built with federal participation. The departments are also usually responsible for the planning and programming of improvements to be made on the various components of the state highway system and for the allocation of state funds to the various local units of government. They are also usually charged with the duty of distributing information regarding highways to the general public, and they may be responsible for motor vehicle registration. In some states the highway agency exercises police functions relative to safety on the highways, usually through the state highway patrol.

2-8 Organization of State Highway Agencies

As might be expected, state highway and transportation departments vary a great deal in their administrative organization. In terms of administrative control, the departments generally fall into one of three classes, those with

1. a single executive
2. a single executive with a board or commission acting in an advisory or coordinate capacity
3. a board or commission

The single executive type of organization usually has a commissioner, director, or secretary who reports directly to the governor. The executive may be elected but is more commonly appointed by the governor.

A large number of state highway organizations are under the administrative control of elected or appointed boards or commissions. The board may have total administrative control and involve itself with the day-to-day management and operations of the department, or it may serve in a purely advisory capacity. In a few states, the routine management functions have been shared by a single executive and a board or commission.

Since the establishment of the U.S. Department of Transportation, most states have created multimodal departments of transportation. Certain of these departments have been organized by transportation mode, as illustrated by Figure 2-2a. Others have been organized by function, with major divisions provided for

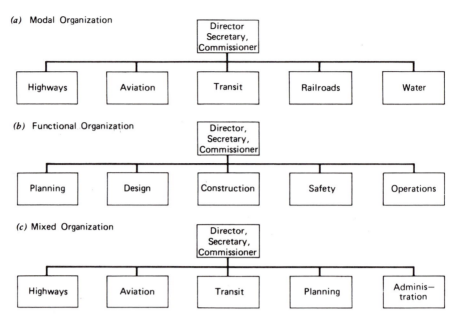

FIGURE 2-2 Basic organizational structures of state transportation departments. (Courtesy Transportation Research Board.)

planning and research, engineering operations, design, and so forth (Figure 2-2*b*). Still others have a mixed organizational structure (Figure 2-2*c*).

An organization chart for a large state highway agency is shown as Figure 2-3.

LOCAL ROAD ADMINISTRATION

As previously pointed out, some 2.9 million miles of highways are currently under the administration of local authorities. Administration of this vast mileage of local roads presents an extremely varied pattern.

2-9 Types and General Features

In most sections of the country the county is the prevailing unit of local road administration. Exceptions to this exist principally in the New England states and Pennsylvania, where the township continues to be the most important unit having responsibility for local roads. In some of these states there are township systems of feeder roads, tributary to the main county routes.

In considering the administration of local roads by the important county unit, differences in practice are also to be noted. In some counties road affairs are administered for the county as a unit; this is the so-called county unit plan. In others, responsibility for roads is divided among various precincts or commissioners' districts, each with its own funds, personnel, and equipment. Extreme variations in area (from a few square miles up to more than 20,000 square miles) and population (from a few hundred or less up to more than 7 million) further account for the various types of county road administrations.

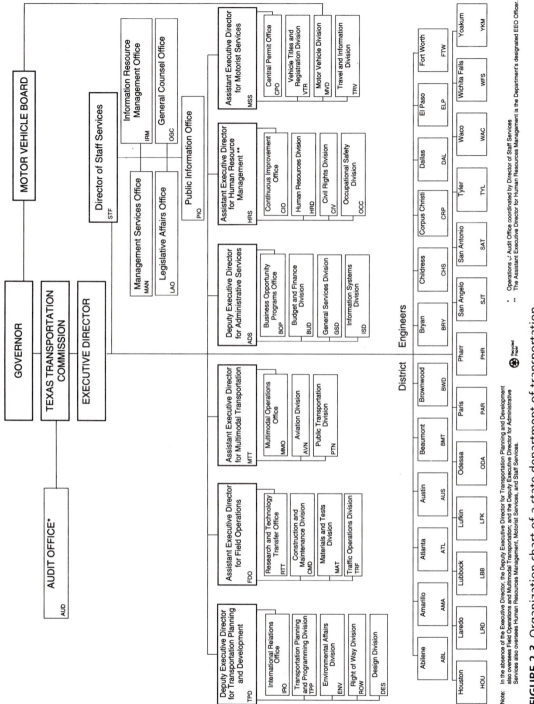

FIGURE 2-3 Organization chart of a state department of transportation.

Note: In the absence of the Executive Director, the Deputy Executive Director for Transportation Planning and Development also oversees Field Operations and Multimodal Transportation; and the Deputy Executive Director for Administrative Services also oversees Human Resources Management, Motorist Services, and Staff Services.

• Operations of Audit Office coordinated by Director of Staff Services
•• The Assistant Executive Director for Human Resources Management is the Department's designated EEO Officer.

 Recycled Paper

It is generally believed that the county-unit plan is the most efficient unit of local road administration. Advantages claimed for the county-unit plan over smaller units of local road administration include the creation of a more efficient and capable organization, a more equitable distribution of the tax burden, higher standards of engineering and construction, and increased ability to cooperate with state highway departments. There has also been some indication that consolidation of the road-building activities of adjacent counties may be advisable in certain instances where the individual counties do not possess sufficient resources to purchase and maintain adequate equipment or to secure necessary engineering services.

County engineering organizations are as varied in character and quality as the county administrative organizations. Generally speaking, the larger and more populous counties have entirely adequate engineering organizations, many of which are on a par with those of the state highway departments. Many other counties have inadequate engineering forces, while hundreds of counties operate with little or no engineering aid.

Large allocations of federal funds for the secondary road program have stimulated local road affairs and led many counties to establish engineering organizations capable of cooperating with the state and federal agencies in the planning, design, and contruction of local roads.

The following criteria have been enumerated for good highway administration at the county level (*2*):

- A road commission whose functions are limited to policy formulation, with an engineer responsible for all management functions.
- Maintenance of adequate and uniform records for guidance in management, for developing and carrying out a long-range road improvement program, and for use in making such reports as the state legislature may direct.
- Cooperation with other county and city agencies in the development of integrated regional highway programs.
- Operation under an annual improvement program, based on a long-range plan and related to a definite annual budget.
- Preparation of construction plans and specifications for all contractual work.
- Maintenance of roads and bridges to uniform standards.
- Extension of contractual highway services to small cities and villages on request.
- Installation of traffic-control and safety measures in accordance with uniform standards.
- Adoption of uniform accounting practices.
- Adoption of standard specifications for purchasing, in accordance with sound business practice.

MUNICIPAL STREET ADMINISTRATION

Administration of the 549,000 miles of local streets and roads in this country is also quite variable in nature. This is not surprising when it is realized that much of this mileage exists in thousands of incorporated communities, the large majority of which have populations of less than 5,000.

Generally speaking, responsibility for city streets is vested in the city engineer, director of public works, city manager, or some similar official, depending on the form of government used. Functions of the city street department or similar body include design and construction of street improvements, formulation of long-range plans for street improvement, the making of traffic surveys, installation of traffic-control and safety devices, and the supervision of maintenance.

As in the case of the counties, street departments in the larger cities operate at a high level of efficiency, performing all necessary engineering functions. In many smaller cities engineering organizations are inadequate or nonexistent. However, it may be noted that there is an increasing appreciation of the desirability of providing engineering control in the planning, design, and construction of city streets, with the result that many municipal engineering organizations have been created or expanded in recent years.

NECESSITY FOR INTERAGENCY COOPERATION

It is recognized that the various components of the highway system are not separate entities but are, or must become, integrated parts of a coordinated system if highway development is to keep pace with the demands of the modern motorist. In recognition of this fact there must be close cooperation among all the units of government concerned with highways and streets, including the federal government, the states, counties, and cities.

In the planning, design, construction, operation, and maintenance of highways, the state highway and transportation departments are the key agencies. They are usually responsible for the distribution of state funds to the various units of local government; they are in a position to supply the technical knowledge and engineering skill so badly needed in many local agencies; they possess facilities for performing special surveys and analyses, such as urban traffic studies; and they possess a wealth of information that should be made available to local units of government for planning and programming purposes. In view of these and other facts, several state highway departments have established special county (secondary road) and urban divisions to cooperate more closely with the appropriate local governments and to further the development of an integrated highway system. In most urban areas cooperative transportation planning agencies have been established.

Planners of important federally assisted highway projects must now coordinate the planning activity through a regional or state clearinghouse (*3*). In metropolitan areas, the clearinghouse is an areawide agency (e.g., a regional planning commission) that has been recognized by the Office of Management and Budget as an appropriate agency to evaluate, review, and coordinate federally assisted projects. In nonmetropolitan areas, the clearinghouse is a comprehensive planning agency designed by the governor or by state law to review and coordinate such activities.

RELATED ORGANIZATIONS

In addition to the governmental agencies described, many other organizations contribute to the development of the highway system. One of the most important of these groups is the American Association of State Highway and Transportation

Officials (AASHTO). The membership of this group consists of the principal executive and engineering officers of the various state highway agencies and the Federal Highway Administration. The principal objective of the association is the fostering of "the development, operation, and maintenance of a nation-wide system of highways to adequately serve the transportation needs of our country." Work of the various committees of the association has resulted in a large number of publications that are widely used by other groups.

The American Road and Transportation Builders Association (ARTBA) is another large and powerful group in the highway field in this country. The membership of ARTBA comprises six divisions, including persons and organizations from every branch of the highway industry and representatives from many foreign countries. As a result of the work of the divisions and the committees included therein, the organization issues technical bulletins and reports at intervals; it also publishes a monthly magazine, *The American Road Builder,* and a weekly newsletter. ARTBA is active in furthering highway legislation and in disseminating information about highways to the general public. ARTBA holds annual and regional meetings, with a road show that is held at periodic intervals.

The Institute of Transportation Engineers (ITE) is a society of transportation professionals with a membership of approximately 11,900. Formerly called the Institute of Traffic Engineers, ITE has a memberhip that consists predominantly of traffic engineers and others interested in vehicular traffic control. The organization meets annually and publishes a number of handbooks and procedural guides, as well as a monthly magazine called the *ITE Journal.*

The Transportation Research Board (TRB), a division of the National Research Council, is a vital factor in the advancement of technical knowledge in transportation. The board does not, in general, perform research itself but has its primary function in the correlation of research efforts of educational institutions, governmental agencies, and industry. The TRB holds an annual meeting and issues many technical publications.

Many other groups are engaged in activities related to highway development, including organizations of highway users, materials producers, equipment manufacturers and dealers, automotive manufacturers, engineering groups, colleges and universities, and many others. Many such groups could be named, and publications of some of them have been cited at various points in this book.

HIGHWAY FINANCE

This section is concerned with the sources and distribution of the money that is obtained for highway purposes; in other words, with the financing of highway systems and their improvement. Attention will be focused principally on current methods of providing money for highway expenditures and the distribution of these sums. Certain general principles of highway finance and taxation will also be presented.

There are two general methods of financing highways. The first of these is the pay-as-you-go method, which involves paying for all highway improvements and the costs of maintaining and operating the highway system from current revenue. This is the method that is currently in use by many governmental agencies, and rightly so in many cases, because of constitutional prohibitions or similar legislative provisions against borrowing for highway purposes. The second method is, of course, that of borrowing money to pay the costs of highway improvement, with the borrowed sums plus interest being repaid over a period of time from future income. This latter method may be generally termed "credit financing." In most cases, it is also "bond financing." Both methods have merit under certain circumstances, and strong advocates of either method may be found among the various authorities in the highway field.

Attention will first be given to the various sources of revenue for highway purposes. These sources may, in many cases, be regarded as sources of current revenue, although they may also be used for debt service, either on obligations incurred in the past or for obligations incurred at the present time and against which some of these revenues may be pledged for future payment. Bond financing is considered separately in Section 2-14.

2-10 Sources of Revenue for Highway Expenditures

The funds necessary for the improvement of the nation's highway system are obtained from a variety of sources. The principal sources are

1. highway user taxes
2. property taxes
3. tolls

Highway User Taxes

Highway user taxes comprise the dominant class of funds for highways at both the state and federal levels. As Table 2-1 indicates, income from federal highway user imposts totaled more than $15 billion in 1992, while state user taxes amounted to more than $30 billion (*4*). A breakdown of state highway user tax revenues for 1992 is given in Table 2-2. User taxes, typically in the form of vehicle fees or "wheel taxes," are also imposed by local governments in a relatively small number of instances.

Highway user taxes are of three general types: (1) fuel taxes, (2) registration taxes and related fees, and (3) special taxes on commercial vehicles.

TABLE 2-1 Revenues Used For Highways in 1992, All Units of Government

Item	*Millions of Dollars*
Highway user tax revenues	
Federal	15,571
State	30,829
Local	1,038
Total	47,438
Tolls	3,449
Property taxes and assessments	4,644
General fund appropriations	12,404
Other taxes and fees	2,886
Interest and other receipts	6,583
Bond proceeds	9,299
Total receipts	86,703

Source: 1993 Highway Statistics, U.S. Department of Transportation, Federal Highway Administration (1994).

TABLE 2-2 State Highway Revenues and Other Highway Receipts, 1993

Item	Millions of Dollars
Motor fuel taxes	22,016
Motor vehicle and motor-carrier taxes	10,727
Tolls	3,147
Appropriations from general funds	1,655
Other state imposts	1,643
Miscellaneous receipts	2,093
Bonds	4,017
Payments from other governments	
Federal Highway Administration	15,821
Other federal agencies	422
Local governments	1,308
Total receipts	62,849

Source: *1993 Highway Statistics,* U.S. Department of Transportation, Federal Highway Administration (1994).

Fuel Taxes Taxes were first levied on motor fuels in this country in 1919 by Oregon and three other states. By 1929, all states levied a gasoline tax, that is, a stated tax on each gallon of gasoline sold as motor vehicle fuel. In 1993, state gasoline taxes ranged from 7.5 to 28 cents per gallon in the United States and averaged about 19 centers per gallon (5). Other motor fuels are similarly taxed. This tax is by far the greatest revenue producer of all highway user taxes, and, in most cases, it is far and away the biggest single source of income for the operation and improvement of the state highway system. Fuel taxes constitute approximately two-thirds of the total obtained from state highway user taxes.

Registration Taxes and Allied Fees Laws pertaining to the registration of motor vehicles have a somewhat longer history than those relating to the imposition of a fuel tax, since the first registration law became effective in New York in 1901. By 1917 all states had enacted laws of this type. Although the procedure of registration was first adopted as a regulatory device, it soon became recognized as a ready source of highway revenue. The practice of graduating registration fees with the weight or capacity of the vehicle also began early in the modern era of highway development in this country, being first occasioned by the great increase in the number of trucks using the highway system after World War I. This practice is now followed in nearly all states. Registration fees are collected principally by the state governments. Several other taxes and fees are closely associated with registration fees and are levied by the various state governments. These special fees include such things as operators' and chauffeurs license fees, fees for certificates of title, title-transfer fees, and fees charged for duplicate license plates. In many states income derived from these latter sources is used primarily for the operation of the highway system rather than for actual construction and maintenance. For example, in some states money derived from operators' and chauffeurs' license fees goes wholly or in part to pay the expense of maintaining a highway police organization.

Special Taxes on Commercial Vehicles The practice of assessing special fees against motor vehicles used for commercial purposes, including both passenger and freight carriers, has developed somewhat more gradually and on a less uniform basis than those taxes that have been previously discussed in this section.

Special fees are levied against this class of motor vehicles in proportion to their use of the highway system and, roughly at least, in proportion to their effect on the physical design of the highway. These imposts have generally taken the form of graduated fees computed on a total mileage basis, on the basis of ton-miles or passenger-miles of travel, or taxes on gross receipts. The special nature of these vehicles is also taken into account by the graduated registration fees previously described.

The general basis on which these taxes are levied seems to be sound, as there is little question that certain classes of commercial vehicles, for example heavy trucks and tractor-trailer combinations, have a disproportionate effect on the design of the highway. The thickness of a pavement that is required to carry heavy trucks may be considerably greater than that required for the adequate support of passenger automobiles. Similarly, a direct effect may be seen in the necessity of providing extra lanes for heavy trucks on long grades. They are often found necessary in highways designed to carry both passenger cars and heavy commercial units in order to provide some relief from congestion. A long and continuing argument has been going on for many years between those who represent the interests of commercial haulers, on the one hand, and highway administrators and engineers, on the other, relative to the imposition and collection of these special taxes.

Property Taxes

Two types of property taxes are of importance when sources of funds for highways are being considered. These are (1) ad valorem property taxes and (2) special assessments against real estate lying contiguous or adjacent to a highway improvement. The first of these types of property tax is extensively used in the financing of highway improvements by local units of government, including counties, townships, incorporated places, and similar administrative units. The second type is now principally used in towns and cities. Neither of these types is an important source of highway revenue to state governments at the present time.

The common concept of the use of general property taxes for highway improvements, which may be levied on real property or on both real and personal property, is that such a procedure is justifiable if the improvements that are to be effected are of general benefit to the community. The same line of reasoning is applicable if the improvements are expected to enhance the value of property located throughout the area. In practice, of course, in many local governmental units, property taxes are not levied specifically for highway improvements and are not earmarked for this purpose. Money derived from this source goes into the general revenue fund and is then budgeted, generally on an annual basis, for highways, education, public health, welfare, and so on.

General property taxes are, of course, levied on the assessed or "fair" value of the land and improvements located thereon or on a similar value of personal property. Although general property taxes have numerous faults, they will almost certainly remain a source of income for highways and other public services for many years to come.

Special assessments are simply what the name implies. They are made directly on the owners of property that lies on or adjacent to a highway or street improvement in accordance with the benefits that are expected to accrue directly to the parcels of land thus affected. As has been indicated, this form of property tax is now used principally in cities and is not in widespread use for the financing of rural roads. It has been most generally used in recent years in the provision of street improvements in new residential areas and is less generally used for reconstruction. Special assessments are usually made on a frontage basis rather than on the property value. The assessments may be designed to recover the entire cost of the improvement over a period of time, or the initial cost may be shared among the property owners and the city government. Assessments may be made on abutting property only or on property located on adjacent side streets and even parallel streets, depending largely on the nature and amount of the benefits that are expected to accrue from the proposed improvement. Assessments against property located on adjacent streets would generally be made on a reduced or sliding scale. As is the case with general property taxes, special assessments have some serious inherent disadvantages. It seems quite certain, however, that this method of financing street improvements will be continued.

Tolls

Funds derived from the assessment of tolls for the use of a particular facility have become an important source of revenue as a result of the construction of a large number of toll roads following World War II. Prior to that time, tolls were generally limited to assessments for particular structures, such as a bridge or tunnel. Tolls are levied by public agencies, including special authorities of various types, to repay costs incurred in the construction of the facility (see Section 2-14 for a discussion of bond financing). As soon as the debt incurred by the construction of the facility has been paid, the facility is made "free." In 1992 toll receipts totaled approximately $3.4 billion.

Miscellaneous Funds

Brief attention may be given at this point to general revenue funds as a source of money for the financing of highway expenditures. Generally speaking, funds from general revenue have not been of major importance in the improvement of highways by the state governments in recent years. In a few states, however, income derived from taxes levied for highway improvement, consisting chiefly of highway user taxes, goes into the general fund from which the funds required for operation and improvement of the state highway system are budgeted periodically. In states where this system is used the amount of money appropriated from the general fund would usually be substantially the same as that which would normally accrue to the state highway department from highway user taxes, so that the end result is about the same. Students of taxation are frequently strong advocates of this method of putting income derived from all tax sources into the general fund, with appropriations being made at intervals for all public services on the basis of need. From the viewpoint of the highway administrator this concept has a serious disadvantage in that highway improvements are best made on a planned, orderly basis with the assurance of a steady and predictable annual income. In 1992 combined receipts from property taxes and general revenues totally more than $17 billion.

Another source of revenue that is worthy of mention here is that derived from parking meters, as used in many towns and cities. Although parking meters were originally envisaged as regulatory devices, the income derived from them has proved to be very substantial and has become quite important in many urban areas. Although parking revenues are closely related to tolls, they are not considered highway-user imposts (*4*).

2-11 Federal Funds

For many years funds collected from certain federal excise taxes (principally on gasoline) went into the general revenue fund of the federal government. Appropriations were then made from the general fund for federal aid to highways and the construction, maintenance, and operation of roads in national parks, national forests, and so on. However, the Highway Act of 1956 changed this system and established the Highway Trust Fund, from which expenditures are made directly for federal aid and for roads on federally controlled lands. Into the Trust Fund go monies received from federal taxes on gasoline and diesel fuel, tires, new trucks, truck-trailers, and buses; there is also a highway use tax on heavy vehicles. Estimated federal taxes paid by private and commercial highway users during 1993 are given in Table 2-3.

2-12 Highway Cost Allocation

One underlying concept should govern all levies for highway purposes. This concept is that the cost of highway improvement should be borne on an equitable basis by the three principal groups who benefit from highway improvement, namely (1) highway users, (2) property owners, and (3) the general public. Every effort should be made to assess the cost of improved highways in an equitable manner against these groups. The effort should be made at every level of government. The accomplishment of the equitable distribution of the tax burden in proportion to benefits received is, of course, much more difficult than the statement of this underlying objective.

Highway cost allocation is a thorny fiscal (and philosophical) problem that

TABLE 2-3 Federal Highway Trust Fund Receipts Attributable to Highway Users in Fiscal Year 1993

Type	Millions of Dollars	
Highway Account		
Gasoline	10,385	
Gasohol	416	
Special Fuels	3,111	
Total motor fuel		13,912
Federal use tax		630
Trucks and trailers		1,199
Tires		304
Total highway account		16,046
Mass Transit Account		
Gasoline	1,537	
Gasohol	132	
Special fuels	323	
Total motor fuel	1,992	
Total mass transit account		1,992
Grand total		18,038

Source: 1993 Highway Statistics, U.S. Department of Transportation, Federal Highway Administration (1994).

has occupied the attention of highway economists for many years. It has been the subject of a number of studies, including at least two required by federal laws (*6, 7*).

Usually the first consideration is the division of cost responsibility among highway users and nonusers. One study recommended that about 8 percent of the cost of the federal-aid highway program be charged to revenue sources other than motor vehicle taxes (i.e., to nonhighway users) (*6*).

Several methods of cost allocation have been proposed, the most prominent of which are the incremental cost method and the differential benefit method.

The incremental cost method is the traditional way of allocating highway user charges. In this approach, each element of highway design affected by the size or weight of vehicles is broken down into a series of additions (increments); the cost of providing each increment is assigned only to those vehicles whose size and weight require them. Thus, all vehicles share in the cost of the basic design, but the cost of each succeeding increment is borne only by the larger vehicles.

The differential benefit method assigns cost responsibility to the various users in direct proportion to the vehicular benefits they receive by their use of the highway system. Four types of vehicular benefits are generally considered: reductions in operating costs, in time costs, in accident costs, and in the strains and discomforts of driving. The last (impedance costs) are normally calculated only for passenger cars.

2-13 Disbursement of Highway Revenues

Table 2-4 presents information relative to the disbursement of highway revenues in 1993 for federal, state, and local governments. Two of the items given in this table deserve additional explanation. These are the items listed as "Interest" and "Bond retirement." These items reflect the money expended in the payment of obligations previously incurred for highway purposes. The total highway debt of the state governments, counties, and municipalities was estimated by the Federal Highway Administration to be nearly $54 billion at the end of 1992; much of this was for toll facilities. Although the sums required for debt service of these obligations represent a sizable portion of the total income for highway purposes, it is generally felt that the various governmental units, taken as a whole, are still

TABLE 2-4 Disbursements For Highways in 1993 For All Units of Government[a]

Item	Federal	State	Local[b]	Total
	Millions of Dollars			
Capital outlay	325	29,874	9,433	39,632
Maintenance	68	9,530	13,362	22,960
Administration and miscellaneous	531	4,656	2,620	7,807
Highway law enforcement and safety	—	3,865	3,178	7,043
Interest	—	2,155	1,464	3,619
Bond retirement	—	2,794	1,992	4,786
Total	924	52,874	32,049	85,847

[a]The disbursements shown do not include $27,158 million federal payments to states and local governments, nor $9,319 million in state grants-in-aid to local governments.

[b]The local disbursements are the amounts for 1992.

Source: 1993 Highway Statistics, U.S. Department of Transportation, Federal Highway Administration (1994).

in a favorable position with regard to the relation between the existing highway debt and the total capital investment that has been made in the various units of the highway system. Several of the toll roads built since World War II have been highly successful from a financial point of view, as are many toll bridges and tunnels.

Another factor in the distribution of highway revenues is diversion. By "diversion" is meant the expenditure of funds collected from highway user taxes for nonhighway purposes. Although these diverted funds are in general expended for very worthwhile purposes, with expenditures of this sort being principally made for education at the present time, the diversion of highway funds is regarded in a very unfavorable light by the large majority of highway administrators and engineers. The principal basis for condemnation of diversion arises from the concept that since highway user tax schedules are predicated on benefits received by the taxpayer in the form of highway improvements, these taxes become unjust and inequitable if funds derived from this source are expended for purposes other than highway improvement. The amount of diversion, expressed in terms of percentage of highway user tax receipts, has decreased substantially in recent years. There is every reason to believe that this trend will continue. A large number of states now have constitutional provisions that prevent the diversion of highway funds.

2-14 Bond Financing for Highways

It sometimes becomes necessary or advisable for a governmental unit to obtain funds for highway improvements by credit financing. That is, the government must make use of its credit in order to borrow money, which is then repaid, with interest, usually over a period of years. Necessary improvements of a highway system or a specific facility may be financed in this manner. By far the greatest amount of credit financing for highways is done through the issue of bonds.

One form of bond that may be used for the financing of highway improvements is the general-obligation bond. This type is backed by the entire faith and credit of the issuing agency. In recent years, general-obligation bonds have been regarded with less favor than in the early period of highway improvement, and the trend has been toward the use of the so-called limited-obligation bond. Limited-obligation bonds are usually secured by the anticipated revenues from a particular source of income. For example, many limited-obligation bonds are now secured by the pledging of future income from gasoline tax receipts to their repayment. Still another form of bond is what may be termed here a revenue bond. A bond issue of this type may be employed, for example, in the construction of a toll road or bridge, with the bond issue secured by the pledging of future tolls to repayment of the bonds. Limited-obligation bonds are also sometimes termed, somewhat loosely perhaps, revenue bonds.

The mechanics of a bond issue and its subsequent repayment may be generally described as follows. The total amount of the bond issue is decided on by the governmental agency involved and then approved by referendum or legislative action. Bonds are then issued and are sold, frequently on a competitive bid basis, to banks, investment companies, or similar organizations, or in some instances to the general public. The amount of interest carried on the bond issue will generally be determined by the type of bond issue, the term of issue, and the credit standing of the issuing agency. The total amount of the bond issue, plus interest, is then repaid by the issuing agency in any one of several ways, as briefly described later. The bonds are said to have been retired when they are repaid.

Highway bonds have been issued for varying periods, with typical periods ranging from 10 to 50 years. Ten years is probably somewhat too short a period of issue for most highway bonds, while a period of 50 years might be regarded as proper under some circumstances in the financing of some relatively long-lived structure, such as road, bridge, or tunnel.

Any one of several general methods may be employed to retire bonds that have been issued for highway improvements. Two of these general methods will be described here. One of these may be termed the "sinking-fund" method and the other the "serial" method. Bonds that are to be retired by the use of a sinking fund are generally term-issue obligations. That is, the entire amount borrowed falls due at the end of a period of time that is fixed when the bonds are issued. Money that will be required to pay the obligation when it falls due or when the bonds mature is provided by setting aside an annual sum toward the payment of the borrowed amount. Payments that are made to the sinking fund must be large enough so that their sum, plus whatever income may be derived from their investment, will be sufficient to retire the bonds when they mature. The sinking-fund method has several inherent disadvantages, chief among which is the temptation that the sinking fund continually presents to public officials over a period of years as it gradually accumulates. This and other factors have generally led to the abandonment of the sinking-fund method in favor of serial bonds. Serial bonds are retired by periodic (annual) payments on the principal plus periodic payments of interest. In their most common form, constant payments are made on the principal each year, with annual interest payments that decrease as time passes. No sinking fund is required when this method of payment is used, and the bonds are completely retired at the end of their term of issue.

There is little doubt that bond financing is entirely proper and desirable as a financing method for highways, at least under certain circumstances. Bond issues of this type may generally be justified by consideration of the benefits expected to emanate from the improvement, particularly the reduced costs of highway transportation. Credit financing may also be resorted to in an effort to bolster the general financial condition of a state through increased employment resulting from construction activity.

The chief disadvantage accruing from bond financing is the added cost occasioned by the payment of interest on the obligation. Other disadvantages that may be associated with bond issues include the generation of a volume of work beyond the capacity of the contractors in the area, with a resultant lack of competitive bidding with accompanying price increases, and the fact that a volume of work may be involved which cannot properly be handled by the engineering force of the agency issuing the bonds.

PROBLEMS

2-1 Determine how state highway affairs are administered in your state. Is the highway agency administered by a single executive or by a commission? Draw an organization chart for the headquarters organization, showing lines of responsibility for the various primary functions of the department. (Consult the annual or biennial reports of the highway department in your college library.)

2-2 How are local road affairs administered in your state? Is the county, township, or some other form of local government responsible for roads? Is there an engineer responsible for such functions as design, construction, and maintenance in the place in which you live or attend school? If so, draw a typical organization chart for this engineering organization.

2-3 Has your state highway department set up special divisions to cooperate with local governmental units in the improvement of secondary roads and urban extensions of state routes? If so, briefly describe the functions of these divisions.

2-4 Prepare a brief report describing an important project now under construction (or recently completed) in your state that required close cooperation among representatives of federal, state, and local highway agencies.

2-5 Prepare a report outlining the organization and functions of the Transportation Research Board.

2-6 Prepare a report describing the organization, membership, and activities of the American Association of State Highway and Transportation Officials. List the current books of standards and policies adopted by this association. What do you think is this group's most valuable contribution to the highway program?

2-7 Determine who is responsible for the planning, design, construction, and maintenance of streets in your community. How much money is expended in a typical year? What are the sources of street funds? Is there a long-range program of street improvement? If so, how did the program come into being? Briefly outline the main points of the program.

2-8 Determine the annual income of the government of your state from direct highway user taxes. Consult the annual report of your state highway agency or some other source, and determine how this money is distributed to the various governmental units in your state.

2-9 Estimate the annual contribution that the owner of an average light passenger car makes in the form of direct highway user taxes levied by your state government. Assume 20,000 kilometers of travel and 10 kilometers to a liter of gasoline. How much does this motorist contribute to the Federal Highway Trust Fund each year?

2-10 Prepare a brief report on a major highway (or street) improvement project, now under way or planned in your state, for which bond financing is to be used. Describe the nature of the project, the total amounts of money involved, and the financing arrangements.

REFERENCES

1. Office of the Federal Register, *U.S. Government Manual, 1993–94,* National Archives and Records Administration, Washington, DC (1993).

2. Highway Study Committee, *Highway Needs in Michigan—An Engineering Analysis,* Michigan Good Roads Federation, Lansing (1948).

3. U.S. Office of Management and Budget, *Intergovernmental Review of Federal Programs: Implementation of Executive Order 12372 Federal Register,* Vol. 48, No. 70 (April 11, 1983).

4. Federal Highway Administration, *1993 Highway Statistics,* U.S. Department of Transportation, Washington, DC (1994).

5. Federal Highway Administration, *Highway Taxes and Fees: How They Are Collected and Distributed,* U.S. Department of Transportation, Washington, DC (1993).

6. *Final Report of the Highway Cost Allocation Study,* House Document No. 72, 87th Congress, 1st Session, U.S. Government Printing Office, Washington, DC (1961).

7. Department of Transportation, *Final Report on Federal Highway Cost Allocation Study,* Report to Congress from the Secretary of Transportation, Washington, DC (May 1982).

CHAPTER 3

TRANSPORTATION PLANNING

Transportation planning can play an important role in an agency's or region's strategy to improve the performance of the transportation system. In its very simplest form transportation planning consists of those activities that collect information on the performance of the existing transportation system; forecast future performance levels given expected changes to key factors such as land use, price of fuel, and growth in employment; and identify possible solutions to expected problems in system performance. At its most complex transportation planning can include a myriad of activities associated with gaining consensus on recommended actions, undertaking numerous technical activities at many different scales of analysis to pinpoint expected problems, and processing large amounts of data associated with system performance and travel behavior.

The purpose of this chapter is to describe the basic characteristics of transportation planning and of the models that are used by planners. The first section presents the basic elements of the transportation planning process. The second section describes the data collection techniques that provide important input into technical approaches toward transportation planning. The third section discusses the different steps of the modeling process and the information they provide to the planning process. The final section describes the evaluation concepts that provide the linkage between technical analysis activities and the ultimate selection of a preferred course of action.

3-1 Basic Elements of Transportation Planning

First and foremost, transportation planning is a process (*1*). This means that it consists of well-defined tasks that must be accomplished before the final set of information is presented to those who must decide which course of action is best for an agency, region, or community. In addition, given that much of the transportation planning activity in the United States occurs in the public sector, many of these tasks are associated with legislative or regulatory requirements that must be satisfied before the results are considered valid. Examples of these tasks include the required consideration of air quality impacts for project in metropolitan areas not in attainment of air quality standards, the involvement of the public throughout the planning process, and the need for the transportation plan and resulting program of action to be financially realistic.

Transportation plans can occur at different scales of application. Many nations, for example, have national planning studies that focus on achieving national goals through changes to the transportation system. In the United States most transportation planning occurs at the state and metropolitan levels. However, no matter at what scale of application, transportation planning usually has similar elements. Figure 3-1 presents a planning framework that shows basic tasks as well as their interrelationship. As shown, transportation planning is primarily a process of producing information that can be used by decision makers to better understand the consequences of different courses of action. The tasks that are part of identifying and assessing these consequences include the following.

Inventory of Facilities

Knowing what your transportation network consists of and the condition and performance of these facilities is an important starting point for transportation planning. Much of the transportation investment that occurs in a state or urban area is aimed at upgrading the physical *condition* of a facility (e.g., repaving a road or building a new bridge) or improving its *performance* (e.g., providing new person-carrying capacity by providing preferential treatment for high occupancy vehicles or building a new road to serve existing demand). Most state transportation agencies have a very extensive inventory of the state's road system that includes such basic information as the number of lanes, type of pavement, the last time the pavement was replaced, the capacity of the road, accident record, and so on (2). Most transit agencies have an inventory of the different assets that

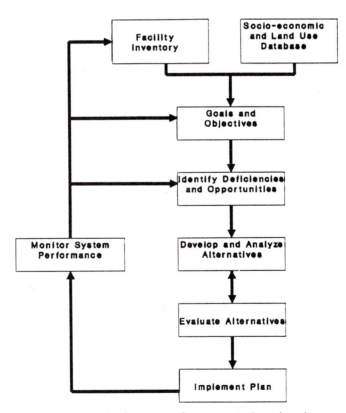

FIGURE 3-1 Basic elements of transportation planning.

constitute a transit system (e.g., buses, stations, shelters, rail cars, etc.). In most cases those transportation engineers and planners involved with the transportation inventory have primary responsibility of keeping it up to date.

Collect and Maintain Socioeconomic and Land Use Data

Transportation planning focuses on the provision of transportation facilities and services to meet existing or expected demand for travel. A fundamental concept in transportation planning that significantly influences the way such demand is estimated is called *derived demand*. Derived demand means that a trip is taken to accomplish some activity at the destination and that the trip itself is simply a means of reaching this activity. Transportation planning must therefore relate tripmaking to the types of activities that occur in a region and also to the characteristics of the tripmaker that will influence the way these trips are made. This is done by combining similar uses of land into a land use category that can then be used in transportation planning to estimate how many trips are attracted to each type of land use (e.g., the number of trips to schools, shopping centers, residential units, office complexes, etc.). Land use maps are most often the source of such information.

The types of data on tripmaker characteristics that are often collected include level of income, number of members in the household, number of autos in the household, number of children, age of head of household, and highest level of education achieved. Each of these socioeconomic characteristics has been shown to affect how many and how trips are made (*3*). Such data are collected from special surveys or, more often, from the U.S. Census that occurs every 10 years.

Define Goals and Objectives

Transportation systems not only provide the means of moving from one location to another, they also relate to much broader issues. For example, investment in transportation facilities has been used to encourage economic development, to reduce the levels of pollution originating from automobiles, to enhance the mobility of the elderly and handicapped, and to revitalize center cities. It is thus very important early in the planning process to establish the specific goals and objectives that the transportation plan and recommended projects will be striving to achieve.

Goals are generalized statements that indicate the desired ultimate achievement of a transportation plan. Examples of goals statements might be, "The transportation system should meet the mobility needs of the population" or "The transportation system should provide enhanced economic development opportunities." *Objectives* are more specific statements that indicate the means by which these goals will be achieved. For example, the goal of meeting the mobility needs of the population could have the following objectives associated with it: "Provide transit service to major markets in the region," "Reduce congestion on major highways," and "Promote bicycle and pedestrian transportation."

The identification of goals and objectives is critical in that they define the evaluation criteria that will be used later in the planning process to assess the relative impacts of alternative projects and strategies. These criteria are often called *measures of effectiveness*. In addition, goals and objectives provide an important linkage to the desires and values of the public that the transportation plan is serving. Thus, the definition of goals and objectives is often subject to intense public involvement, with numerous opportunities provided for input and ultimate sign-off by appointed or elected officials (*4*).

Identify System Deficiencies or Opportunities

Transportation planning identifies and prioritizes those elements of the transportation system where problems exist today or where problems will exist in the future given growth in travel. In addition, transportation planning can also identify areas where significant problems do not exist today, but where changes to the system can provide opportunities for enhanced efficiency of operation. The methods used to identify these deficiencies and opportunities can vary widely. In some cases, large scale transportation network models are used to estimate the future traffic volumes and to then compare them to the capacity of the road network to handle such volumes. The volume-to-capacity (V/C) ratio has served as the major means of identifying the location of system deficiencies. However, the Intermodal Surface Transportation Efficiency Act (ISTEA) of 1991 has introduced into transportation planning the concept of *performance measures,* which can be much broader than the traditional V/C ratios. Such performance measures are targeted at specific elements of the transportation system, and when a certain threshold is reached, corrective action needs to be taken. Example performance measures are listed in Table 3-1.

Develop and Analyze Alternatives

Once the planning process has identified areas where improvements are needed, transportation planners define different strategies that could solve the problem. Historically, these strategies have focused on improvements to highways, such as adding new lanes, improving traffic control through signals or signing, or improving traffic flow through channelization. Today, however, planners are looking at a wide variety of strategies that can be used to solve the transportation problem. These strategies can include reducing the demand for transportation through flexible working hours, increasing average vehicle occupancy through such things as carpools or transit use, or raising the price of travel through the use of tolls.

More recently, the application of advanced transportation technologies to the operation of a road system, known as *intelligent transportation systems,* has become an important strategy in many cities. Such systems might include network surveillance through video cameras, centralized control centers that can reroute traffic around incidents, and dynamic traffic control devices that provide coordinated traffic signal timings to maximize the amount of traffic that can flow through a set of intersections.

Evaluate Alternatives

Evaluation brings together all the information gathered on individual alternatives and provides a framework to compare the relative worth of the alternatives. This evaluation process most often relies on the different measures of effectiveness that relate to the goals and objectives defined at the beginning of the planning process. In addition, evaluation includes methods for comparing in an analytical way the relative value of the alternatives. One of the most used approaches is

TABLE 3-1 Candidate Performance Measures

Time-Related Measures
Average travel speed
Average travel time
Travel time contours
Origin-destination travel time
Percent travel time under delay conditions
Percent of time average speed below threshold value
Volume Measures
VMT/lane mile
Traffic volume
Congestion Indices
Roadway congestion index
Excess delay
Delay Measures
Delay/trip
Delay/VMT
Minute-miles of delay
Delay due to construction/incidents
Level-of-Service Measures
Lane-miles at/of LOS "X"
VHT/VMT at/of LOS "X"
Predominant intersection LOS
Number of congested intersections
Vehicle Occupancy/Ridership Measures
Average vehicle ridership
Persons/vehicle

VMT = Vehicle Miles Traveled
LOS = Level of Service

the benefit/cost ratio, which compares the alternatives on the basis of discounted benefits and costs. The benefit/cost ratio is a means of identifying the most economically efficient alternative by defining benefits and costs in terms of dollars. Many measures of effectiveness, however, cannot be defined in dollar terms, and thus effective evaluation must include many different means of presenting relevant information to those who select the preferred alternative (5,6).

Implement Plan

The major product of the transportation planning process is the transportation plan. However, in the United States, another major product of transportation planning is the programming document. At the regional level, this document is called the *transportation improvement program (TIP)* and lists all the transportation projects that will be implemented over the next few years in the region. If federal funds are to be used in the implementation of a project, the project must be listed in the TIP. Programming the projects that result from a planning effort is thus a very important step. Because the number of projects recommended by various planning efforts often exceeds the amount of funding that is available, the programming process must rely on a process of prioritizing projects. This prioritizing can be based on subjective assignment of weights to specific types of projects (e.g., safety projects receive a weight of 0.6, while capacity expansion projects receive a weight of 0.3), or the priorities can be assigned on the basis of a benefit/cost methodology (e.g., those projects having the best benefit/cost outcome are programmed first).

Monitor System Performance

Transportation planning is a process that continually examines the performance and condition of the transportation system to identify where improvements can be made (7). Therefore, some means of monitoring system performance is necessary to systematically identify areas where these improvements might occur. This system monitoring has traditionally been based on traffic volumes obtained from traffic counters placed at strategic locations in the network (i.e., increasing traffic volumes indicate congestion). More recently, transportation planning has begun to incorporate the results of management systems that are aimed at specific network issues. For example, most state transportation agencies have pavement management systems that continually monitor the condition of road pavements, predict likely deterioration, and estimate when new pavement surfaces should be provided. Other types of management systems now being used by many transportation agencies include bridge management systems, safety management systems, public transportation management systems, intermodal management systems (which focus on the transfer of people and goods from one mode to another), and congestion management systems.

3-2 Data Collection

As indicated in the previous section transportation planning depends on quality data to provide input into the many different steps in the planning process. Not surprisingly, in many cases the cost of data collection is the biggest share of the budget for planning studies. Transportation planners use a variety of methods for collecting the data needed for input into the planning process (8). Some of these methods are described in the following paragraphs.

Road-Use Studies

Highway planners perform road-use studies to determine the relative use of various parts of a highway system. Information may be obtained by personal interviews with a representative sample of the registered motor vehicle operators within the highway jurisdiction. Drivers are asked to state the total mileage driven in one year and the proportion of that mileage that was driven on various classifications of streets and highways. The planners can use such information to determine the proportionate use made of the various roads by urban, suburban, and rural residents. By correlation with financial data, planners may then compare the benefits received with the revenues contributed by various groups of citizens. Many states use such information to more equitably distribute the costs of highway transportation among the various benefiting groups in accordance with the concepts presented in Chapter 2.

Pavement-Life Studies

State highway agencies have been conducting pavement-life studies since the mid-1930s. The primary objective of such studies is to determine the average rate of retirement and the estimated average service life for each type of pavement. Roads are usually retired by resurfacing, reconstructing, or abandoning them. Information is also assembled on costs of construction, maintenance, and depreciation.

With this information, planners may estimate the amount and cost of replacements that will be required in the future. They can program construction and reconstruction operations on the basis of anticipated future revenue receipts. Highway engineers use pavement-life data in economic calculations, as the previous chapter indicated.

Traffic Volume Studies

Engineers and planners have been conducting traffic volume studies for more than 60 years. The procedures used in such studies are well documented elsewhere (*7,8*) and will be only briefly described in this section. Traffic volume studies provide highway officials with essential information on the number of vehicles using the highway system. Such information is needed for "the determination of design standards, the systematic classification of highways and the development of programs for improvement and maintenance" (*9*). From traffic volume studies, planners may estimate the vehicle-miles of travel (VMT) or vehicle-kilometers of travel (VKT) on the various classes of rural and urban roadways. Such travel data indicate a measure of service provided by the system and facilitate the appraisal of safety programs and the development of equitable highway finance and taxation programs.

The fundamental measure for traffic volume studies is annual average daily traffic (*ADT*). By using empirically based relationships, *ADT* values can easily be converted to peak hourly volumes.

Because of the wide variability of traffic flows during different times and at different locations, it is practically impossible to arrive at a true accounting of the traffic volumes on each section within a highway system. Even if comprehensive and continuous monitoring of traffic were possible, the costs of maintaining such a counting program would be prohibitive. Traffic volume studies are therefore carried out on a sample basis.

Many highway agencies have established statewide or areawide traffic counting programs. In such programs, traffic volumes are measured by means of a large number of short counts covering an entire state or area. The locations for the counts are called *coverage stations*. Coverage stations are typically spaced at intervals of 3 to 8 km (2 to 5 miles) along rural highways and about 1.6 km (1 mile) apart along urban highways and streets. Measurements at coverage stations are typically recorded on portable counters actuated by pneumatic detectors. At coverage stations, the traffic volume is not monitored continuously but is sampled over time. The duration and frequency of such counts generally depend on the class of the street or highway and the amount of traffic it serves. Typically, coverage counts consist of a single nondirectional 24- or 48-hr count repeated at 4-year intervals.

In order to convert coverage counts to estimates of average daily traffic, traffic volumes are measure at a small number of strategically selected *control stations*.

At those locations, traffic counts are usually made over a 7-day period to measure hourly and daily traffic patterns of variation. In addition, 24-hr nondirectional counts are normally made at control stations on a monthly basis to measure the seasonal variation. On the basis of temporal patterns of variation measured at the control stations, the sample coverage counts can be adjusted by correction factors to provide estimates of *ADT* throughout the system.

In a number of highway jurisdictions, permanent counting stations have been established at a small number of selected locations. At those locations, traffic volumes are measured continuously, providing an accurate measure of the hourly, daily, and seasonal patterns of variation. Generally, automatic traffic detection and recording equipment is employed at continuous counting stations, and in some states traffic volume data are transmitted to a central computer over telephone lines for processing and analysis (*10, 11*).

Traffic counts made with automatic devices generally must be supplemented by manual surveys to record information on the character of traffic. Such surveys are often undertaken at weighing or loadometer stations. There, information is collected on the type of vehicle; its length, width, and height; the rated capacity; and the axle and gross loads. Additional information on the type of commodity carried, the origin, the destination, and other pertinent facts may also be noted. When all the information of the traffic survey has been assembled, the expansion factors have been determined, and the annual average 24-hr traffic has been computed, traffic maps and flow maps can be prepared. A base map previously prepared by the road inventory survey is used for the traffic map. The numerical traffic volumes for each segment of the road are usually placed on this map. For more rapid visualization, the volume of traffic may be shown on the traffic flow map by bands proportional in width to the volume of traffic passing through the various points at which base station counts were taken. Figure 3-2 shows a typical traffic flow map.

Travel Surveys

Travel surveys identify where and when trips begin and end, the trip purpose, and the mode of travel, as well as certain social and economic characteristics of the tripmaker. Such surveys may also determine the types of land use at trip termini, automobile occupancy, and, for freight, the type of commodity transported. Travel surveys serve as a fundamental source of data for urban transportation studies and are being increasingly used in statewide transportation studies (*10, 11*).

By means of travel surveys, planners attempt to obtain a sample cross section of all travel on a typical day within a specified area. By expanding the data, they are able to produce an estimate of the average travel demand on the transportation system at the time the survey was conducted.

There are four general classifications of travel surveys:

1. household travel surveys
2. roadside surveys
3. modal surveys
4. goods movement surveys

Household Travel Surveys These surveys are especially useful for determining the number and characteristics of person trips or auto-driver trips made by

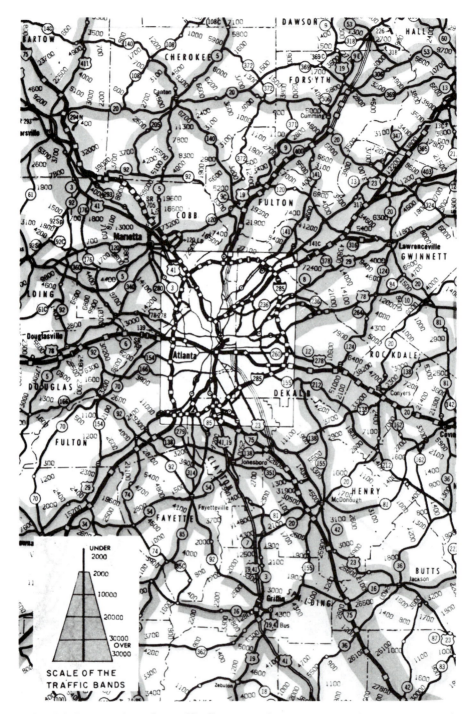

FIGURE 3-2 A portion of a traffic flow map. (Courtesy Georgia Department of Transportation.)

residents within a specified study area. By such surveys, planners may collect extensive data on socioeconomic characteristics of households and relate such information to household travel. Such relationships are extremely valuable for making forecasts of future travel by persons who reside in the study area (*12*). Household travel surveys include home interview, telephone, and mail surveys.

The home interview method involves identifying a sample of all the households within the survey area, arranging for interviews, and questioning respondents about trips made by persons in their respective households the previous day. The sample size varies with the population of the survey area, ranging from 20 percent in areas with a population less than 50,000 to 4 percent in areas with over 1 million. Home interviews will usually be preferred over other household survey methods if a long or complicated survey instrument is to be used. Higher response rates may be expected in home interview surveys than in telephone or mail surveys. On the other hand, home interview surveys are usually more expensive than other methods.

The Roadside Survey Method This method consists of stopping vehicles and asking the drivers for information on trip origin, destination, and purpose. Typically, all the vehicles passing the roadside survey station are counted, but only a sample of the drivers (e.g., 20 percent) are interviewed. Alternative roadside survey methods involve handing out voluntary return postcard questionnaires to drivers or recording license numbers and mailing questionnaires to the vehicle owners. The stations where the counts are made are usually located on all the main approach highways to the survey area. If the stations are located along an imaginary line encircling a survey area, they are referred to as "cordon stations," and the counts are known as "cordon counts." Roadside surveys are also used to intercept and interview drivers of vehicles crossing selected boundary lines (i.e., screen lines) within a state or region.

Roadside surveys are especially useful in recording relatively long and infrequent trips, including those made by persons who reside outside the survey area. Such surveys provide only limited information about the nature of the trips surveyed and practically nothing about the social and economic characteristics of the tripmaker. Roadside surveys must be carefully planned and executed to prevent intolerable congestion, especially along heavily traveled routes.

Modal Surveys In a large number of states where multimodal transportation agencies have been established, special surveys may have to be made of travel by bus, rail, and air transportation. Typically, modal surveys are made by interviewing passengers at terminals or while on board public carriers. To a lesser extent, telephone interviews and surveys of license plates at terminals have been used to collect modal travel data. The U.S. Department of Transportation has published detailed guidelines on statewide modal surveys (*12*) and urban mass transportation travel surveys (*13*) to which the reader should refer for additional information.

Goods Transportation Surveys The transportation of goods within a state or urban area is complex, involving several transport modes, a combination of line-haul and distribution traffic, and a variety of commodity types and users. The impact of goods movement on the functioning of a transportation system is significant, and planners should endeavor to incorporate goods transportation planning into the existing planning process. Unfortunately, planning procedures for goods transport analyses have not kept pace with those for passenger transportation.

Because of the competitive nature of the freight industry, freight carriers may be reluctant to respond to mailed questionnaires. For this reason, experience suggests that the personal visit interview technique is the most effective means of gathering confidential information from transportation industry sources.

In certain instances, goods transportation surveys may be limited to the collection of origin-destination information for trucks by truckload type and weight. On the other hand, comprehensive multimodal goods transportation surveys may be desired in which detailed shipping information is sought for specific commodities, including consignee, shipper, carrier, commodity type, origin, destination, weight, value, time of origin and destination, and shipping rate. It may be possible to collect a great deal of terminal facility data on a confidential basis from terminal operators, including the number of movements per day (by mode type) and the class and volume of cargo handled during average and peak periods. Additional information on goods transportation surveys is given by Ref. 14.

Parking Surveys   The purpose of a parking survey is to determine the parking habits and requirements of motorists and the relation of these factors to other uses of existing parking facilities.

A parking survey should be designed so that the information collected will provide data on: (1) the location, kind, and capacity of existing parking facilities, (2) the amount of parking space needed to serve present demands, (3) the approximate location of possible additional parking facilities, and (4) the legal, administrative, financial, and economic aspects of parking facilities.

The first phase of the parking survey is to make an inventory of all available parking facilities. This includes curb and off-street parking areas such as parking lots, garages, and service stations. Any physical and legal restrictions are also noted. The theoretical capacity of a parking lot or garage is usually determined by dividing the gross area by the area needed by one car, which varies from about 23 to 37 m^2 (250 to 400 ft^2), depending on the type of facility and the design. (See Chapter 9.) Twenty-two lineal feet is usually used in determining theoretical curb capacities. Calculation of the theoretical capacity does not imply that more cars cannot be accommodated. Some cars are parked for short periods of time, and their spaces then become available to others. Cooperation of the owners of off-street parking facilities must be obtained in order to determine the actual capacities or numbers of cars that are accommodated for periods of the survey so that the results can be correlated with capacities of other facilities. Additional information on the duration of parking is also to be obtained.

The determination of actual capacities of curb parking facilities is more readily done. Parking, when permitted, is usually for 1- or 2-hr periods. Therefore, the number of space-hours available is determined from the inventory of the curb space available. It is assumed that parking is for the legal time specified. Studies are also made to ascertain the extent of illegal parking and its effect on the total effective capacity of curb parking facilities.

In order to determine the number of cars that may utilize available parking spaces in a central area, a cordon line is usually drawn around the area to be studied, and cordon counts are made of all vehicles entering and leaving it. The counts are made manually so that a motor-vehicle classification can be made. Automatic recorders are often used to supplement the manual counts and to serve as a basis of control for abnormal conditions.

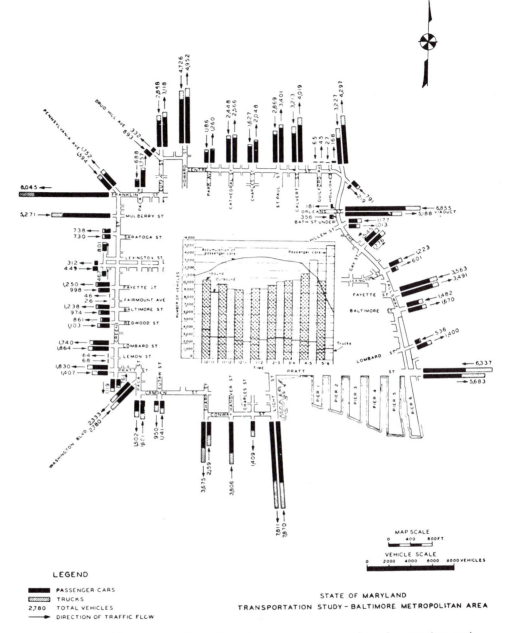

FIGURE 3-3 Parking survey of number of passenger cars and trucks entering and leaving the downtown area on a weekday between 10:00 A.M. and 6:00 P.M.

Motorists who are interviewed at all parking facilities within the cordon area are asked for their home address, destination, and the purpose of the trip. Figure 3-3 shows the results of a parking survey in the city of Baltimore.

The information received in a parking survey may point to a number of recommendations for the relief of congestion and for general improvement in traffic facilities. Such recommendations might, for example, include the provision of additional parking facilities, either publicly or privately owned, better enforcement of parking regulations to make available space at the curb now being used by motorists parking illegally, changes in zoning regulations, and so forth.

3-3 Use of Data for Origin-Destination Analysis

Perhaps the best illustration of how the different data collection methods discussed in the previous section are used in a coordinated way can be found in the development of an origin-destination database. Determining the origins and corresponding destinations of the trips made in a study area is a critical step in the transportation planning process. This so-called O-D database provides some indication of where the majority of trips are coming from and where they are going to and in which transportation corridors these trips are likely to travel. Much of the data collection effort in a typical planning study is, in essence, designed to determine these origin-destination patterns.

The first step in this determination is to establish a system of designating origins and destinations. This is done by dividing the study area into geographic units called traffic analysis zones (TAZ), which are representative of the types of land use found in each particular area. The boundaries of these zones should, whenever feasible, coincide with census tract boundaries and with jurisdictional boundaries. The zones should also be small enough so that there are a large number of *inter*zonal trips being made in the study area. The FHWA, for example, suggests that the *intra*zonal trips should not be more than 10 to 15 percent of the total trips being made in each zone (*15*). Figure 3-4 illustrates the concept of the zonal system that is basic to any transportation planning effort. As shown, the study area is divided into analysis zones internal to the study and zones that represent origins or destinations outside the study area. The cordon line represents the imaginary boundary between the internal and external zones.

Comprehensive urban O-D studies usually involve the use of a combination of the survey procedures described previously: household travel surveys, roadside surveys, modal surveys, and goods movement surveys.

Such surveys consist of two parts:

1. an external survey involving the collection of travel data on vehicles crossing the external cordon line
2. an internal survey that focuses on the travel habits of persons who live within the area bounded by the external cordon line

The external survey data are collected along the external cordon line. Directional traffic counts are normally obtained for every street and highway that crosses the cordon line. Origin-destination data are collected by roadside surveys performed on all major routes that cross the cordon line.

The internal survey consists of the following subdivisions: (1) home interviews (trips by all individuals), (2) truck studies, (3) taxi studies, and (4) public transit information.

Interviews are made on a predetermined schedule, and forms are provided for tabulating the necessary information. The information required is for the day preceding the home interview. This information includes the number of occupants in a household, occupation, number making the trip, use of private or public transportation, origin and destination, purpose of the trip, and so forth.

In selecting the sample, use is made of certain available basic records. These may include (1) Sanborn maps, a series of copyrighted maps prepared for the use of insurance companies, assessors, and so forth, which show all structures

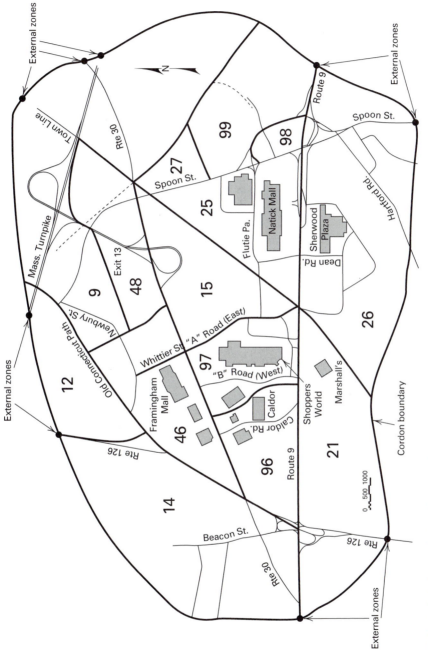

FIGURE 3-4 Example traffic analysis zones.

together with other information, (2) city directories, (3) Bureau of the Census "Block Statistics," which show blocks and block numbers, and (4) land-use maps that may be available.

Truck and taxi information is generally available from the records of the companies that operate them. Usually 20 percent of the trucks and 50 to 100 percent of the taxis registered in an area are sampled. Public transit information is usually obtained directly from the transit companies. This should include location of existing routes, schedule of operations, and the total number of passengers carried on an average weekday during the period of the survey.

The completeness with which trips are reported in the internal survey may be checked by three methods: (1) use of control points for comparison, (2) a screen-line comparison, and (3) a cordon-line comparison. Not more than three or four control points should be selected for use in any one study. Control points are usually viaducts, bridges, underpasses, or other points of constriction through which large volumes of traffic pass. The average traffic volume and classification at each control point must be determined by manual counts made during the survey so that the expanded counts obtained from home interviews can be compared with the actual ground counts at these points.

Screen lines are natural barriers such as rivers or railroads; they usually have a limited number of crossings at which ground counts can be made at a minimum expense. The purpose of the screen line is to divide the area of the internal survey into two parts in order to determine the number of vehicles moving from one part to the other. A comparison can then be made between the number of trips having their origin on one side and destination on the other, as actually counted, and the number of trips as determined from the expanded interviews.

A cordon-line comparison deals with passenger-car trips by residents of the internal area and truck trips by trucks registered in the area, but only those trips that cross the cordon line. The total number of such trips recorded in the external survey can be compared with the total number recorded in the internal survey.

In many recent surveys conducted in various municipalities by the method just described, about 90 percent of the traffic passing control points and screen lines during 16 hrs of operation of external stations has been accounted for by the expanded interview data. The remaining 10 percent has been assumed to be made up of cars circulating in search of a parking place, of unimportant short trips not reported in interviews, and of trips by persons living outside the area and not intercepted at the external cordon stations. These trips are considered to be of minor importance as far as the principal purposes of the survey are concerned.

Figure 3-5 shows a comparison of the number of vehicles counted at control points with the number passing those points as determined from the interviews.

For cities with a population of less than 5,000 it may not be necessary to conduct an internal survey. It may be possible to collect sufficient O-D information by means of roadside interviews at the external cordon. In slightly larger cities (with populations in the range of 5,000 to 50,000), in lieu of an internal survey, roadwise surveys may be performed along two cordon lines, one at the edge of the city and one at the fringe of the central business district.

Origin-destination data have traditionally been displayed in the form of desire line maps. Desire lines indicate the desires of vehicle users as direct lines of travel from one point to another within a given area, assuming that direct routes are available. The information that these lines represent is especially helpful to plan-

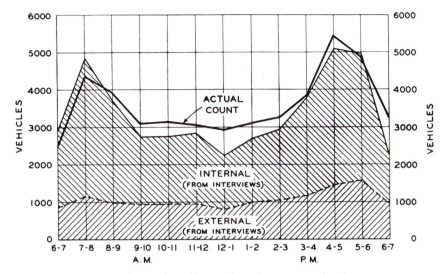

FIGURE 3-5 Comparison of traffic passing three control points as reproduced from interviews and as actually counted at those points. (Courtesy
Federal Highway Administration.)

ning and design engineers when routes for new arterial improvements are to be
selected.

Plans for improvement should be keyed to the major demands of traffic, as
indicated by the desire lines with the greatest widths in a plot of the type shown
in Figure 3-6. Assuming that this can be done, the routes thus improved should
best meet the desires of the largest number of vehicle users.

3-4 Transportation Systems Modeling

Transportation planning by its very definition is concerned with future travel
demands and putting in place the facilities and services that will accommodate
these demands. The challenge to transportation planners is to make reliable
forecasts of traffic demand that reflect the effects of changes in population, social,
and economic conditions as well as changes in the transportation network. Figure
3-7, for example, illustrates the relationship between the many different factors
that influenced travel demand in the United States over the last decades. Unless
reliable forecasts of future traffic that take these types of factors into account
are made, transportation officials run the risk of building facilities that will either
receive little use or be prematurely overloaded.

The recognized components of future travel demand include:

1. *Existing traffic.* Traffic currently using an existing highway that is to be
   improved.
2. *Normal traffic growth.* Traffic that can be explained by anticipated growth
   in state or regional population or by areawide changes in land use.
3. *Diverted traffic.* Traffic that switches to a new facility from nearby roadways.
4. *Converted traffic.* Traffic changes resulting from change of mode.
5. *Change of destination traffic.* Traffic that has changed to different desti-

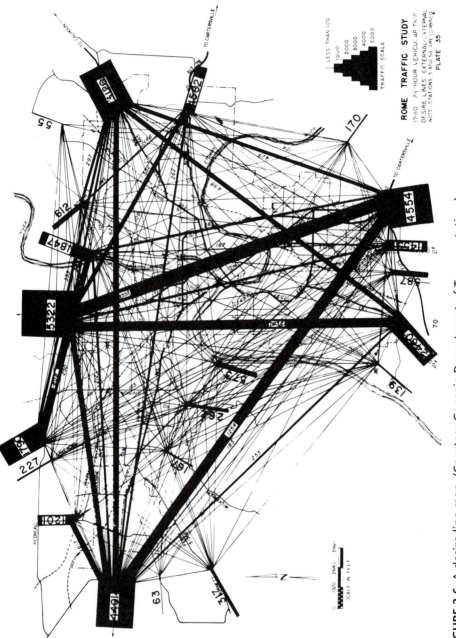

FIGURE 3-6 A desire line map. (Courtesy Georgia Department of Transportation.)

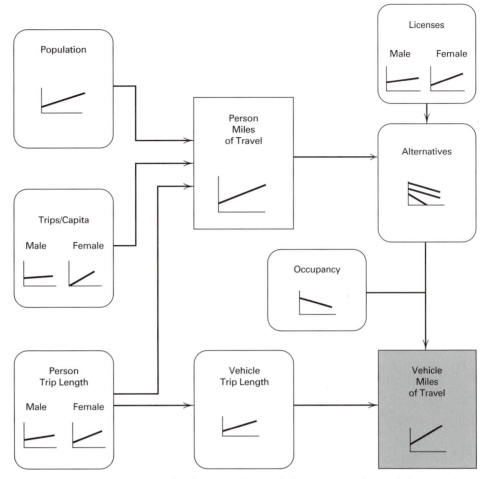

FIGURE 3-7 Factors in growth of personal travel. (SOURCE: Federal Highway Administration, *Travel Behavior Issues in the 90's, 1990 Personal Transportation Survey.*)

nations, where such change is attributable to the attractiveness of the improved transportation and not to changes in land use.

6. *Development traffic.* Traffic due to improvements on adjacent land in addition to the development that would have taken place had the new or improved highway not been constructed.

7. *Induced traffic.* Traffic that did not previously exist in any form but results when new or improved transportation facilities are provided.

The models that are used to predict future travel demand must account for these different components of the likely traffic flow that will occur given future conditions.

Basic Concepts in Transportation Systems Modeling

Most transportation models are based on some fundamental assumptions and approaches that heavily influence how these models are used by transportation planners and engineers (*1, 16, 17*). The most important are as follows:

Tripmaking is a Function of Land Use   The concept of derived demand discussed earlier is a critical point of departure in model development. Most models are developed on the assumption that tripmaking is related to the types of land use at the origin and destination ends of the trip (*18*).

Trips Are Made for Different Purposes   Trips are made to accomplish different objectives. In the morning, a large percentage of travel in urban areas consists of individuals going to work or going to school. In midday, the types of trips might include shopping, personal business and recreational. Because modeling is based on the types of activities and land uses found at both ends of the trip, the modeling process often treats these trips separately and then aggregates them at the end for an estimate of the total number of trips on the network.

Trips Are Made at Different Times of the Day   The basis for most modeling is the determination of origin and destination patterns in the study area. These patterns will clearly differ by time of day in that different types of trips are being made at different times of the day. Thus, modeling is often done with an origin-destination trip table (a matrix that indicates the trip patterns) that represents defined time periods. It is not uncommon, for example, to have traffic estimates for the morning peak three hours, the afternoon peak three hours, the midday time period (e.g., 10:00 A.M. to 2:00 P.M.), and for the entire day.

Travelers Often Have Different Options Available to Them   Not only is it important to know the origin-destination trip patterns in a study area, but travelers often have the option of making this trip with different transportation modes and/or on different routes in the network. For example, in an urban area, a trip from one zone to another might have available the following options for making this trip: single-occupant automobile, carpool, bus transit, or rail transit. If the single occupant or carpool modal option is chosen, the traveler(s) could choose different routes on the highway network to reach their destination.

Trips (and thus the Characteristics of Travel) Are Made to Minimize the Level of Inconvenience Associated with Reaching a Destination   The choice of mode, for example, is often based on the relative travel times associated with reaching a destination by each mode or by how expensive the use of each mode is to the user. The choice of paths through a network, for example, is often assumed to be based on the minimum amount of time it takes to go from origin to destination.

Transportation Networks and Traffic Analysis Zones Are the Basis of Systems Modeling   Given the numerous origins and destinations in a study area, and given the different paths that can be used to reach destinations, transportation models must be able to represent the land use and transportation network characteristics that are fundamental to trip estimation. As noted earlier, traffic analysis zones are used to represent the land use and demographic data that influence tripmaking. The transportation system is often pictured as a network of *nodes* and *links* that conceptually represent intersections and roadways (*19*). Figure 3-8 shows a representation of a transportation network.

Transportation modeling consists primarily of four steps. These will be denoted with the following variables:

Trip Generation (T_i):   The number of trips produced in traffic analysis zone *i*.

Trip Distribution (T_{ij}):   The number of trips produced in zone *i* and attracted to zone *j*.

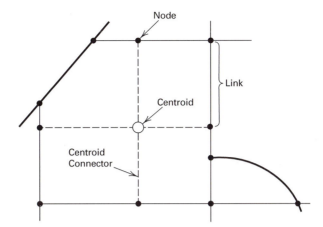

FIGURE 3-8 Representation of a transportation network. (SOURCE: Federal Highway Administration, *Traffic Assignment.*)

Mode Split (T_{ijm}):   The number of trips produced in zone *i* and attracted to zone *j* traveling by mode *m*.

Trip Assignment (T_{ijmr}): The number of trips produced in zone *i* and attracted to zone *j* traveling by mode *m* over route *r*.

Trip Generation

The trip generation step in transportation modeling relates the number of trips being produced from a zone or site by time period to the land use and demographic characteristics found at that location. A necessary input step into trip generation is to have some indication of what the land use and demographic characteristics are likely to be. For future conditions, special models are used to estimate population, number of dwelling units, auto ownership, income, employment, retail sales, and the like that will likely characterize the future conditions for this zone. In many cases, and especially for regional or statewide planning, these future estimates are provided by economic or demographic planners rather than by transportation professionals.

Studies have shown that the rate of tripmaking is closely related to three characteristics of land use: (1) intensity of land use (e.g., dwelling units per acre, employees per acre, etc.), (2) character of land use (e.g., average family income, car ownership, etc.), and (3) location relative to major economic activities (e.g., closeness to downtown). The different methods thus relate tripmaking rates to these types of variables.

There are many different methods that can be used to estimate the trip-producing activity for particular zones. The three most common methods include trip rates from national or local studies, cross classification analysis, and regression equations.

Trip Rates from National/Local Sources Planners often undertake special studies to determine the number of trips associated with different types of land uses. For example, traffic counts can be taken at the driveways of stores or restaurants to count the number of vehicles attracted to these locations. The

Institute of Transportation Engineers (ITE) publishes the *Trip Generation Handbook,* which is a compilation of trip rate studies that have occurred throughout the United States *(20)*. In the absence of local information, the *Handbook* is often used as a source of information on trip rates.

Cross Classification Analysis If good data are available from a study area to provide information on the relationship between socioeconomic variables and tripmaking, cross classification analysis can be used to develop relevant trip rates. An illustration of cross classification analysis is shown in Figure 3-9. In this case data were available from the Census on the number of persons per household and household income and the corresponding number of trips made per household type. By estimating the number of future households and applying the calculated trip rates, the future number of trips produced in this zone can be calculated.

Regression Analysis Given the high correlations that typically exist between trip rates and socioeconomic variables, many models use least-squares regression equations to estimate trip production per zone. The ITE *Handbook* also provides regression equations for estimating trip productions for certain types of land use. Typical regression equations might include the following:

Household Size	Auto Ownership					
	0		1		2+	
	HH	Trips	HH	Trips	HH	Trips
1	1,200	2,520	2,560	6,144	54	130
2	874	2,098	3,456	9,676	5,921	20,165
3+	421	1,137	2,589	8,026	8,642	33,704

Number of trips per household size by auto ownership, obtained from regional study

Household Size	Auto Ownership		
	0	1	2+
1	2.1	2.4	2.4
2	2.4	2.8	3.4
3+	2.7	3.1	3.9

Trip rates obtained from previous matrix

Household Size	Auto Ownership		
	0	1	2+
1	25	125	3
2	32	175	254
3+	10	89	512

Forecasted number of households in study zone by autoownership and size

Household Size	Auto Ownerships			
	0	1	2+	
1	52	300	7	
2	77	490	864	Total
3	27	276	2001	4,094

Forecasted number of trips in zone determined by multiplying trip rates by number of households in category

FIGURE 3-9 Cross classification analysis

$$T_i = 0.34(P) + 0.21(DU) + 0.12(A)$$
$$A_j = 57.2 + 0.87(E)$$

where T_i = Total number of trips produced in zone i
A_j = Total number of trips attracted to zone j
P = Total population for zone i
DU = Total number of dwelling units for zone i
A = Total number of automobiles in zone i
E = Total employment in zone j

Trip Distribution

The major product of the trip distribution step in transportation modeling is the *trip table,* an origin-destination matrix that shows the number of trips originating in the study zones and where these trips are destined to. Two major methods of producing such trip tables are the gravity model and the Fratar method.

The Gravity Model The gravity model is so named because of its similarity to Newton's law of gravitation. Employed first for sociological and marketing research, the gravity model began to be used for transportation studies in the early 1950s. Since that time, the model has been slightly modified and has become the predominant technique for trip distribution. The original version of the model, which was introduced by Voorhees (*21*), was of the form:

$$T_{ij} = \left[\frac{\dfrac{A_j}{(D_{ij})^n}}{\dfrac{A_1}{(D_{i1})^n} + \dfrac{A_2}{(D_{i2})^n} + \cdots + \dfrac{A_m}{(D_{im})^n}} \right] P_i \qquad (3\text{-}1)$$

where T_{ij} = trips from zone i to zone j for a specified purpose
P_i = total trips produced at zone i for the specified purpose
A_j = a measure of attraction of the jth zone for trips of this purpose
D_{ij} = distance from zone i to zone j
n = some exponent that varies with trip purpose

Consider the following numerical example. Given a residential zone that produces a total of 110 shopping trips per day, distribute these trips to shopping centers 1, 2, 3 in accordance with the gravity model. Distances between zones are shown on the sketch. The value of *n* in the gravity model is 2. Use the amount of commercial floor space within the destination zone as the measure of attractiveness:

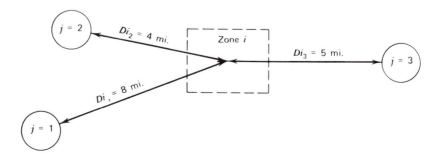

Shopping Center	Floor Space (thousand ft²)
1	184
2	215
3	86

$$\text{Trips from zone } i \text{ to zone } 1 = \frac{\dfrac{184}{(8)^2}}{\dfrac{184}{(8)^2} + \dfrac{215}{(4)^2} + \dfrac{86}{(5)^2}} \times 110 = 16$$

$$\text{Trips from zone } i \text{ to zone } 2 = \frac{\dfrac{215}{(4)^2}}{\dfrac{184}{(8)^2} + \dfrac{215}{(4)^2} + \dfrac{86}{(5)^2}} \times 110 = 75$$

Similarly, the trips from zone i to zone 3 $= 19$

Total trips $= 110$

The gravity model has been modified in recent years to reflect research and experience with the model. It has been found that decreases in travel propensity are more closely related to travel time than to distance. In addition, it has been established that the exponent of travel time, n, varies not only with trip purpose but also with trip length. Trip distribution analyses are therefore usually stratified according to travel time t with different calibrated values of the exponent being determined for a given city and trip length. Furthermore, to facilitate efficient computer use of gravity models, it is now the practice to represent the effect of spatial separation on travel between zones in the form of travel time factors

$$F_t = \frac{C}{t^n}$$

where C is a constant. Instead of a surrogate measure of attractiveness such as commercial floor space use or number of employees, actual zonal total trip attractions are used in the equation. Current gravity models permit an analyst to make adjustment to allow for special social or economic conditions by choice of socioeconomic adjustment factor.

Currently, the recommended formulation of the gravity model is

$$T_{ij} = \frac{A_j F_{t_{ij}} K_{ij}}{\sum\limits_{\text{all zones}} A_x F_{tij} K_{ix}} \times P_i \tag{3-2}$$

where $F_{t_{ij}}$ = travel time factor for travel time between zones i and j $= \dfrac{C}{t_{ij}^n}$

K_{ij} = socioeconomic adjustment factor between zones i and j

A_j = total attractions at zone j

The Fratar Method   Proposed by T. J. Fratar in 1954, this method is designed to compute trip interchanges where there is nonuniform growth within various sections of a study area. This method and variations on it are called "growth factor" methods. A growth factor for a particular zone is simply the ratio of expected future traffic to the existing traffic emanating from the zone. According to the Fratar method, future travel patterns between zones are determined by the present travel patterns and growth factors at the destination zone. The method is an iterative factoring process in which the number of future origins at each zone is held constant. It is analogous to the Hardy Cross method of successive approximations for moment distribution in indeterminate structures. In recent years, the method has been principally used to predict trip interchanges between external sections of a study area. It has also been employed for statewide transportation studies. For further information on the Fratar method, the reader should consult Ref. 22.

Mode Split Models

Mode split models are primarily oriented toward predicting the percentage of individuals who will choose one mode over others for making a particular trip. A great deal of research has been undertaken to better specify models that correctly reflect the individual's decision-making process in making this choice. Empirical evidence indicates that the following factors influence mode choice (*19*):

1. type of trip (e.g., trip purpose, time of day)
2. characteristics of the tripmaker (e.g., income, age, auto ownership)
3. characteristics of the transportation system (e.g., relative travel times for the modes available to make the trip)

The most recent advances in transportation modeling have focused on what are called *individual choice models,* which relate the choice of one mode over another to the utility associated with each mode. The utility or level of attractiveness of each mode can be defined with a variety of variables relating to travel costs, travel time, reliability, and so on. Different model formulations have been used to predict mode choice based on different modal utilities, but the most common formulation is called the *logit model* and is of the following form.

$$P_{it} = \frac{e^{U_{it}}}{\sum\limits_{\text{All } j} e^{U_{jt}}} \tag{3-3}$$

where P_{it} = probability of individual t choosing mode i
U_{it} = utility of mode i to individual t
U_{jt} = utility of mode j to individual t

An example of this type of mode split model follows. Assume that we know there are 1,000 trips being made between zones i and j (which we obtained from trip distribution). There are three modes available to make this trip. The utility of the individual modes is defined as

$$U_{\text{auto}} = 1.0 - 0.1 \, (TT_{\text{auto}}) - 0.05 \, (TC_{\text{auto}})$$
$$U_{\text{bus}} = -0.1 \, (TT_{\text{bus}}) - 0.05 \, (TC_{\text{bus}})$$
$$U_{\text{walk}} = -.05 - 0.1 \, (TT_{\text{walk}})$$

where TT = Travel time by mode in minutes
TC = Travel cost by mode in dollars

Assume that we know that the travel time for auto is 5 minutes, for bus 15 minutes, and for walking 20 minutes. The corresponding costs are $0.60 for auto and $0.50 for bus. Substituting these numbers into the utility equations results in the following estimates of modal utilities:

$$U_{auto} = 0.47 \qquad U_{bus} = -1.525 \qquad U_{walk} = -2.5$$

Using equation 3-3, we find the following probabilities associated with the use of each mode:

$$P_{auto} = \frac{e^{0.47}}{e^{0.47} + e^{-1.525} + e^{-2.5}} = 0.842$$

$$P_{bus} = \frac{e^{-1.525}}{e^{0.47} + e^{-1.525} + e^{-2.5}} = 0.114$$

$$P_{walk} = \frac{e^{-2.5}}{e^{0.47} + e^{-1.525} + e^{-2.5}} = 0.043$$

Given the 1,000 trips between these two zones, one would predict that 842 would use the automobile, 114 would use the bus, and 43 would walk.

In some cases, such as small urban areas, the mode split step is not used in the modeling process in that there are few alternative modes. The auto is the only way of making a trip. In larger urban areas, however, mode split is an important step in the modeling process.

Trip Assignment

The final step in transportation modeling is to assign the trips to paths in the network. The most important concept in this trip assignment process is that travelers will choose a path that minimizes travel time from origin to destination. Trip assignment models, therefore, are based on minimum time algorithms that identify the minimum time paths through networks. Common to all assignment approaches is the existence of link performance functions that relate travel time to factors such as volume and roadway capacity variables (*23*). Figure 3-10 graphically illustrates the concept of a link performance function. In this figure the time associated with travel on a link in the network is related in a nonlinear way with the amount of traffic on that link. As shown, the closer this volume approaches to the capacity of the roadway, the greater travel time is associated with traversing this link. In practical terms this relationship means that as a roadway becomes congested, it will take a longer time to travel the link distance.

Several approaches toward network assignment are commonly used by planners. All of them use the minimum time path concept in producing estimated volumes for each link in the network.

All-Or-Nothing Assignment This assignment procedure estimates the shortest time path between each zonal pair in the system based on uncongested speeds and assigns all of the volume making these trips to the shortest path. Therefore, if there are 1,000 trips going from zone *i* to zone *j* and the shortest path consists of a route defined by a set of links, all 1,000 trips would be assigned to each link. Clearly, such an assignment ignores the effect congestion has on individual's choice of route, and in general, all-or-nothing assignments are only used in those situations where uncongested conditions are expected to occur.

Capacity Restraint Assignment Capacity restraint assignment recognizes that the travel time on a link will clearly be related to the amount of traffic using that

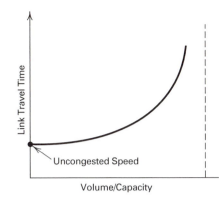

FIGURE 3-10 Graphical representation of link performance.

link as illustrated in Figure 3-10. Therefore, capacity restraint assignment begins with some increment of the total trips making a trip between an origin and destination and assigns this increment to the shortest time path. The travel times for each link are then recalculated using an equation that represents the nonlinear function in Figure 3-10, and another increment of trips is assigned to the shortest time path using the revised time estimates. In this way the shortest time path between an origin and destination could change over several iterations as congestion builds up on specific links, thus making another route the shortest time path. For example, 800 of the 1,000 trips discussed earlier could be assigned to one path because it represented the shortest time path, but 200 of the trips might be assigned to another path because at some point in the assignment process the path followed by the 800 trips was no longer the shortest time path.

Stochastic Equilibrium Assignment Stochastic assignment assumes that several routes between an origin and destination might be perceived by travelers as having equal times or otherwise be equally attractive to the traveler. As a result these routes might be equally used by the travelers. In this case route choice probabilities are calculated often using a similar concept of the logit model formulation presented above and then used to estimate from a probabilistic perspective the number of trips that will take different routes. This type of assignment is not as readily used in practice, although it is likely that in the future some form of stochastic assignment will be the norm.

The result of the assignment step in modeling is an estimate of the volumes that will be found on each link in the network. These estimated volumes can then be used to determine where deficiencies exist in the system (e.g., by comparing the volumes to existing capacity). The resulting volumes can also be compared to other assignments that are undertaken with different input assumptions to determine the effects of different policies. For example, a transportation model could be used to determine the effect of reducing the number of trips originating at certain locations through the use of parking restraints or carpool incentives. The impact on the transportation system of such policies would be estimated by running the model with reduced trip generation numbers and comparing the resulting link assignments with nonadjusted trip generation numbers.

Although the examples in this section are primarily oriented toward highway planning, the same approach toward modeling is used for transit planning. In this case the network is the transit network and the assignment process is oriented toward which transit route will most likely be used by those who will use transit.

3-5 Plan and Project Evaluation

The previous section discussed the modeling methodology that typifies transportation planning efforts. The output of these models, in essence, is numbers that show what each alternative is likely to mean in terms of estimated volumes or riders. These estimated volumes can be related to a wide variety of other issues that might be important in selecting which alternative to proceed with. For example, a reduction in auto volumes could be related to decreases in air pollutants. Thus, volume estimates are critical elements in the evaluation of transportation plans and projects. However, as was indicated earlier in this chapter, the evaluation process is often much broader than simply estimates of volumes. In most cases, the evaluation of alternatives includes a large number of measures of effectiveness, many of which are not related to estimates of volumes. The evaluation process therefore necessarily incorporates information from a wide variety of sources.

Table 3-2 shows typical criteria used to assess different impacts of alternatives being considered. As shown, there are impacts associated with estimated travel volumes, but there are also a large number of impact categories that go beyond transportation impacts. In particular, there are impact categories associated with land use, the environment, distribution (or equity) of benefits, and financial issues. In most cases, the method of portraying the evaluation results is in the form of a matrix with the different measures of effectiveness along one axis and the alternatives under consideration along another. As can be seen in Table 3-2, some of the measures of effectiveness can have numerical values associated with the impact (e.g., number of trips, transit boardings, air quality emissions, costs, etc.). However, it is quite common for some measures of effectiveness to be "measured" with subjective assignments of the degree to which the alternative achieves the stated measure. For example, compatibility with land use plans can be measured by subjectively assigning high, medium, or low to this particular impact category. This approach to measuring impacts is very common for those types of issues that are important to the decision makers but for which there is no quantitative measurement.

Another set of measures that is becoming increasingly important in evaluation is the financial feasibility of the proposed actions. For many years, transportation plans were developed with little consideration for the level and source of funding to implement the recommended actions. Today, transportation plans and programs are to be financially realistic, which means that actions recommended should be accompanied by a strategy for funding their implementation. Such funding might include traditional sources such as government transportation funds, but increasingly is including innovative sources of funding such as tolls, impact fees, and private sector contributions.

A final characteristic of successful plan evaluation is the strategy for public involvement. Opportunities for public input and comment should be provided throughout the plan development process. However, such opportunities are especially important during the evaluation phase. This is the first time in the planning process that sufficiently detailed information is available for the public to understand what is being proposed. Public involvement strategies include a wide variety of techniques. The most common technique is the public hearing, which

TABLE 3-2 Measures of Effectiveness Used in Evaluation

General Category	*Typical Criteria*
1. Transportation system performance	Number of trips by mode Vehicle miles traveled Congestion Peak hour congestion Transit boardings Highway level of service
2. Mobility	Mobility options Improved movement of people
3. Accessibility	Percent within 30 minutes, etc. Transit and highway speeds
4. System development, coordination, and integration	Terminal transitions Transportation system development Regional importance Projects in existing plans
5. Land use	Compatibility with land use plans Growth inducement
6. Freight	Reduced goods movement costs
7. Socioeconomic	Homes or businesses displaced Maximize economic benefit Historic impacts Construction employment
8. Environmental	Air quality Sensitive areas Natural environment
9. Energy	Energy consumption
10. Safety	Annual accidents by mode Safety ratings
11. Equity	Equity of benefit and burden
12. Costs	Capital costs Operating costs
13. Cost effectiveness	Annualized costs per trip or mile FTA (UMTA) index
14. Financial arrangements	Funds required Funding feasibility—build/operate Public/private sources
15. Institutional factors	Ease of staging and expansion Nonimplementing agency support
16. Other	Fatal flaw Right of way opportunities Enforcement Recreation

Source: Transportation Research Board, "Multimodal Evaluation in Passenger Transportation," National Cooperative Highway Research Program Synthesis 201, Washington, DC (1994).

for many types of project planning is required by law. Other techniques include newspaper supplements, public meetings, special television shows, voter surveys, brochures and pamphlets, focus groups, and design workshops. The results of this public review and input are incorporated into the overall evaluation process and often presented to decision makers when the preferred alternative is being selected. For controversial plans or projects, the level of public involvement activities can be quite high.

PROBLEMS

3-1 Assume that the following goals have been established for a transportation planning study.

Goal 1:   The transportation system should provide mobility for all segments of the population.

Goal 2:   The transportation system should minimize impact on the natural environment.

Define at least three objectives for each goal that could be used to achieve the stated purpose. For each objective, define a measure of effectiveness that could be used to measure the degree to which the objective is achieved.

3-2 The regional planning agency has adopted persons per vehicle as a performance measure that can be used to measure the level of success of regional policies. These policies are designed to increase vehicle occupancy so that traffic congestion will be reduced. Given the types of data collection techniques discussed in this chapter, outline a data collection strategy that would provide input into this performance measure.

3-3 Given the following data from your local planning agency, use cross classification analysis to determine the number of trips that will occur in the future if the estimated number of future households is as shown.

Household Income (rows), columns grouped by vehicles 1, 2, 3+:

Household Income	1 HH	1 Trips	2 HH	2 Trips	3+ HH	3+ Trips
Low	350	1,190	5,640	23,124	4,230	19,458
Medium	675	2,498	6,950	31,275	9,641	47,241
High	540	2,106	2,420	11,616	3,202	17,291

Current Tripmaking

Household Income	Household Size 1	2	3+
Low	105	275	430
Medium	220	1,222	2,412
High	90	120	250

Future Zonal Households

3-4 Referring to the sketch in section 3-4, calculate the interzonal trips due to 450 work trips produced at zone *i*. There are 750 attractions at zone 1, 400 attractions at zone 2, and 300 attractions at zone 3. The exponent of travel time is 0.6 and the travel times are 9 minutes to zone 1, 5 minutes to zone 2, and 7 minutes to zone 3. Assume all socioeconomic adjustment factors are equal to 1.0.

3-5 A calibration study resulted in the following utility equation for different modes in a particular city:

$$U_k = a_k - (0.25)(X_1) - (0.032)(X_2) - (0.015)(X_3) - (0.002)(X_4)$$

where a_k = mode specific constant
X_1 = access plus egress time in minutes
X_2 = waiting time in minutes
X_3 = line-haul time in minutes
X_4 = out-of-pocket cost in cents

For a particular origin-destination pair, the forecasted number of trips is 5,000. For this particular trip, there are two modes available, bus and auto.

a) If the characteristics of these two modes are as follows, how many trips will be taken by bus? by car?

	X_1	X_2	X_3	X_4
Automobile	5	0	20	100
Bus	10	15	40	50

Assume that the mode specific constant is equal to -0.12 for the automobile and -0.56 for bus.

b) Assume that a new mode, rapid transit, is to be introduced into the market between these two zones. The characteristics of this new service are as follows:

$$X_1 = 10 \qquad X_2 = 5 \qquad X_3 = 30 \qquad X_4 = 75$$

Assume that from experience in other cities that the mode specific constant for rapid transit is -0.41. What will be the modal shares of the 5,000 trips between these two zones for all three modes.

c) Assume that the city council wants to increase ridership on transit in this new three-mode system. They are considering one of two actions: lower the fare on bus and rapid transit to a flat $0.25 for all trips, or place a surcharge of $2.00 on all cars parking in these zones. Which policy would you recommend to achieve the council's objective?

3-6 Obtain a transportation improvement program (TIP) for a metropolitan area with which you are familiar. What types of projects will be implemented in that region over the next several years? What are the different types of funding sources?

REFERENCES

1. Meyer, M., and E. Miller, *Urban Transportation Planning: A Decision-Oriented Approach,* McGraw-Hill, New York (1984).

2. Federal Highway Administration, *Highway Performance Monitoring System Field Manual,* U.S. Department of Transportation (1987).

3. Federal Highway Administration, *Travel Behavior Issues in the 90's,* U.S. Department of Transportation, Report FHWAQ-PL-93-012 (July 1992).

4. Wachs, M., B. M. Hudson, and J. Schofer, "Integrating Localized and Systemwide Objectives in Transportation Planning," *Traffic Quarterly* (April 1974).

5. Thomas, E., and J. Schofer, "Strategies for the Evaluation of Alternative Transportation Plans," *National Cooperative Highway Research Program Report 96,* Transportation Research Board (1970).

6. Rutherford, S., "Multimodal Evaluation of Passenger Transportation," *National Cooperative Highway Research Program Synthesis of Highway Practice 201,* Transportation Research Board (1994).

7. Meyer, M., "Monitoring System Performance: A Foundation for TSM Planning," *Special Report 192,* Transportation Research Board (1980).

8. Institute of Transportation Engineers, "The Age and Status of Transportation Planning Databases," Report of ITE Technical Council Committee 6F-42, *ITE Journal* (March 1993).

9. Federal Highway Administration, *Guide to Urban Traffic Counting,* Report FHWA/PL/81/019, U.S. Department of Transportation (September 1981).

10. American Association of State Highway and Transportation Officials, *Guidelines for Traffic Data Programs,* Washington, DC (1992).

11. Mahmassani, H., et al, "Survey Approach for Study of Urban Commuter Choice Dynamics," *Innovations in Travel Survey Methods, Transportation Research Record 1412,* Transportation Research Board (1993).

12. DiRenzo, J., *Travel Survey Procedures for Statewide Transportation Planning,* Federal Highway Administration, U.S. Department of Transportation (1976).

13. Urban Mass Transportation Administration, *Transit Data Collection Design Manual,* Report DOT-I-85-38, U.S. Department of Transportation (June 1985).

14. Ogden, K. W., *Urban Goods Movement,* University Press, Cambridge, England (1992).

15. Ismart, D., *Calibration and Adjustment of System Planning Models,* Federal Highway Administration, Report FHWA/HEP/23/1-91(2M)E (December 1990).

16. Stopher, P., and A. Meyburg, *Urban Transportation Modeling and Planning,* D. C. Heath (1975).

17. Ortuzar, J., and L. Willumsen, *Modelling Transport,* 2nd ed., John Wiley & Sons (1994).

18. Stover, V., and F. Koepke, *Transportation and Land Development,* Institute of Transportation Engineers (1988).

19. Newell, G., *Traffic Flows on Transportation Networks,* MIT Press, Cambridge, MA (1980).

20. Institute of Transportation Engineers, *Trip Generation Handbook,* Washington, DC (1992).

21. Voorhees, A. M., "A General Theory of Traffic Movement," *Proceedings of The Institute of Traffic Engineers,* Washington, DC (1955).

22. Fratar, T. J., "Vehicular Trip Distribution by Successive Approximations," *Traffic Quarterly* Vol. VIII, No. 1, pp. 53–65 (1954).

23. May, A., *Traffic Flow Fundamentals,* Prentice Hall, Englewood Cliffs, NJ (1990).

CHAPTER 4

HIGHWAY EVALUATION

In the United States up until about 1950, highway engineers often were able to select desirable highway projects and plan, design, and construct them primarily on the basis of engineering and travel demand factors. Engineers, and the public as well, understood the need for extensive improvements to the highway system and recognized the considerable benefits from such projects.

During the 1950s and 1960s, thousands of miles of new highways were constructed, many in heavily populated areas, causing social disruption, public hostility, and, in certain instances, militant opposition. Some segments of the public expressed concerns about possible environmental and ecological harm from highway construction. The partial fulfillment of the need for new highways and changes in public attitudes have brought about higher standards for economic, social, and environmental review.

Highway agencies are increasingly focusing attention on reconstruction and improvement of existing highway systems rather than building new facilities. Now new highway systems and improvements to existing facilities are being subjected to greater scrutiny, and formal evaluation of the economic impact as well as social and environmental impacts of highway projects is becoming commonplace.

In this chapter we will examine some of the techniques and procedures for making such evaluations. As is the case with many of the subjects covered in this text, space will not permit a detailed discussion of all aspects of highway evaluation. It will be necessary to simplify some of the material presented, with emphasis being placed on certain principles of fundamental importance.

ECONOMIC ANALYSIS FOR HIGHWAY IMPROVEMENTS

Although economic analysis for highway projects was proposed as a decision-making tool in the mid-nineteenth century, the technique was adopted by highway officials much later. About 1920 the first adequate motor vehicle running cost data appeared in the literature, but economic analyses for highways did not gain wide acceptance until after the publications of the AASHTO "Red Book" (*1*) in 1952. The Red Book was updated in 1960, and others (*2–5*) have since updated highway user cost factors and refined economic analysis methodology. A Stanford Research Institute report (*6*) prepared for the National Cooperative Highway Research Program served as the basis for the 1977 edition of the AASHTO Red Book (*6*). That document, which contains unit highway user cost factors based on 1975 levels of vehicle performance characteristics and prices, remains the primary reference source for economic analyses in the United States. It serves as the primary basis for the material in the following paragraphs.

Economic studies for highway purposes are done principally for one or more of the following reasons:

- to determine feasibility of a project
- to compare alternative locations
- to evaluate various features of highway design, for example, the type of surface to be used
- to determine priority of improvement
- to allocate responsibility for the costs of highway improvement among the various classes of highway users (and nonusers, in some cases)
- Occasionally, to compare proposals for highway improvement with proposals for other public projects such as education

Before describing analytical concepts and approaches for economic studies, let us briefly consider the costs and benefits of highway transportation.

4-1 Costs of Highway Transportation

Highway transportation cost is defined as the sum of the highway investment cost, the maintenance and operating costs, and the highway user costs. Where bus transit is present, the transportation cost includes the transit capital costs, the costs of operating the transit system, and the costs that accrue to transit users (fares and travel time costs). Examples of highway transportation costs are given in Table 4-1.

The highway investment cost is the cost of preparing a highway improvement for service including the cost of rights-of-way, engineering design, construction, traffic control devices, and landscaping. Maintenance cost is the cost of preserving a highway and its appurtenances and keeping the facility in serviceable condition. Operating costs include the costs of traffic control, lighting, and the like.

TABLE 4-1 Examples of Highway Transportation Costs

Type of Cost	*Examples*
Highway investment cost	Engineering design, right-of-way, grading and drainage, pavement
Highway maintenance cost	Mowing, care of roadside parks, lighting
Highway user cost	
(a) Motor vehicle operating cost	Fuel, lubrication, tires
(b) Travel time	Total vehicle-hours of travel times unit value of time
(c) Accident costs	Estimated accident rate times unit accident cost
Transit capital costs	Terminals, shops, administrative offices
Transit operating costs	Drivers' wages, operating and maintenance cost of buses
Transit user costs	
(a) Transit money costs	Sum of fares paid, auto running costs to get to terminal
(b) Transit travel time costs	Time waiting for bus times unit value of time Time riding bus times unit value of time

Transit capital costs are the investments in fixed facilities for bus transit such as terminals, shops, bus parking facilities, and passenger benches and shelters. Special highway facilities for buses such as busways may also be included. Reference 6 treats the costs of bus vehicles as an operating rather than a capital cost. Other transit operating costs include drivers' wages, bus fuel and tires, bus maintenance, and the costs of management, administration, and insurance.

Transit user cost consists of the sum of bus fares paid to make a transit trip, the passenger car running costs experienced by users in getting to a stop or terminal, and the costs of waiting and walking (6).

Highway User Costs

A major component of the costs of highway transportation, highway user costs include motor vehicle operating costs, the value of travel time, and traffic accident costs. Usually, only those operating costs that depend on the mileage of travel are included in highway economic analyses, including the cost of fuel, tires, engine oil, maintenance, and a portion of depreciation. Registration and parking fees, insurance premiums, and the time-dependent portion of depreciation may be excluded when estimating the reduction in user costs due to a highway improvement. Anderson et al. (6) have published a set of nomographs by which estimates of operating costs can be made. Examples of these nomographs and a discussion of their use are given in Section 4-4.

The value of travel time is the product of the total vehicle-hours of travel (by vehicle type) and the average unit value of time. The magnitude of travel time depends on average running speed and the number and duration of stops. Studies have indicated that the perceived value of travel time is sensitive to trip purpose and time savings per trip. Generally, travelers place little value on time savings less than 5 min. For passenger cars, Ref. 6 recommends a travel time value of $3.00 per vehicle per hour based on a value of $2.40 per person-hour and 1.25 persons per vehicle. The report recommends time values of $7.00 per hour for single-unit trucks and $8.00 per hour for combination trucks based on truck driver wages and fringe benefits.

The accidents that occur every year on the streets and highways of the United States represent a great economic loss besides the pain and suffering they cause. Total economic loss resulting from motor vehicle accidents in 1992 was estimated to be $156.6 billion (7). This total included losses incurred in fatal accidents, accidents involving nonfatal personal injury, and those involving only property damage. It follows that from an economic standpoint highway improvements that result in decreased accident rates result in monetary saving to the road user and the public in general.

The National Safety Council issues periodic reports (8) on the costs of vehicular accidents. Such costs include wage and productivity losses, medical expenses, administrative expenses, motor vehicle damages, and employer costs. In 1992 the costs of all these items for each death, injury, and property damage accident were:

Death	$880,000
Nonfatal disabling injury	$ 29,500
Property damage accident (including minor injuries)	$ 6,500

For each death in motor vehicle accidents the National Safety Council (8) estimates that there are approximately:

	Urban Areas	*Rural Areas*
Nonfatal injuries per death	97	31
Property damage accidents per death	418	100

To account for savings in accident costs in highway economic studies, accident rates before and after the proposed improvement should be estimated. In many instances, the analyst is able to rely on empirical data that will likely account for regional differences in vehicle mix, driver behavior, and climate. General guidance on typical accident rates for various classes of roadways is given by Table 4-2. A number of researchers (*9–11*) have also developed multiple regression equations by which estimates of accident rates can be made on the basis of roadway and traffic characteristics.

4-2 Benefits of Highway Improvements

It is obvious that many benefits result from highway improvement or, to put it more broadly, from improved highway transportation. Some of these benefits are direct and readily apparent; others are indirect and more difficult to discern. Likewise, some benefits may be readily evaluated in terms of dollars and cents; others defy evaluation in this fashion, although they are nonetheless as real and lasting as monetary returns. Sometimes, of course, the consequences of a given highway improvement are not entirely beneficial; for example, construction of an urban expressway may produce a decline in ridership on a commuter railroad, and public funds spent from the general treasury for highways cannot be used, say, for education. Such consequences should be considered in comprehensive economic analyses.

For purposes of discussion it is convenient to group highway benefits into the following categories: (1) direct benefits that result from a reduction in highway user costs and (2) indirect benefits, including benefits to adjacent property and to the general public. The most quantifiable and, to the analyst, the most significant highway benefits are those that result from a reduction in user costs. As the previous section implies, such benefits result from decreased operating costs, higher operating speeds, fewer delays, and decreased accident losses. The procedure for estimating such benefits are described later in this chapter.

Those benefits that accrue to the owner of property located adjacent to or in the vicinity of improved highways are fairly simple to observe and delineate but are somewhat more difficult to evaluate. In the vast majority of cases highway improvements result in an increase in the value of land located adjacent to or served by the highway involved. A simple case of direct property benefit is that of a farm or a number of farms located on an unimproved earth road that is rough, muddy in wet weather and dusty in dry. Any degree of improvement of a road of this sort would be of direct benefit to the property owners located along it since they would be provided with easier access to market, schools, and recreational facilities. Similar direct benefits may accrue to city dwellers who live on an improved residential street. In many other cases, property benefits may be more difficult to discern and to evaluate.

Benefits of improved highway transportation that are enjoyed by the general public may well be regarded as the most important of the three classes of benefits that have been listed. These benefits include many of the advantages commonly

TABLE 4-2 Accident Rates by Road Type

	Rate (Number per MVKM[a])			
	Fatal Accidents	*Injury Accidents*	*Property Damage Only Accidents*	*Total Accidents*
Rural				
No access control				
2 lane	0.043	0.58	0.86	1.49
4 or more lanes, undivided	0.029	0.55	1.21	1.80
4 or more lanes, divided	0.039	0.48	0.78	1.30
Partial access control				
2-lane expressway	0.032	0.32	0.47	0.83
Divided expressway	0.024	0.27	0.47	0.77
Freeway	0.016	0.17	0.30	0.49
Urban				
No access control				
2 lanes	0.028	0.94	2.10	3.07
4 or more lanes, undivided	0.025	1.32	2.79	4.13
4 or more lanes, divided	0.017	1.03	1.98	3.02
Partial access control				
2-lane expressway	0.021	4.13	0.65	1.08
Divided expressway	0.014	0.67	1.27	1.95
Freeway	0.007	0.25	0.63	0.89
Suburban				
No access control				
2 lanes	0.030	0.78	1.59	2.41
4 or more lanes, undivided	0.023	0.98	2.06	3.06
4 or more lanes, divided	0.019	0.68	1.39	2.09
Partial access control				
2-lane expressway	0.060	0.51	0.88	1.45
Divided expressway	0.037	0.51	0.80	1.34
Freeway	0.009	0.20	0.46	0.66
Statewide				
No access control				
2 lanes	0.041	0.63	1.03	1.70
4 or more lanes, undivided	0.025	1.17	2.47	3.65
4 or more lanes, divided	0.019	0.93	1.80	2.75
Partial access control				
2-lane expressway	0.033	0.36	0.50	0.87
Divided expressway	0.023	0.38	0.67	1.08
Freeway	0.009	0.22	0.53	0.76

[a]Number per million vehicle kilometers.

Source: Adapted from American Association of State Highway and Transportation Officials, *A Manual on User Benefit Analysis of Highway and Bus-Transit Improvements* (1977).

associated with modern civilization, including such things as efficient fire and police protection; rapid postal deliveries; improved access to educational, recreational, and social facilities; decreased costs of commodities; improved national defense; and so on ad infinitum. The general advantages of improved highways are so interwoven with every phase of modern life in this country that it is virtually impossible to imagine what existence would be like without them.

Although it is apparent that indirect benefits are of great importance, these benefits cannot usually be evaluated except in a general sense and are therefore seldom used in economic calculations

4-3 Present Value Concepts

In economic studies it is important to recognize the time value of money. Because of the existence of interest, a quantity of money is worth more now than the prospect of receiving the same quantity at some future date. Therefore, in order to compare costs and benefits of highway improvements on a sound basis, they must be converted to equivalent values at some common date. This procedure, which is known as discounting, is accomplished by using a suitable interest rate in accordance with established principles of compound interest and present value concepts. Such concepts are more fully described in standard textbooks on economic analysis. For example, see Thuesen and Fabrycky (*12*).

The present value (also termed present worth) P of some future single payment F can be calculated by the following equation:

$$P = \frac{F}{(1 + i)^n} \tag{4-1}$$

where i = interest or discount rate per period
      n = number of interest periods, usually years

Alternatively, present worth can be determined by use of present worth factors (*PW*) such as those given in Table 4-3.

EXAMPLE 4-1   In 20 years the residual or salvage value of a certain highway will be $250,000. Determine the present value of that sum, using an interest rate of 6 percent.

By Eq. 4-1,

$$\text{present value} = \frac{\$250,000}{(1 + 0.06)^{20}} = \$77,951$$

The same result is obtained by multiplying $250,000 by the appropriate present worth factor *PW* from Table 4-3.

The present worth P_s of a series of uniform annual end-of-period payments A can be determined by the following equation:

$$P_s = A \left[\frac{(1 + i)^n - 1}{i(1 + i)^n} \right] \tag{4-2}$$

or

$$P_s = A(SPW) \tag{4-3}$$

where SPW = series present worth factor (Table 4-3)

TABLE 4-3 Present Worth (*PW*) and Series Present Worth (*SPW*) Compound Interest Factors for $1 and Various Interest Rates

Years	4%		6%		8%		10%	
n	*PW*	*SPW*	*PW*	*SPW*	*PW*	*SPW*	*PW*	*SPW*
5	0.8219	4.4518	0.7473	4.2124	0.6806	3.9927	0.6209	3.7908
10	0.6756	8.1109	0.5584	7.3601	0.4632	6.7101	0.3855	6.1446
15	0.5553	11.1184	0.4173	9.7122	0.3152	8.5595	0.2394	7.6061
20	0.4564	13.5903	0.3118	11.4669	0.2145	9.8181	0.1486	8.5136
25	0.3751	15.6221	0.2330	12.7834	0.1460	10.6748	0.0923	9.0770
30	0.3083	17.2920	0.1741	13.7648	0.0994	11.2578	0.0573	9.4269
40	0.2083	19.7928	0.0972	15.0463	0.0460	11.9246	0.0221	9.7791

Source: American Association of State Highway and Transportation Officials, *A Manual on User Benefit Analysis of Highway and Bus-Transit Improvements* (1977).

EXAMPLE 4-2

The user benefits for a certain highway are estimated to be a uniform $85,000 per year. Determine the present worth of those benefits assuming an interest rate of 8 percent and an analysis period of 25 years.

By Eq. 4-2,

$$P_s = \$85,000\left[\frac{(1 + 0.08)^{25} - 1}{0.08(1.08)^{25}}\right] = \$907,358$$

In many instances, the annual values of user benefits (or highway costs) change from year to year. In those cases, the present worth cannot be calculated by Eq. 4-2 or 4-3 but can be computed for each year of the analysis period by Eq. 4-1 and then summed. An equation given in Ref. 6 makes it possible to closely approximate the present value of a series of annual values based on the average annual rate of growth of the values. The equation, which follows, is based on the assumption that the annual values increase (or decrease) at approximately an equal annual percentage. The present worth factor

$$PW_g = \frac{e^{(r - i)n} - 1}{r - i} \tag{4-4}$$

where i = assumed interest rate (decimal fraction)
   r = rate of growth of annual value (percent per annum/100)
   n = analysis period (years)

The average rate of growth r can be approximated on the basis of estimates of 2 years' values:

$$r = \frac{\ln(\alpha)}{Y} \tag{4-5}$$

where α = future annual value estimate/first year value estimate
   Y = number of years between estimates

EXAMPLE 4-3

A certain highway has user benefits of $17,500 during the first year that uniformly increase at a rate of 4 percent each succeeding year. Determine the present value of the benefits, given a 12-year analysis period. Use an interest rate of 6 percent.

By Eq. 4-4, the present worth factor,

$$PW_g = \frac{e^{(0.04 - 0.06)12} - 1}{0.04 - 0.06} = 10.67$$

present worth $= 10.67 \times \$17,500 = \$186,725$

4-4 Economic Analysis Techniques

Economic calculations intended to demonstrate economic justification for a particular project, to permit comparison of alternative schemes or locations, to determine priority of improvement, and so on, may be carried out by one of several methods. Methods used in engineering economy studies include (1) benefit-cost ratio, (2) net present value, (3) comparison of annual costs, and (4) determination of the interest rate at which alternatives are equally attractive. The benefit-cost ratio (*B/C*) and net present value (*NPV*) methods, described here, are commonly used by highway engineers and administrators. For additional information on the latter two techniques the reader is referred to standard textbooks on economic analysis.

Benefit-Cost Ratio

Traditionally, in the benefit-cost ratio method, analysis was made by comparing the relations of the *annual* road user costs to the *annual* highway costs for logical alternatives in location and design. The following formulation of the benefit-cost ratio, which utilizes present values, is now recommended for all highway and transit applications as an approximate indicator of project desirability (6). The benefit-cost ratio

$$\frac{B}{C} = \frac{PV(\Delta U)}{PV(\Delta I) + PV(\Delta M) - PV(\Delta R)} \tag{4-6}$$

where $PV =$ present value of the indicated amount or series

$\Delta U =$ user benefits, the reductions in highway or transit user costs due to the investment (costs without the improvement less costs with the improvement)

$\Delta M =$ increase in annual maintenance, operating, and administrative cost due to the investment (costs with the improvement less costs without the improvement)

$\Delta R =$ increase in residual value due to the project at end of project life

$\Delta I =$ increased investment costs due to the project

Net Present Value Method

In the net present value method, the procedure is to reduce all benefits and costs to their present value in accordance with the principles given in Section 4-3. The net present value *NPV* is defined as the difference between the present values of the benefits from the project and the costs of developing the project. The net present value may be computed as follows:

$$NPV = PV(\Delta U) + PV(\Delta R) - PV(\Delta I) - PV(\Delta M) \tag{4-7}$$

Stated another way:

$$NPV = \frac{B_1 - C_1}{(1 + i)} + \frac{B_2 - C_2}{(1 + i)^2} + \cdots + \frac{B_n - C_n + R}{(1 + i)^n} \quad (4\text{-}8)$$

where B_1 and C_1 are the benefit and cost for the first year, B_2 and C_2 for the second year, and so forth; R is the residual value, i is the discount rate, and n is the analysis period.

Decision Rules in Economic Analysis

Rules of application of the *B/C* ratio and the net present value techniques depend on whether there are budgetary constraints and whether the projects under consideration are independent or mutually exclusive. Consider first the case where there are no budgetary constraints and independent projects are being compared; that is, the selection of one project would not preclude the selection of other projects. In this instance all projects with positive net present values or benefit-cost ratios greater than 1 would be economically feasible.

Where the investment budget is constrained,* the analyst should select the combination of projects that produces a maximum net present value but does not exceed the available budget. Alternatively and approximately, projects may be selected in order of decreasing *B/C* ratios, adding projects until the budget is exhausted.

In cases in which several mutually exclusive locations or designs are being compared, the alternative with the highest net present value would be chosen. If benefit-cost ratios are used to select the preferred project, the selection must be made in increments. Beginning with the lowest-cost alternative having a *B/C* ratio greater than 1, each increment of additional cost is justified only if the *incremental B/C* ratio is greater than 1. This is illustrated by the following example.

EXAMPLE 4-4   Three alternative designs ($X1$, $X2$, and $X3$) are being considered for a major interchange. For each of the alternatives, the present values are shown below for user benefits, investment costs, maintenance costs, and the residual value. By means of Eqs. 4-6 and 4-7, the *B/C* ratios and *NPV*s are computed as indicated. Which is the preferred alternative?

On the basis of net present values, alternative $X2$ is chosen.

To make the selection by the *B/C* ratio method, it is necessary to use incremental ratios, that is, based on the differences in costs of pairs of alternatives. Alternative $X1$, which has the lowest investment cost, is tentatively selected. The next most costly alternative, $X2$, is tested by calculation of the *B/C* ratio for the increment $X2 - X1$. Since that ratio is greater than 1, alternative $X2$ is selected to replace $X1$. Alternative $X3$ is then tested by the incremental *B/C* ratio for $X3 - X2$. Since that ratio is less than 1, alternative $X3$ is not justified, and $X2$ is selected.

---

*A linear or dynamic programming procedure is recommended for those situations where both the investment budget and the future maintenance and operating budget are constrained (*6*).

	X1	*X2*	*X3*	*(X2 − X1)*	*(X3 − X2)*
$PV(\Delta U)$	95	127	128	32	1
$PV(\Delta I)$	12	13	20	1	7
$PV(\Delta M)$	7	20	17	13	−3
$PV(\Delta R)$	2	1	3	−1	2
B/C ratio	5.6	4.0	3.8	2.1	0.5
NPV	78	95	94	17	−1

4-5 Measurement of User Benefits

One of the most difficult aspects of economic analyses for highways is the measurement of highway user benefits. First, consider highways that do not serve transit vehicles. The unit highway user cost for a given section of highway *HU* expressed in dollars per thousand vehicles is the sum of the basic section costs, accident costs, and delay costs:

$$HU = (B + A)L + T + D \qquad (4\text{-}9)$$

where B = basic section costs, consisting of the unit cost (time value and vehicle running costs) associated with vehicle flow and the basic geometrics (grades and curves) of the analysis section
  A = unit accident costs in the analysis section
  L = analysis section length in miles*
  T = transition costs—additional unit user time and running costs incurred due to changes in speeds between accident sections
  D = additional unit time and running costs caused by delays at intersections, at traffic signals, stop signs, or other traffic control devices.

The highway user benefit ΔU in Eq. 4-6 is the difference in highway user costs for any two alternatives:

$$\Delta U = HU_0 - HU_1 \qquad (4\text{-}10)$$

Highway user costs are determined largely on the basis of empirical data, and Ref. 6 provides data for the estimation of the basic section costs, transition costs, and delay costs for various highway and traffic conditions.

The revised manual includes nomographs that give one-way basic section costs for three classes of vehicles (passenger cars, single unit trucks, and 3-S2 combination trucks) using four types of highways (freeways, multilane highways, two-lane highways, and arterials). One of the 12 nomographs that shows the basic section costs for passenger cars on multilane highways is reproduced as Figure 4-1.

To use Figure 4-1, the nomograph is entered at the lower left with average running speed or with the volume-capacity ratio for the representative hour of operation for the anaysis section. Travel time, the inverse of running speed, can be read on the left ordinate of the lower left graph. A horizontal line at the level

---

*Conventional U.S. units are used in this and the following sections in order to be consistent with available nomographs.

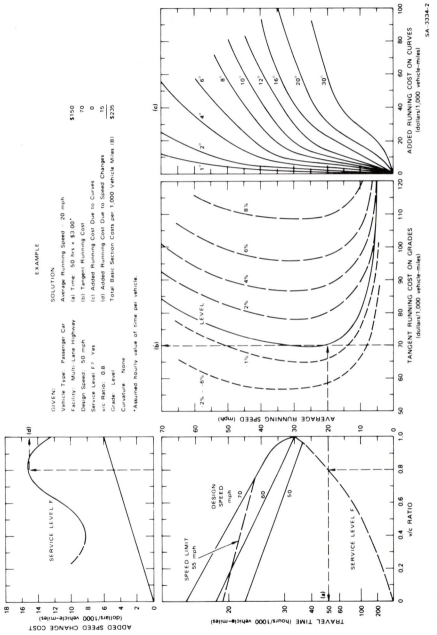

EXAMPLE

GIVEN: SOLUTION

Vehicle Type: Passenger Car   Average Running Speed - 20 mph
Facility: Multi-Lane Highway   (a) Time: 50 hrs x $3.00* $150
Design Speed: 50 mph   (b) Tangent Running Cost   70
Service Level F?   Yes   (c) Added Running Cost Due to Curves   0
v/c Ratio: 0.8   (d) Added Running Cost Due to Speed Changes   15
Grade: Level   Total Basic Section Costs per 1,000 Vehicle Miles (B) $235
Curvature   None

*Assumed hourly value of time per vehicle.

SA-3334-2

FIGURE 4-1 Basic section costs (*B*) for passenger cars on multilane highways.

of average running speed is then projected to the right to determine the tangent running cost (along a specified grade) and added running costs due to curves. Finally, added running costs due to speed change cycles are determined by entering the upper left graph with the volume-capacity ratio and reading the added cost on the ordinate. The added costs due to speed change cycles are relatively small except under level of service *F* (or queuing) conditions. An example of the use of the nomograph is shown on the figure.

The basic section costs given by Figure 4-1 apply only to paved roadways. The costs of operating vehicles on gravel and stone roadway surfaces can be estimated by multiplying the basic section costs from the nomographs (e.g., Figure 4-1) by an adjustment factor. For passenger cars, such conversion factors range from about 1.1 for a speed of 8 km/hr (5 mph) to about 1.6 for a speed of 96 km/hr (60 mph) (*2*). Larger values are recommended for single-unit trucks and truck combinations.

As Section 4-1 details, unit accident costs can be estimated on the basis of empirical accident rates that reflect regional effects of vehicle mix, driver behavior, roadway type, and climate, or can be estimated from the average rates (Table 4-2) and costs (by accident type) previously given.

Empirical studies have made it possible to estimate the costs of speed changes when vehicles pass between sections with different physical or traffic characteristics. These section transition costs, symbolized by the letter *T* in Eq. 4-9, depend on the magnitude of average speed change. Figure 4-2 provides a means of estimating the transition costs for one-way traffic based on the speeds on adjacent sections. Reference 6 gives adjustment factors that can be applied to the graph values to estimate transition costs for single unit and combination trucks.

Perhaps the most complex aspect of determination of unit highway user costs for a highway section is the estimation of the costs of intersection delay (referred to as *D* in Eq. 4-9). Two nomographs have been published by Anderson et al. (*6*) to determine intersections delay costs: one to estimate the unit costs of stopping at a traffic signal and the other to estimate the unit costs of idling. These nomographs are reproduced as Figures 4-3 and 4-4. The figures require data on the following traffic and signalization parameters for the intersection being studied.

1. *Green-to-cycle time ratio* λ. The ratio of effective green time of the signal (generally taken to be the actual green time) to the cycle length of the signal, both expressed in the same unit of time (usually seconds).

2. *Saturation flow s.* The approach volume in vehicles per hour of green that is found for the intersection when the load factor is 1.0 and the appropriate adjustment factors are applied. In the absence of Highway Capacity Manual (*13*) solutions, recommended values for saturation flow are 1700 to 1800 vehicles per hour times the number of approach lanes.

3. *Capacity c.* Where the Highway Capacity Manual is used, capacity is the service volume of the approach at a load factor of 1.0. It is also equal to the saturation flow times the green-to-cycle time ratio.

4. *Degree of saturation* χ. The ratio of the volume of traffic approaching the intersection (usually in vehicles per hour) to the capacity of the intersection, in the same units.

5. *Approach speed.* Also termed "midblock speed," this is the average running speed at which the signalized intersection is approached by the vehicle stream.

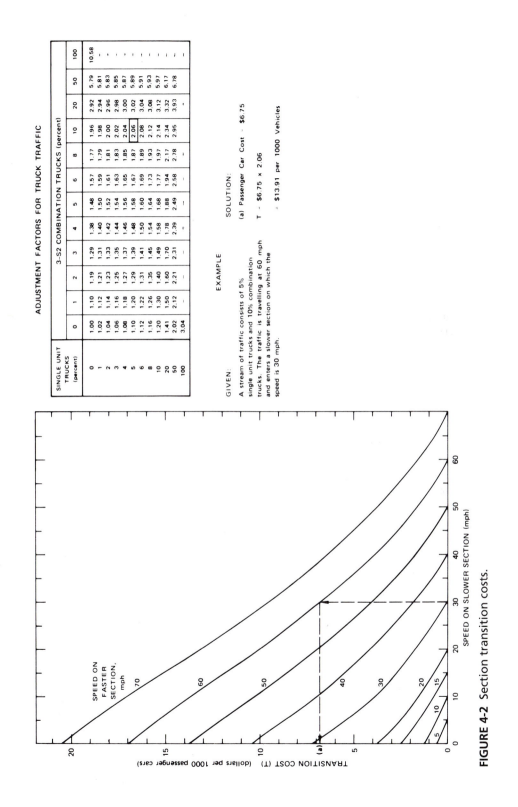

ADJUSTMENT FACTORS FOR TRUCK TRAFFIC

SINGLE UNIT TRUCKS (percent)	3-S2 COMBINATION TRUCKS (percent)											
	0	1	2	3	4	5	6	8	10	20	50	100
0	1.00	1.10	1.19	1.29	1.38	1.48	1.57	1.77	1.96	2.92	5.79	10.58
1	1.02	1.12	1.21	1.31	1.40	1.50	1.59	1.79	1.98	2.94	5.81	-
2	1.04	1.14	1.23	1.33	1.42	1.52	1.61	1.81	2.00	2.96	5.83	-
3	1.06	1.16	1.25	1.35	1.44	1.54	1.63	1.83	2.02	2.98	5.85	-
4	1.08	1.18	1.27	1.37	1.46	1.56	1.65	1.85	2.04	3.00	5.87	-
5	1.10	1.20	1.29	1.39	1.48	1.58	1.67	1.87	2.06	3.02	5.89	-
6	1.12	1.22	1.31	1.41	1.50	1.60	1.69	1.89	2.08	3.04	5.91	-
8	1.16	1.26	1.35	1.45	1.54	1.64	1.73	1.93	2.12	3.08	5.93	-
10	1.20	1.30	1.40	1.49	1.58	1.68	1.77	1.97	2.14	3.12	5.97	-
20	1.41	1.50	1.60	1.70	1.78	1.88	1.94	2.17	2.34	3.32	6.17	-
50	2.02	2.12	2.21	2.31	2.39	2.49	2.58	2.78	2.95	3.93	6.78	-
100	3.04	-	-	-	-	-	-	-	-	-	-	-

EXAMPLE

GIVEN:

A stream of traffic consists of 5% single unit trucks and 10% combination trucks. The traffic is travelling at 60 mph and enters a slower section on which the speed is 30 mph.

SOLUTION:

(a) Passenger Car Cost - $6.75

T - $6.75 x 2.06

= $13.91 per 1000 Vehicles

FIGURE 4-2 Section transition costs.

SPEED ON FASTER SECTION, mph

70 60 50 40 30 20 15 10 5

SPEED ON SLOWER SECTION (mph)

TRANSITION COST (T) (dollars per 1000 passenger cars)

ADJUSTMENT FACTORS FOR PERCENT TRUCKS IN TRAFFIC STREAM

TIME COST

APPROACH SPEED (mph)	SINGLE UNIT TRUCKS (percent)	3 S-2 COMBINATION DIESEL TRUCKS (percent in traffic stream)				
		0	5	10	20	100
5-20	0	1.00	1.15	1.30	1.61	
	5	1.07	1.22	1.37	1.67	
	10	1.13	1.28	1.43	1.74	
	20	1.26	1.41	1.57	1.87	
	100	2.31				4.03
21-40	0	1.00	1.25	1.51	2.01	
	5	1.10	1.35	1.60	2.11	
	10	1.20	1.45	1.70	2.21	
	20	1.40	1.65	1.90	2.41	
	100	2.99				6.05
41-60	0	1.00	1.41	1.82	2.63	
	5	1.11	1.56	1.93	2.74	
	10	1.22	1.61	2.04	2.85	
	20	1.44	1.85	2.26	3.07	
	100	3.20				9.17

RUNNING COST

APPROACH SPEED (mph)	SINGLE UNIT TRUCKS (percent)	3 S-2 COMBINATION DIESEL TRUCKS (percent in traffic stream)				
		0	5	10	20	100
5-20	0	1.00	1.35	1.70	2.40	
	5	1.08	1.43	1.78	2.49	
	10	1.16	1.51	1.86	2.57	
	20	1.32	1.68	2.03	2.73	
	100	2.62				8.02
21-40	0	1.00	1.35	1.71	2.41	
	5	1.07	1.42	1.78	2.48	
	10	1.14	1.49	1.84	2.55	
	20	1.27	1.63	1.98	2.69	
	100	2.37				8.07
41-60	0	1.00	1.35	1.70	2.39	
	5	1.06	1.41	1.76	2.45	
	10	1.12	1.47	1.82	2.51	
	20	1.24	1.59	1.94	2.63	
	100	2.21				7.96

EXAMPLE

GIVEN:

Volume: 480 vehicles/hr
Saturation Flow: 1,600 vehicles/hr
Signal Cycle Time: 60 sec
Effective Green Time: 30 sec
Intersection Approach Speed: 30 mph
5% Single Unit Trucks
5% 3-S2 Combination Trucks

SOLUTION:

$\lambda = 30/60 = 0.5$
Capacity of Approach $= 0.5 \times 1600 = 800$
$\lambda = 480/800 = 0.6$
(a) Average Stops per Vehicle (per Signal): 0.71
(b) Stopping Delay per Signal: 2.5 hrs
(c) Cost of Stopping: $10.30
Time Cost: $2.5 \times \$3.00^* \times 1.35^\dagger$ $10.13
Running Cost: $\$10.30 \times 1.42^\dagger$ 14.63
Total Cost Due to Stopping per 1,000 vehicles per Signal (excludes idling) $24.76

*Assumed hourly value of time per passenger car.
†Adjustment factors for trucks in traffic stream.

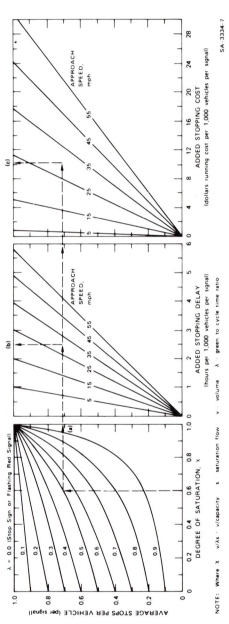
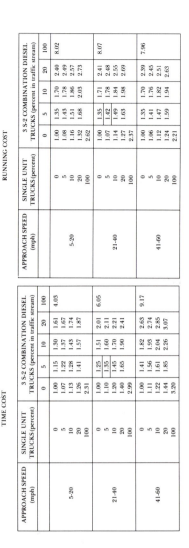

NOTE: Where X $v/\lambda s \cdot v/capacity$ s = saturation flow v = volume λ = green to cycle time ratio

SA-3334-7

FIGURE 4-3 Costs due to stopping at intersections (excludes idling).

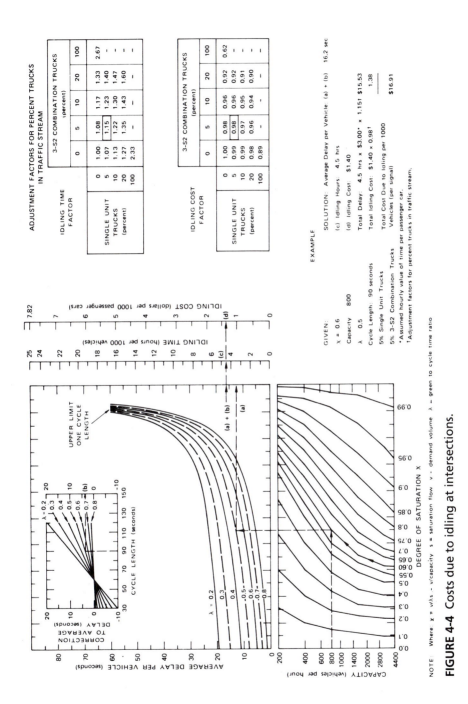

ADJUSTMENT FACTORS FOR PERCENT TRUCKS
IN TRAFFIC STREAM

IDLING TIME FACTOR

SINGLE UNIT TRUCKS (percent)	3-S2 COMBINATION TRUCKS (percent)				
	0	5	10	20	100
0	1.00	1.08	1.17	1.33	2.67
5	1.07	1.15	1.23	1.40	-
10	1.13	1.22	1.30	1.47	-
20	1.27	1.35	1.43	1.60	-
100	2.33	-	-	-	-

IDLING COST FACTOR

SINGLE UNIT TRUCKS (percent)	3-S2 COMBINATION TRUCKS (percent)				
	0	5	10	20	100
0	1.00	0.98	0.96	0.92	0.62
5	0.99	0.98	0.96	0.92	-
10	0.99	0.97	0.95	0.91	-
20	0.98	0.96	0.94	0.90	-
100	0.89	-	-	-	-

EXAMPLE

GIVEN:

$x = 0.6$
Capacity 800
λ 0.5
Cycle Length: 90 seconds
5% Single Unit Trucks
5% 3-S2 Combination Trucks

SOLUTION: Average Delay per Vehicle (a) + (b) 16.2 sec

(c) Idling Hours: 4.5 hrs
(d) Idling Cost $1.40

Total Delay: 4.5 hrs × $3.00* × 1.15† $15.53
Total Idling Cost: $1.40 × 0.98† 1.38
Total Cost Due to Idling per 1000
Vehicles (per signal) $16.91

*Assumed hourly value of time per passenger car.
†Adjustment factors for percent trucks in traffic stream.

NOTE Where $x = v/\lambda s$ - v/capacity s = saturation flow v - demand volume λ - green to cycle time ratio

FIGURE 4-4 Costs due to idling at intersections.

83

The examples that accompany Figures 4-3 and 4-4 illustrate their use. Adjustment factors are provided in the manual to account for the effects of single-unit trucks and combination trucks.

One-Way Versus Two-Way Traffic

The costs given by the highway user nomographs relate to one-way traffic on a highway facility. In the usual case where the improvements affect both sides of the roadway and the traffic is balanced in direction over the course of a typical day, the benefits that are calculated from the one-way analysis should be multiplied by 2.

Effects of Induced and Diverted Traffic

Highway economic analyses should account for changes in traffic volume that are (1) induced by the improvement, that is, trips that would not be made without the improvement, and (2) diverted from other less attractive routes to the new facility. It is therefore recommended that the difference in user costs be multiplied by the average of the traffic volume with and without the improvement.

$$\text{user benefits} = (U_0 - U_1)\left[\frac{(V_0 + V_1)}{2}\right] \qquad (4\text{-}11)$$

where U_0 = user cost per unit of traffic (vehicles or trips) without the improvement
U_1 = user cost per unit of traffic with the improvement
V_0 = level of traffic without the improvement
V_1 = level of traffic with the improvement, including any induced or diverted traffic (V_1 may be either larger than V_0, as shown, or smaller, as when traffic is diverted from a link to a new parallel facility)

To properly account for the benefits to both induced and diverted traffic as well as the continuing traffic, the equation should be applied to each link in the highway network being analyzed and summed over all links (6).

EXAMPLE 4-5

Highway Economic Analysis Calculations The following example, adapted from Ref. 6, illustrates application of some of the data and procedures presented earlier in the chapter.

A highway agency is considering eliminating (straightening) an *S*-shaped curve on a multilane highway. Total construction cost for the improvement is estimated at $2 million and will be incurred at the inception of the project. Annual maintenance costs are anticipated to be $2,000 less after the improvement because the elimination of the curve reduces pavement and guardrail damage.

The *S*-curve is actually made up on two continuous 8° curves (radii = 218 m) in reverse alignment to each other. Since the curves are the same degree of curvature, they can be analyzed together, and since traffic can be assumed to be uniform over the entire affected roadway, the analyst may view the improvement in terms of a single analysis section, designated *ab*. In analyzing the two curves together, it will be assumed that the average speed over the curves is the same and that there are no vehicular speed changes in passing from one curve to the other. If such assumptions cannot be made, separate analysis sections should be designated for each of the curves.

TABLE 4-4 Facility Data

			Basic Section Data			
Alternative	Section	Length (miles)[a]	Lanes (one way)	Grade	Curvature	Design Speed (mph)[a]
0	ab	0.20	2	0%	8°	50
1	ab	0.16	2	0%	0°	60

[a]1 mile = 1.6093 km.

The characteristics of the existing route (alternative 0) and the proposed route (alternative 1) are summarized in Table 4-4. Note that the length of the section *ab* includes both 8° curves. Note also that after the improvement, the design speed of the straightened section will be 60 mph rather than 50 mph.

The analyst has obtained the traffic estimates with and without the proposed improvement for the 1st and the 20th year (the study years) of the 25-year analysis period. It is assumed that level of service *F* conditions will not occur during the analysis period and that the hourly variation of traffic is small enough that explicit consideration of peaking characteristics is unnecessary. Hence the analysis can be performed essentially on an *ADT* basis. This entails utilizing traffic volume data for a typical hour of the day in each of the study years. Such traffic data are summarized in Table 4-5, where typical one-way hourly volume is defined as *ADT* divided by 18 (the number of hours assumed, for analysis purposes, in a day). Note that a small increment of traffic is estimated to be induced as a result of the proposed improvement.

Note also that capacity in year 20 is somewhat below that in year 1. This is to reflect normal deterioration of the roadway over the years.

There are so few trucks that the traffic stream will be considered to consist entirely of passenger cars—otherwise, the percentage of trucks by type would also be shown in Table 4-5.

Average running speeds (determined from Figure 4-1 as a function of *v/c*) are also noted in Table 4-5. Since speeds on the analysis section will change as a result of the improvement, transition costs between sections will change; thus speed on the previous and subsequent sections must be specified (as in Table 4-5).

TABLE 4-5 Traffic Data

				Number of Vehicles			Average Running	Speed (mph)[a] on:	
Alternative	Section	Study Year	Period	One-Way Volume	One-Way Capacity	v/c Ratio	Speed (mph)[a]	Previous Section	Next Section
0	ab	1	Typical hour	1900	3800	0.50	36	41	50
1	ab	1	Typical hour	2000	3800	0.53	41	41	50
0	ab	20	Typical hour	2900	3500	0.83	30	34	45
1	ab	20	Typical hour	3000	3500	0.86	34	34	45

[a]1 mile = 1.6093 km.

Calculation of User Costs

The above traffic and facility data enable calculation of user costs on the existing route and the proposed route, as shown in Table 4-6. Unit highway user costs (*HU*) are calculated from Eq. 4-9 by using basic section costs (*B*), section length (*L*), and transition costs (*T*). Intersection delay costs are not relevant in this example. Accident costs are not considered in this example, but if accident rates were known and changes in accidents were discernible as a result of this improvement, their unit cost would be added to *B* before multiplication by *L*. The numbers for *B* and *T* are derived from Figures 4-1 and 4-2, using the facility and traffic data in Tables 4-4 and 4-5. A unit value of time of $3 per hour per vehicle is assumed.

Determination of *B* and *T* for the existing route (alternative 0) in year 1 is further described as follows. Entering the lower left-hand (speed-flow) chart of Figure 4-1 with a *v/c* ratio of 0.50 and a design speed of 50 mph yields an average running speed of approximately 36 mph and a corresponding travel time of 27.8 hr per thousand vehicle-miles. Tangent running cost at 36 mph on a level (0%) grade is approximately $71 per thousand vehicle-miles (KVM). Added running cost due to an 8° curve is (from Figure 4-1) approximately $20 per KVM. Added running cost due to speed changes at a *v/c* ratio of 0.50 is approximately $3 per KVM. Total running cost per KVM on the existing route is thus about $94 (71 + 20 + 3). Adding this to a total time value of about $83.40 (27.8 × $3 per hour) per KVM yields a total unit basic section cost (*B*) for the existing route in year 1 of $177.40 per KVM.

Entering Figure 4-2 with a speed of 41 mph on the "faster" previous section and 36 mph on the section being analyzed yields an added one-way transition cost of approximately $1 per thousand vehicles (KV). Similarly, for the transition *from* the section, with an assumed speed of 50 mph, the transition cost, from Figure 4-2, is approximately $2.50 per KV. Total one-way section transition cost is thus $3.50 per KV.

Similar utilization of Figures 4-1 and 4-2 applied to the proposed route in year 1 and both the existing and proposed routes in year 20 yields corresponding values for *B* and *T*. The absence of curves under alternative 1 does the most to reduce the *B* values relative to the alternative 0, while reduction in the magnitude of speed changes between sections reduces the *T* values for alternative 1 relative to alternative 0. Multiplying basic section costs by section length (*L*) and adding

TABLE 4-6 Calculation of *HU*

Alternative	*Section*	*Year*	*Period*	*(B × L)ᵃ*	*+T*	*= One-Way HU per KVᵇ*
0	*ab*	1	Typical hour	177.4 × 0.2	3.5	$38.98
1	*ab*	1	Typical hour	148.7 × 0.16	1.7	$25.49
0	*ab*	20	Typical hour	186.90 × 0.2	3.0	$40.38
1	*ab*	20	Typical hour	165.00 × 0.16	2.0	$28.40

ᵃ*B* and *T* are in dollars; *L* is in miles; 1 mile = 1.6093 km.
ᵇ*KV* = thousand vehicles.

section transition costs results in the one-way highway user costs (*HU*) per KV in Table 4-6.

Calculation of User Benefits

Benefits in the representative 1st and 20th years can be calculated from the above user cost (*HU*) estimates by applying Eq. 4-11, as demonstrated in Table 4-7.

Computation of Present Values and Measures of Economic Desirability

Using an assumption that benefits grow at an equal annual percentage rate, the benefits for the two representative years may be converted into an estimate of the present value of all benefits by the application of Eqs. 4-4 and 4-5. Using the ratio α of benefits in the 20th year to those in the 1st year ($232,190/172,827 = 1.34$), the rate of growth of annual value $r = \ln(\alpha)/19 = 0.0155$. For an assumed 25-year analysis period and a 4 percent discount rate, a present value factor PW_g of approximately 18.7 is obtained by Eq. 4-4. The present value of benefits is then $PW_g \times$ first-year benefits or

$$18.7 \times 172,827 = \$3,231,865$$

Since the improvement affects traffic symmetrically in both directions, this one-way benefit estimate can be multiplied by 2 to yield the total present value of the project's benefits, $6.46 million.

A complete project analysis also requires conversion of project costs and related items into present values. Project costs include the capital cost of construction and the change in maintenance expenses incurred on the section as a result of the improvement. A present value adjustment for the residual or scrap value of the project at the end of the 25-year analysis period is also necessary. As noted, the total construction cost of $2 million will be incurred at the inception of the project and therefore has a present value of $2 million. Also as noted, annual maintenance costs are anticipated to be $2,000 less after the improvement. The project is estimated to have a residual value of roughly $500,000 in its 25th year. Again using a discount rate of 4 percent, the stream of annual maintenance savings is converted to present value by using the appropriate series present worth factor in Table 4-3. The present value of the residual value in the 25 years is also calculated from Table 4-3. The conversion of all project-related costs to present value is summarized in Table 4-8.

TABLE 4-7 User Benefit Calculations[a]

Section	Year	Period	Days per year	×	Hours per Day in Period	×	$(HU_0 - HU_1) \times (V_0 + V_1)/2$	=	One-Way Annual Benefit
ab	1	Typical hour	365		18		$(38.98 - 25.49) \times (1.9 + 2.0)/2$		$172,827
ab	20	Typical hour	365		18		$(40.38 - 28.40) \times (2.9 + 3.0)/2$		$232,190

[a]Unit user costs (*HU*) are expressed in units of dollars per thousand vehicles, and traffic volumes (*V*) are expressed in units of thousands of vehicles.

TABLE 4-8 Calculation of Present Value of Project-Related Costs

Item		Present Value
Construction costs		$2,000,000
Changes in annual maintenance expenditures	$-$2,000 $\times$ 15.62 $=$	$-$31,244[a]
Residual value in 25 years	$-$500,000 $\times$ 0.375 $=$	$-$187,558[a]
Present value of all "costs"		$1,781,198

[a]Since incremental maintenance costs are decreased and residual value is a positive amount, these quantities are expressed as negative "costs."

The net present value of the project is the difference between the present value of benefits and the present value of costs:

$$\text{net present value} = \text{present value of benefits} - \text{present value of costs}$$
$$= \$6,463,730 - \$1,718,198$$
$$= \$4,745,532$$

The net present value of $4,745,532 indicates that the project is economically desirable. According to Eq. 4-6, the benefit-cost ratio is

$$6,463,730 \div (2,000,000 - 31,244 - 187,558) \text{ or } 3.62$$

Since the benefit-cost ratio is greater than 1.0, the project is economically desirable. However, if there are budget constraints, the decision maker should investigate other projects or project alternatives for their relative economic attractiveness before making a decision to implement the project. In any event, the decision maker should also investigate any significant community or noncostable benefits or costs such as air and noise pollution that might be associated with the project before a final decision is made.

4-6 Cost Updating Procedures

It is noted that the nomographs shown as Figures 4-1, 4-2, 4-3, and 4-4 are based on price levels prevailing in January 1975. It is suggested that user costs be reviewed for currentness about once every two years and appropriately updated. Reference 6 recommends a cost updating procedure that employs multipliers based on commonly available consumer and wholesale price indexes published by the U.S. Bureau of Labor Statistics. It is noted, however, that these index values published in the AASHTO reference (6) have now been indexed to a later time (1982–84 = 100).

Table 4-9 gives a list of consumer price indexes and wholesale price indexes (now termed producer price indexes), updated to May 1994. The table also shows the January 1975 values as well as those values that have been adjusted for rebasing. These indexes make it possible to update the user costs from the nomographs for inflation.

A series of updating formulas was published in the AASHTO manual (6) that accounts for the various components of user costs and recognizes that these components may escalate at different rates. These formulas have been rearranged and are shown in Table 4-10. The formulas provide a weighted average multiplier that accounts for the varying rates of inflation. In the first formula in Table 4-10, for example, fuel costs account for 28 percent of total passenger car running costs

TABLE 4-9 Consumer and Wholesale Price Indexes* for January 1975 and May 1994

Index	*(1)* January 1975 From Reference 6	*(2)* January 1975 Value Adjusted for Rebasing	*(3)* May 1994 Index
CPI, total	156.1	52.1	147.5
CPI, private transportation	142.2	48.0	130.0
C_f, gasoline, regular and premium	160.2	42.3	96.0
C_o, motor oil, premium	152.8	47.7	96.5
C_t, tires, new, tubeless	124.9	68.4	99.7
C_m, auto repairs and maintenance	170.1	51.7	149.7
C_d, automobiles, new	123.4	60.8	135.7
WPI, industrial	167.5	53.6	119.8
P_{fg}, gasoline, single unit trucks	198.9	30.6	61.2
P_{fd}, diesel fuel, combination trucks	237.6	29.8	55.0
P_o, motor oil for trucks	151.5	51.2	127.7
P_t, truck and bus tires	152.0	58.0	95.1
P_{sd}, motor trucks, <10,000 lbs	148.9	55.8	158.1
P_{ld}, motor trucks, >10,000 lbs	148.9	49.3	141.9

*The Wholesale Price Index (WPI) is now referred to as the Producer Price Index (PPI).

Note: The letter *C* refers to the consumer price index; the letter *P* refers to the producer price index.

Source: U.S. Bureau of Labor Statistics.

on level tangents; motor oil accounts for 1 percent; tires account for 5 percent, and so on. In these formulas, for a given index i, the variable C_i represents the ratio of the value of the present consumer price index to the adjusted value of the January 1975 index. Similarly,

$$P_i = \frac{\text{(Current Producer Price Index)}_i}{\text{(Adjusted Wholesale Price Index for January 1975)}_i}$$

The use of these formulas for updating costs is illustrated in Example 4-6.

EXAMPLE 4-6 **Updating User Costs** Using the indexes from Table 4-9 and the appropriate formula from Table 4-10, update the tangent running cost for the highway described in Example 4-5 to May 1994.

Solution From Figure 4-1, the tangent running cost for Example 4-5 was found to be $71 per thousand vehicle-miles (KVM). From Table 4-10 (a), the updating formula is:

$$M = 0.28\,C_f + 0.01\,C_o + 0.05\,C_t + 0.27\,C_m + 0.39\,C_d$$

$$M = 0.28\frac{96.0}{42.3} + 0.10\frac{96.5}{47.7} + 0.05\frac{99.7}{68.4} + 0.27\frac{149.7}{51.7} + 0.39\frac{135.7}{60.8}$$

$$M = 2.38$$

The updated tangent running cost = $2.38 \times \$71 = \$169$ per KVM.

4-7 Cost and Benefits of Bus Transit Improvements

In economic analyses involving bus transit the user benefits should include reductions in highway user costs (as previously described) as well as reductions in

TABLE 4-10 (a) Multiplier Formulas for Updating Running Costs of Passenger Cars

Condition for Use	Updating Formula
General and level tangents	$M = 0.28\ C_f + 0.01\ C_o + 0.05\ C_t + 0.27\ C_m + 0.39\ C_d$
Positive grades	$M = 0.40\ C_f + 0.02\ C_o + 0.05\ C_t + 0.16\ C_m + 0.37\ C_d$
Negative grades	$M = 0.17\ C_f + 0.03\ C_o + 0.06\ C_t + 0.20\ C_m + 0.54\ C_d$
Excess curve cost	$M = 0.22\ C_f + 0.01\ C_o + 0.77\ C_t$
Speed change and stopping cost	$M = 0.35\ C_f + 0.01\ C_o + 0.41\ C_t + 0.02\ C_m + 0.21\ C_d$
Idling cost	$M = 0.83\ C_f + 0.01\ C_o + \qquad\qquad 0.05\ C_m + 0.11\ C_d$

Source: Adapted from American Association of State Highway and Transportation Officials, *A Manual on User Benefit Analysis of Highway and Bus-Transit Improvements,* Washington, DC (1977).

TABLE 4-10 (b) Multiplier Formulas for Updating Running Costs of Single Unit Trucks

Condition for Use	Updating Formula
General and level tangents	$M = 0.40\ P_f + 0.01\ P_o + 0.04\ P_t + 0.30\ C_m + 0.25\ P_d$
Positive grades	$M = 0.54\ P_f + 0.01\ P_o + 0.01\ P_t + 0.24\ C_m + 0.20\ P_d$
Negative grades	$M = 0.21\ P_f + 0.03\ P_o + 0.04\ P_t + 0.33\ C_m + 0.39\ P_d$
Excess curve cost	$M = 0.24\ P_f + 0.01\ P_o + 0.74\ P_t + 0.01\ C_m$
Speed change and stopping cost	$M = 0.36\ P_f + \qquad\qquad 0.48\ P_t + 0.04\ C_m + 0.12\ P_d$
Idling cost	$M = 0.80\ P_f + 0.01\ P_o + \qquad\qquad 0.04\ C_m + 0.15\ P_d$

Source: Adapted from American Association of State Highway and Transportation Officials, *A Manual on User Benefit Analysis of Highway and Bus-Transit Improvements,* Washington, DC (1977).

TABLE 4-10 (c) Multiplier Formulas for Updating Running Costs of 3-S2 Combination Trucks

Condition for Use	Updating Formula
General and level tangents	$M = 0.31\ P_f + 0.01\ P_o + 0.11\ P_t + 0.37\ C_m + 0.20\ P_d$
Positive grades	$M = 0.61\ P_f + 0.01\ P_o + 0.04\ P_t + 0.25\ C_m + 0.09\ P_d$
Negative grades	$M = \qquad\qquad 0.03\ P_o + 0.14\ P_t + 0.50\ P_m + 0.33\ P_d$
Excess curve cost	$M = 0.07\ P_f + 0.01\ P_o + 0.90\ P_t + 0.02\ C_m$
Speed change and stopping cost	$M = 0.20\ P_f + \qquad\qquad 0.73\ P_t + 0.02\ C_m + 0.05\ P_d$
Idling cost	$M = 0.64\ P_f + 0.01\ P_o + \qquad\qquad 0.14\ C_m + 0.21\ P_d$

Source: Adapted from American Association of State Highway and Transportation Officials, *A Manual on User Benefit Analysis of Highway and Bus-Transit Improvements,* Washington, DC (1977).

bus user costs. Bus user costs consist of the costs of time (including in-vehicle travel time and the time for waiting, walking, and transferring) and the out-of-pocket costs to the user such as bus fares and the costs of getting to and from the bus stops. It is recommended that different costs be specified for the time spent in the transit vehicle and the time spent out of the vehicle. Reference 6 recommends that the costs assigned for walking, waiting, and transferring be 1.5 to 2.0 times that for traveling in the transit vehicle.

The change in transit system costs is the difference in operating the system before and after the improvement adjusted to reflect any change in income from transit fares. If changes in fare revenues are not also accounted for in cost calculations, a simple lowering of the fares might be interpreted to yield a net benefit when in fact the user benefits may be largely offset by losses in bus company revenues (*6*).

Research has shown that the annual cost of operating a transit system is most closely related to the following variables: (1) annual bus miles produced by the transit system, (2) bus size, (3) bus driver wages, and (4) average service speed. Anderson et al. (*6*) give an empirical equation by which annual transit operating costs may be estimated.

The present value of user benefits for transit improvements should be compared with the present value of changes in transit system costs. The user benefits must exceed the changes in the system costs (both expressed as present values) if the project is to be judged economically feasible.

The material presented here on economic analyses involving bus transit is admittedly sketchy. For further information on this subject the reader should consult Ref. 6.

SOCIAL AND ENVIRONMENTAL CONSEQUENCES OF HIGHWAYS

In recent years there has been increasing public awareness of the impact of highways and other public projects on the environment. Reflecting this concern, Congress has passed several laws that have brought about fundamental changes in the highway evaluation process and the relationship between highway investment and environmental quality. In the rest of this chapter some of these laws and certain of the procedures and techniques employed to evaluate social and environmental impacts are described. Generally speaking, the evaluation guidelines described in this chapter are mandatory for federally assisted highway improvements and are increasingly being used for projects constructed from state or local funds.

4-8 Environmental Legislation

Environmental legislation is important to the highway engineering process not only because of the types of impacts that must be considered in the project development process, but also because it outlines which agencies have approval power of project implementation. Several key concepts should be kept in mind during the project development process. First, many different types of social and environmental impacts must often be considered by the highway designers. These impacts, whose mandatory consideration is often the result of laws targeting that particular impact, can include such things as air quality, coastal zone management, endangered species, energy conservation, farm land protection, fish and wildlife preservation, floodplain considerations, hazardous waste, historic preservation, noise levels, socioeconomic impacts, aesthetic or visual concerns, water quality, wetlands, and wild/scenic rivers.

Second, different federal and state agencies often have an important role in approving project development where environmental impacts are present. For example, the U.S. Coast Guard is involved in bridge projects over navigable waters, the U.S. Army Corps of Engineers has jurisdiction over all wetlands, the state historic preservation officer is part of the process of approving projects that are near historic properties, and so on. Highway engineers spend a great deal of time interacting with the representatives of such agencies to ensure that their concerns are addressed in the engineering process.

Third, although these agencies are often part of the project development process, many are also required to issue permits that allow the environmental impacts to occur with appropriate levels of mitigation. For example, the U.S.

Army Corps of Engineers must issue a Section 404 permit if wetlands are to be affected by a particular project. Thus, the project development process is not really over until these permits are granted.

Finally, an important element of environmental legislation is that not only are impacts to be identified, but strategies to mitigate these impacts are also to be part of the project design. Mitigation can range from avoiding the impact completely (e.g., by relocating the highway alignment) or complete compensation for affecting a site (e.g., building new wetlands off site that replace those impacted). The highway engineering process is thus involved not only with the design of the project itself, but also with the migration strategies that are necessary to allow the project to proceed.

The most significant legislative provisions relating to evaluation of social and environmental consequences of highways were contained in the Department of Transportation Act of 1966, the National Environmental Policy Act of 1969, the Clean Air Act of 1963 (and subsequent acts dealing with air pollution), the Uniform Relocation Assistance and Real Property Acquisitions Policies Act of 1970, and the Federal-Aid Highway Act of 1970.

The Department of Transportation Act, passed in 1966, declared it to be the national policy that a special effort should be made to preserve the natural beauty of the countryside. Section 4(f) of that act stated that the Secretary of Transportation shall not approve any program or project that requires any land from a public park, recreation area, wildlife and waterfowl refuge, or historic site unless there is no feasible and prudent alternative to the use of such land. It further stated that transportation programs that require land from such areas must include all possible planning to minimize environmental harm. Thus, in those cases where such lands are to be impacted, highway engineers must conduct a Section 4(f) study to determine likely impacts and possible mitigation strategies.

The National Environmental Policy Act of 1969 declared it to be a national policy that the federal government would use all practicable means and measures "to create and maintain conditions under which man and nature can exist in productive harmony." Section 102 of the act, quoted in part below, established the requirement for environmental impact statements.

> Sec. 102. The Congress authorizes and directs that, to the fullest extent possible: (1) the policies, regulations, and public laws of the United States shall be interpreted and administered in accordance with the policies set forth in this Act, and (2) all agencies of the Federal Government shall—
>
> (A) utilize a systematic, interdisciplinary approach which will insure the integrated use of the natural and social sciences and the environmental design arts in planning and in decisionmaking which may have an impact on man's environment;
>
> (B) identify and develop methods and procedures, in consultation with the Council on Environmental Quality established by title II of this Act, which will insure that presently unquantified environmental amenities and values may be given appropriate consideration in decisionmaking along with economic and technical considerations;
>
> (C) include in every recommendation or report on proposals for legislation and other major Federal actions significantly affecting the quality of the human environment, a detailed statement by the responsible official on—
>
> (i) the environmental impact of the proposed action,
>
> (ii) any adverse environmental effects which cannot be avoided should the proposal be implemented,

(iii) alternatives to the proposed action,

(iv) the relationship between local short-term uses of man's environment and the maintenance and enhancement of long-term productivity, and

(v) any irreversible and irretrievable commitments of resources which would be involved in the proposed action should it be implemented.

The Clean Air Act of 1963 encouraged increased state and local programs for the control of air pollution. The act also provided for the development of air quality standards. In 1965, Congress passed the Motor Vehicle Control Act, which initiated controls on motor vehicle manufacturers to require the installation of air pollution control devices on all new vehicles. The most recent amendment to the Clean Air Act occurred in 1990, when Congress greatly expanded the linkage between transportation investment and air quality concerns. In essence, the 1990 amendments required that no highway investment could occur in an urban area that is in nonattainment of federally mandated air quality standards if that project was not in conformance with the area's air quality plan. The interpretation of this conformity requirement is that any highway project that will result in increased vehicle miles traveled (VMT) will not meet this test. In many of the largest U.S. urban areas, therefore, highway planning and project design is very much linked to the air quality planning process.

In 1970, Congress passed the Uniform Relocation Assistance and Real Property Acquisitions Policies Act in an effort to establish uniform treatment of all people affected by public projects financed with federal funds. The act sets forth standard policies and procedures for the determination of just compensation, for negotiation with property owners, for taking possession of property, and, when required, for instituting formal condemnation proceedings.

As was indicated in Chapter 1, the Federal-Aid Highway Act of 1970 required the Secretary of Transportation to publish guidelines designed to ensure that possible adverse economic, social, and environmental effects are properly considered during the planning and development of federal-aid highway projects.

4-9 Federal Highway Environmental Policies

As a means of implementing the various environmental laws, the Federal Highway Administration has issued regulations (*14, 15*) that specify the process by which highway agencies plan and develop highway projects of any federal-aid highway system. The policy of the FHWA is, to the fullest extent possible, to coordinate all environmental investigations, reviews, and consultations into a single process. The agency further intends that decisions on proposed actions be based on a balanced consideration of the need for safe and efficient transportation and of national, state, and local goals to protect and enhance the environment, conserve energy, and revitalize urban areas.

Other matters of FHWA environmental policy include

1. involvement of the public in the planning process
2. use of a systematic interdisciplinary approach in the planning development process
3. incorporation of measures to mitigate adverse impacts from proposed projects

4-10 Documentation for Environmental Analyses

The Federal Highway Administration defines three classes of action that prescribe the level of documentation required in the environmental planning process (*14*):

1. *Class I* actions, which require the preparation of an environmental impact statement (EIS). Examples of class I projects include controlled access freeways; projects with four or more lanes on a new location; and major projects whose construction involves a large amount of demolition, extensive displacement of individuals or businesses, or substantial disruption of local traffic patterns.

2. *Class II* actions (categorical exclusions), which do not require an environmental impact statement or environmental assessment. Such projects include planning and technical studies that do not fund the construction of facilities; construction of bicycle and pedestrian lanes, paths, and facilities; and reconstruction of an existing bridge structure on essentially the same alignment or location.

3. *Class III* actions, in which the significance of the impact is not clearly established. Such actions require the preparation of an *environmental assessment* (EA) concisely describing the environmental impacts of the proposed work and its alternatives. This assessment serves as a basis for deciding the nature of further environmental analysis and documentation. If the FHWA determines that a proposed project will not have a significant impact on the environment, a statement to that effect is prepared. This document, which is called a finding of no significant impact (FONSI), must be made available to the public on request.

4-11 Environmental Impact Statements

An environmental impact statement must be prepared for all federal actions that will have a "significant impact" on the environment. For federally supported highway projects, the FHWA determines the need for an EIS based on preliminary environmental studies or when the review of the environmental assessment and the comments received indicate that the expected impacts may be significant.

The environmental impact statement is a written statement containing an assessment of the anticipated significant effects that the proposed action may have on the quality of the environment. The purpose of the EIS is to ensure that careful attention is given to environmental matters and that such matters are appropriately considered in the highway agency's decision. The contents of an environmental impact statement are briefly outlined in Table 4-11.

Environmental impact statements are proposed in two stages. The highway agency prepares and circulates a draft statement in a prescribed fashion to public agencies, private groups, and individuals for review and comments. After receiving reactions to the draft EIS and making appropriate modifications, the agency then submits a final EIS to the regional or Washington headquarters of FHWA for review and approval.

As noted earlier, a section 4(f) statement is required by Section 4(f) of the Department of Transportation Act. The statement relates to the proposed use of publicly owned land from a park, recreation area, or wildlife refuge. In order to use such lands for highway purposes, the 4(f) statement must declare that

TABLE 4-11 Contents of an Environmental Impact Statement

1. Description of the proposed action and alternatives considered
   Location, type, and length of facility, termini, number of lanes, right-of-way width
   Other design features such as general horizontal and vertical alignment, structures, etc.
   Deficiencies of existing facilities, anticipated benefits
2. Land use planning
   Description of planning process for the area
3. Probable impact of proposed action on the environment
   Natural, ecological, scenic resource impacts
   Relocation of individuals and families
   Social impacts
   Air quality impacts
   Noise impacts
   Water quality impacts
   Construction impacts
4. Alternatives to the proposed action
5. Probable adverse environmental effects that cannot be avoided
6. The relationship between local short-term uses of man's environment and the maintenance and enhancement of long-term productivity
7. Irreversible and irretrievable commitments of resources
8. The impact on properties and sites of historic and cultural significance
9. Summary of coordination and public and minority involvement

1. there is no feasible and prudent alternative to the use of such land
2. the proposed program includes all possible planning to minimize harm to the section 4(f) land resulting from such use

In the planning and development of complex highway projects, the FHWA encourages the preparation of tiered environmental impact statements. The first tier of such statements focuses on broad issues such as mode choice, general location, and areawide air quality and land-use implications of alternative transportation improvements. The second tier is site-specific and describes detailed project impacts and measures to mitigate harmful impacts.

4-12 Coordination and Public Involvement

State highway agencies conduct public hearings to afford interested persons an opportunity to participate in the process of determining the need for and the location of highways. For federal-aid highways, one or more pubic hearings are normally required for proposed projects

1. if substantial amounts of right-of-way are required
2. if there is a substantial change in the layout of connecting roadways or the facility being improved
3. if there is a substantial adverse effect on abutting property
4. if the property otherwise has significant social, economic, or environmental effects

At such hearings, a representative of the state highway agency explains the project's purpose and need; its major design features; the social, economic, and environmental impacts; and measures planned to mitigate such effects. The agency must also accept oral and written statements from the public and consult with citizens at various stages of federally supported projects.

In the early stages of the development of all important federal-aid highway projects, the responsible state highway agency must send notice of the project and available plans to a state or metropolitan area clearinghouse. The clearinghouse then informs other interested agencies and invites their comments on the proposed work. The same process is used for annual programs of proposed future projects. The coordination process was formerly known as the "A-95 notification," named after a U.S. Office of Management and Budget circular that required it. It is now called the "Intergovernmental Review of Federal Programs." Various federal agencies have published rules for intergovernmental review in the *Federal Register,* implementing Executive Order 12372 (*16*).

HIGHWAY NOISE PLANNING

In the next several sections some of the common environmental impacts of highways will be described. The first impact to be considered is perhaps the most troublesome to control: highway noise. First, noise characteristics and measurement will be described, and then some remedial programs to reduce noise levels and mitigate their harmful effects will be examined.

4-13 Noise Characteristics and Measurement

Noise is defined as excessive or unwanted sound. It is unwanted because it annoys, interferes with conversation, disturbs sleep, and, in the extreme, is a danger to public health. Sound, whether noisy or noiseless, is produced by vibrations in air, water, steel, or other substance. When an object vibrates, it produces rapid small-scale variations in the normal atmospheric pressure. This disturbance is propagated from the source in a repetitive spherical pattern at a speed (in air) of approximately 1100 ft/sec (340 m/sec). It may be reflected, partially absorbed, or attenuated before reaching an eardrum to produce a sensation of sound.

Noise is characterized by its sound level, its frequency spectrum, and its variation over time. The term *sound level* refers to a physical measure that corresponds to the hearer's subjective conception of loudness. It is a function of the magnitude of the pressure fluctuations about the ambient barometric air pressure. One can speak of the strength of these fluctuations in terms of several variables, the most common being *sound intensity* and *sound pressure.*

Sound intensity (also called sound power density) is the average rate of sound energy transmitted through a unit area perpendicular to the direction of sound propagation, typically measured in picowatts (10^{-12} watt) per square meter. The human ear can detect sound intensities as weak as 1 picowatt and tolerate intensities as high as 10^{13} picowatts. Because of the difficulties of dealing with such a large range of numbers, a logarithmic measure called the decibel (dB) is used to describe sound level. The sound intensity, expressed in decibels, is

$$\text{sound intensity} = 10 \log_{10}\left(\frac{I}{I_0}\right) \qquad (4\text{-}12)$$

where I = sound intensity (picowatts/m²)

I_0 = 1 picowatt/m², a standard reference intensity representing approximately the weakest audible sound

Because no instrument is available for directly measuring the power level of a source, sound pressure, which is usually proportional to the square root of sound power, is used as a measure of the magnitude of a sound disturbance (*17*). The sound pressure, in decibels, is

$$\text{sound pressure} = 10 \log_{10}\left(\frac{P}{P_0}\right)^2$$

$$= 20 \log_{10}\left(\frac{P}{P_0}\right)$$

(4-13)

where P = root-mean-square sound pressure, typically expressed in units of newtons per square meter

P_0 = 20 micronewtons/m² or 0.0002 dyne/cm², a standard reference pressure corresponding to the weakest audible sound

Sound level is measured by a sound level meter, which consists essentially of a microphone that converts the pattern of sound pressure fluctuations into a similar pattern of electrical voltage, amplifiers, and a voltage meter that is normally calibrated to read in decibels. For practical purposes, the decibel scale ranges from 0, the threshold of hearing, to about 140 dB, the onset of pain. For every increase of about 10 dB, there is a doubling of the sound's apparent loudness.

The apparent loudness of a sound also depends on the *frequency* of the sound. Frequency is the rate of occurrence of the sound pressure fluctuations, commonly expressed in cycles per second or hertz (Hz). The frequency determines the pitch of the sound; the higher the frequency, the higher the pitch. The normal human ear can hear sounds with frequencies from about 20 to 20,000 Hz but is more sensitive to sounds in the middle- to high-frequency range.

> Most noises are made up of a mixture of components having different frequencies: the sound of a diesel tractor/trailer at high speed on the freeway combines the high-pitched singing of the tires and the low-pitched roar of the engine and the exhaust, both of which the ear readily distinguishes. . . . A flute, on the other hand, if played softly, makes an almost pure tone containing only a single prominent frequency. Depending on how the components of a noise are distributed in frequency, our ears make a subjective judgment of "quality." (*17*)

A great many scales have been used to express noise levels. One of the simplest and most straightforward noise measurement techniques consists of measuring the *overall sound pressure level* that is related to the total sound energy over the audible frequency range. However, the unweighted overall sound pressure level is not strongly correlated with a hearer's subjective response to the noise. As indicated earlier, the human ear tends to be more sensitive to sounds with relatively high frequencies.

The *A-weighted sound level* was devised to more closely represent a person's subjective response to sounds. In the *A*-weighted filter network, the lower frequencies are deemphasized in a manner similar to that of human hearing. The *A*-weighted sound level, measured in decibels (dBA), is the generally accepted scale for measuring highway transportation noises.

The extent to which a person is annoyed by noise also depends to a considerable degree on its variation over time. The temporal effect is manifested in the duration of a single noise event as well as its repetitiveness and time of occurrence. Clearly, the longer a noise lasts and the more frequently it is repeated, the greater the interruption of human activity and the more pronounced the annoyance.

One factor that gives a good representation of both the magnitude and the fluctuation pattern of noise is the 10th percentile noise level, L_{10}, that is, the noise level exceeded 10 percent of the time (*18*). With reference to Figure 4-5, the L_{10} noise level is that value for which the sum of the deltas equals 10 percent of the total time being considered, T_T.

Another descriptor of traffic noise is the equivalent sound level, L_{eq}, the average acoustic intensity over time. It represents the sound level that, if held constant over a specified time period, would contain the same total acoustic energy as the actual varying levels of the traffic noise during the same period (*19*). L_{eq} is considered more reliable than L_{10} for noise studies along low-volume roadways (*20*).

For obvious reasons, noise tends to be more annoying during evening and nighttime hours than during the day. There is a greater need for quiet during sleeping hours, and during those periods when there is less noise from other sources, peak traffic noises tend to be more disturbing than even higher levels during daylight hours.

4-14 Noise Studies and Standards

It is suggested that a noise study be prepared during the location stage of all alternative lines under consideration for each highway project. Early in the design process, the present and future uses of land adjacent to the highway should be determined and noise-sensitive areas should be identified.

Existing noise levels should be determined by field measurements at representative times and locations, with special emphasis on potentially noise-sensitive times and areas. There is considerable variation in types of equipment and levels

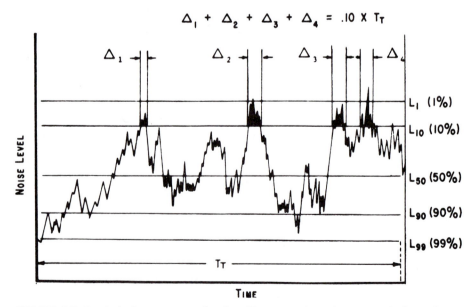

FIGURE 4-5 Statistical measures of noise. (Courtesy American Association of State Highway and Transportation Officials.)

of sophistication and effort employed by the various agencies for noise studies. Detailed procedure for making such measurements are given in Ref. 19.

The prediction of future noise levels is a complex task in which the effects of a large number of parameters should be considered, including traffic volumes and composition, accelerations and speeds, roadway gradients, topography, the presence and types of barriers, and the distance between the highway and the affected area. Some guidance on the predication of future noise levels is given by Refs. 19 and 20.

AASHTO (*18*) has established standards for evaluation of the impact of highway-generated noise on nearby land uses or activities. The standards provide a basis for the selection of appropriate noise attenuation measures. The following exterior design noise levels are recommended for the land uses and activities indicated (*20*):

Category	Description	Noise Level (dBA) L_{eq}	L_{10}
A	Lands on which serenity and quiet are of extraordinary significance and serve an important public need and where the preservation of those qualities is essential if the area is to continue to serve its intended purpose	57	60
B	Picnic areas, recreation areas, playgrounds, active sports areas, parks, residences, motels, hotels, schools, churches, libraries, and hospitals	67	70
C	Developed lands, properties, or activities, not included in categories A and B above	72	75
D	Undeveloped lands	—	—
E	Residences, motels, hotels, public meeting rooms, schools, churches, libraries, hospitals, and auditoriums	52	55

Unusual conditions may, of course, require special analyses that could lead to the establishment of design noise levels different from those indicated above.

4-15 Noise Attenuation

Highway engineers can lessen the impact of highway noise on nearby human activities in three general ways:

1. by providing a buffer zone between the highway and adjacent land activities
2. by modifying the horizontal or vertical alignment of the highway
3. by providing noise shielding

Since noise tends to spread out uniformly as it travels away from the source, the sound level decreases at a rate of about 6 dB for each doubling of distance. One approach to noise attenuation, therefore, is to provide a buffer zone between the highway and the users of adjacent land. The acquisition of additional right-of-way width may, of course, be expensive and not be justified for the purpose of noise attenuation alone. However, it may be possible to use such buffer areas for other purposes such as for bicycle paths, pedestrian walkways, traffic safety recovery zones, and so forth.

Designers may greatly decrease the noise impact of new highways by locating the alignment so as to avoid noise-sensitive areas. It may also be possible to lower noise levels along adjacent land areas by utilizing independent alignment for opposing traffic lanes and by taking advantage of natural features for noise attentuation (*19*). Other possible desirable modifications to the highway alignment include depressing or elevating the roadway section and reducing the highway grade.

A variety of noise shields or barriers have been used to lessen the impact of noise on adjacent land, including earth berms, timber walls, concrete and masonry walls, metal walls, vegetative screens, and stucco and chain link fences. Experience indicates that a well-designed barrier will provide a noise attenuation of about 10 dBA.

Some general principles for noise barrier design are outlined below.

1. Barriers should be high enough and long enough to intercept the line of sight between the source and the observer. It is especially important that the barrier be high enough to shield truck exhaust noise. To accomplish this without aesthetic sacrifice, some agencies have constructed combination barriers consisting of an earth berm that supports some type of vertical wall.

2. To be most effective, barriers should be constructed so as to cast a big sound "shadow." This generally means that the barrier should be placed close to either the source or the receiver. It should be remembered that noise barriers are fixed objects that, if placed close to the traffic lanes, may be unsafe. For this and other reasons, noise barriers are frequently located near the right-of-way line.

3. Barriers should be constructed of a dense material, and there should be no air paths through or under the barrier that could constitute noise leaks.

4. In the interests of community acceptance, barriers should be made as attractive as possible. They should be compatible with local architecture and blend into the forms of the landscape.

Reference 21 gives additional information on the selection, design, and construction of highway noise barriers.

HIGHWAY AIR QUALITY

Air pollution may constitute the most serious environmental impact caused by highway transportation. It is a complex problem, and its evaluation and control may require the involvement of highly trained environmental specialists. A detailed treatment of the subject is well beyond the scope of this textbook. This section will include a brief description of the air pollution problem and ways in which highway engineers may lessen its impact. For additional information, it is suggested that the reader consult the FHWA publication *Fundamentals of Air Quality (22),* which served as the principal source of reference for the following material.

4-16 Air Pollution

Air pollutants are natural and man-made contaminants in the atmosphere. They may be grouped into five major classes:

1. carbon monoxide (CO)
2. hydrocarbons (HC) and photochemical oxidants
3. nitrogen oxides (NO_x)
4. particulate matter
5. sulfur dioxide (SO_2)

Carbon monoxide, the most widely distributed and the most commonly occurring air pollutant, is a colorless, odorless, highly poisonous gas that results from the incomplete combustion of carbonaceous fuels. Gaseous organic compounds of carbon and hydrogen (hydrocarbons) and oxides of nitrogen are also emitted during the combustion process. While prevailing concentrations of hydrocarbons do not appear to be detrimental to human health, certain of these substances may react with nitrogen oxides to produce harmful pollutants. Ozone and other oxidizing agents are formed when hydrocarbons and nitrogen oxides are exposed to sunlight. These photochemical oxidants can cause irritation of the respiratory and alimentary systems and can damage vegetation, metals, and other materials. There is also some evidence that long-term exposure of humans to nitrogen dioxide, even in low concentrations, may contribute to chronic respiratory diseases.

Particulate matter is any solid or liquid matter dispersed in air under standard conditions of temperature and pressure. Particulate matter discharged into the atmosphere may be in the form of dust, soot, fly ash, and so on. Generally, particles released into the atmosphere from man-made sources are 1 to 10 μm in size.

Sulfur dioxide is generated by the combustion of sulfur-bearing fuels and by certain industrial processes that use sulfur-bearing raw materials. Motor vehicles are not a major source of sulfur dioxide pollution.

Highway vehicles are the most important source of air pollution in the United States. They are the principal source of carbon monoxide and hydrocarbons and account for nearly half of the total nitrogen oxides emitted each year. In addition, motor vehicles are the chief source of particulate lead in the atmosphere.

Highway air pollution results principally from emissions from internal combustion and diesel engines. Internal combustion engines produce emissions that originate from three basic sources within the vehicle system (*22*):

1. *Exhaust emissions.* Pollutants that are present in the exhaust gas as it is discharged into the atmosphere.
2. *Evaporative emissions.* Vapors lost directly to the atmosphere from the fuel tank, carburetor, or any other part of the system.
3. *Crankcase blow-by.* Gases and vapors that under pressure escape from the combustion chamber, pass the engine pistons, and enter the crankcase.

The diesel engine emissions problem has not yet been well defined or quantified. The most apparent source of atmospheric pollution from a diesel engine is the smoke and odor from its exhaust. In addition, diesel engines are known to emit significant concentrations of unburned hydrocarbons, nitric oxides, and oxygenated compounds (*22*). Studies have shown that the concentration of emissions from diesel engines varies widely depending on engine speed and load.

4-17 Air Quality Standards and Pollution Modeling

The U.S. Environmental Protection Agency (EPA) has the overall responsibility for the establishment of national ambient air quality standards. Such standards have been established and published in the *Federal Register.* Enforcement of these standards is the responsibility of state air pollution control agencies approved by the EPA.

To thoroughly assess the impact of a proposed highway improvement, it is necessary to estimate the quantities of pollutant concentrations within the study area both before and after development. Such an assessment depends heavily on a thorough statistical analysis of past meterological data and the evaluation of the effects of existing and forecast traffic volumes. The evaluation process is a complex one for which two types of computer programs or models have been developed: (1) emission models and (2) dispersion models.

Emission models provide estimates of quantities of various types of pollutants based on the amount of travel, driving patterns, and emission rates for various classes of vehicles. Dispersion models use data on source emissions from emission modeling, and, with additional data on meteorological conditions, geographic boundaries, and so forth, compute the dispersion of pollutants over the area of interest for specified periods of time.

4-18 Control of Highway Air Pollution

Programs to control air pollution from motor vehicles have concentrated primarily on modification of engine designs to meet federally prescribed maximum levels of emissions. However, motor vehicle emissions are influenced not only by engine operating characteristics but also by a wide variety of transportation system characteristics. The following paragraphs, which are quoted directly from Ref. 22, describe ways of lessening the impact of air pollution by appropriate planning, design, and operation of the highway facilities.

> *Planning.* Air quality is affected by both mobile emission sources such as automobiles and aircraft, and stationary emission sources such as factories and power plants. In theory, there should be an optimum location pattern for these various activities that would minimize ground level concentrations of pollutants. Different urban forms, such as urban sprawl, a centralized core, or satellite cities, affect air quality by requiring different transportation systems and energy expenditure. Actual development of these urban forms, however, which approximates theoretical forms, is generally limited to new towns. This may allow incorporation of air quality planning considerations into the basic layout of a proposed new community.
>
> More commonly, planning studies are conducted within the context of an existing pattern of development. Planning for improvements to highway and transit networks should be coordinated to permit opportunities for transfer between public transit facilities and private automobiles. This planning strategy is the most direct method for reducing highway-related air pollution in cities, since transit vehicles can serve high-density urban core areas more effectively than automobiles. This method of reducing private vehicle travel, however, is probably the most difficult to implement since it reduces an individual's travel flexibility.

Planning for highway improvements can provide for increased capacity to reduce congestion, higher speed links in the road network, and shorter travel distances. The first two planning strategies tend to reduce emissions of CO and HC due to smoother traffic flow and higher and more uniform average speeds. The third item reduces all emissions. It should be recognized that these are long-term approaches to reducing emissions, with implementation periods on the order of five to twenty years.

Design. The physical layout of highway facilities also affects vehicle emissions. In general, highway design features which reduce congestion and increase operating speeds will also contribute to reduced emissions of air pollutants.

Number of travel lanes. The number of lanes is an obvious factor that affects the potential of total pollutants which may be emitted in a given corridor. With a greater number of lanes, the roadway capacity is increased, resulting in a potential for higher traffic densities and therefore higher pollutant emissions.

Lane capacity. All roadway lanes are limited in their ability to accommodate vehicular traffic, and emissions are dependent on the number of vehicles utilizing each lane. One of the many variables that enters into the evaluation of lane capacity is driver behavior. Lane width affects capacity on two-lane and multi-lane highways and at intersections, because drivers tend to encroach on adjacent lanes (or totally avoid a lane) when lane width is narrow or conditions make a lane appear narrow. Such behavior causes inefficient lane usage and lower operating speeds, and therefore, a possible reduction in capacity. Closely related to lane width is the presence of lateral obstructions, such as median barriers, curbs, and bridge abutments. The presence of a single lateral restriction can cause a bottleneck, leading to alternate deceleration and acceleration, and the products of increased emissions. The absence of adequate shoulders along the highway or on structures can also create bottlenecks at times of accidents or breakdowns, even under moderate traffic loadings.

Horizontal and vertical alignment. Emission rates are also affected by features such as horizontal and vertical curves, due to varying vehicle power requirements. Emissions from all types of vehicles vary whenever there is a deviation from steady-state speed or power. Network capacity can be decreased by long, steep grades, though truck climbing lanes can help alleviate this problem.

Consideration must also be given to the varying effects of at-grade, elevated, or depressed roadway sections. Though these configurations do not affect emission rates, they do influence the dispersion capabilities of a given roadway section. In general, ground-level concentrations downwind of an elevated roadway will be less than those from an at-grade section. The dispersion capabilities are more limited in a cut section, where the potential exists for pollutants becoming trapped and accumulating.

Operating speeds. As noted previously, vehicle pollutant emissions are speed-dependent. As operating speeds for a given roadway segment increase, emissions of carbon monoxide and hydrocarbons decrease, while those for oxides of nitrogen increase. Changes in speed contribute to higher emissions, since emissions of HC and NO_x in the acceleration mode are significantly higher than those at a steady-state speed.

Interchanges and intersections. The existence of interchanges and ramps affects operating speeds. Directional freeway interchanges generally allow smoother flow and require fewer speed changes, and therefore cause lower emissions than cloverleaf or diamond interchanges.

At-grade intersections are significant generators of increased emissions, due chiefly to changes in operating mode brought about by vehicles decelerating, idling, and accelerating. Generally, any strategy used to improve traffic flow also contributes to a decrease in the total emission burden.

Operation. Transportation system management strategies can contribute to reductions in pollutant emissions by ensuring a smooth flow of traffic on arterials and freeways. For example, a system of synchronized traffic signals contributes to lower emissions by decreasing vehicle time delays and increasing travel speeds. By replacing stop signs with yield signs, or allowing right turns during the red phase of the cycle, average vehicle speeds are increased, resulting in lower emissions of CO and HC. On freeways, ramp metering and variable message information signs can reduce congested conditions, and thereby also reduce emissions.

These different ways of lessening the air pollution from mobile sources provide a good point of departure for considering the implications of highway design on air quality. In recent years, the transportation profession has focused even more attention on understanding the basic phenomena associated with automobile pollution (*23*). It is likely in future years that important changes will occur in the way air quality impacts are modeled and in the strategies that are considered by highway engineers to minimize the impact of automobile emissions.

4-19 Impacts on Wetlands

An increasing concern in recent years for the preservation of the ecology has presented serious challenges to highway engineers, especially when proposed highway alignments come near wetlands. The U.S. Army Corps of Engineers and the U.S. Environmental Protection Agency define wetlands as follows (*24*):

> Those areas that are inundated or saturated by surface or ground water at a frequency and duration sufficient to support, and that under normal circumstances do support, a prevalence of vegetation typically adapted for life in saturated soil conditions. Wetlands generally include swamps, marshes, bogs and similar areas.

As can be seen in this definition, the three most important factors for defining wetlands are the type of vegetation, the type of soil, and the hydrology that exist at the potential wetland location. Once the wetlands have been identified and the boundaries defined, the next step in the analysis process is to determine what impact will occur given a proposed highway alignment. The importance of wetlands lies in the functions they serve in the ecosystem. Most of the analysis procedures that are used by environmental specialists to identify wetlands relate to the following functions and corresponding values associated with these functions (*25, 26, 27*).

Ground water recharge	High export of organic material
Ground water discharge	Wildlife diversity/abundance
Floodflow alteration	Aquatic diversity/abundance
Sediment stabilization	Recreation
Sediment/toxicant retention	Uniqueness/heritage
Nutrient removal/transformation	

In addition to an assessment of the impact on these functions, many analysis procedures incorporate what is referred to as a "fatal flaw" determination. In essence, this determination identifies potential impacts so detrimental to the wetland that significant mitigation measures (including not building the road in this location) must be undertaken. To give some indication of what type of impact is considered a fatal flaw in a proposed highway design, the following characteristics of wetlands that are considered to have such significant levels of impact are listed from Ref. 26. The term in parentheses following each impact characteristic refers to the wetland function affected.

1. Do any federal or state endangered or threatened species use the wetland area regularly? (uniqueness/heritage)
2. Is the wetland owned by an organized conservation group or public agency for the primary purpose of preservation, ecological enhancement, or low-intensity recreation? (uniqueness/heritage)
3. Is the wetland included in a statewide listing of historical or archeological sites? (uniqueness/heritage)
4. Does the wetland have ecological or geological features consistently considered by regional scientists to be unusual or rare for wetlands in that region? (uniqueness/heritage)
5. Does the wetland represent most or all of this wetland type in this locality? (all functions)
6. Have substantial public or private expenditures been made to create, restore, protect, or ecologically manage the wetland? (uniqueness/heritage)

The analysis procedure lists 25 other questions pertaining to the potential impact on the wetland (not as fatal flaw impacts) that the analyst needs to answer to determine the overall impact. If a proposed highway alignment will affect any of the functions of a wetland, the design must incorporate strategies to mitigate the expected impact. The first mitigation option is to avoid the impact by changing the alignment of the highway or not building the road in the first place. In many cases, this option is not feasible. Given that unavoidable losses of wetlands will likely occur with the construction of a project, several strategies for compensating for their loss should be considered by designers. These strategies include

Wetland replacement: Wetlands are constructed at other locations to compensate for those impacted at the construction site. The replaced wetland should not necessarily be of the same type as that which was lost. Different types of wetlands might, in fact, offer improvements for fish and wildlife or for the control of water quality, flooding, and shore erosion (*28*).

Wetland restoration: Wetlands that are impacted by construction are restored to their original or enhanced function. This is often used to mitigate the impact of temporary construction staging sites or other construction-related locations that will not be permanently part of the highway project itself.

Wetland enhancement: Wetlands either in the right-of-way or replacement wetlands beyond the project boundaries are designed to improve the value of the wetland over that which the original wetland possessed. In most cases, this enhancement relates to fish and wildlife or major improvements to the social values associated with a wetland.

As shown in Table 4-12, the types of impacts and associated mitigation strategies that are associated with highway projects occur at all phases of project development, construction, and operations/maintenance (*29*). Pursuant to Section 404 of the Clean Water Act, a permit is required from the U.S. Corps of Engineers for the discharge of dredge or fill material into all waters of the United States. Thus, for many of the activities that highway engineers deal with in the design

TABLE 4-12 Highway Wetland Impacts and Potential Mitigation Strategies

Project Phase	General Ecological Concerns and/or Impacts	Types of Mitigation and Enhancement Measures
Predesign	Loss of highly productive areas Loss of oligotrophic water supplies	Careful selection of corridors and alignments
Design	Changes in water quality Interruption of fish migration Enhanced eutrophication Changes in hydrology	Design of alignments, culverts Design erosion control measures Specify types of materials Specify provisions in construction contract
Construction	Loss and/or disruption of aquatic species Loss of aquatic habitat Loss of water quality in receiving waters	Implementation of design features Minimize encroachment and alteration of aquatic habitats Implementation of erosion control Careful timing of construction Careful management of construction camp and site
Operation and maintenance	Accidental spillage of potential toxicants and irritants Runoff of terrestrial herbicides Changes in aquatic productivity due to road runoff	Implement plans for potential spill Regulate application of herbicides in right-of-way Implement plans for corrective actions

Source: Erickson, P., et al., *Highways and Ecology: Impact Assessment and Mitigation,* Federal Highway Administration, Report FHWA RWE/OEP-78-2, March 1978.

and operation of highways, strong communication with the Corps of Engineers is necessary.

In many cases, state transportation agencies have guidelines that are used for each type of mitigation strategy. Figure 4-6 shows the wetland construction guidelines for the Minnesota Department of Transportation. As can be seen in these guidelines, the major thrust in the design of replacement wetland is to make them as natural as possible and to design the wetland for the basic function(s) it will serve in the ecosystem.

OTHER HIGHWAY SOCIAL AND ENVIRONMENTAL IMPACTS

As noted in Section 4-8, a large number of environmental impacts must be considered as part of the highway project development process. Unfortunately, space does not allow a detailed examination of each impact category. It is important to realize, however, that the success of highway design, and indeed of

RETENTION POND TO MITIGATION POND

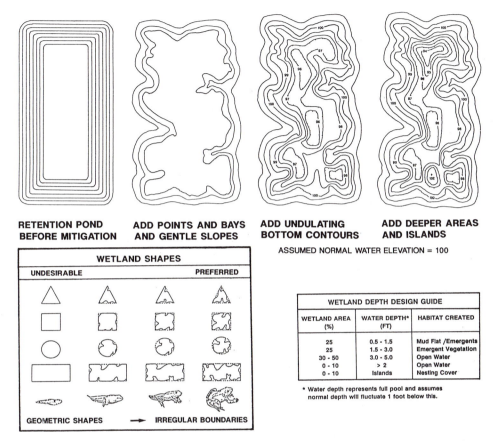

| RETENTION POND BEFORE MITIGATION | ADD POINTS AND BAYS AND GENTLE SLOPES | ADD UNDULATING BOTTOM CONTOURS | ADD DEEPER AREAS AND ISLANDS |

ASSUMED NORMAL WATER ELEVATION = 100

WETLAND SHAPES

UNDESIRABLE   PREFERRED

GEOMETRIC SHAPES ⟶ IRREGULAR BOUNDARIES

WETLAND DEPTH DESIGN GUIDE		
WETLAND AREA (%)	WATER DEPTH* (FT)	HABITAT CREATED
25	0.5 - 1.5	Mud Flat /Emergents
25	1.5 - 3.0	Emergent Vegetation
30 - 50	3.0 - 5.0	Open Water
0 - 10	> 2	Open Water
0 - 10	Islands	Nesting Cover

* Water depth represents full pool and assumes normal depth will fluctuate 1 foot below this.

FIGURE 4-6 Wetland construction guidelines. (Source: Minnesota Department of Transportation, *Environmental Services Section, Wetland Construction Guidelines,* [April 1990].)

any type of infrastructure, is necessarily tied to the effective handling of the associated social and environmental impacts. Each of the impact categories has analysis methodologies that are considered standard for highway applications. And in most cases environmental specialists are part of the design team to take responsible charge in dealing with these impacts. Quite simply, highway engineering is not just laying out an alignment from point A to point B. The process of design is as much an understanding of the social and environmental context of the highway, and in dealing with the resulting impacts, as it is providing the most efficient layout of basic engineering design.

PROBLEMS

4-1 Determine the basic section costs for passenger cars using a multilane highway under the following conditions:

design speed = 100 km/hr (62 mph)
volume/capacity ratio = 0.7
level of service = B
grade = 2%
curvature, R = 291 m (955 ft)

4-2 Estimate the one-way section transition costs for the highway described in Problem 4-1 given that the average running speed on an adjacent section is 40 km/hr (25 mph).

4-3 The volume of traffic along a certain intersection approach is 800 passenger cars per hour. Determine the costs due to stopping (excluding idling) at the intersection under the following conditions:

saturation flow = 1650 vehicles/hr
signal cycle time = 60 sec
effective green time = 33 sec
approach speed = 70 km/hr (44 mph)

4-4 For the conditions described in Problem 4-3, determine the costs due to idling.

4-5 A certain highway project is planned that would have an initial investment cost of $1.2 million. The user benefits for the facility (in excess of maintenance costs) are estimated to be $87,000 per year over its useful life of 20 years. At the end of the 20-year period, the residual (salvage) value would be $200,000. On the basis of present worth concepts, should the project be built? Assume an interest (discount) rate of 9 percent. What is the benefit-cost ratio?

4-6 A highway agency is considering two alternative energy attenuation systems, described below, to lessen the severity of crashes into bridge columns and other fixed object hazards. System A has low initial cost but must be repaired extensively after each "hit." System B has high initial cost but needs little maintenance. It is expected that such a device will be hit once every two years. On the basis of present worth concepts, which system is preferred? Use an annual interest rate of 10 percent.

	System A	System B
First cost	$3,500	$8,500
Routine annual maintenance	$ 500	$ 300
Repair cost after each hit	$1,200	None
Life	20 yrs	20 yrs
Salvage value	$1,000	$2,500

4-7 The benefits from a certain highway project are estimated to be $15,000 in the first year and $30,000 in the 21st year. Assuming a uniform growth rate and an interest rate of 9 percent, calculate the present value of benefits over a 20-year analysis period.

REFERENCES

1. American Association of State Highway Officials, *Road User Benefit Analysis for Highway Improvements,* Washington, DC (1952).

2. Winfrey, Robley, *Economic Analysis for Highways,* International Textbook, Scranton, PA (1968).

3. Claffey, P. J., *Running Costs of Motor Vehicles as Affected by Road Design and Traffic,* National Cooperative Highway Research Program (NCHRP) Report No. 111, Washington, DC (1971).

4. Winfrey, R., and C. Zellner, *Summary and Evaluation of Economic Consequences of Highway Improvements,* NCHRP Report No. 122, Washington, DC (1971).

5. Curry, D. A., and D. G. Anderson, *Procedures for Estimating Highway User Costs, Air Pollution, and Noise Effects,* NCHRP Report No. 133, Washington, DC (1972).

6. Anderson, D. G., D. A. Curry, and R. J. Pozdena, *User Benefit Analysis for Highway and Bus Transit Improvements,* Final Report NCHRP Project 2-12, Stanford Research Institute, Menlo Park, CA (published as *A Manual on User Benefit Analysis of Highway and Bus—Transit Improvements, 1977* by the American Association of State Highway and Transportation Officials).

7. *Accident Facts, 1993 Edition,* National Safety Council, Chicago (1993).

8. *Estimating the Cost of Accidents 1992,* National Safety Council Bulletin, Chicago (1992).

9. Cribbins, P. D., J. M. Arey, and J. K. Donaldson, Effects of Selected Roadway and Operational Characteristics on Accidents on Multi-Lane Highways, *Highway Research Record No. 188* (1967).

10. Dart, Olin K., and Lawrence Mann, Jr., Relationships of Rural Highway Geometry to Accident Rates in Louisiana, *Highway Research Record No. 312* (1970).

11. Kihlberg, J. K., and K. J. Tharp, Accident Rates as Related to Design Elements of Rural Highways. NCHRP Report No. 47, Washington, DC (1968).

12. Thuesen, G. J., and W. J. Fabrycky, *Engineering Economy,* 8th ed., Prentice Hall, Englewood Cliffs, NJ (1993).

13. *Highway Capacity Manual,* Transportation Research Board Special Report No. 209 (1994).

14. *Code of Federal Regulations,* Title 23, Part 771.

15. *Federal-Aid Highway Program Manual,* Vol. 7, Chap. 7, Federal Highway Administration, Washington, DC.

16. Intergovernmental Review of Federal Programs, Executive Order 12372, *Federal Register,* Vol. 48, No. 16, January 24, 1983.

17. Schultz, T. J., Bolt Benanek and Newman, Inc., *Noise Assessment Guidelines: Technical Background,* Prepared for U.S. Department of Housing and Urban Development, Washington, DC (1972).

18. *Guide on Evaluation and Attenuation of Traffic Noise,* American Association of State Highway and Transportation Officials, Washington, DC (1974).

19. *Guide on Evaluation and Abatement of Traffic Noise 1993,* American Association of State Highway and Transportation Officials, Washington, DC (1993).

20. *Highway Traffic Noise in the United States: Problem and Response,* Federal Highway Administration, Washington, DC (August 1994).

21. Snow, C. H., *Highway Noise Barrier Selection, Design, and Construction Experiences, A State of the Art Report—1975,* Implementation Package 76-8, Federal Highway Administration, Washington, DC (1976).

22. *Fundamentals of Air Quality,* Implementation Package 76-5, prepared for the Federal Highway Administration by Greiner Engineering Sciences, Inc., Baltimore, MD (1976).

23. Wholley, T., ed., *Transportation Planning and Air Quality II,* American Society of Civil Engineers, 1994.

24. U.S. Army Corps of Engineers, *Recognizing Wetlands,* undated.

25. Cowardin, L., et al., *Classification of Wetlands and Deepwater Habitats of the United States,* Fish and Wildlife Service, Department of the Interior, Report FWS/OBS-79/31, 1992.

26. Adams, P. et al., *Wetland Evaluation Technique (WET),* Volume II, Federal Highway Administration, Report FHWA-IP-88-029, October 1987.

27. Fish and Wildlife Service, Environmental Protection Agency, Department of the Army, and Soil Conservation Service, *Federal Manual for Identifying and Delineating Jurisdictional Wetlands,* January 1989.

28. Garbisch, E., *Highways and Wetlands: Compensating Wetland Losses,* Federal Highway Administration, August 1986.

29. Erickson, P., et al., *Highways and Ecology: Impact Assessment and Mitigation,* Federal Highway Administration, Report FHWA-RWE/OEP-78-2, March 1978.

CHAPTER 5

DRIVER, PEDESTRIAN, AND VEHICLE CHARACTERISTICS

The highway engineer must design for a wide range of vehicle operating charac-
teristics and allow for great differences in driver and pedestrian characteristics.
Most highway facilities must be designed to accommodate the smallest subcom-
pact automobile as well as the largest tractor-trailer truck. In many instances the
design must also accommodate motorcycle and bicycle users and pedestrians. It
should be remembered that within each class of users there is great variability:
differences in vehicle sizes, weights, and operating characteristics; differences in
pedestrian walking speeds; and differences in drivers' and pedestrians' ability to
comprehend and react to highway features and traffic events. In the first part of
this chapter we will examine some of the driver and pedestrian characteristics that
are of consequence and interest to the designer. In the latter paragraphs, we will
discuss certain physical characteristics of passengers cars and trucks that have a
bearing on the design process.

DRIVER CHARACTERISTICS

Drivers in the United States are licensed to drive under laws and regulations of
the various states. There is a great deal of variability among the states in licensing
standards and presumably in driving skills, especially for beginning drivers. The
vast majority of U.S. drivers have received no formal driver training but were
taught by family or friends or were self-taught.

Despite efforts to improve driver licensing procedures spanning more than 80
years, many observers doubt that driver licensing makes much of a contribution
to safe motor-vehicle operation (1). State reports, spot checks, and special surveys
have suggested that "on a given day there are at least 10 million unlicensed
drivers on the road" (2). A report from one state indicated that 9.6 percent of
drivers involved in fatal accidents were unlicensed at the time of the crash (2).

In 1992 there were over 173 million licensed drivers in the United States,
approximately 0.73 licensed drivers for every registered motor vehicle. About 92
percent of males and 82 percent of females who are old enough to drive have a
driver's license (3).

About 5.1 percent of licensed drivers are 19 years of age and under and about
8.9 percent are aged 70 and over. The number of elderly drivers has increased
dramatically in recent years. In 1940 drivers aged 65 years and older comprised
only 3 percent of all licensed drivers; by 1992 drivers of that age bracket comprised 13.9 percent of the total. The number of elderly drivers increased from

1.36 million in 1940 to 5.71 million in 1959 (*4*) to 24.14 million in 1992 (*3*). As Figure 5-1 illustrates, the very young and the very old drivers have the highest accident rates.

It is important that highway engineers keep in mind that street and highway facilities must be designed to accommodate drivers with a wide range of ages and skills—the young and the elderly, and the novice as well as the experienced professional.

5-1 The Human Sensory Process

A driver's decisions and actions depend principally on information received through the senses. This information comes to the driver through the eyes, ears, and the sensory nerve endings in the muscles, tendons, joints, skin, and organs. In general order of importance, the senses most used by drivers are

1. visual (sight)
2. kinesthetic (movement)
3. vestibular (equilibrium)
4. auditory (hearing)

Researchers report that about 90 percent of the information a driver receives is visual (*5*). Visual information may come to a driver by means of foveal or peripheral vision. In foveal vision, images are concentrated in a small area of the eye near the center of the retina at which visual perception is most acute. At a given moment a person's sharpest vision is concentrated within a cone with a central angle of about 3°. For most persons, visual acuity is reasonably sharp within a conical angle up to about 10°. Beyond this, a person's vision is less well defined. A person with normal vision can perceive peripheral objects within a cone having a central angle ranging up to about 160° (*6*).

A driver increases the amount of visual information received by movement of the head and eyes. The driver's eyes search and scan the field of view, moving the area with sharpest acuity. The direction of foveal vision repeatedly shifts and

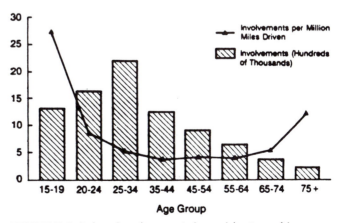

FIGURE 5-1 Driver involvements in accidents and involvement rates by age. (Source: Transportation Research Board, *Transportation in an Aging Society: Improving Mobility and Safety for Older Persons*, Special Report 218 [1988].)

the eyes fixate, each such event requiring respectively 0.15 to 0.33 sec and 0.1 to 0.3 sec (6).

Mourant and Rockwell (7) found that experienced drivers scanned a wider range of horizontal fixation locations than novice drivers. The average horizontal fixation range for the experienced group varied from 30° to 48°, depending on the type of driving task. The novice drivers concentrated their eye fixations in a smaller area and looked closer in front of the vehicle and more to the right. By pursuit eye movements, the novice drivers tended to observe the guardrail and lane edges continuously for periods of about 1.0 sec to gain information on lateral position. Similar findings have been reproted for drivers who have been deprived of sleep (7).

Visual acuity declines and the field of vision narrows with advancing age, especially when lighting conditions are poor (4). Older drivers also tend to have more difficulty in judging distances and distinguishing colors than young drivers.

Drivers receive information by kinesthesia, the sensation of bodily position, presence, or movement that results primarily from stimulation of sensory receptors in muscles, tendons, and joints. Lateral or longitudinal accelerations exert forces on the driver that are transmitted by the seats, steering wheel, brake pedal, arm rest, and so forth. Proprioceptors in the muscles and joints are stimulated by the tendency of the body to move and shift as a result of the forces. This provides a feedback that may cause a driver to brake, slow down, or take some other action. This phenomenon has been termed "driving by the seat of the pants."

Drivers also receive messages from the vestibular nerve within the inner ear. Three fluid-filled semicircular canals in the inner ear enable a person to recognize the direction of movement and maintain balance or equilibrium. Movement of the head or body causes the fluid in the canals to move, affecting delicate fibrous cells located at one end of each of the canals. Impulses from the vestibular nerve are then transmitted to the brain, which, in turn, sends messages to the muscles, which take action to maintain balance.

The kinesthetic and vestibular senses provide important information to a driver about forces associated with change of direction, steering, braking, vibrations, and stability of the vehicle (8).

Sounds of horns or skidding tires may alert a driver to an impending collison, and engine noise aids the driver as well as pedestrians in judging vehicle speeds. When traveling along a curve or turning at an intersection, tire noises may indicate to the driver a need to slow down. Highway engineers employ sound in the form of special pavement textures, raised pavement markers, or "rumble strips" to signal to drivers to slow down or to avoid certain paved areas. Sounds within a vehicle such as loud conversation or radio music may mask out the sounds of sirens, bells, or other important warning devices.

Drivers may detect a fire or engine malfunction by their sense of smell. Hunger, thirst, or discomfort may cause drivers to change their trip pattern or stop. Drivers may also occasionally receive information through the sensory nerve endings in the head, skin, and internal organs and experience sensations of touch, coldness, warmth, fatigue, or pain.

5-2 Driver Perception and Reaction

Driver perception-reaction time is defined as the interval between seeing, feeling, or hearing a traffic or highway situation and making initial response to what has been perceived. Take, for example, a simple braking situation. The "clock" for perception-reaction time begins when the object or condition first becomes visible, and it stops when the driver's foot touches the brake.

Traditionally, perception-reaction time has been couched in terms of perception, intellection, emotion, and volition, termed PIEV time (6). Perception is the process of forming a mental image of sensations received through the eyes, ears, or other parts of the body. It is recognizing and becoming aware of the information received by the senses. Intellection is another word for reasoning or using the intellect. Emotion is the affective and subjective aspect of a person's consciousness. It has to do with how a person feels about a situation. More often than not, emotion is detrimental to safe motor vehicle operation. Volition is the act of making a choice or decision.

Highway engineers now generally accept the notion that there are at least four steps or phases in information processing that occur from the presentation of a stimulus until the driver responds. What is of significance to the designer is that each of these stages requires time. Olson (9) lists the stages and explains:

1. Detection. Perception-response time begins when some object or condition of concern enters the driver's field of vision. In the example given earlier (of a driver being confronted with debris that fell off a truck beyond the crest of a hill), perception-response time began when it first became possible to see the debris on the road over the hill crest. This first step concludes when the driver develops a conscious awareness that "something" is present. The something may be within the field of view of the driver some time before it is detected. Hence, there is the potential for a significant delay between the presentation of the stimulus and its being detected. Typically, once detection occurs the eyes will be moved to bring that which has been detected into focus on the central portion of the retina so that the next step can be undertaken.

2. Identification. In this step sufficient information is acquired about the object or condition to be able to reach a decision as to what action, if any, is required. Identification need not be complete in detail. For example, it is not necessary to know if the object ahead is a truck or boulder, it is enough to know that the lane is blocked. If that which is being identified is moving, or capable of movement, it is also necessary to determine what it is doing. This may require estimates of speed and trajectory.

3. Decision. At this point the operator must decide what action is appropriate. Assuming some action is decided upon, the choices come down to change in speed and/or direction.

4. Response. In this step commands are issued by the motor center of the brain to the appropriate muscle groups to carry out the required actions.

Perception-response time typically ends when the driver begins to turn the steering wheel or presses on the brake pedal. However, certain situations call for more complex maneuvers that can be time consuming. An example occurs where a lane closure forces the driver to search for a gap in an adjacent lane, adjust his/her speed accordingly, and move into the gap. Where such a situation exists, it is appropriate to extend the overall perception-response time to include maneuver time.

Perception-reaction times vary from less than 0.5 sec to 3 or more sec. Greenshields (10) has indicated that, in general:

1. The speed of all forms of reaction varies from one person to another and from time to time in the same person.
2. Reaction time changes gradually with age, very young and very old people being slower in their reactions.

3. People generally react more quickly to very strong stimuli than to weak ones.

4. Complicated situations take longer to react to than simple ones.

5. A person's physical condition affects his reactions. For example, fatigue tends to lengthen a person's reaction time.

6. Distractions increase the time of all reactions except the reflex.

Johansson and Rumar (*11*) measured brake-reaction times for a group of 321 drivers who had some degree of braking expectation. The median brake-reaction time was 0.66 sec and the range was from 0.3 to 2.0 sec. The frequency distribution of the brake-reaction times is shown in Figure 5-2. The researchers performed a second experiment to see how brake-reaction time in response to a completely unexpected signal compares with such time in response to a somewhat antici-pated signal. They reported that unexpected brake-reaction times were larger by a factor of 1.35.

Elderly drivers have more difficulty than the young in perceiving traffic situ-ations and tend to pause longer between successive acts. Data presented by Marsh (*4*) suggest that on the average, simple reaction times for persons aged 65 are about 16 percent higher than for those aged 20; complex reaction times are about 33 percent higher for the older subjects. However, because of the wide variability among drivers, "chronological age is not a sound criterion for appraisal of driving competence" (*4*).

As Figure 5-3 illustrates, a person's response time depends on the sense stim-ulated. Generally, a person responds quickest to stimuli of touch and hearing and requires only slightly more time for response to visual stimuli. Response times for the kinesthetic and vestibular stimuli are considerably longer.

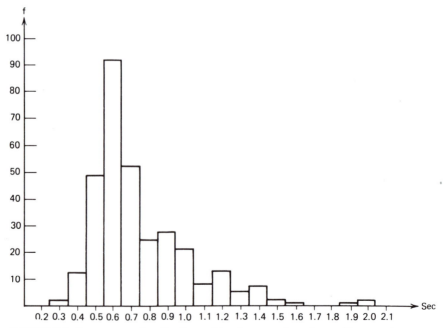

FIGURE 5-2 Frequency distribution of brake-reaction times. (Source: Johansson, Gunnar, and Rumar, Kare, Driver's Brake Reaction Times, *Human Factors* [1971].)

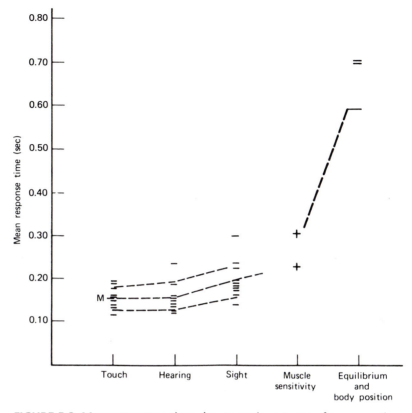

FIGURE 5-3 Mean response time due to various types of sensory stimulus. (Source: Forbes, T. W. and Katz, N. S., Summary of Human Engineering Data and Principles Related to Highway Design and Traffic Engineering Problems; sketch courtesy of the Institute of Transportation Engineers.)

Figure 5-4 shows that response time increases with the complexity of a situation, as does the error rate.

Perception-reaction time tends to be detrimentally affected by fatigue, the use of alcohol or other drugs, and certain physiological and psychological conditions.

For the purpose of computing stopping sight distances, AASHTO recommends the use of a brake-reaction time of 2.5 sec (*12*). However, stopping sight distances based on this reaction time are often inadequate when drivers must make complex or instantaneous decisions, when information is difficult to perceive, or when unexpected or unusual maneuvers are required (*12*). In such circumstances a longer sight distance, called a *decision sight distance,* is recommended. Decision sight distances have been developed by engineers from empirical data and are introduced in Chapter 7.

5-3 Driver Information Needs

Given the ever increasing demands of the driving task, it is important that highway and traffic engineers understand the driver's need for information and the means of its transmission. Research was undertaken in the late 1960s by AIL, a division of Cutler-Hammer, to identify types of information needed by drivers

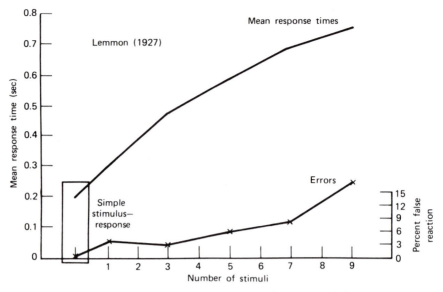

FIGURE 5-4 Increase in mean response time and errors with increasing number of stimuli. (Source: Forbes, T. W. and Katz, N. S., Summary of Human Engineering Research Data and Principles Related to Highway Design and Traffic Engineering Problems; sketch courtesy of the Institute of Transportation Engineers.)

and the principal factors and interactions that affect the reception and use of this information. The results of one aspect of that work have been described in a paper by Allen, Lunenfeld, and Alexander (*13*), which forms the primary basis for the material presented in the following paragraphs.

In order to make sound decisions, a driver must receive reliable and understandable information to reduce uncertainty. A driver gathers information from a variety of sources, uses it as a basis for decision making, and then translates those decisions into actions to control the vehicle (*14*).

As Table 5-1 indicates, the various activities involved in driving a motor vehicle may be grouped into three categories or levels of performance: (1) the control level, (2) the guidance level, and (3) the navigation level.

The control level of driver performance is related to the physical operation of the vehicle. It includes lateral control that is maintained by steering and longitudinal control that is exercised by accelerating and decelerating. To maintain vehicle control, the driver requires information on the vehicle's position and orientation with respect to the road as well as feedback on the vehicle's response to braking, accelerating, and steering. Such information is received primarily through the visual, kinesthetic, and vestibular senses.

> Guidance level of driver performance refers to the driver's task of selecting a safe speed and path on the highway. While performance at the control level is overt action, performance at the guidance level is a decision process. The driver must evaluate the immediate situation, make appropriate speed and path decisions, and translate these decisions into control actions needed to survive in the traffic stream. Activities include decisions relating to lane positioning, overtaking, and passing. . . .

> The driver is constantly time-sharing activities at the guidance level with those at the control and navigation levels. Furthermore, two or more guidance level processing

TABLE 5-1 A Framework for Driving Task Conceptualization

Subtask Category	Related to	Example of Sources of Information	Importance of Information	Likely Consequence of Failure
Control (Microperformance)	Physical operation of vehicle Steering control Speed control	Road edges Lane divisions Warning signs Kinesthesia	Highest	Emergency situation or crash
Guidance (Situational performance)	Selecting and maintaining a safe speed and path	Road geometry Obstacles Traffic conditions Weather conditions	Intermediate	Emergency situation or crash
Navigation (Macroperformance)	Route following Direction finding Trip planning	Experience Directional signs Maps Touring service	Lowest	Delay, confusion, or inefficiencies

requirements may occur simultaneously or in close enough time proximity to require simultaneous action, which may not necessarily be compatible. This points to the importance of experience and prior knowledge throughout the driving task, since the driver may be required to make several decisions under heavy time pressure. (*14*)

The navigation level of driver performance refers to the driver's ability to plan and execute a trip. It includes trip preparation and planning and direction finding. The former activities may be performed before the trip begins, while the latter occurs en route. Information at the navigation level comes from maps, directional signs, landmarks, and from the driver's memory based on previous experience.

Allen et al. (*13*) have listed five basic principles for the systematic presentation of information required by drivers:

1. First things first—primacy.
2. Do not overload—processing channel limitations.
3. Do it before they get on the road—*a priori* knowledge.
4. Keep them busy—spreading.
5. Do not surprise them—expectancy.

Primacy

The concept of primacy is based on the realization that at any given instant certain driver information is more important than other information. In evaluating the primacy of information, the effect of failure to receive that information is considered. Lack of certain types of information may lead to an emergency situation or a crash; lack of other types of information may only cause momentary confusion or delay. Avoiding a car emerging from a driveway (a control subtask) is more important than deciding whether to pass a slow-moving vehicle (a guidance subtask). Either of these subtasks would take precedence over observance of directional signs (a navigation subtask).

In the guidance category, where failures tend to be catastrophic, the question is one of degree. Generally, colliding with a guardrail is less serious than colliding with a bridge pier. Colliding with a vehicle moving in the same direction is less serious than a head-on collision. . . .

A second aspect of primacy judgment is based on distance criterion. A hazard close by would have a higher primacy than a more distant hazard of the same or similar type. . . . Assessing the primacy (both distance and severity potential) of a single hazard is a task that drivers probably do well. Assessing the relative primacy of two competing hazards in a dynamic situation is a difficult task that is more prone to driver error. (*14*)

Processing Limitations

As the complexity of a driving situation increases, so does the driver's response time. Obviously, if a driver is confronted with several complex situations within a short time span, a point may be reached when the driver is unable to receive and process the amount of information required for error-free response. Information needed by the driver should be presented systematically, requiring a series of simple choices rather than a smaller number of complex choices.

A Priori Knowledge

Drivers should be given as much information as possible before they get on the road. To the extent feasible, *a priori* information should be communicated to drivers on distances and directions to destination points, average travel times, highway types and design features, and other such information that will facilitate the planning of trips.

Spreading

During the course of trips, drivers may be confronted with highly variable demands for their attention, too much information being provided in certain areas and too little in others. Efforts should be made to spread the information load in order to avoid boredom of drivers in certain areas and overloading of their processing capability in others. This is accomplished by performing a primacy analysis and transferring the less important information from the high-information peaks to the low-information valleys. Ideally, the goal of good highway design should be to confront the driver with only one decision at a time.

Expectancy

A driver's readiness to respond to roadway conditions, traffic situations, or information systems is referred to as *driver expectancy*. Based on previous experience, drivers are conditioned to expect certain characteristics of road design and operation. For example, drivers generally expect exit and entrance ramps along freeways to be on the right-hand side. Drivers do not generally expect lane drops or other changes in the roadway cross section. At locations where driver expectancies are violated, drivers may require a longer response time and will have a greater tendency to make an inappropriate or hazardous response. Although expectancies occur at all levels of the driving task, their importance is probably greatest at the guidance level.

PEDESTRIAN CHARACTERISTICS

The pedestrrian is the major user of the roadway system; when the system fails, he is a major victim. Approximately 7,000 pedestrians are killed in the United

States each year, and about 70,000 are injured. The problem is most serious in urban areas, where pedestrian fatalities comprise about 28 percent of all traffic deaths.

Certain segments of the pedestrian population—notably the very young and the very old—are either unaware of rules of safe pedestrian behavior or unresponsive to efforts to enforce pedestrian traffic regulations. For example, Wiener (*15*) studied the responses of elderly pedestrians to an education and enforcement program directed toward eliminating jaywalking behavior. He concluded:

> A large percentage of elderly pedestrians are totally unresponsive to campaigns, legal sanction, enforcement symbols, or traffic regulation. No amount of campaigning will reach this group, for through lifelong habits, ignorance of traffic hazards and regulations, senility or willful neglect, their behavior is unmodifiable by the threat of sanctions. To make cities safer for pedestrians, officials must look for means which do not require their conscious cooperation.

Engineers are challenged to design safe and convenient pedestrian facilities that will function well even for those persons who willfully or ignorantly disobey rules of safe walking behavior. Such facilities serve small children as well as elderly and physically handicapped persons, including those who are blind or not ambulatory.

Designers of pedestrian facilities require knowledge of the space requirements and walking speeds of individual walkers as well as an understanding of traffic flow charcteristics of groups of pedestrians. The former subjects are treated in the following paragraphs, while the latter is discussed in Chapter 7.

5-4 Space Needs for Pedestrians

The design of pedestrian facilities depends to a large extent on the shoulder breadth and body depth of the human body. One study of fully clothed male laborers indicated, for the 95th percentile, a shoulder breadth of 579 mm (22.8 in.) and a body depth of 330 mm (130 in.) (*16*). An elliptical shape with a 610-mm (24-in.) major axis and a 457-mm (18-in.) minor axis has been used for determining the capacity of subway cars (*16*). The ellipse, encompassing an area of 0.21 m² (2.3 ft²), is slightly larger than the area occupied by the 95th percentile male. It allows for body sway, the tendency to avoid bodily contact with others, and for packages or personal articles that many pedestrians carry. It should be emphasized that the 457- by 610-mm (18- by 24-in.) ellipse is useful primarily for the determination of space needs or capacities for elevators or other conveyances or locations where the pedestrians are standing rather than walking. For design of sidewalks or other pedestrian corridors, one needs to be concerned with the dynamic spatial requirements for avoiding collisions with other pedestrians.

> The space required for locomotion may be divided into a pacing zone, the area required for foot placement, and the sensory zone, the area required by the pedestrian for perception, evaluation, and reaction. The length of the pacing zone is dependent on the age, sex, and physical condition of the pedestrian, and has been shown to have a direct linear relationship with speed. Both the pacing and sensory zones can be affected by external influences such as terrain and traffic conditions. Pedestrian pacing lengths may be physically measured, but sensory zone requirements are comprised of many human perceptual and psychological factors (*16*).

Research has shown that one pedestrian following another prefers to leave an average distance-spacing between himself and the lead pedestrrian of about

2.4 m (8 ft). This corresponds to an average time-spacing of about 2 sec between pedestrians walking in line.

In a remarkably efficient manner, pedestrians monitor the varying speeds and angles of other approaching pedestrians and make adjustments to their pace and speed in order to avoid collisions.

5-5 Walking and Running Speeds

Empirical research has shown that under free-flow conditions pedestrian walking speeds tend to be approximately normally distributed. Under such conditions, pedestrian walking speeds range from about 0.6 to 1.8 m/sec (2.0 to 6.0 ft/sec). Typical mean walking speeds are in the 1.2 to 1.4 m/sec (4.0 to 4.5 ft/sec) range. A pedestrian walking rate of 4.0 ft/sec is generally assumed for the timing of pedestrian traffic signals. In areas where there are large numbers of elderly pedestrians, AASHTO (*12*) recommends the use of a 0.9 m/sec (3.0 ft/sec) walking rate. The presence of significant numbers of handicapped persons would also dictate the use of a lower rate of movement.

Free-flow walking speeds vary with the pedestrian's age and sex as well as trip purpose. Grades less than about 5 percent seem to have little effect on walking speed, and pedestrians carrying baggage tend to walk about as fast as those without baggage (*16*).

Walking speeds decrease with increases in pedestrian density. Empirical studies indicate that for an average of 2.3 m² (25 or more ft²) per pedestrian, walking speeds are only slightly affected by pedestrian conflicts (*16*). As the available space per pedestrian drops below 2.3 m², the average walking speed decreases sharply. Traffic flow characteristics for pedestrian flow are further examined in Chapter 7.

Speeds beyond about 143 m/min (7.8 ft/sec) may be construed as running (*16*). The fastest movement afoot is the 100 yard dash in 9.1 sec, a speed of 10 m/sec (33 ft/sec).

VEHICLE CHARACTERISTICS

During the decade 1983 to 1993 automobile registrations in the United States increased 15 percent, while the number of registered trucks increased 26 percent (*17*). By 1993 there were over 145 million passenger cars, 46 million trucks, and 658,000 buses (*17*). In 1990 the United States had about 33 percent of the world's passenger cars and 32 percent of the world's trucks and buses. These percentages reflect U.S. ownership rates of 0.58 cars per person and 0.76 vehicles per person. U.S. ownership rates are exceeded only in the tiny European republics of Andorra and San Marino and in the principality of Monaco.

About 20 percent of the automobiles in use in the United States are less than 3 years old, while 30 percent are 10 years or older. The average age of automobiles in use in the United States is about 7.9 years, while the average age of trucks is 8.1 years.

Retail sales of automobiles in the United States totaled 8.51 million in 1993, including 1.78 million imports. These sales included, by size category, the following percentages: small 32.8 percent, middle 43.3 percent, large 11.1 percent, and luxury 12.8 percent. Table 5-2 gives a more detailed breakdown of domestic car sales for 1992.

TABLE 5-2 Ward's Domestic Car Market Segmentation, 1992

Category	Example	Calendar Year Sales	Percent Subgroup
Lower small	Saturn	565,033	32.3
Upper small	Escort	1,060,281	60.7
Small specialty	Eclipse	122,377	7.0
Total small		1,747,691	100.0
Lower middle	Grand Am	956,689	30.1
Upper middle	Taurus	1,920,861	60.5
Middle specialty	Mustang	297,706	9.4
Total middle		3,175,256	100.0
Large	LeSabre	613,079	81.5
Large specialty	Thunderbird	138,743	18.5
Total large		751,822	100.0
Lower luxury	Roadmaster	86,947	14.5
Middle luxury	Town Car	399,302	66.4
Upper luxury	Seville	41,152	6.8
Luxury specialty	Eldorado	52,037	8.6
Luxury sport	Corvette	22,350	3.5
Total luxury		601,788	100.0

Source: Ward's Automotive Yearbook, Ward's Communications, Inc., Detroit, MI (1993).

Total retail sales of new trucks in the United States in 1993 exceeded 5.6 million. Table 5-3 gives a breakdown of U.S. retail sales of trucks by gross vehicle weight rating and body type.

Vehicle Type and Size

The type and size of motor vehicles influence clearances for bridges, tunnels, and grade separation structures and the geometric characteristics of streets, roads, and parking facilities. Street and highway facilities are used by a variety of vehicles ranging from motorcycles* to articulated truck units up to 23 m (75 ft) in length and weighing over 45,359 kg (100,000 lb).

Motorcycles are small motorized two- or three-wheeled vehicles that can transport one or two people. Since motorcycles can operate satisfactorily on any street or highway suitable for cars and trucks, they are given little consideration in highway design.

Passenger cars and light trucks are the dominant source of personal transportation. These vehicles typically weigh about 680 to 1,800 kg (1,500 to 4,000 lb) and accommodate 2 to 6 passengers.

Single unit trucks have the cargo and power unit mounted on a common frame. These trucks range from 2-axle, 6-tire vehicles weighing about 4,536 kg (10,000 lb) to 3- or 4-axle vehicles weighing up to 18,144 kg (40,000 lb). Single unit trucks are most commonly used for hauling freight relatively short distances.

Combination trucks have a power unit or tractor plus one or more trailers.

---

*Bicyclists also may use public streets and highways, except for Interstate and other controlled-access highways. The design of special bicycle facilities is covered in Chapter 7.

TABLE 5-3 U.S. Retail Sales of Domestic Light Trucks by Gross Vehicle Weight Rating and Body Type, 1993

Gross Vehicle Weight/Type	Sales
Trucks with GVW 2,721 kg (6,000 lbs) or less	
Utility	720,526
Compact pickup	962,084
Van	17,699
Minivan	69,608
Conventional pickup	660,994
Station wagon (on truck chassis)	320,720
Mini passenger carrier	1,002,388
Passenger carrier	200
Trucks with GVW 2,722-4,536 kg (6,000–10,000 lbs)	
Utility	59,993
Van	279,090
Van cutaway chassis	44,968
Conventional pickup	646,961
Station wagon (on truck chassis)	114,699
Passenger carrier	59,625
Multi-stop	27,049
4,537-6,350 kg (10,101–14,000 lbs)	
Conventional pickup	13,826
6,531-14,968 kg (14,001–33,000 lbs)	129,063
14,969 kg (33,000 lbs) and over	157,886
Total domestic	5,287,379
Total imports	393,616
Total U.S. retail sales	5,680,995

Source: AAMA Motor Vehicle Facts and Figures '94, American Automobile Manufacturers Association, Detroit, MI (1994).

The tractor and trailer are joined at a hitch point and can be separated when not in use. The tractor operating alone is referred to as a "bobtail truck." Combination trucks may have two general types of trailers: a *semitrailer,* which has one or more axles near its rear but no front axles; or a *full trailer,* which has one or more axles at both its front and rear ends.

The big trucks that operate on U.S. highways may be seen in various configurations, as Figure 5-5 illustrates. The most common of the big trucks is the tractor-semitrailer combination, which has 5 axles, 3 on the tractor and 2 on the semitrailer. Such trucks are commonly referred to as "18-wheelers." These trucks typically have 14.6-m (48-ft) trailers.

Twin-trailer and triple-trailer configurations are becoming more common on American highways. *Twin-trailer trucks* consist of a tractor plus an 8.5-m (28-ft) semitrailer followed by another 8.5-m (28-ft) semitrailer. Such trucks are sometimes referred to as "western doubles." *Rocky Mountain double trucks* have a tractor plus a 13.7-m (45-ft) semitrailer plus an 8.5-m (28-ft) full trailer. *Turnpike double trucks* consist of a tractor plus a 13.7-m (45-ft) semitrailer followed by a

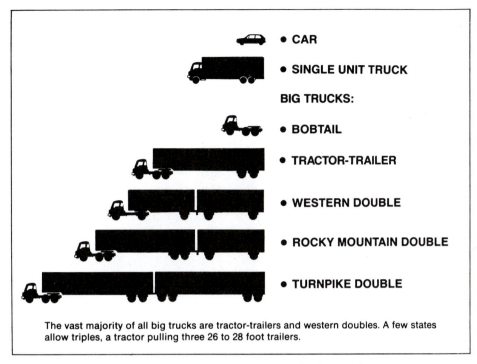

CAR

SINGLE UNIT TRUCK

BIG TRUCKS:

BOBTAIL

TRACTOR-TRAILER

WESTERN DOUBLE

ROCKY MOUNTAIN DOUBLE

TURNPIKE DOUBLE

The vast majority of all big trucks are tractor-trailers and western doubles. A few states allow triples, a tractor pulling three 26 to 28 foot trailers.

FIGURE 5-5 Various classes of vehicles that use U.S. highways. (Courtesy Insurance Institute for Highway Safety.)

13.7-m (45-ft) full trailer. A few states allow *triples,* a tractor pulling three 8.5-m (28-ft) trailers.

A *van* is a passenger vehicle mounted on an automobile or light truck chassis that is capable of transporting 6 to 15 people.

Buses usually transport more than 15 passengers. A *minibus* is a vehicle less than 7.6 m (25 ft) in length that has a capacity of 16 to 25 people. *Single unit transit buses* measure 2.4 to 2.6 m (8.0 to 8.5 ft) in width and 9.1 to 12.2 m (30.0 to 40.0 ft) in length and are capable of transporting up to 53 seated passengers. *Articulated transit buses* are 2.6 m (8.5 ft) in width and as much as 18.3 m (60 ft) in length, with a seated passenger capacity of 73 passengers. *Intercity buses* are most commonly 2.6 m (8.5 ft) in width and 12.2 m (40 ft) long and have a seated capacity of 54 passengers.

Recreational vehicles have a variety of configurations, including passenger cars with trailers, pickup trucks with camper bodies, and motor homes. Although recreational vehicles are operationally similar to some passenger cars, light trucks and buses, they often perform poorly because they are more heavily loaded that other vehicles of similar size (*18*).

Since the highway system must be designed to accommodate all classes of vehicles, the designer requires information on maximum and minimum dimensions—for example, the class of vehicle with the greatest overall width, the longest turning radius, the smallest dimension from the bottom of the bumper to the ground, and so forth. Table 5-4 gives maximum and minimum dimensions of standard American passenger cars manufactured in 1994. These dimensions are

TABLE 5-4 Maximum and Minimum Dimensions of Standard American Passenger Cars Manufactured in 1994*

Dimension	Minimum (inches)	Maximum (inches)	Minimum (mm)	Maximum (mm)
Overall length	142.5	225.1	3,620	5,718
Overall width	64.2	79.8	1,631	2,027
Overall width, doors open	127.8	173.6	3,246	4,409
Overall height	44.0	65.9	1,118	1,674
Front overhang distance	30.9	47.6	785	1,209
Rear overhang distance	27.4	59.6	696	1,514
Wheelbase length	90.7	121.5	2,304	3,086
Front tread width	53.7	62.8	1,364	1,595
Rear tread width	52.8	64.1	1,341	1,628
Bottom of front bumper to ground	5.1	17.6	130	447
Bottom of rear bumper to ground	7.3	18.5	185	470
	(feet)	*(feet)*	*(meters)*	*(meters)*
Turning diameter, outside front				
Wall to wall	17	46	5.2	14.0
Curb to curb	16	25	4.9	13.7
Turning diameter, inside rear				
Wall to wall	9	28	2.7	8.5
Ramp breakover angle, in degrees	8.0	26.0		

*Maximum and minimum values are for those vehicles included in the source document.

Source: Vehicle Dimensions, 1994 Model Year, American Automobile Manufacturers Association, Detroit, MI (1994).

considered to be representative of the total population of such cars since the dimensions have changed little during the past decade.

The maximum size of trucks and buses is established by state laws, and there is considerable variation among the states regarding the maximum allowable vehicle weights and dimensions. The range of state limitations is given in Table 5.5. The maximum weight and width of vehicles that use the Interstate System are set by federal law. Title 23 of the United States Code specifies a maximum width of 102 in. for such vehicles. The code specifies a maximum overall gross vehicle weight for the Interstate System of 80,000 lb, including all enforcement tolerances. The law also specifies a limit of 20,000 lb for any one axle, including tolerances, and gives a formula for calculating the maximum overall gross weight on any group of two or more consecutive axles.

Vehicle Performance Characteristics

Vehicle performance limitations and capabilities affect required lengths of speed-change lanes, weaving lanes, and no-passing zones, and the design of traffic control systems. If highway systems are to operate safely and efficiently, highway and trafic engineers need to understand performance characteristics of the various vehicles that use such systems. In the following paragraphs, we will describe some of the more important vehicle operating characteristics.

TABLE 5-5 Summary of State Limitations on Vehicle Weights and Dimensions[a]

	Minimum in Any State	*Maximum in Any State*
Width (m)	2.6	2.7
Height (m)	4.0	4.4
Length (m)		
Tractor-semitrailer combinations		
Single unit truck	12.2	18.3
Semitrailer on Interstate and		
national network	14.6	18.3
Semitrailer off national network	13.7	18.3
Overall combination length, other roads	16.8	26.8
Twin Combinations		
Semitrailer or trailer on Interstate and		
national network	8.5	29.0
Twin combination length on other roads	8.5	24.7
Straight truck + trailer	15.2	25.9
Weight (kg)		
Axle limits		
Single	9,072	10,206
Tandem	15,422	19,958
Tridem	15,422	29,938
Maximum gross weight, Interstate	36,288	53,071
Maximum gross weight, other roads	33,240	53,071

[a]This table is shown in U.S. units in Appendix A.

Source: Summary of Size and Weight Limits, American Trucking Association, Inc., Alexandria, VA, January 1994.

5-7 Resistances

A vehicle's motion tends to be retarded by at least five types of resistance:

1. inertia resistance
2. grade resistance
3. rolling resistance
4. curve resistance
5. air resistance

Inertia Resistance

It will be recalled from the study of physics that inertia is the tendency of a body to resist acceleration: the tendency to remain at rest or to remain in motion in a straight line unless acted upon by some force. The force, F_i, required to overcome a vehicle's inertia is described by the familiar relationship

$$F_i = ma \text{ (newtons)} \tag{5-1}$$

where　m = vehicle mass (kg)
　　　　a = acceleration (m/sec^2)

When acting to move a vehicle forward, this force is positive. When the force is slowing or stopping a vehicle, it and the corresponding acceleration (deceleration) are negative.

Grade Resistance

Grade resistance is the component of the gravitational force acting down a frictionless inclined surface. (See Figure 5-5.) This force

$$F_g = m\,g\,\sin\theta \text{ (newtons)} \qquad (5\text{-}2)$$

where m = vehicle mass (kg)
   g = acceleration due to gravity (9.8 m/sec)
   θ = angle of incline (degrees)

For even the steepest highway gradients encountered in practice, $\sin\theta \approx \tan\theta$, and for practical purposes, the resistance is

$$F_g = m\,g\,(G/100) \text{ (newtons)}$$
$$G = \text{gradient (percent)}$$

Rolling Resistance

A vehicle does not operate on a smooth, frictionless surface. There is resistance to motion as the tires roll over irregularities in the surface and push through mud, sand, or snow. This resistance, termed rolling resistance, includes that caused by the flexing of the tires and the internal friction of the moving parts of the vehicle. Rolling resistance is higher on low-quality pavement surfaces, and it increases with increase in vehicle speed. Table 5-6 shows typical rolling resistances as determined from empirical tests.

Curve Resistance

As was stated earlier, once a vehicle is set in motion, it tends to remain in motion in a straight line unless acted on by some force. The forces changing the direction of a vehicle are imparted through the front wheels. Components of these forces tend to impede a vehicle's forward motion. Curve resistance then is the force required to cause a vehicle to move along a curved path. It is a function of the radius or degree of curvature and the vehicle speed. Typical curve resistances for passenger cars are shown in Table 5-7.

Air Resistance

Air resistance includes the force required to move air from a vehicle's pathway as well as the frictional effects of air along its top, sides, and undercarriage. It is

TABLE 5-6 Rolling Resistances of Passenger Cars on Various Surfaces and at Different Speeds (Newtons/Metric Ton)

	Uniform Speed (km/hr)				
	32	48	64	80	96
Smooth pavement	123	132	142	152	167
Badly broken and patched pavement	142	167	196	250	—
Dry, well-packed gravel	152	172	245	304	—
Loose sand	172	196	279	373	—

Sources: Adapted from Paul J. Claffey, Vehicle Operating Characteristics, *Transportation and Traffic Engineering Handbook*, Institute of Transportation Engineers, Arlington, VA (1975) and Douglas W. Harwood, Traffic and Vehicle Operating Characteristics, *Traffic Engineering Handbook*, Institute of Transportation Engineers, Arlington, VA (1992).

TABLE 5-7 Typical Curve Resistances of Passenger Cars on High-Type Road Surfaces (Newtons)

Radius of Curve (m)	Speed (km/hr)	Resistance (N)
350	80	80
350	96	160
175	48	80
175	64	240
175	80	480

Source: Adapted from Paul J. Claffey, Vehicle Operating Characteristics, *Transportation and Traffic Engineering Handbook,* Institute of Transportation Engineers, Arlington, VA (1975).

a function of the frontal cross-sectional area of the vehicle and the square of the vehicle speed. The air resistance can be estimated by the following equation:

$$F_a = C_D A \frac{\rho v^2}{2} \text{ (newtons)} \tag{5-3}$$

where C_D = aerodynamic drag coefficient
A = frontal cross-sectional area (m²)
ρ = air density (kg/m³), typically 1.29 kg/m³
v = vehicle velocity (m/sec)

Power Requirements

Power is defined as the rate at which work is done. The power P required to overcome the various resistances and to propel a vehicle may be computed by the following equation:

$$P = R v \text{ (watts)} \tag{5-4}$$

where R = sum of the various resistances
= $F_i + F_g + F_r + F_c + F_a$ (newtons)
v = vehicle velocity (m/sec)

It should be emphasized that the power computed by equation 5-4 is only that required for a propulsion. It does not include the power required for a vehicle's automatic transmission, power steering, air conditioner, and other such accessories.

5-8 Acceleration and Deceleration Performances

A motor vehicle moves according to fundamental laws of motion; relationships among distance, time, velocity, and uniform acceleration are given by the following equations:

$$v_f = v_0 + at \tag{5-5}$$
$$d = v_0 t + \tfrac{1}{2}at^2 \tag{5-6}$$
$$v_f^2 = v_0^2 + 2ad \tag{5-7}$$

where v_f = final velocity (m/sec)
v_0 = initial velocity (m/sec)
a = acceleration or deceleration rate (m/sec²)
t = time (sec)
d = distance (m)

Maximum acceleration rates vary with the size of the vehicle and its operating speed. Vehicles are capable of greatest acceleration at lowest speeds. From a standing start to a speed of 25 km/hr, maximum acceleration values range from about 0.9 m/sec² (2.9 ft/sec²) for tractor-semitrailer trucks up to about 4.5 m/sec² (14.7 ft/sec²) for large cars (*19*). For a speed change of 0 to 48 km/hr (30 mph), typical maximum accelerations are

For tractor-semitrailer truck	0.5 m/sec² (1.5 ft/sec²)
For large car	3.1 m/sec² (10.3 ft/sec²)
For high-performance sports car	4.3 m/sec² (14.2 ft/sec²)

Figure 5-6 shows typical maximum accleration rates for 16 km/hr (10 mph) increases at various running speeds along level roads. Wide variations are shown for various vehicle types and running speeds.

Without braking, a vehicle will decelerate when the driver releases the accelerator due to the "drag" of the engine, air resistance, grade resistance, and so forth.

A passenger car operating in the range of 80 to 100 km/hr (50 to 60 mph) will decelerate about 0.9 m/sec² (3.0 ft/sec²) without braking; in the range of 32 to 48 km/hr (20 to 30 mph), an automobile will decelerate about 0.45 m/sec² (1.5 ft/sec²).

Under normal braking conditions the levels of deceleration developed usually do not reach the limit of a vehicle's braking capability nor that of the pavement-tire interface. The magnitude of decelerative force developed under normal or nonskid braking depends on the force the driver applies to the brake pedal. That

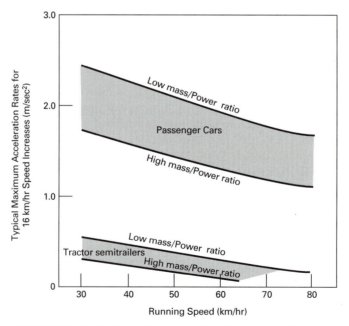

FIGURE 5-6 Typical maximum acceleration rates of passenger cars and tractor semitrailers for 16 km/hr (10 mph) speed increases at various running speeds.

force, in turn, is related to driver comfort. Observed normal deceleration rates for passengers cars (with braking) are about 1.5 m/sec^2 (4.8 ft/sec^2) in the range of 48 to 112 km/hr (30 to 70 mph) and about 2.0 m/sec^2 (6.7 ft/sec^2) in the range of 0 to 48 km/hr (0 to 30 mph).

PROBLEMS

5-1 A driver in a vehicle traveling 95 km/hr shifts her eyes from left to right and focuses on construction activities along the right shoulder. Estimate the distance in meters the vehicle travels as the driver's eyes shift and fixate.

5-2 A driver in a vehicle traveling 40 mph shifts his eyes from right to left and focuses on a child about to dart across the street. Estimate the distance in feet the vehicle travels as the driver's eyes shift and fixate.

5-3 In an intersection collision one of the vehicles leaves 30 m of skid marks. A skid mark analysis indicates that the vehicle was traveling 75 km/hr at the onset of braking. Assuming an average driver (i.e., median brake-reaction time), estimate the distance from the point of impact to the vehicle position when the driver initially reacted.

5-4 In an intersection collision one of the vehicles leave 160 ft of skid marks. A skid mark analysis indicates that the vehicle was traveling 50 mph at the onset of braking. Assuming an average driver (i.e., median brake-reaction time), estimate the distance from the point of impact to the vehicle position when the driver initially reacted.

5-5 Estimate the power required to accelerate a 1,350 kg vehicle traveling 48 km/hr up a 5.0 percent grade at a rate of 1.8 m/sec^2. The vehicle has a frontal cross-sectional area of 1.9 m^2. The roadway has straight alignment and a badly broken and patched asphalt surface. Assume the drag coefficient = 0.3.

5-6 On the basis of Figure 5-6 prepare a table showing the typical maximum acceleration rates in mph/sec for a low mass/power ratio at 20, 30, 40, and 50 mph.

5-7 A high power/mass ratio passenger car enters an acceleration lane at 40 km/hr and merges into a traffic lane at 70 km/hr. Estimate the desired length of the acceleration lane.

5-8 Prepare a report on the changes in the numbers and sizes of large trucks using U.S. highways during the past 20 years. How have these trucks impacted traffic congestion, safety, and durability of highway pavements? Give specific (quantitative) evidence of such impacts.

5-9 Plot the brake reaction times shown in Figure 5-2 as a cumulative distribution on normal probability paper. On the basis of that data, discuss the reasonableness of selecting the "average" driver's brake reaction time for design purposes. What would you recommend as a reasonable percentile for brake reaction for design? Take into account that many situations that confront drivers are unexpected.

REFERENCES

1. Goldstein, Leon G., *On the Future of Driving Licensing and Driver License Research,* Transportation Research Board Special Report No. 151 (1974).

2. Kerrick, John C., *What Have We Learned to Date?* Transportation Research Board Special Report No. 151 (1974).

3. Federal Highway Administration, *Highway Statistics 1992,* U.S. Department of Transportation, Washington, DC (1993).

4. Marsh, Burton W., Aging and Driving, *Traffic Engineering,* Arlington, VA (November 1960).

5. Dewar, Robert, Driver and Pedestrian Characteristics, *Traffic Engineering Handbook,* Institute of Transportation Engineers, Prentice Hall, Englewood Cliffs, NJ (1992).

6. Matson, Theodore M., Wilbur S. Smith, and Frederick W. Hurd, *Traffic Engineering,* McGraw-Hill, New York (1955).

7. Mourant, Ronald R., and Thomas H. Rockwell, Strategies of Visual Search by Novice and Experienced Drivers, *Human Factors,* Vol. 19, No. 4, pp. 325–335 (1972).

8. Greenshields, Bruce D., The Driver, *Traffic Engineering Handbook,* John E. Baerwald (ed.), Institute of Traffic Engineers, Washington, DC (1955).

9. Olson, Paul L., Driver Perception Response Time, *Motor Vehicle Accident Reconstruction: Review and Update,* Society of Automotive Engineers, Inc., Warrendale, PA (1989).

10. Greenshields, Bruce D., Reaction Time in Automobile Driving. *J. Appl. Psych.,* Vol. 20, No. 3, p. 355 (1936).

11. Johansson, Gunnar, and Kare Rumar, Driver's Brake Reaction Times, *Human Factors,* Vol. 13, No. 1, pp. 23–27 (1971).

12. *A Policy on Geometric Design of Highways and Streets.* American Association of State Highway and Transportation Officials, Washington, DC (1994).

13. Allen, T. M., H. Lunenfeld, and G. J. Alexander, Driver Information Needs, *Highway Research Record 366* (1971).

14. Alexander, Gerson J., and Harold Lunenfeld, *Positive Guidance in Traffic Control,* U.S. Department of Transportation, Washington, DC (1975).

15. Wiener, Earl L., The Elderly Pedestrian: Response to an Enforcement Campaign, *Research Review* (December 1968).

16. Fruin, John J., *Pedestrian Planning and Design,* Metropolitan Association of Urban Designers and Environmental Planners, New York (1971).

17. *AAMA Motor Vehicle Facts and Figures 94,* American Automobile Manufacturers Association, Detroit, MI (1994).

18. Harwood, Douglas W., Traffic and Vehicle Operating Characteristics, *Traffic Engineering Handbook,* Institute of Transportation Engineers, Prentice Hall, Englewood Cliffs, NJ (1992).

19. Claffey, Paul J., Vehicle Operating Characteristics, *Transportation and Traffic Engineering Handbook,* Institute of Transportation Engineers, Arlington, VA (1975).

20. Wilson, Ernest E., Deceleration Distances for High Speed Vehicles, *Proceedings, 20th Annual Meeting of the Highway Research Board,* pp. 393–397, Washington, DC (1940).

CHAPTER 6

TRAFFIC CHARACTERISTICS

A knowledge of traffic characteristics is useful to the highway engineer in developing highway and transportation plans, performing economic analyses, establishing geometric design criteria, selecting and implementing traffic control measures, and evaluating the performance of transportation facilities. Dozens of measures have been employed to describe the quality and quantity of traffic flow. In this chapter information is presented on those flow characteristics that fundamentally bear on the planning, design, and operation of highway and transport facilities: traffic speed, travel time, volume, and density. In a section on highway capacity we will consider ways of estimating the ability of various highway facilities to accommodate traffic flow. Finally, we will describe the nature and severity of the highway accident problem and examine the causes of traffic crashes.

TRAFFIC FLOW CHARACTERISTICS

6-1 Speed and Travel Time

Speed of travel is a simple and widely used measure of the quality of traffic flow. Basically, speed is the total distance traversed divided by the time of travel. Speed is commonly expressed in kilometers (miles) per hour or meters (feet) per second. Its reciprocal, travel time, is usually expressed in units of minutes per kilometer (mile).

There are three basic classes or measures of speed of travel:

1. spot speed
2. overall speed
3. running speed

Spot speed is the "instantaneous" speed of a vehicle as it passes a specified point along a street or highway. There are, of course, practical difficulties in measuring instantaneous speeds since, by definition, speed is the distance traveled divided by the travel time. Spot speeds may be determined by manually measuring the time required for a vehicle to traverse a relatively short specified distance. A variety of electromechanical and electronic devices are commonly employed to measure spot speeds. Such devices typically involve some sort of vehicle detectors (e.g., pneumatic tubes) that actuate and stop a timing mechanism, the time of travel or speed being printed on a tape or recorded on a graph. Radar meters have also been widely used by traffic engineers and enforcement officers to measure spot speeds.

The average of a series of measures of spot speeds can be expressed in two ways, as a *time-mean speed* and a *space-mean speed.* Time-mean speed is the arithmetic mean of speeds of all vehicles passing a point during a specified interval of time. The time-mean speed is

$$\bar{u}_t = \frac{\sum_{i=1}^{n} u_i}{n} \tag{6-1}$$

where u_i = observed speed of *i*th vehicle
    n = number of vehicles observed

The space-mean speed is the arithmetic mean of speeds of vehicles occupying a relatively long section of street or highway at a given instant. It is the average of vehicle speeds weighted according to how long they remain on the section of road. The space-mean speed is

$$\bar{u}_s = \frac{nd}{\sum_{i=1}^{n} t_i} \tag{6-2}$$

where d = length of roadway section
    t_i = observed time for the *i*th vehicle to travel distance d

Space-mean speed and time-mean speed are not equal. In fact, Wardrop (*1*) has shown that

$$\bar{u}_t = \bar{u}_s + \frac{\sigma_s^2}{\bar{u}_s} \tag{6-3}$$

where σ_s^2 = variance of the space distribution of speeds

For general-purpose usage, no distinction is normally made between time-mean and space-mean speeds. For theoretical and research purposes, the type of mean should be specified.

Overall speed and *running speed* are speeds over a relatively long section of street or highway between an origin and a destination. These measures are used in travel time studies to compare the quality of service between alternative routes. Overall speed is defined as the total distance traveled divided by the total time required, including traffic delays. Running speed is defined as the total distance traveled divided by the running time. The running time is the time the vehicle is in motion; time for stop-delays is excluded.

Overall and running speeds are normally measured by means of a test vehicle that is driven over the test section of roadway. The driver attempts to travel at the average speed of the traffic stream or to "float" in the traffic stream, passing as many vehicles as pass the test vehicle. A passenger uses a stopwatch to record time of travel to various previously chosen points along the course. Distances are usually recorded by the vehicle's odometer. The test drive is repeated several times and the average travel time is used to compute the overall and running speeds.

Spot speeds vary with time, location, and environmental and traffic conditions. Since 1942 the average speed on main rural highways has generally increased, rising from 64 km/hr (40 mph) in 1951 to 96 km/hr (60 mph) in 1968. Following a petroleum embargo and the subsequent imposition of a nationwide speed limit, the average speed on main rural highways decreased to 90 km/hr (55.7 mph) in 1983 but has risen gradually since that time. In 1992, the average speed on rural interstate highways was about 95 km/hr (59 mph), but 17.5 percent of the drivers were exceeding 105 km/hr (65 mph).

Speeds vary with the quality of traffic service, being generally higher along expressways and other well-designed facilities and during times when traffic congestion is not a factor. Oppenlander (2) found that mean spot speeds along two-lane rural highways were positively related to late width and minimum sight distance and negatively related to degree of curvature, gradient, and the number of roadside establishments per mile of highway.

At a given time and location, speeds are widely dispersed and can generally be represented by a normal probability distribution. As Figure 6-1 illustrates, the range of speeds decreases with increase in traffic volume.

6-2 Traffic Volume and Rate of Flow

Traffic volume is defined as the number of vehicles that pass a point along a roadway or traffic lane per unit of time. A measure of the quantity of traffic flow, volume is commonly measured in units of vehicles per day, vehicles per hour, vehicles per minute, and so forth.

Two measures of traffic volume are of special significance to the highway engineer: average daily traffic (ADT) and design hourly volume (DHV). The average daily traffic is the number of vehicles that pass a particular point on a roadway during a period of 24 consecutive hours averaged over a period of 365 days.

ADT is a fundamental measurement of traffic that is used for the determination of the vehicle-kilometers (or vehicle-miles) of travel on the various categories of highway systems. Vehicle-kilometers (or vehicle-miles) are important for the development of highway financing or taxation schedules, the evaluation of safety programs, and as a measure of the service provided by a highway transportation system.

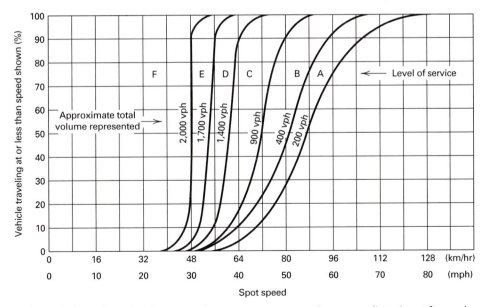

FIGURE 6-1 Typical distribution of passenger car speeds in one direction of travel under ideal uninterrupted flow conditions on freeways and expressways. (Courtesy Transportation Research Board.)

It is not feasible to make continuous counts 365 days a year along every section of a highway system. Average daily traffic values for many road sections are therefore based on a statistical sampling procedure described in Chapter 3.

The design hourly volume is a future hourly volume that is used for design. It is usually the 30th-highest hourly volume of the design year. Traffic volumes are much heavier during certain hours of the day or year, and it is for these peak hours that the highway is designed.

It has been found that, for the United States as a whole, traffic on the maximum day is approximately 233 percent of the annual average daily traffic, and traffic volume during the maximum hour is approxiamtely 25 percent of the annual average daily traffic. In order to design a highway properly, it is necessary to know the capacity that must be provided in order to accommodate the known traffic volume.

The relation between peak hourly flows and the annual average daily traffic on rural highways is shown in Figure 6-2. Experience has indicated that it would be uneconomical to design the average highway for an hourly volume greater than that which will be exceeded during only 29 hr in a year. The hourly traffic volume chosen for design purposes, then, is that occurring during the 30th-highest hour.

An approximate value of the 30th-highest hour can be obtained by applying

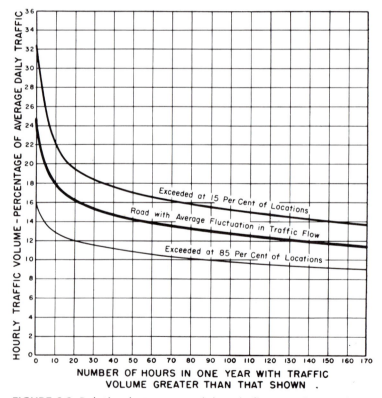

FIGURE 6-2 Relation between peak hourly flows and annual average daily traffic on rural highways. (Courtesy Federal Highway Administration.)

an empirically based percentage to the future *ADT*. The 30th-highest hour, as a percentage of the average daily traffic, ranges from 8 to 38 percent, with an average for the United States of 15 percent for rural locations and 12 percent for urban locations.

Early studies of U.S. traffic indicated that the relationship between the 30th-highest hour and the annual average daily traffic remained unchanged from year to year. However, later studies suggest that the 30th-highest hour factor has a tendency to decline slightly with the passing of time. If this trend continues, appropriate adjustments will have to be made in the design hourly volume for any future year.

On a given roadway, the volume of traffic fluctuates widely with time. Figures 6-3, 6-4, and 6-5 illustrate the variations in volume that occur with time of day, day of the week, and season of the year. These variations tend to be cyclical and to some extent predictable. The nature of the pattern of variation depends on the type of highway facility. Urban arterial flow is characterized by pronounced peaks during the early morning and late afternoon hours, due primarily to commuter traffic. The peaking pattern is not generally evident on weekends, and such facilities experience lowest flows on Sundays. Rural highways tend to experience less pronounced daily peaks, but they may accommodate heaviest traffic flows on weekends and holidays because of recreational travel. Highway facilities generally must accommodate heaviest flows during the summer months. Peaks typically occur during July or August. As might be expected, the seasonal fluctuations are most pronounced for rural recreational routes.

The term *rate of flow* accounts for the variability or the peaking that may occur during periods of less than 1 hr. The term is used to express an equivalent hourly rate for vehicles passing a point along a roadway or for traffic during an interval less than 1 hr, usually 15 min (*3*).

The distinction between volume and rate of flow may be illustrated by an example. Suppose the following traffic counts were made during a study period of 1 hr:

Time Period	Number of Vehicles	Rate of Flow (vehicles/hr)
8:00–8:15	1,000	4,000
8:15–8:30	1,100	4,400
8:30–8:45	1,000	4,000
8:45–9:00	900	3,600
Total	4,000	

The total volume is the sum of these counts or 4,000 vehicles/hr. The rate of flow varies for each 15-min period and during the peak period is 4,400 vehicles/hr. Note that 4,400 vehicles did not actually pass the observation point during the study hour, but they did pass *at that rate* for one 15-min period.

Consideration of peak rates of flow is of extreme importance in highway capacity analyses.* Suppose the example roadway section is capable of handling a maximum rate of only 4,200 vehicles/hr. In other words, its capacity is 4,200

---

*Highway capacity analyses are discussed in more detail in Section 6-6.

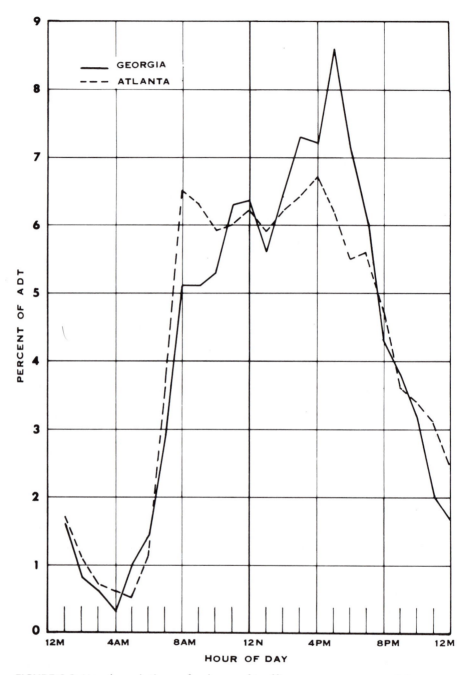

FIGURE 6-3 Hourly variations of volume of traffic on an average weekday. (Courtesy Georgia Department of Transportation.)

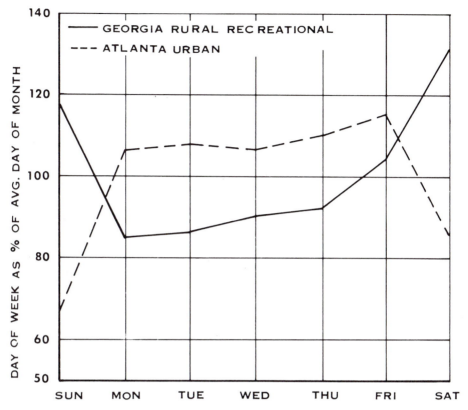

FIGURE 6-4 Traffic volume fluctuation by day of week. (Courtesy Georgia Department of Transportation.)

vehicles/hr. Since the peak rate of flow is 4,400 vehicles/hr, an extended breakdown in the flow would likely occur even though the volume, averaged over the full hour, is less than the capacity.

The *Highway Capacity Manual* (3) uses a *peak hour factor* to relate peak rates of flow to hourly volume. The peak hour factor is defined as the ratio of total hourly volume to the maximum rate of flow within the hour. If there was no variability in flow rate during the hour, the peak hour factor would be 1.00. Typical peak hour factors for 2-lane roadways range from about 0.83 to 0.96.

6-3 Traffic Density

Traffic density, also referred to as traffic concentration, is defined as the average number of vehicles occupying a unit length of roadway at a given instant. It is generally expressed in units of vehicles per mile. As Section 6-5 indicates, traffic density bears a functional relationship to speed and volume. Density has not been extensively employed in the past by highway and traffic engineers to describe traffic flow; however, it is now recommended as the basic parameter for describing the quality of flow along freeways and other multilane highways. It has also been the focus of a number of theoretical and analytical studies.

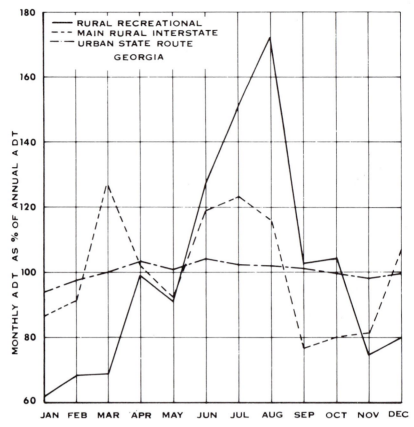

FIGURE 6-5 Monthly variation in traffic volume. (Courtesy Georgia Department of Transportation.)

6-4 Spacings and Headways

There are many situations that engineers encounter for which it is necessary to consider the behavior of individual vehicles in the traffic stream rather than the average traffic stream characteristics. Such situations include calculating the probability of delay and average delay for vehicles or pedestrians crossing a traffic stream and predicting the length of waiting lines at toll booths, traffic signals, and entrances to parking facilities. Two measures are of fundamental importance in such calculations: spacing and time-headway between successive vehicles. The spacing is simply the distance between successive vehicles, typically measured from front bumper to front bumper. It is the reciprocal of density. Time-headway is the time between the arrival of successive vehicles at a specified point, and it is the reciprocal of volume.

For many light traffic situations, traffic can be described by the Poisson probability distribution. The equation for the Poisson distribution is

$$P(x) = \frac{m^x e^{-m}}{x!} \tag{6-4}$$

where $P(x)$ = the probability that exactly x randomly arranged vehicles will be observed in a unit length of road, or the probability of arrival of exactly x vehicles in a unit length of time

$$m = Vt/3{,}600 = \text{the average number of vehicles arriving in an interval of length } t$$

V = traffic volume (vehicles/hr)

t = length of time interval (sec)

EXAMPLE 6-1 **The Poisson Distribution** Consider the following example, which has been taken from Ref. 4. The number of vehicles arriving along a Los Angeles street was recorded for each of 120 30-sec intervals. During 9 of the intervals, no vehicles arrived; during 16 intervals, exactly one vehicle arrived; and so forth. See Table 6-1. The average number of arrivals is

$$m = \frac{Vt}{3600} = \frac{368(30)}{3600} = 3.067$$

$$P(0) = \frac{(3.067)^0 e^{-3.067}}{0!} = 0.047$$

$$P(1) = \frac{(3.067)^1 e^{-3.067}}{1!} = 0.143$$

Other probabilities are shown in Table 6-1. The theoretical frequencies corresponding to the probability $P(x_i)$ are

$$F_i = 120\, P(x_i)$$
$$F_0 = 120(0.047) = 5.6 \text{ intervals, etc.}$$

Corresponding to the Poisson counting distribution for free-flow conditions is the negative exponential gap or headway distribution. The equation is

TABLE 6-1 Example Data for the Poisson Distribution

Traffic Arrivals: 30-Sec Intervals
(Durfee Avenue, Northbound)

Number of cars per interval X_i	Observed frequency[a] f_i	Total cars observed f_iX_i	Probability of X_i, $P(X_i)$	Theoretical frequency F_i
0	9	0	0.047	5.6
1	16	16	0.143	17.2
2	30	60	0.219	26.3
3	22	66	0.224	26.9
4	19	76	0.172	20.6
5	10	50	0.105	12.6
6	3	18	0.054	6.5
7	7	49	0.023	2.8
8	3	24	0.009	1.1
≥9	1	9	0.002	0.4
Total	120	368	1.000	120.0

$$m = \frac{368}{120} = 3.067 \qquad e^{-m} = 0.047$$

$$F_i = 120\,(P(X_i))$$

[a]Data by courtesy of Los Angeles County Road Department.

Source: Poisson and Other Distributions in Traffic, Eno Foundation for Transportation, Saugatuck, CT (1971).

$$P(g \geq T) = e^{-VT/3600} \qquad (6\text{-}5)$$

where $P(g \geq T)$ = the probability of a headway equal to or greater than T sec

The value e is the Naperian logarithmic base and V and T are as previously defined.

EXAMPLE 6-2   **The Negative Exponential Distribution**   Given the conditions described in Example 6-1, estimate the percentage of a randomly chosen group of gaps that will be greater than 10 sec.

According to the negative exponential distribution,

$$P(g \geq T) = e^{-368(10)/3600} = 0.36$$

Thirty-six percent of the gaps will be greater than 10 sec.

It should be emphasized that the Poisson and negative exponential distributions are applicable to random or free-flow traffic situations. More complex distributions have been proposed to describe a wider variety of traffic flow conditions. References 5 and 6 give detailed descriptions of many counting and headway distributions.

6-5 Speed-Volume-Density Relationships

The relationship among traffic speed, volume, and density is shown by the fundamental equation

$$q = k\bar{u}_s \qquad (6\text{-}6)$$

where q = average volume of flow (vehicles/hr)
   k = average density or concentration (vehicles/mile)
   $\bar{u}_s$ = space-mean speed (mph)

Although a number of theoretical and analytical speed-density relationships have been published, the exact shape of the $k - \bar{u}_s$ curve has not been conclusively established. A model proposed by Greenshields (6) assumed a linear relationship between speed and density. With that assumption, a parabolic volume-density model results. The Greenshields model is illustrated by Figure 6-6 along with the corresponding relationships between speed and volume and volume and density. The curves represent a hypothetical situation with a maximum (or mean-free) speed of 80 km/hr (50 mph) and a maximum density of 109 veh/km (175 vehicles/mile).

Let us consider three key points on the curves shown in Figure 6-6. Point A represents a low-density, high-speed situation. Low volumes exist because few vehicles are on the road. Point B is the point of maximum flow. For this condition, intermediate levels of speed and density occur. Point C represents the worst possible type of flow situation, with maximum density occurring and speeds (and volume) approximating zero.

Empirical research has indicated that speed decreases exponentially with increases in density. Figure 6-7 shows empirical data collected in the Lincoln Tunnel to which a logarithmic flow-concentration curve has been fitted. The equation is

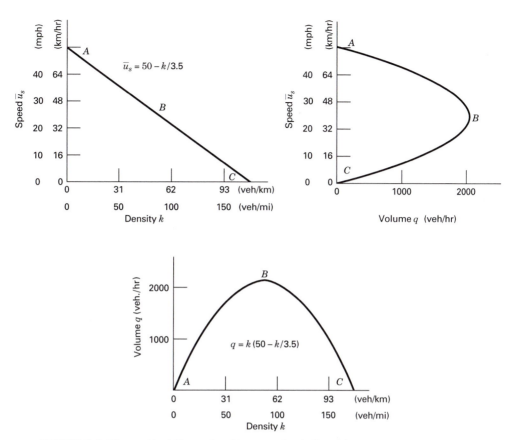

FIGURE 6-6 Theoretical flow, density, speed relationships.

$$q = u_m k \ln\left(\frac{k_j}{k}\right) \tag{6-7}$$

where u_m = speed at maximum flow
k_j = maximum density

For the data represented by Figure 6-7, u_m = 17.2 vehicles/hr and k_j = 142 veh/km (228 vehicles/mi).

6-6 Highway Capacity

Even with ideal roadway conditions (e.g., a test track) traffic volume tends to reach a maximum point at a relatively low speed. (See Figure 6-8.) This phenomenon, which is puzzling to the casual observer, results from the fact that spacing allowed by the average driver when trailing another vehicle increases nonlinearly with increases in speed (7). Stated another way, traffic density decreases exponentially with increases in speed, as Figure 6-9 illustrates. Referring to Eq. 6-6, increasing speed tends to increase volume, but that effect is offset by concomitant decreases in density.

Another factor limiting the number of vehicles passing a point is interference between vehicles in the traffic stream. This effect is especially noticeable along

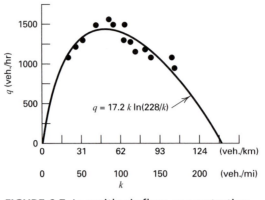

FIGURE 6-7 Logarithmic flow-concentration diagram. (Courtesy Transportation Research Board.)

two-lane roads. With a low traffic volume, the vehicle operator has a wide latitude in selecting the speed at which he or she wishes to travel. As traffic volume increases, the speed of each vehicle is influenced in a large measure by the speed of the slower vehicles. As traffic density increases, a point is finally reached where all vehicles are traveling at the speed of the slower vehicles. This condition indicates that the ultimate capacity has been reached. The capacity of a highway

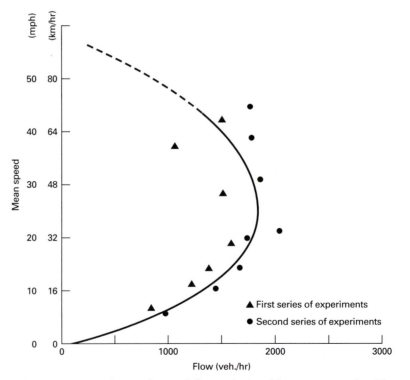

FIGURE 6-8 Experimental speed-flow relationship on test track with straight roadway section. (Courtesy Transportation Research Board.)

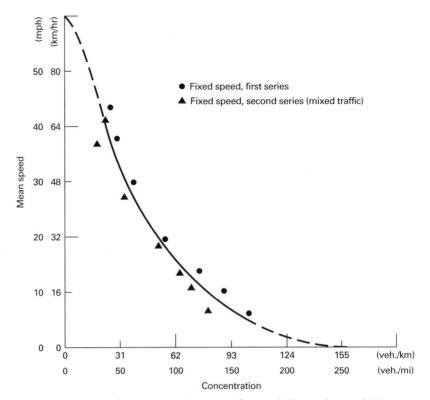

FIGURE 6-9 Speed-concentration curve for straight track at British Road Research Laboratory. (Courtesy Transportation Research Board.)

is therefore measured by its ability to accommodate traffic and is usually expressed as the number of vehicles that can pass a given point in a certain period of time at a given speed.

Unlike speed on two-lane highways, speed on freeways tends to be insensitive to flow over a broad range of flows. Along a 112 km/hr (70 mph) freeway, for example, speed tends to remain constant for flows up to 1,300 passenger cars per hour per lane (pcphpl). Along a freeway with a free flow speed of 88 km/hr (55 mph), average speed is independent of flow up to 1,750 pcphpl. (See Figure 6-10.)

Roadways, of course, are not ideal, and in highway capacity estimates allowances must be made for prevailing roadway and traffic conditions that inhibit the ability of a road to accommodate traffic. In determining highway capacities for uninterrupted flow conditions, the general procedure, described below, is to begin with capacities corresponding to ideal conditions and to apply appropriate empirically based adjustments for prevailing roadway and traffic conditions. The *capacity* of a given section of roadway, either in one direction or in both directions for a two-lane or three-lane roadway, may be defined as the maximum hourly rate at which vehicles can reasonably be expected to traverse a point or uniform section of a lane or roadway during a given time period under prevailing roadway, traffic, and control conditions (*3*).

Although the maximum number of vehicles that can be accommodated remains fixed under similar roadway and traffic conditions, there is a range of lesser volumes that can be handled under differing operating conditions. Operation at

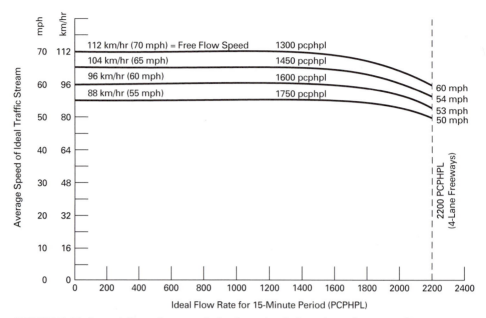

FIGURE 6-10 Speed-flow characteristics for a basic four-lane freeway. (Courtesy Transportation Research Board.)

capacity provides the maximum volume, but as both volume and congestion decrease there is an improvement in the level of service.

Level of service is a qualitative measure that describes operational conditions within a traffic stream and their perception by drivers and/or passengers (*3*). Six levels of service, A through F, define the full range of driving conditions from best to worst, in that order. These levels of service qualitatively measure the effect of such factors as travel time, speed, cost, and freedom to maneuver, which, in combination with other factors, determine the type of service that any given facility provides to the user under the stated conditions. With each level of service, a *service flow rate* is defined. It is the maximum volume that can pass over a given section of lane or roadway while operating conditions are maintained at the specified level of service.

Level A represents free flow at low densities with no restrictions due to traffic conditions. Level B, the lower limit of which is often used for the design of rural highways, is the zone of stable flow with some slight restriction of driver freedom. Level C denotes the zone of stable flow with more marked restriction on the driver's selection of speed and with reduced ability to pass. The conditions of Level D reflect little freedom for driver maneuverability, and, while the operating speeds are still tolerable, this region approaches the condition of unstable flow. Low operating speeds and volumes near or at capacity indicate level of service E, which is the area of unstable flow. Level F is the level of service provided by the familiar traffic jam, with frequently interruptions and breakdown of flow, as well as volumes below capacity coupled with low operating speeds.

The operational parameters used to determine the level of service differ with the type of facility. These parameters or "measures of effectiveness" describe the operating quality for each type of facility. Table 6-2 sets out the measures of effectiveness that best define the level of service for each facility type.

TABLE 6-2 Measures of Effectiveness for Level of Service Definition

Type of Facility	Measure of Effectiveness	Units
Uninterrupted Flow Facilities		
Freeways		
Basic freeway segments	Density	Passenger cars per mile per lane
Weaving areas	Average travel speed	Miles per hour
Ramp junctions	Flow rate	Passenger cars per hour
Multilane highways	Density	Passenger cars per mile per lane
Two-lane highways	Percent time delay	Percent
	Average travel speed	Miles per hour
Interrupted Flow Facilities		
Signalized intersections	Average individual stopped delay	Seconds per vehicle
Unsignalized intersections	Average total delay	Seconds per vehicle
Arterials	Average travel speed	Miles per hour
Transit	Load factor	Persons per seat, vehicles per hour, people per hour
Pedestrians	Space	Square feet per pedestrian

Source: *Highway Capacity Manual,* Transportation Research Board Special Report No. 209 (1994).

The Highway Capacity Manual

The *Highway Capacity Manual* (*3*) is the authoritative guide for the performance of highway capacity analyses. The manual reflects over 40 years of comprehensive research by a number of research agencies. This document was prepared under the guidance of the Transportation Research Board's Committee on Highway Capacity and Quality of Service.

The procedures described in the *Highway Capacity Manual* cover a wide range of facilities, including streets and highways as well as facilities for transit, pedestrians, and bicyclists. As Table 6-2 illustrates, capacity analyses are performed for two general categories of facilities, those with *uninterrupted flow* and those with *interrupted flow*. Uninterrupted flow facilities include two-lane highways, freeways, and other multilane highways. The traffic flow conditions on such facilities result from interactions among vehicles in the traffic stream as well as between vehicles and the physical and ambient characteristics of the roadway.

Interrupted flow facilities generally have traffic control devices that cause periodic interruptions of traffic flow. Examples of such facilities are signalized and two-way and four-way STOP-controlled intersections. The capacity of interrupted flow facilities is limited not only by the physical features of the roadway space but also by the portion of time that is available for various components of traffic flow.

Factors Affecting Capacity, Service Flow Rate, and Level of Service

Many of the procedures described in the *Highway Capacity Manual* are based on simple tables or graphs for specified standard conditions, which must be adjusted to account for prevailing conditions different from those specified. The conditions so defined are often described in terms of ideal conditions.

Ideal conditions for uninterrupted flow facilities include

- 3.6-m (12-ft) lane widths
- 1.8-m (6-ft) clearance between the edge of travel lanes and the nearest roadside obstructions
- all passenger cars in the traffic stream
- a driver population comprised predominantly of regular and familiar users of the facility

An ideal signalized intersection approach has

- 3.6-m (12-ft) lane widths
- level grade
- no curb parking allowed on the intersection approaches
- all passenger cars in the traffic stream
- no turning movements at the intersection
- intersection located outside the central business district
- green signal available at all times

Since prevailing conditions are seldom ideal, computations of capacity, service flow rate, or level of service must be adjusted to account for departures from the ideal. Prevailing conditions may be grouped into three categories: roadway, traffic, or control conditions.

Roadway factors include

- the type of facility and its development environment
- lane widths
- shoulder widths and/or lateral clearances
- design speed
- horizontal and vertical alignments

Traffic conditions refer to the types of vehicles using the facility and how the traffic flow is distributed by lane use and direction. It is well known that larger and heavier vehicles have an adverse effect on traffic flow in a number of ways. Capacity analyses account for the impacts of trucks, buses, and recreational vehicles, utilizing the results of empirical studies that have quantified these effects.

In addition to the distribution of vehicle types, the effects of two other traffic characteristics on capacity, service flow rates, and level of service must be considered. Directional distribution has a significant impact on the operation of two-lane rural highways. Capacity generally declines as the directional split becomes unbalanced. For multilane highways, capacity analysis procedures focus on a single direction, that of the peak rate of flow. Lane distribution must also be considered for multilane facilities. Normally, the shoulder lane of a multilane facility carries less traffic than other lanes, and analysis procedures must account for these differences.

For interrupted flow facilities, the control of the time available for movement of each traffic flow is of utmost importance in determining capacity, service flow rates, and level of service. Of greatest importance is the control of such facilities by the traffic signal. Signalized operations are affected by the type of signal in use, the phasing, the allocation of green time, and the length of cycle. Stop and yield signs also affect capacity, but in a different way. The stop or yield sign assigns the right-of-way permanently to a major street, while a signal assigns designated times when specific movements are permitted.

A complete treatment of highway capacity analyses is beyond the scope of this book. Two examples, one each for uninterrupted and interrupted flow, are given to illustrate the general approach used for such analyses.

General Methodology for Capacity Analyses for Freeway Segments

With controlled access, limited ramp locations, and no fixed interruptions of traffic flow, freeways operate under the purest form of uninterrupted flow. A capacity analysis of a freeway segment begins with a recognition that a maximum service flow rate per lane exists for each level of service. The various levels of service for freeways are defined by the density, expressed in pasenger cars per mile per lane. Suggested maximum service flow rates, MSF_i, are shown in Table 6-3.

These rates, which represent ideal conditions, must be adjusted to account for any prevailing conditions that are not ideal and to reflect the number of lanes in one direction on the freeway segment. The service flow rate for the specified level of service and prevailing conditions is computed by using correction factors, as the following equation demonstrates:

$$SF_i = MSF_i \times N \times f_{\text{w}} \times f_{\text{HV}} \times f_{\text{P}} \qquad (6\text{-}8)$$

where SF_i = service flow rate for level of service i under prevailing roadway and traffic conditions for N lanes in one direction (vehicles/hr)

N = number of lanes in one direction of the freeway

f_{w} = factor to adjust for the effects of restricted lane widths and/or lateral clearances

f_{HV} = factor to adjust for the effect of heavy vehicles (trucks, buses, and recreational vehicles) in the traffic stream

f_{P} = factor to adjust for the effect of driver population

TABLE 6-3 Level of Service Criteria for Basic Freeway Sections

Level of Service	Maximum Density (pc/mi/ln)	Minimum Speed (mph)	Maximum Service Flow Rate (pcphpl)	Maximum v/c Ratio
Free-Flow Speed = 70 mph				
A	10.0	70.0	700	0.318/0.304
B	16.0	70.0	1,120	0.509/0.487
C	24.0	68.5	1,644	0.747/0.715
D	32.0	63.0	2,015	0.916/0.876
E	36.7/39.7	60.0/58.0	2,200/2,300	1.000
F	varies	varies	varies	varies
Free-Flow Speed = 60 mph				
A	10.0	60.0	600	0.272/0.261
B	16.0	60.0	960	0.436/0.417
C	24.0	60.0	1,440	0.655/0.626
D	32.0	57.0	1,824	0.829/0.793
E	41.5/46.0	53.0/50.0	2,200/2,300	1.000
F	varies	varies	varies	varies

Note: For split values, first value applies to four-lane freeways, the second to six- and eight-lane freeways.

Source: Adapted from *Highway Capacity Manual,* Transportation Research Board, 1994.

The factor f_w is based on the lane width, the number of freeway lanes, the distance to roadside obstructions, and whether obstructions are present on both sides of the freeway. Table 6-4 gives recommended adjustment factors for a range of lane widths and lateral clearances.

The factor f_{HV} is used to account for the effects of trucks, buses, and recreational vehicles in the traffic stream. The factor depends on the terrain, specifically the length and magnitude of up grades, and the mix of vehicles in the traffic stream. Heavy vehicles, because of their restricted maneuverability, reduce the number of vehicles that a facility can handle. This reduction is represented by the term *passenger-car equivalent,* which indicates the equivalent number of passenger cars that have been displaced by the presence of each truck, bus, or recreational vehicle. Table 6-5 gives passenger-car equivalents for stated percentages of trucks and buses operating on specific freeway grades. The table applies when the grade is

<3% grade and >1/2 mile in length or
>3% grade and >1/4 mile in length

The *Highway Capacity Manual* (3) gives a procedure for estimating the passenger-car equivalents of heavy vehicles operating on an extended length of freeway that contains a number of upgrades, downgrades, and level segments. This procedure, not included here, allows an analyst to calculate passenger-car equivalents along an extended segment where no one grade is long enough to have significant effect on the overall operation of the highway segment.

Passenger-car equivalents for various percentages of recreational vehicles on specific freeway grades are given in Table 6-6.

By using passenger-car equivalents such as those shown in Tables 6-5 and 6-6, the adjustment factor for heavy vehicles can be computed by the following equation:

$$f_{HV} = 1/[1 + P_T(E_T - 1) + P_R(E_R - 1)] \qquad (6\text{-}9)$$

where f_{HV} = adjustment factor for the combined effect of trucks, recreational vehicles, and buses on the traffic stream

E_T, E_R = passenger-car equivalents for trucks/buses and recreational vehicle, respectively

P_T, P_R = proportion of trucks/buses and recreational vehicles, respectively, in the traffic stream

TABLE 6-4 Adjustment Factors for Restricted Lane Width and/or Lateral Clearance

Distance from Traveled Way to Obstruction (ft)*	Lane Width (ft)*			
	Obstructions on One Side		Obstructions on Two Sides	
	≥12 ft	10 ft	≥12 ft	10 ft
≥6 ft	1.00	0.90	1.00	0.90
4 ft	0.99	0.89	0.98	0.88
2 ft	0.97	0.88	0.95	0.86
0 ft	0.92	0.84	0.86	0.78

*Interpolation may be used for lane width and/or distance from traveled way to obstruction.

Source: Adapted from *Highway Capacity Manual,* Transportation Research Board, 1994.

TABLE 6-5 Passenger-Car Equivalents for Trucks and Buses on Specific Upgrades

Grade (%)	Length (Miles)	E_T						
	Percent Trucks and Buses	2	4	5	6	8	10	15
<2	All	1.5	1.5	1.5	1.5	1.5	1.5	1.5
2	0–¼	1.5	1.5	1.5	1.5	1.5	1.5	1.5
	¼–½	1.5	1.5	1.5	1.5	1.5	1.5	1.5
	½–¾	1.5	1.5	1.5	1.5	1.5	1.5	1.5
	¾–1	2.5	2.0	2.0	2.0	1.5	1.5	1.5
	1–1½	4.0	3.0	3.0	3.0	2.5	2.5	2.0
	>1½	4.5	3.5	3.0	3.0	2.5	2.5	2.0
3	0–¼	1.5	1.5	1.5	1.5	1.5	1.5	1.5
	¼–½	3.0	2.5	2.5	2.0	2.0	2.0	2.0
	½–¾	6.0	4.0	4.0	3.5	3.5	3.0	2.5
	¾–1	7.5	5.5	5.0	4.5	4.0	4.0	3.5
	1–1½	8.0	6.0	5.5	5.0	4.5	4.0	4.0
	>1½	8.5	6.0	5.5	5.0	4.5	4.5	4.0
4	0–¼	1.5	1.5	1.5	1.5	1.5	1.5	1.5
	¼–½	5.5	4.0	4.0	3.5	3.0	3.0	3.0
	½–¾	9.5	7.0	6.5	6.0	5.5	5.0	4.5
	¾–1	10.5	8.0	7.0	6.5	6.0	5.5	5.0
	>1	11.0	8.0	7.5	7.0	6.0	6.0	5.0
5	0–¼	2.0	2.0	1.5	1.5	1.5	1.5	1.5
	¼–⅓	6.0	4.5	4.0	4.0	3.5	3.0	3.0
	⅓–½	9.0	7.0	6.0	6.0	5.5	5.0	4.5
	½–¾	12.5	9.0	8.5	8.0	7.0	7.0	6.0
	¾–1	13.0	9.5	9.0	8.0	7.5	7.0	6.5
	>1	13.0	9.5	9.0	8.0	7.5	7.0	6.5

Note: If the length of grade falls on a boundary, apply the longer category; interpolation may be used to find equivalents for intermediate percent grades.

Source: Adapted from *Highway Capacity Manual,* Transportation Research Board, 1994.

It is known that certain types of traffic, such as weekend and recreational traffic, use freeways less efficiently than weekday or commuter traffic. Little research has been done on this effect, but it is known that capacities tend to be lower on weekends, particularly in recreational areas. Lacking local data, considerable engineering judgment must be exercised in making an adjustment for the character of the traffic stream. The *Highway Capacity Manual* suggests that an adjustment factor f_P ranging from 0.75 to 1.00 be used to account for this effect.

EXAMPLE 6-3 Capacity of a Basic Freeway Segment (a) Determine the service flow rate with level of service C for a section of a 4-lane freeway (2 lanes in each direction) with 3.6-m (12-ft) lanes and obstructions 0.9 m (3 ft) from the traveled pavement on one side of the roadway. The section has a 4 percent gradient 1.3 km (0.8 mi) long. It is to accommodate 10 percent heavy trucks and buses and 2 percent recreational vehicles. Based on local studies, an adjustment factor for the driver

TABLE 6-6 Passenger-Car Equivalents for Recreational Vehicles on Specific Upgrades

Grade (%)	Length (Miles)	E_R						
Percent Recreational Vehicles		*2*	*4*	*5*	*6*	*8*	*10*	*15*
≤2	All	1.2	1.2	1.2	1.2	1.2	1.2	1.2
3	0–½	1.2	1.2	1.2	1.2	1.2	1.2	1.2
	> ½	2.0	1.5	1.5	1.5	1.5	1.5	1.2
4	0–¼	1.2	1.2	1.2	1.2	1.2	1.2	1.2
	¼–½	2.5	2.5	2.0	2.0	2.0	2.0	1.5
	> ½	3.0	2.5	2.5	2.0	2.0	2.0	2.0
5	0–¼	2.5	2.0	2.0	2.0	1.5	1.5	1.5
	¼–½	4.0	3.0	3.0	3.0	2.5	2.5	2.0
	> ½	4.5	3.5	3.0	3.0	3.0	2.5	2.5

Note: If the length of grade falls on a boundary, apply the longer category; interpolation may be used to find equivalents for intermediate percent grades.

Source: Adapted from *Highway Capacity Manual,* Transportation Research Board, 1994.

population, f_P, of 0.90 is indicated. The design speed is 112 km/hr (70 mph). (b) What is the capacity of the segment?

Solution to Part (a). From Table 6-3, the maximum service flow rate for LOS C is 1,664 pcphpl. The service flow rate for LOS C is computed by Eq. 6-8.

From Table 6-4, the factor to adjust for the effects of lane width and lateral clearance is $f_W = 0.98$.

To adjust for the effect of heavy vehicles, passenger-car equivalents for trucks/buses and recreational vehicles are obtained from Tables 6-5 and 6-6, respectively:

$$E_T = 5.5$$
$$E_R = 3.0$$

By Eq. 6-9,

$$f_{HV} = 1/[1 + 0.10(5.5 - 1) + 0.02(3.0 - 1)] = 0.67$$
$$f_P = 0.90 \text{ (given)}$$

By Eq. 6-8, the service flow rate in one direction

$$SF_C = 1,664 \times 2 \times 0.98 \times 0.67 \times 0.90 = 1,943 \text{ veh/hr}$$

Solution to Part (b). The capacity corresponds to LOS E and from Table 6-3 is 2,200 passenger cars/hr per lane (ideal conditions). For the prevailing conditions, the capacity is

$$SF_E = 2,200 \times 2 \times 0.98 \times 0.67 \times 0.90 = 2,600 \text{ veh/hr}$$

General Methodology for Capacity Analysis for Signalized Intersection

The capacity of a signalized intersection is highly dependent on the type of signal control being used. A wide variety of equipment and control schemes may be used for such intersections. The capacity of a signalized intersection is therefore far more variable than that of other types of facilities, where capacity depends primarily on the physical geometry of the roadway.

In intersection analysis the concepts of capacity and level of service are not as strongly correlated as they are for other types of facilities. The *Highway Capacity Manual* (3) therefore recommends that separate analyses be used to determine the capacity and level of service for a signalized intersection.

The capacity of signalized intersections should be defined for each approach.* Intersection approach capacity is the maximum rate of flow that may pass through the intersection by that approach under prevailing roadway, traffic, and signalization conditions. To account for peaking, the rate of flow is usually measured or projected for a 15-min period, and capacity is expressed in vehicles per hour (3).

Operational analysis of a signalized intersection requires detailed information on the roadway, the signal system, and the traffic at the intersection. Required information on the roadway includes approach grades, the number and width of lanes, parking conditions, and existence and lengths of exclusive turning lanes.

Complete information is needed on signalization, including the phase plan, cycle length, green times, type of signal operation (actuated or pretimed), and existence of push-button pedestrian-actuated phases. (These and other traffic control concepts are discussed in more detail in Chapter 12.)

Capacity analyses of signalized intersections, require detailed information on traffic conditions, including

1. traffic volumes for each movement on each approach
2. percentage of heavy vehicles
3. volume and pattern of pedestrian traffic
4. rate of parking maneuvers within the vicinity of the intersection
5. number of local buses picking up or discharging passengers at the intersection

In addition, information is needed on the shape of the arrival curve distribution and its relationship to the signal operation. This factor describes the platooning effect in arriving flows. It affects the average stopped delay of vehicles passing through the intersection, which defines the level of service. The *Highway Capacity Manual* (3) categorizes arrival distributions by defining five arrival types, defined as follows:

Type 1 is the worst platoon condition, defined as a dense platoon arriving at the intersection at the beginning of the red phase.

Type 2 is a generally unfavorable platoon condition, which may be a dense platoon arriving during the middle of the red phase or a dispersed platoon arriving throughout the red phase.

Type 3 refers to totally random arrivals that are widely dispersed throughout the red and green phases.

Type 4 is a moderately favorable platoon condition, defined as a dense platoon

---

*It may also be desirable to evaluate separately the capacity of designated lanes or lane groups such as those serving a particular movement or set of movements.

arriving during the middle of the green phase or a dispersed platoon arriving throughout the green phase.

Type 5 is the most favorable platoon condition, defined as a dense platoon arriving at the beginning of the green phase.

The *Highway Capacity Manual* (3) relates the arrival types to a platoon ratio, which is defined by

$$R_P = PVG/PTG \qquad (6\text{-}10)$$

where R_P = platoon ratio
PVG = percentage of all vehicles in the movement arriving during the green phase
PTG = percentage of the cycle that is green for the movement

The relationship between arrival type and the platoon ratio is shown in Table 6-7.

The capacity of signalized intersections is based on the concept of a *saturation flow rate*. The *Highway Capacity Manual* (3) defines saturation flow rate as the maximum rate of flow that can pass through an intersection approach or lane group under prevailing roadway and traffic conditions, assuming that the approach or lane group has 100 percent of real time available as effective green time. Saturation flow rate is expressed in units of vehicles per hour of effective green time.

The capacity of a given lane group or approach may be calculated by the equation

$$c_i = s_i \times (g/C)_i \qquad (6\text{-}11)$$

where c_i = capacity of lane group or approach i (vehicles/hr)
s_i = saturation flow rate for lane group or approach i (vehicle/hr of green)
$(g/C)_i$ = green ratio for lane or approach i

The computation of the saturation flow rate begins with the selection of an ideal saturation flow rate, usually taken to be 1,900 passenger cars per hour of green time per lane. This value is then adjusted to account for the various prevailing conditions. The equation for estimating saturation flow rate is

$$s = s_0 N f_w f_{HV} f_g f_p f_{bb} f_a f_{RT} f_{LT} \qquad (6\text{-}12)$$

where s = saturation flow rate for the subject lane group, expressed as a total for all lanes in the lane group under prevailing conditions (vehicles/hr green)

TABLE 6-7 Relationship Between Arrival Type and Platoon Ratio

Arrival Type	Range of Platoon Ratio, R_R	Progression Quality
1	≤0.5	Very poor
2	>0.50 and ≤0.85	Unfavorable
3	>0.85 and ≤1.15	Random arrivals
4	>1.15 and ≤2.00	Highly favorable
5	>2.00	Exceptional

Source: Highway Capacity Manual, Transportation Research Board, 1994.

S_0 = ideal saturation flow rate per lane, usually 1,900 passenger cars per hour of green time per lane

N = number of lanes in the lane group

f_w = adjustment factor for lane width; 12-ft lanes are standard; given in Table 6-8

f_{HV} = adjustment factor for heavy vehicles in the traffic stream, given in Table 6-9

f_g = adjustment factor for approach grade, given in Table 6-10

f_p = adjustment factor for existence of a parking lane adjacent to the lane group and parking activity in that lane, given in Table 6-11

f_{bb} = adjustment factor for blocking effect of local buses stopping within the intersection area, given in Table 6-12

f_a = adjustment factor for area type, given in Table 6-13

f_{RT} = adjustment factor for right turns in the lane group

f_{LT} = adjustment factor for left turns in the lane group

TABLE 6-8 Adjustment Factor for Lane Width

Lane width, ft	8	9	10	11	12	13	14	15
Lane width factor, f_w	0.867	0.900	0.933	0.967	1.000	1.033	1.067	1.100

Source: Highway Capacity Manual, Transportation Research Board, 1994.

TABLE 6-9 Adjustment Factor for Heavy Vehicles

Percent heavy vehicles, % HV	0	2	4	6	8	10	15	20	25
Heavy vehicle factor, f_{HV}	1.000	0.980	0.962	0.943	0.926	0.909	0.870	0.833	0.800

Source: Highway Capacity Manual, Transportation Research Board, 1994.

TABLE 6-10 Adjustment Factor for Grade

	Downhill			*Level*		*Uphill*	
Grade, %	−6	−4	−2	0	+2	+4	+6
Grade factor, f_g	1.03	1.02	1.01	1.00	0.99	0.98	0.97

Source: Highway Capacity Manual, Transportation Research Board, 1994.

TABLE 6-11 Adjustment Factor for Parking, f_p

Number of Lanes in Lane Group	*No Parking*	*Number of Parking Maneuvers per Hour, N_m*				
		0	*10*	*20*	*30*	*40*
1	1.000	0.900	0.850	0.800	0.750	0.700
2	1.000	0.950	0.925	0.900	0.875	0.850
3	1.000	0.967	0.950	0.933	0.917	0.900

Source: Highway Capacity Manual, Transportation Research Board, 1994.

TABLE 6-12 Adjustment Factor for Bus Blockage, f_{bb}

Number of Lanes in Lane Group	Number of Buses Stopping per hour, N_B				
	0	*10*	*20*	*30*	*40*
1	1.000	0.960	0.920	0.880	0.840
2	1.000	0.980	0.960	0.940	0.920
3	1.000	0.987	0.973	0.960	0.947

Source: Highway Capacity Manual, Transportation Research Board, 1994.

TABLE 6-13 Adjustment Factor for Area Type

Type of Area	Factor f_a
Central business district	0.90
All other areas	1.00

Source: Highway Capacity Manual, Transportation Research Board, 1994.

Adjustment factors for turning movements are not included here but may be found in Ref. 3.

The level of service for signalized intersections is defined in terms of delay. Specifically, level of service is based on the average stopped delay per vehicle for a 15-min analysis period, as specified in Table 6-14.

The *Highway Capacity Manual* (*3*) includes an empirical formula for the estimation of average stopped delay. The estimated stopped delay depends on the traffic volume, the capacity, the cycle length, and the effective green time for the lane group being considered. The first term in the formula accounts for the *uniform delay,* assuming uniform arrivals and stable flow. It includes a "delay adjustment factor" that accounts for quality of progression and type of controller. The second term in the delay equation accounts for the *incremental delay,* that is, the delay due to nonuniform arrivals and cyclic failures. This term includes an "incremental delay calibration term," which accounts for the arrival type and degree of platooning. Further discussion of the calculation of stopped delay is beyond the scope of this book but may be found in Ref. 3.

TABLE 6-14 Level of Service Criteria for Signalized Intersections

Level of Service	Stopped Delay per Vehicle (sec)
A	<5.0
B	>5.0 and ≤15.0
C	>15.0 and ≤25.0
D	>25.0 and ≤40.0
E	>40.0 and ≤60.0
F	>60.0

Source: Highway Capacity Manual, Transportation Research Board, 1994.

EXAMPLE 6-4 **Capacity of a Signalized Intersection** (a) Determine the capacity of the south approach lanes for a signalized intersection given the following conditions:

> 2 approach lanes, 3.4 m (11 ft) wide
> 10 percent heavy vehicles
> 2 percent downgrade along the south approach
> location in the central business district
> no buses stopping in the intersection
> no parking permitted
> through traffic only
> cycle length = 60 sec
> green ratio = 0.45

(b) If the average stopped delay is 12 sec, what is the level of service?

> **Solution to Part a.** By Eq. 6-12, and using Tables 6-8 through 6-13,

$$s = 1,900 \times 2 \times 0.967 \times 1.01 \times 0.90$$
$$s = 3,036 \text{ vehicles/hr of green}$$

By Eq. 6-11,

$$c_i = 3,036 \times 0.45 = 1,366 \text{ vehicles/hr}$$

> **Solution to Part b.** From Table 6-14, the approach operates at level of service B.

THE HIGHWAY ACCIDENT PROBLEM

Since 1900 more than 2 million deaths have resulted from motor vehicle accidents, far exceeding the number of Americans killed in all of the nation's wars combined. In 1992 motor vehicle accidents in the United States produced 40,300 deaths, 2.2 million disabling injuries, and economic losses of $156.6 billion (8). Traffic accidents are the fourth leading cause of death in the country, exceeded only by heart disease, cancer, and stroke. For persons between the ages of 1 and 24, motor vehicle accidents are the leading cause of death.

The highway accident problem is a complex one that is viewed differently by different groups. The police and human factors specialists tend to see it largely as an enforcement problem. Highway engineers view it in terms of needed improvements in highway design and traffic control. Motor vehicle manufacturers and the medical community recognize the need to "package" vehicle occupants in more crash-resistant vehicles. Emergency medical specialists see the needs for better first aid of the of the injured and faster access to hospital trauma units. Because of these different perspectives, there are differences of opinion concerning the programs of action most needed to reduce highway losses.

6-7 Accident Loss Factors

It is helpful to think of an accident as an event comprised of nine parts, called crash factors. One may visualize an accident in three phases: precrash, at-crash, and postcrash. Within each phase, one or more human, vehicular, and environmental factors may contribute to the initiation of the crash or alter its consequences. Examples of the nine classes of loss factors are shown in Table 6-15.

These factors may be arranged in the form of a nine-cell matrix. Countermeasures designed to reduce highway deaths and injuries may be focused on any one or a combination of the cells that comprise the loss factor matrix. Precrash countermeasures are concerned with the prevention of accidents. At-crash countermeasures attempt to reduce injuries by controlling the energy of crashes. Postcrash programs are designed to limit the losses by improving first-aid and other emergency procedures. This part of the chapter is concerned primarily with precrash factors, that is, those aspects of the driver, the vehicle, and the roadway environment that explain the occurrence of the accident event.

6-8 The Nature of Vehicular Crashes

One of the most striking features of scientific accident causation data is the wide variety of human, vehicular, and roadway factors that contribute to vehicular crashes. Many accidents—perhaps one-third to one-half of all crashes—are uncomplicated events that can be explained by the presence of a single human, vehicular, or environmental factor. These factors range from simple lapse of driver attention to the relatively rare case of a driver who uses the automobile as an instrument of suicide.

Many crashes, however, are complex events that are triggered by two or more factors. For example, consider a crash involving a driver driving a vehicle with worn tires at an excessive speed along a wet and slippery roadway. The combined presence of these factors could cause a crash. Removal of any of the factors—the excessive speed, the slippery roadway, or the worn tires—might have prevented the crash from occurring.

For many years, traffic safety specialists have maintained that 90 percent of all accidents are due to improper driving. It appears that this conclusion, which is based on analyses of police accident reports, oversimplifies a complex problem and exaggerates the role of driver behavior. Nevertheless, multidisciplinary research has indicated that human factors are the most prevalent type of contributing factors. It is believed, however, that not more than half of traffic accidents can be attributed to human factors alone.

Research has shown that a substantial percentage of crashes results from a

TABLE 6-15 Examples of Nine Classes of Accident Loss Factors

Designation	Example
Precrash	
Human	Driver fell asleep
Vehicle	Brake failure
Environment	Slippery roadway surface
At-crash	
Human	Seat belts improperly worn
Vehicle	Structural weakness of side of vehicle
Environment	Unyielding signpost near pavement
Postcrash	
Human	Bystanders took improper first-aid action
Vehicle	Vehicle not equipped with a fire extinguisher
Environment	Emergency telephone not available

variety of driver malfunctions and conditions. Driver inattention or distraction, excessive speed, and violation of other safe driving practices appear to be recurring contributing factors. Many crashes can be explained by some form of driver impairment such as fatigue, emotional stress, physical illness, or failure to wear needed corrective lenses. By far the most prevalent form of driver impairment is alcohol intoxication.

Vehicular factors that contribute to crashes are more commonly classified as vehicular conditions (e.g., slick tires) rather than malfunctions or catastrophic failures due to poor design. Defective brakes (especially gross brake failure) and tires (especially inadequate tread depth and underinflation) are frequently cited as contributors to crashes. Vehicular factors are rarely solely responsible for an accident but usually occur jointly with human and/or environmental factors.

Environmental factors commonly interact with some other factor (human or vehicle) to produce an accident. Although a crash may result from a natural disaster or from a malfunction in the highway or its appurtenances, these events are extremely rare. Typical exampes of environmental factors that contribute to crashes are wet roadways, missing or improper traffic control derives, and poor geometric design.

6-9 Conceptualization of an Accident

Multidisciplinary research has shown that traffic accidents are caused by diverse human, vehicular, and environmental factors that often interact in a complex way to trigger the initiation of the event. A fairly wide tolerance is permissible (not accident causative) in any one component of the human/machine/environment system. Even a drunk may successfully manipulate an automobile under certain conditions, such as down a wide street with which he or she is familiar and on which there are no other vehicles or roadside hazards. Conversely, a skilled driver may avoid an unexpected hazard presented either by vehicle failure or by the environment. Blumenthal (9) has conceptualized the variability in tolerence in terms of driver performance and system demands, both of which fluctuate unpredictably over a period of time. As Figure 6-11 illustrates, accidents occur at

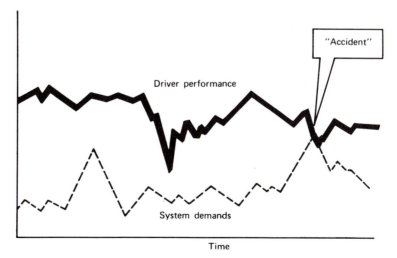

FIGURE 6-11 A concept of a traffic accident.

some point in time when the system demands exceed the performance capability of the driver.

6-10 Current Accident Trends and Concerns

In recent years significant advances have been made in improving highways for safety and in designing more crashworthy vehicles. Numerous and expensive programs aimed at improving the driving function have been sponsored by the federal government and implemented by the state governments. Strong evidence exists that these initiatives have yielded positive results. This is best demonstrated by the fatality rate, which fell to an all-time low of 1.1 per 100 million vehicle-kilometers (1.8 per 100 million vehicle-miles) of travel in 1992, approximately 65 percent of the rate for 1982 (*8*).

In fact, according to the Insurance Institute for Highway Safety (IIHS), most categories of motor vehicle deaths have declined (*10*). In 1993 pedestrian deaths had declined 30 percent and motorcycle deaths had gone down 53 percent, compared to 1980. Bicyclist deaths had decreased 20 percent since 1975, while large truck occupant deaths had declined 55 percent after peaking in 1978. Passenger vehicle occupant deaths declined but at a lower rate than other categories, approximately 3 percent during the period 1975–1993.

The death rate per million registered passenger vehicles 1 to 3 years old decreased from 258 in 1978 to 157 in 1993. Rates decreased for every type and size of passenger vehicle, from the smallest cars to the largest pickups and utility vehicles (*10*).

By the mid-1990s, traffic safety specialists were focusing attention on four areas of special concern: (1) alcohol impairment, (2) large trucks, (3) roadside hazards, and (4) the effects of vehicular crashes on the elderly.

Alcohol Impairment

The National Highway Traffic Safety Administration (NHTSA) reported that alcohol was involved in 45 percent of fatal crashes in 1992 (*11*). The incidence of alcohol impairment in fatal crashes tends to be greatest in late night and early morning accidents, when 80 to 90 percent of all fatally injured drivers are reported to have been drinking. It is estimated that alcohol is involved in about 19 percent of serious injury accidents and about 6 percent of property-damage accidents (*8*).

Under strong pressure from citizens' groups and the general public, legislatures in many states have increased the penalties for driving under the influence of alcohol. As a result, alcohol impairment in 1993 was not as serious as it had been in 1980. In that year about half of all fatally injured drivers had blood alcohol concentrations (BACs) of 0.10 percent or more; by 1993 only 36 percent of fatally injured drivers had BACs of 0.10 percent or more (*10*).

Large Trucks

For many years researchers have understood the relationships between vehicle size and injury severity. Clear evidence exists that when two vehicles collide, occupants of the larger vehicle are far less likely to be seriously injured or killed than those traveling in the smaller vehicle. The hazards associated with vehicle size are most prominent in collisions that involve a large truck and an automobile. In such crashes, automobile occupants are 30 times as likely to be killed as occupants of the truck. Deregulation of the trucking industry and the relaxation of restrictions on truck weight and size are likely to increase accident losses. Combination trucks with 5 axles or more accounted for 8 percent of the traffic volume on rural interstates in 1970; those vehicles accounted for 14 percent of the total volume in 1992. These larger vehicles have increased not only in number but also in size. Loadings on the rural interstates quadrupled between 1970 and 1992 (*12*).

Highway and traffic engineers face the difficult challenge of providing a safe driving environment for drivers of an ever-widening array of vehicle types and sizes.

Roadside Hazards

Many crash deaths involve collisons with objects fixed in place along the roadside, such as trees and utility poles. Deaths in such crashes decreased from a peak of 15,232 in 1980 to 11,300 in 1993, but during this time roadside hazard crash deaths accounted for a steady 25–30 percent of all motor vehicle deaths (*10*). The pattern of such crashes also remained virtually unchanged. Nearly all such crashes involve one vehicle, about a third involve the vehicle rolling over, and about a third involve occupant ejection (*10*). Approaches to ameliorating the roadside hazard problem are discussed in Chapter 8.

Traffic Safety and the Elderly

People over the age of 65 represent a small proportion of those who are involved in crashes; however, when they are involved, they are at greater risk of death or serious, prolonged injury than are young people. Older drivers are more likely to be driving the struck rather than the striking vehicle in multiple-vehicle accidents and tend to be involved in fewer single-vehicle accidents than younger drivers. Efforts to decrease crash-related injuries and deaths among the elderly should therefore be directed to providing better protection for them in the crash and providing improved emergency medical services and hospital care during the postcrash period.

PROBLEMS

6-1 Three cars travel over a 60-m section of highway at constant speeds of 19, 20, and 25 m/sec. Compute the time-mean and space-mean speeds for this condition.

6-2 Three cars travel over a 200-ft section of highway at constant speeds of 35, 40, and 45 ft/sec. Compute the time-mean and space-mean speeds for this condition.

6-3 For the conditions described in Problem 6-1, estimate the variance about the space-mean speed.

6-4 For the conditions described in Problem 6-2, estimate the variance about the space-mean speed.

6-5 The estimated future average daily traffic for a rural highway is 8,000. Assume that the relationship between peak hourly flows and *ADT* is as shown in Figure 6-2. Estimate the design hourly volume. What would be the effect of using the 10th-highest hour for design purposes? The 50th-highest hour?

6-6 Arrivals at a parking lot are assumed to follow the Poisson distribution. The average arrival rate is 2.8 per minute. What is the probability that during a given minute no cars will arrive? What percentage of the time will no cars arrive?

6-7 A street has an hourly volume of 360 vehicles. A pedestrian requires a gap of at least 10 sec to cross. Assuming headways can be described by a negative exponential distribution, what is the probability that the pedestrian will have to wait to cross? Comment on how the negative exponential distribution fails to describe headways realistically.

6-8 Compute the slope of the volume-density curve shown in Figure 6-6 at points *A, B,* and *C*. What are the units of the slope? Discuss the meaning of the slope of this curve. (See Ref. 6.)

6-9 Compute the density corresponding to the maximum volume for the *q-k* curve shown in Figure 6-7.

6-10 Estimate the service flow rate for level of service B for a 6-lane freeway (3 lanes in each direction) with a 3.7-m (12-ft) lines and obstructions 1.8 m (6 ft) from the edge of pavement on 1 side only. The section is to accommodate 12 percent heavy trucks and buses and 2 percent recreational vehicles. Assume the adjustment factor for the driver population, f_P, is 0.95. The design speed is 120 km/hr (75 mph), and the section has a plus 3 percent grade that is 1.1-km (0.7-mile) long.

6-11 Estimate the capacity of the freeway segment described in Problem 6-10.

6-12 Estimate the capacity of the approach lanes of a signalized intersection located in the central business district given the following conditions:

2 approach lanes 3.0-m (10 ft) wide

8 percent heavy vehicles

+4 percent grade along the approach

30 buses stopping per hour

20 parking maneuvers per hour

through traffic only

cycle length = 65 sec

green ratio = 0.50

6-13 If the average stopped delay per vehicle for the intersection described in Problem 6-12 is 20 sec, what is the level of service?

6-14 Consult with the state office of highway safety and prepare a report on the approximate annual costs of the following types of safety countermeasures. Describe quantitatively the benefits of these programs.

(a) Driver education

(b) Motor vehicle inspection

(c) Traffic enforcement

REFERENCES

1. Wardrop, J. G., Some Theoretical Aspects of Road Traffic Research, *Proceedings of the Institution of Civil Engineers,* Pt. 2, Vol. I, pp. 325–362 (1952).

2. Oppenlander, Joseph C., Multivariate Analysis of Vehicular Speeds, *Highway Research Record No. 35,* p. 63 (1963).

3. *Highway Capacity Manual,* Special Report 209, Third Edition, Transportation Research Board, Washington, DC (1994).

4. *Poisson and Other Distributions in Traffic,* Eno Foundation for Transportation, Westport, CT (1971).

5. *Traffic Flow Theory, A Monograph,* Transportation Research Board Special Report No. 165 (1975).

6. Greenshields, B. D., A Study of Traffic Capacity, *Proceedings of the Highway Research Board* Vol. 14, pp. 448–477 (1934).

7. *Highway Capacity Manual,* Bureau of Public Roads, U.S. Department of Commerce, Washington, DC (1950).

8. *Accident Facts, 1993 Edition,* National Safety Council, Chicago, IL (1993).

9. Blumenthal, M., Problem Definition: The Driving Task in the System Context, presented at the American Psychological Association, San Francisco (1968).

10. *Status Report,* Vol. 29, No. 9, Insurance Institute for Highway Safety, Arlington, VA (August 20, 1994).

11. *Fatal Accident Reporting System 1992,* National Highway Traffic Safety Administration, Washington, DC (1993).

12. *Highway Statistics 1992,* Federal Highway Administration, Washington, DC (1993).

CHAPTER 7

GEOMETRIC DESIGN OF HIGHWAYS

This chapter describes the criteria, standards, and engineering procedures used to design principal elements of the highway alignment, highway cross sections, and adjacent roadside environment. Development of a comprehensive highway design focuses on the establishment of a travel lane configuration, alignment location, and all dimensions related to the highway cross section. A three-dimensional physical location is determined through calculation of a horizontal and vertical alignment of the highway centerline, based on a variety of operational considerations. The results of these activities are referred to as the geometric design *and represent all visible features of a highway or street. The first and major portion of the chapter deals with the design of motor-vehicle facilities. Specific design elements are described and discussed with respect to design methodology. The chapter concludes with a brief discussion of design for bicycle and pedestrian facilities.*

DESIGN STANDARDS

Geometric design practices of the state highway and other designing agencies are not entirely uniform on a national basis. A considerable variation exists in the laws of the various states, which serves to limit the size and weight of motor vehicles. Differences also exist in the financing ability of the various governmental agencies, and these among other policy issues significantly influence the designer's decisions and modify the implementation of wholly uniform design standards. Differences in experience and the interpretation of research also contribute to variation in design practices. Furthermore, differences in local conditions among regional factors such as terrain, weather conditions, and available construction materials affect standards and design practices on a state-by-state basis.

These differences are allowed and tolerated by the Federal Highway Administration as unavoidable and are accepted when approval is requested on plans developed by the various states for federally funded highway improvements. The strongest force tending to standardize these differences lies in the numerous and diverse technical committees of the American Association of State Highway and Transportation Officials (AASHTO). All state highway agencies and the Federal Highway Administration hold membership in this association and join in the deliberations of its technical committees. Upon approval by a required majority, a standard is declared adopted and becomes, in effect, a guide for all members of the association.

Each state highway agency typically develops its own versions of specifications, standard drawings, and plan preparation procedures based on the national guidelines adopted by AASHTO. The establishment and distribution of requirements for the design of highway facilities serves to ensure that uniform roadways are designed and constructed on a statewide basis. This statewide uniformity facilitates the understanding of engineering designs by road building contractors and significantly contributes to the implementation of consistent construction methods.

DESIGN CRITERIA

7-1 Design Controls and Criteria

The elements of design are influenced by a wide variety of design controls, engineering criteria, and project-specific objectives. Such factors include the following:

- functional classification of the roadway
- projected traffic volume and composition
- required design speed
- topography of the surrounding land
- capital costs for construction
- agency funding mechanisms
- human sensory capacities of roadway users
- vehicle size and performance characteristics
- traffic safety considerations
- public involvement, review, and comment
- environmental considerations
- right-of-way impacts and costs

These considerations are not, of course, completely independent of one another. The functional class of a proposed facility is largely determined by the volume and composition of the traffic to be served. It is also related to the type of service that a highway will accommodate and the speed that a vehicle will travel while being driven along a highway. For a given class of highway, the choice of design speed is governed primarily by the surrounding topography, regional importance within the larger highway network, magnitude of related construction impacts, and capital costs associated construction of the highway project. Once a design speed is chosen, many of the elements of design (e.g., horizontal and vertical alignment, shoulder width, and side slopes) may be established on the basis of fundamental human sensory capabilities, vehicle performance, and other related operating characteristics described in Chapter 5.

The design features of a highway influence all visible features and directly affect its capacity and traffic operations, its safety performance, and its social acceptability to highway users, owners of abutting land, and the general public.

Of all the factors that are considered in the design of a highway, the principal design criteria are traffic volume, design speed, vehicle size, and vehicle mix. Each of these criteria is discussed in more detail along with other related design considerations in the following paragraphs.

7-2 Relationship of Traffic to Highway Design

The major traffic elements that influence highway design are

- average daily traffic (ADT)
- design hour volume (DHV)
- directional distribution (D)
- percentage of trucks (T)
- design speed (V)

As indicated in Chapter 6, ADT is a fundamental measure of traffic flow. To the designer, the most significant measure of traffic volume is the design hour volume, a two-way value, which may be determined by multiplying the ADT by a percentage representative of the amount of traffic occurring during the peak hour during an average weekday. This percentage, K, is typically 8 to 12 percent for urban facilities and 12 to 18 percent for rural facilities. The directional distribution (D) is the one-way volume in the predominant direction of travel, expressed a percentage of the volume in the two-way design hour volume. For rural roads, D ranges from 55 to 80 percent and typically is about 67 percent.

Composition of traffic (T) is usually expressed as the percentage of trucks (exclusive of light delivery trucks) present in the traffic flow during the design hour. That percentage typically varies from about 5 to 10 percent. In urban areas, the percentage of trucks traveling within the overall flow of traffic during peak hours tends to be considerably less than percentages on a daily basis (*1*).

> The objective in highway design is to create a highway of appropriate type with dimensional values and alignment characteristics such that the resulting design service flow rate is at least as great as the traffic flow rate during the peak 15-minute period of the design hour, but not enough greater as to represent extravagance or waste. Where this objective is accomplished, a well balanced, economical highway system will result (*1*).

Roadway Capacity

It will be recalled from Chapter 6 that the traffic volumes that can be served at each level of service are referred to as "service volumes." Once a level of service has been chosen for a particular project design, the corresponding service volume logically becomes the design service volume. This implies that if the traffic volume using this facility exceeds that value, the operating conditions will be inferior to the level of service for which the roadway was designed (*1*). The level of service appropriate for the design of various types of highways located within representative surrounding terrain conditions is shown in Table 7-1.

The configuration and number of travel lanes to be provided for a highway design have a significant impact on the level of service. An evaluation of the estimated volume of traffic anticipated to occur during the design year is useful in determining the number of travel lanes required to achieve the desired level of service operation. A more extensive capacity analysis, based on data and procedures given in Ref. 2, may be required to identify potential bottlenecks and to facilitate the design of intersections, freeways, and grade-separated interchanges.

TABLE 7-1 Guide for Selection of Design Levels of Service

| Highway Type | Type of Area and Appropriate Level of Service | | | |
	Rural Level	Rural Rolling	Rural Mountainous	Urban and Suburban
Freeway	B	B	C	C
Arterial	B	B	C	C
Collector	C	C	D	D
Local	D	D	D	D

Note: General operating conditions for levels of service (source: Ref. 2):

A—free flow, with low volumes and high speeds.

B—reasonably free flow, but speeds beginning to be restricted by traffic conditions.

C—in stable flow zone, but most drivers restricted in freedom to select their own speed.

D—approaching unstable flow; drivers have little freedom to maneuver.

E—unstable flow; may be short stoppages.

Source: From *A Policy on Geometric Design of Highways and Streets,* copyright 1990. American Association of State Highway and Transportation Officials, Washington, DC. Used by permission.

Design Speed

The assumed design speed for a highway may be considered as "the maximum safe speed that can be maintained over a specified section of highway when conditions are so favorable that the design features govern" (*1*). The choice of design speed will depend primarily on the surrounding terrain and the functional class of the highway. Other factors determining the selection of design speed include traffic volume and composition, costs of right-of-way and construction, and aesthetic considerations.

Design speeds typically range from 30 to 120 km/hr (20 to 75 mph), and intermediate values are chosen in increments of 10 km/hr; 40, 50, 60 km/hr, and so forth. Where feasible, a constant design speed should be used in the design of a highway of substantial length. Where changes in terrain or other conditions dictate a change in design speed, such change should be made over a sufficient distance to permit drivers to change speed gradually. The changes in design speed should, of course, be indicated by appropriate traffic control devices.

Design Designation

A tabulation of the major traffic controls used for the design of a highway are collectively referred to as the *design designation*. Examples of such designations are shown below. The tabulation on the left is for a two-lane highway; the traffic data on the right for a multilane highway.

			Control of access = full
ADT (1995)	= 2,500	*ADT* (1995)	= 10,200
ADT (2015)	= 5,200	*ADT* (2015)	= 21,200
DHV	= 720	*DHV*	= 2,950
D	= 65%	*D*	= 60%
T	= 12%	*T*	= 8%
V	= 100 km/hr	*V*	= 110 km/hr

The highway design designation is normally placed on the cover sheet of the plan set for proposed improvements in order to indicate to all those handling and reviewing the plans the traffic values used in establishing the basis for the project design.

7-3 Design Vehicle

The dimensions of the motor vehicles that will utilize the proposed facility also influence the design of a roadway project. The width of the vehicle naturally affects the width of the traffic lane; the vehicle length has a bearing on roadway capacity and affects the turning radius; the vehicle height affects the clearance of the various structures. Vehicle weight, as mentioned previously, affects the structural design of the roadway.

The American Association of State Highway and Transportation Officials (*1*) recommends that 15 design vehicles be used for determining the controls for geometric design of a highway. These vehicles range in size from the passenger car to a large semitrailer-full trailer combination. Normally the design engineer will select for design the largest vehicle that is expected to use the roadway facility in significant numbers on a daily basis during the design year. The dimensions of these design vehicles are given in Table 7-2 and are exemplified by Figure 7-1, for a WB-50 design vehicle.

Vehicle performance for the 15 classes of motor vehicles also has a considerable impact on the design of a highway facility. Acceleration and deceleration rates are important parameters taken into account in the development of several numerical equations discussed later in this chapter. One obvious consideration of particular significance when designing a two-lane roadway is an evaluation of truck slow-down speeds. It is highly desirable to limit the speed differential between trucks and passenger cars to a desirable maximum of 15 km/hr (10 mph). Factors affecting truck slow-downs are the steepness of a roadway upgrade and the total length of the upgrade. This relationship is shown in Figure 7-2.

CROSS SECTION ELEMENTS

One of the most important parts of developing a roadway design focuses on the selection and configuration of the elements that comprise the roadway cross section. Once defined, these elements are often collectively referred to as the *typical sections* or *templates* when used within the context of most computer design programs. Specifically, aspects of the cross section directly relate to the number of travel lanes to be provided and the width and location of shoulders, medians, slopes, embankments, and ditches. Each of these elements will be discussed along with other related design considerations in the following paragraphs.

7-4 Highway Travel Lanes

The width of the surfaced road and number of lanes should be adequate to accommodate the type and volume of traffic anticipated and the assumed design speed of vehicles. Roads presently in use have traditionally been separated into generalized categories that include two-lane, three-lane, multilane undivided, multilane divided, and limited-access highways.

TABLE 7-2 Design Vehicle Dimensions

Design Vehicle Type	Symbol	Overall			Overhang		WB_1	WB_2	S	T	WB_3	WB_4
		Height	Width	Length	Front	Rear						
Passenger car	P	1.3	2.1	5.8	0.9	1.5	3.4					
Single Unit truck	SU	4.1	2.6	9.1	1.2	1.8	6.1					
Single Unit bus	BUS	4.1	2.6	12.1	2.1	2.4	7.6					
Articulated bus	A-BUS	3.2	2.6	18.3	2.6	2.9	5.5		1.2[a]	6.1[a]		
Combination trucks												
Intermediate semitrailer	WB-12	4.1	2.6	15.2	1.2	1.8	4.0	8.2				
Large semitrailer	WB-15	4.1	2.6	16.7	0.9	0.6	6.1	9.1				
Double bottom semitrailer—												
full trailer	WB-18	4.1	2.6	19.9	0.6	0.9	3.0	6.1	1.2[b]	1.6[b]	6.4	
Interstate semitrailer	WB-19*	4.1	2.6	21.0	1.2	0.9	6.1	12.8				
Interstate semitrailer	WB-20**	4.1	2.6	22.5	1.2	0.9	6.1	14.3				
Triple semitrailer	WB-29	4.1	2.6	31.0	0.8	1.0	4.1	6.3	1.0[d]	1.8[d]	6.6	6.6
Turnpike double semitrailer	WB-35	4.1	2.6	35.9	0.6	0.6	6.7	12.2	0.6[c]	1.8[c]	13.4	
Recreational vehicle												
Motor home	MH		2.4	9.1	1.2	1.8	6.1					
Car and camper trailer	P/T		2.4	14.9	0.9	3.0	3.4	6.1	1.5			
Car and boat trailer	P/B		2.4	12.8	0.9	2.4	3.4	4.6	1.5			
Motor home and boat trailer	MH/B		2.4	16.1	1.2	2.4	6.1	4.6	1.8			

* = Design vehicle with 14.6 m trailer as adopted in 1982 STAA (Surface Transportation Assistance Act).
** = Design vehicle with 16.2 m trailer as grandfathered in 1982 STAA (Surface Transportation Assistance Act).
a = Combined dimension 7.3, split is estimated
b = Combined dimension 2.9, split is estimated
c = Combined dimension 2.4, split is estimated
d = Combined dimensions 2.8, split is estimated
WB_1, WB_2, WB_3, WB_4, are effective vehicle wheelbases
S is the distance from the rear effective axle to the hitch point
T is the distance from the hitch point to the lead effective axle of the following unit

Source: From *A Policy on Geometric Design of Highways and Streets 1994*, copyright 1994. American Association of State Highway and Transportation Officials, Washington, DC. Used by permission.

THIS TURNING TEMPLATE SHOWS THE TURNING PATHS OF THE AASHTO DESIGN
VEHICLES. THE PATHS SHOWN ARE FOR THE LEFT FRONT OVERHANG AND THE
OUTSIDE REAR WHEEL. THE LEFT FRONT WHEEL FOLLOWS THE CIRCULAR CURVE,
HOWEVER, ITS PATH IS NOT SHOWN.

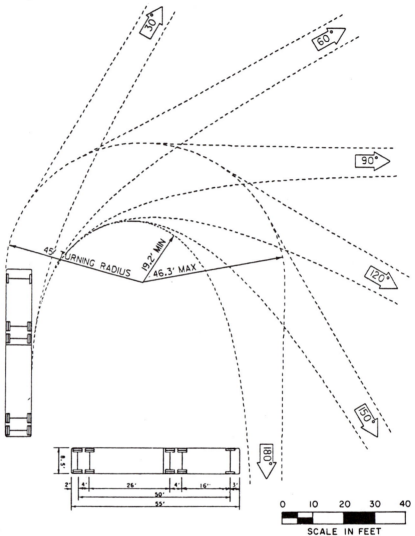

Source: Texas State Department of Highways and Public Transportation

FIGURE 7-1 Minimum turning path for WB-50 design vehicle. (Source: *A Policy on Geometric Design of Highways and Streets,* copyright 1990, American Association of State Highway and Transportation Officials, Washington, DC. Used by permission.)

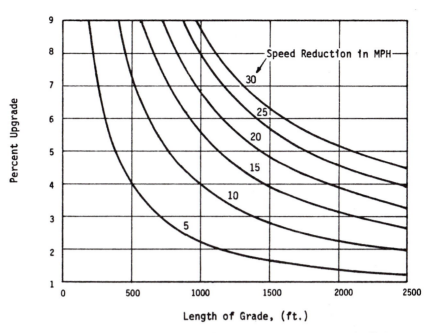

FIGURE 7-2 Critical length of grade for typical heavy truck—300 lb/hp, entering speed = 88 km/hr (55 mph). (Source: *A Policy on Geometric Design of Highways and Streets,* copyright 1990, American Association of State Highway and Transportation Officials, Washington, DC. Used by permission.)

Two-Lane Highways

The bulk of our highway system is composed of two-lane highways, and over 90 percent of the rural roads in the United States are of this type. These two-lane roads vary from low-type roads, which follow the natural ground surface, to high-speed primary highways with paved surfaces and stabilized shoulders. As traffic density, vehicle speeds, and truck widths have increased, two-lane highways have also increased in width from 4.8 m (16 ft) to the current recommended value of 7.2 m (24 ft) with 3-m (10-ft) stabilized or paved shoulders on either side along primary routes. Where lighter traffic volumes are expected, the American Association of State Highway and Transportation Officials recommends a range of minimum surface widths, based on a combination of the design speed and traffic volume magnitude.

The surrounding terrain and ability to pass or overtake a slower-traveling vehicle play a large role in the operational adequacy of this type of roadway facility. Consider a two-lane highway in a rural portion of Florida with long straight sections and no significant uphill grades and carrying a two-way volume of approximately 500 vehicles per hour. Now consider a similar two-lane highway in the mountains of Colorado, where passing is restricted and truck speeds are adversely affected by steep grades. The same type of highway in Florida will provide a better level of service, given an equivalent amount of traffic volume, primarily due to the surrounding conditions.

Three-Lane Highways

Many three-lane roads were built in previous years and are still in use. Their great advantage seemed to stem from an operational improvement over the two-lane road, with only a moderate increase in construction and right-of-way costs. The center lane is either used as two-way center left-turn lane or alternates in the uphill direction as a directional passing lane. The three-lane road does appear to have a high accident rate, and construction of these types of roads is no longer recommended except in special cases.

Four-Lane Highways

On four-lane highways traffic flows in opposite directions on each pair of lanes, and passing is accomplished within the lanes of forward movement of traffic and not in the lanes of opposing traffic. The amount of interaction between opposing traffic flows is substantially reduced, except in the case of left-turning vehicles. A feeling of freedom from opposing traffic generally results in smoother operation and an increase in the capacity per lane over that of two- and three-lane highways. Four-lane highways provide at least four times the capacity of a two-lane highway for the same assumed design speed.

Although the four-lane highway is the basic multilane type, traffic volume sometimes warrants the use of highways with six, eight, or more lanes, particularly in highly developed urban areas. Eight-lane highways, with four lanes in each direction, are typically the upper limit for multilane highways due to a variety of problems related to multiple lane changes, which adversely affect capacity and safety. If more than four directional travel lanes are required, channelization, frontage roads, or other directional traffic flow measures are often added to the highway to improve the efficiency and traffic-handling capacity of the facility.

The undivided multilane highway does, however, appear to have an accident rate higher than that of the two-lane highway. This would appear to be due to higher traffic volumes with traffic frictions from cars traveling both in the same direction and opposing directions and from conflicts with left-turning vehicles.

It is generally agreed that for rural roads in locations where design speeds are high, and when traffic volumes are sufficient to require multilane construction, traffic separation is desirable.

7-5 Divided Highways

In order to provide positive protection against conflict of opposing traffic, highways are frequently divided by a median strip. (See Figure 7-3.) On such highways, lane widths should be a minimum of 3.6 m (12 ft), with 4-m (13-ft) lanes provided where many large truck combinations are anticipated. It is highly desirable that all multilane highways should be divided. The width of these median strips varies from 1.2 m (4 ft) to 18 m (60 ft) or more. A median strip less than 1.2 m (4 ft) to 1.8 m (6 ft) in width is considered to be little more than a centerline stripe, and its use, except for special conditions, should be discouraged. Where narrow medians must be used, many agencies install median barriers to physically separate opposing flows of traffic and minimize the potential for head-on collisions. Medians of 4.2 m (14 ft) to 4.8 m (16 ft) have been used and are sufficient to provide most of the separate advantages for opposing traffic while permitting the inclusion of a median lane at crossroads for left-turn movements; however, medians 4.8 m (16 ft) to 18 m (60 ft) wide and greater are now recommended when surrounding conditions permit.

The median should also be of sufficient width to maintain vegetation and to support low-growing shrubs that reduce the headlight glare of opposing traffic.

FIGURE 7-3 A divided highway in Connecticut. (Courtesy Connecticut Department of Transportation.)

Median strips at intersections should receive careful consideration and should be designed to permit necessary turning movements, which may require single or even dual left turn lanes. Many agencies design the width of medians so that additional travel lanes can be added in the future, if required, within the limits of the median. For rural divided highways a median width of 13.4 m (44 ft) is considered desirable to provide one 3.6 m (12 ft) travel lane in each direction, while leaving a 6-m (20-ft) wide median strip.

Divided highways need not be of a constant cross section. The median strip may vary in width; the roads may be at different elevations; and superelevation may be applied separately on each set of lanes. In rolling terrain, substantial savings may be effected in construction and maintenance costs by this variation in the design. This type of design also tends to eliminate the monotony of a constant width and equal grade alignment.

Where it is necessary to narrow the median strip, or where intersections make it desirable to widen the median strip in tangent alignment, the change may be effected by reverse curves of 1750-m radius (1° curvature), which can be provided without superelevation or transitions. Where such changes in width on curves are desirable, they should be accomplished if possible by changing the curvature of one or both sides of the roadway alignment.

7-6 Limited-Access Highways

A very important feature of the design of a multilane highway is the control of access from adjacent property.

A limited-access highway may be defined as a highway or street especially designed for through traffic, to which motorists and owners of abutting properties have only restricted right of access. Limited- or controlled-access highways may consist of freeways that are open to all types of traffic or parkways from which all commercial traffic is excluded (*1*). Most of the present expressway systems in the United States have been developed as freeways.

Limited-access highways may be elevated, depressed, or at grade. Many examples of the various types may be found in the United States in both rural and urban areas.

The control of access is attained by limiting the number of connections to and from the highway, facilitating the flow of traffic by separating cross traffic with overpasses or underpasses, and eliminating or restricting direct access by owners by abutting property through the use of frontage roads, which connect to the limited access facility at consistently spaced grade-separated interchanges.

The design of limited-access routes should provide adequate width of right-of-way, adequate landscaping, prohibition of outdoor advertising on the controlled access proper, and control of abutting service facilities such as gas stations, parking areas, and other roadside appurtenances.

In urban areas, the design of a limited-access facility is usually accompanied by the design of frontage roads, parallel to the facility, which serve local traffic and provide access to adjacent land. Such roads may be designed for either one-way or two-way operation. Reasonably convenient connections should be provided between through-traffic lanes and frontages. In general, desirable spacing of access points or interchanges along limited access facilities is 1500 m (1 mile) or greater in urban areas and 4 to 7 km (3 to 5 miles) in rural areas.

7-7 Pavement Crowns

Another element of the highway cross section is the pavement crown, which is the raising of the centerline of the roadway above the elevation of the pavement edges. Pavement crowns have varied greatly throughout the years. On the early low-type roads, high crowns were necessary for good drainage and were commonly constructed at a 4 percent slope rate or more ($\frac{1}{2}$ in. or more per foot). With the improvement of construction materials, road-building techniques, and equipment innovations that permit closer control, pavement crowns have been decreased. Present-day high-type pavements with good control of drainage now have crowns as low as 1 percent slope rate ($\frac{1}{8}$ in. per foot). Low crowns are satisfactory when little or no settlement of the pavement is expected and when the drainage system is of sufficient capacity to quickly remove the water from a traffic lane to prevent a motor vehicle from hydroplaning. When four or more traffic lanes are used, it is desirable to provide a higher rate of crown on the outer lanes in order to expedite the flow of water from the pavement into the gutter or onto adjacent unpaved shoulders.

Crowns may be formed by intersecting tangent lines or by curved lines that emanate from the road centerline. In the latter case, circular arcs of long radii are used, as well as parabolic arcs. It makes little difference which is employed, but the parabolic arc lends itself better to making computations for the initial offsets or coordinates in constructing templates or in setting grades.

7-8 Shoulders

Closely related to the lane width is the width of the shoulders. It is necessary to provide shoulders for safe operation and to allow the development of full traffic capacity. Well-maintained, smooth, firm shoulders increase the effective width of the traffic lane as much as 0.6 m (2 ft), as most vehicle operators drive closer to the edge of the pavement in the presence of adequate shoulders. To accomplish their purpose, shoulders should be wide enough to permit and encourage vehicles to leave the pavement when stopping. The greater the traffic volume, the greater is the likelihood of the shoulders being put to emergency use.

A usable outside shoulder width of at least 3 m (10 ft) and preferably 3.6 m (12 ft) clear of all obstructions is desirable for all heavily traveled and high-speed highways. Inside shoulders often are not as wide. In mountainous areas, where the extra cost of providing shoulders of this width may be prohibitive, or on low-type highways, a minimum width of 1.2 m (4 ft) may be provided; a width of 1.8 to 2.4 m (6 to 8 ft) is preferable. Under these conditions, however, emergency parking pull-outs should be provided at proper intervals. For areas of terrain where guardrails or other vertical elements (such as retaining walls) are required, an additional 0.6 m (2 ft) of shoulder widening should be provided.

The slope of the shoulder should be greater than that of the pavement. A shoulder with a high-type surfacing should have a slope of at least 3 percent ($\frac{3}{8}$ in. per foot). Sodded shoulders may have a slope as high as 8 percent (1 in. per foot) in order to efficiently carry water away from the pavement.

7-9 Guardrail

Guardrail should be provided where fills are over 2.4 m (8 ft) in height, when shoulder slopes are greater than 4:1, in locations where there is sudden change in alignment, and where a greater reduction in speed is necessary. In locations with deep roadside ditches, steep banks, or other right-of-way limitations, it is often necessary to steepen the side slopes and to require the use of guardrail. Where guardrail is used, the width of the shoulders is increased approximately 0.6 m (2 ft) to allow space for placing the posts.

Various types of guardrail are in use at the present time. The most important of these are the W-beam guardrail, the cable guardrail, and the box beam guardrail. These may be installed on wood, steel, or concrete posts. Some agencies have successfuly used the so-called weak post system. In this system, the posts collapse when struck, and the rail deflects and absorbs the energy of the impact. When strong posts are used, the rail should be blocked out or supported away from the post to minimize pocketing of the vehicle.

7-10 Curbs, Curb and Gutter, and Drainage Ditches

The use of curbs is generally confined to urban and suburban roadways. The design of curbs varies from a low, flat, angle-type to a nearly vertical barrier-type curb. In areas where sidewalks are not provided, curbs adjacent to traffic lanes should be low in height and constructed with a flatter vertical angle so as not to create an obstruction. The face of the curb should be no steeper than 45° so that vehicles may drive over the curb without difficulty. This type of curb design is generally referred to as a mountable curb.

Curbs at parking areas and adjacent to sidewalks should be 150 to 200 mm (6 to 8 in.) in height, with a curb face that is nearly vertical. Clearance should be sufficient to clear passenger car fenders and bumpers and to permit the opening of car doors without scraping. Storm water drainage and the ability to accommodate curb inlets will also affect the shape and height of the curb. From a pedestrian's viewpoint, curbs should be limited to one step in height. When a barrier curb is used, it should be offset a minimum of 3 m (10 ft) from the edge of the traffic lane. Figure 7-4 shows some commonly utilized typical curb sections (*1*).

Drainage ditches should be located and shaped to avoid creating a hazard to traffic safety. Under normal conditions, ditches should be low enough to drain the water from under the pavement. A broad, flat, rounded ditch section has been found to be safer than a V-type ditch, which also may be subject to undesirable erosive hydraulic action. The longitudinal gradient or slope of the ditch may vary greatly from that of the adjacent roadway pavement profile grade line.

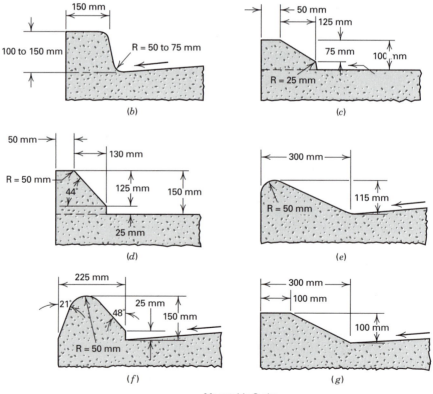

FIGURE 7-4 Typical highway curb sections. (Source: *A Policy on Geometric Design of Highways and Streets,* copyright 1994, American Association of State Highway and Transportation Officials, Washington, DC. Used by permission.)

7-11 Slopes

Side slopes and back slopes may vary considerably depending on soil character-istics and the geographic location of the highway. Well-rounded flat slopes present a pleasing appearance and are most economical to build and maintain. Side slopes of 4:1 are used a great deal in both cut and fill sections up to about 3 m (10 ft) in depth or height, but where the height of cut or fill does not exceed 1.8 m (6 ft), a maximum side slope of 6:1 is recommended. Extremely flat slopes are sometimes used in low-lying areas. As previously mentioned, where guardrail is used, slopes may be as high as 1.5:1. Slopes as high as 1:1 are generally not satisfactory and typically exhibit an expensive long-term maintenance problem. In certain fill sections, special slopes may be built with riprap, mechanically stabilized fabric applications, reinforced concrete cribbing, and various types of retaining walls.

The back slopes in cut areas may vary from 6:1 to vertical in rock sections and to 1.5:1 in normal soil conditions. In general, 2:1 is most commonly used for a back slope in order to remain within the right-of-way width and to reduce the impact on adjacent properties. It is sometimes advisable to have back slopes as flat as 4:1 when side borrow is needed. Slope transtions from cuts to fills should be gradual and should extend over a considerable length of the roadway.

7-12 Right-of-Way

The right-of-way width for a two-lane highway on secondary roads with an annual average daily traffic volume of 400 to 1000 vehicles, as recommended by the American Association of State Highway and Transportation Officials, is 20 m (66 ft) minimum and 25 m (80 ft) desirable. Along the Interstate Highway system minimum widths vary, depending on local conditions, from 46 m (150 ft) without frontage roads and 76 m (250 ft) with frontage roads, to 60 to 90 m (200 to 300 ft) for an eight-lane divided highway without frontage roads. On high-type two-lane highways in rural areas a minimum width of 30 m (100 ft), with 37 m (120 ft) desirable, is recommended. A minimum width of 45 m (150 ft) and a desirable width of 76 m (250 ft) are recommended for divided highways.

Sufficient right-of-way should be acquired in order to avoid the expense of purchasing developed property or the removal of other physical encroachments from the highway right-of-way. A wide section of right-of-way must be given careful consideration for a balanced design. The selection of a width based on minimum or desirable dimensions is typically established with respect to facility type and surrounding conditions. A typical design is illustrated in Figure 7-5.

ROADWAY ALIGNMENT

An ideal and most desirable roadway is one that generally follows the existing natural alignment of the countryside. This is the most economical type of highway to construct, but certain aspects of the design that must be maintained may prevent the designer from following this undulating surface without making considerable adjustments in both the vertical and horizontal directions.

The designer must produce an alignment in which conditions are consistent and uniform to help reduce problems related to driver expectancy. Sudden changes in alignment should be connected with long sweeping curves, and short sharp curves should not be interspersed with long curves of small curvature. The

Typical Cross-Sections

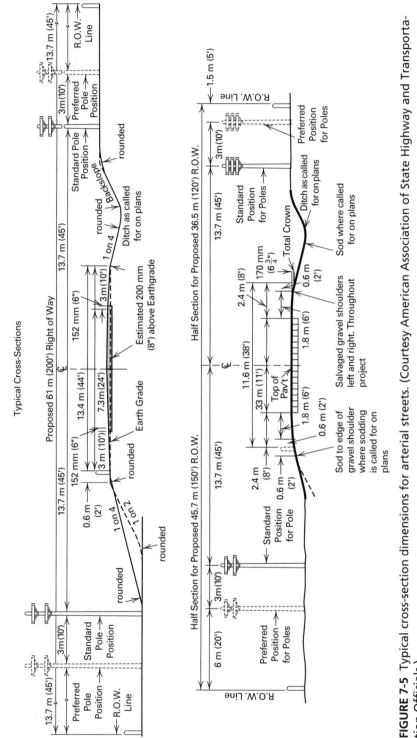

FIGURE 7-5 Typical cross-section dimensions for arterial streets. (Courtesy American Association of State Highway and Transportation Officials.)

ideal highway location is one with consistent alignment, where both vertical grade and horizontal curvature receive consideration and are configured to satisfy limiting design criteria. The optimal final alignment will be that in which the best balance between grade and curvature is achieved.

Terrain has considerable influence on the final choice of alignment. Generally, the topography of the surrounding area is fitted into one of three classifications: level, rolling, or mountainous.

In level country, the alignment is in general limited by considerations other than grade, that is, cost of right-of-way, land use, waterways requiring expensive bridging, existing cross roads, railroads, canals, power lines, and subgrade conditions or the availability of suitable borrow material.

In rolling country, grade and curvature must be carefully considered and to a certain extent balanced. Depths of cut and heights of fill, drainage structures, and number of bridges will depend on whether the route follows the ridges, the valleys, or a cross drainage alignment.

In mountainous country, grades provide the greatest problem, and, in general, the horizontal alignment or curvature is controlled by maximum grade criteria.

7-13 Horizontal Alignment

The physical location of a highway for design and construction is developed and dimensioned with respect to a calculated centerline that is expressed in terms of bearings, distances, curvature, transitions, stations, and offsets. This numerically derived centerline information is collectively referred to as the *geometric alignment.*

Tangent or straight sections of the highway are connected with circular curves to create a flowing and smooth alignment. Most high-speed highways have long tangent distances that are connected by very gradual long sweeping curves. The horizontal alignment is developed to accommodate a given design speed, and typically the alignment dimensions and distances are tabulated in a manner that facilitates construction staking as conducted by a field surveying crew.

7-14 Circular Curves

In the past, traditional U.S. units have focused on describing circular curves in terms of the degree of curvature, which is defined as the central angle subtended by an arc of 100 ft. This is known as the *arc definition.* Implementation of the metric system for construction of highways specifies that all curves be defined based on the length of the radius in meters and in effect drops the degree of curvature term from use in defining circular curves. For the purpose of creating a more useful reference both systems will be presented.

Using the traditional U.S. units and the arc definition, the circumference of a circle bears the same relationship to 360° as a 100-ft arc does to the degree of curve D. Therefore,

$$\frac{2\pi R}{360°} = \frac{100 \text{ ft}}{D} \quad \text{and} \quad R = \frac{5729.58}{D} \tag{7-1}$$

Numerous books on railroad location define the degree of curvature as the central angle subtended by a chord of 100 ft, and some highway departments have also followed this procedure in the past. In the use of long sweeping circular curves for highway design, arc or chord measurements can be considered alike for all curves less than 4° without appreciable error. From a careful examination of tabulated curve data, it has been determined that the following chords may be assumed to be equal to the arc length without appreciable error:

100-ft chords up to 4° curvature

50-ft chords up to 10° curvature

25-ft chords up to 25° curvature

10-ft chords up to 100° curvature

The use of degree curvature and chord lengths in the design of circular curves primarily stems from the methods used in surveying to locate and stake out the highway curves for construction. Typically a surveyor would place his instrument at a point on the tangent where the curve begins and then incrementally turn angles equivalent to D and measure calculated chord distances until the entire length of the curve was located. The length of radius is not as useful in field stake-out activities because the center of the curve is typically located a considerable distance from the construction area.

Under the guidelines of the metric system, circular curves are defined in terms of the radius length R in meters. Other important variables relate to the intersection angle or central angle Δ, the tangent distance T and the length of curve L. A number of other useful variables relating to a circular curve can also be calculated for use in design evaluations. Once two elements of a circular curve are determined, such as R and Δ, all other variables can be calculated. Figure 7-6 identifies a simple highway curve with variable definitions and basic equations, shown in both traditional U.S. units and the metric system. Since the definition of a curve is based on $2\pi R$, once the radius is determined in the desired units (meters or feet), no other conversions are required and all equations can be used in a similar fashion for both the traditional U.S. unit system and the metric system.

Several of the equations pertain to locating a specific point along the curve, given that two of the variables that define the curve are established. Equations related to point location along the curve are very useful during both design and contruction layout activities.

In general, the sharpest desirable curve utilized in open country ranges from a radius of 350 m to 250 m, which equates to a similar range in degree of curvature of 5 to 7°. For convenience, Table 7-3 provides a direct comparison between the curve radius in meters and the degree of curvature. In mountainous areas, desirable curves should be designed with at least a 175-m radius or should not exceed a 10° curvature. It is preferable when surrounding conditions permit to use a radius of curve for highway alignment that exceeds the shortest radius allowed for a given design speed. With respect to this objective, many states limit the radius of curvature to 585 m (3° curvature) along principal highways. However, exceptions to these limitations will occur and subsequently may affect the design speed and posted speed of the proposed highway.

Several variations of the circular curve deserve consideration when developing the horizontal alignment for a highway design. When two curves in the same direction are connected with a short tangent, this condition is known as a "broken back" arrangement of curves. This type of alignment should be avoided except where very unusual topographical or right-of-way conditions dictate otherwise. Highway engineers generally consider the broken back alignment to be unpleasant and awkward and prefer spiral transitions or a compound curve alignment with continuous superelevation for such conditions.

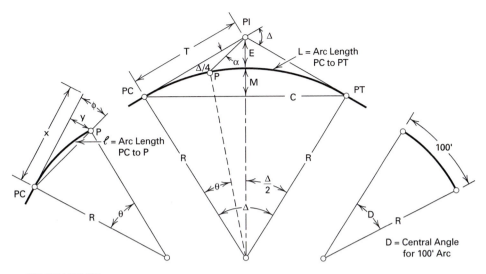

VARIABLES

PC = Point of curvature (Beginning of curve)
PT = Point of tangency (End of curve)
PI = Point of intersection
Δ = Central angle
L = Length of curve (PC to PT) *(2)*
l = Length of arc (PC to P) *(2)*
θ = Central angle for arc length *l*
T = Tangent length (PC to PI & PT to PI) *(2)*
φ = Deflection angle at PC between tangent and chord for P
α = Deflection angle at PI between tangent and line from PI to P
x = Tangent distance from PC to P *(2)*
y = Tangent offset P *(2)*

PI = Point of intersection
D = Degree of curvature *(1)*
R = Radius of curve *(2)*
E = External distance *(2)*
M = Middle ordinate *(2)*
C = Chord length *(2)*

CIRCULAR CURVE EQUATIONS

$$D = \frac{5729.57795}{R} \quad \text{(arc def.)}$$

$$L = \frac{2\pi R \Delta}{360}$$

$$l = \frac{100\,\theta}{D}$$

$$T = R \tan \frac{\Delta}{2}$$

$$E = R\left(\sec \frac{\Delta}{2} - 1\right)$$

$$M = R\left(1 - \cos \frac{\Delta}{2}\right)$$

$$C = 2\,R \sin \frac{\Delta}{2}$$

$$\phi = \frac{\theta}{2} = \frac{lD}{200}$$

For any tangent distance x,

$$y = R - [R^2 - x^2]^{1/2}$$

For any arc length *l*,

$$x = R \sin \theta$$

$$y = R\,(1 - \cos \theta)$$

NOTES: (1) This variable used only for curve definition in traditional US units.
(2) Units for these variables can be expressed in either meters or feet.

FIGURE 7-6 Properties of a simple circular curve.

TABLE 7-3 Metric Curve Radius and Degree of Curvature

Metric Radius (meters)	Traditional U.S. Units Degree of Curvature, D
3,495	0°30'
1,750	1°00'
1,165	1°30'
875	2°00'
700	2°30'
585	3°00'
500	3°30'
440	4°00'
350	5°00'
295	6°00'
250	7°00'
220	8°00'
195	9°00'
175	10°00'
160	11°00'
145	12°00'
135	13°00'
125	14°00'
110	16°00'
95	18°00'
85	20°00'
80	22°00'

Figure 7-7 identifies elements of a typical compound highway curve with variable definitions and basic equations developed for a larger and smaller radius curve, based on the assumption that the radius dimensions R_L and R_S and central angles Δ_L and Δ_S are given or have been previously determined. Compound curves should also be used with caution. Generally speaking, where rugged topography or restricted right-of-way make their use necessary, the radius of the flatter circular arc, R_L, should not be more than 50 percent greater than the radius of the sharper curve, R_S, that is, $R_L < 1.5\ R_S$.

Another important variation of the circular highway curve is the use of reverse curves, which are adjacent curves that curve in opposite directions. The alignment illustrated in Figure 7-8, which shows a point of reverse curvature, PRC, and no tangent separating the curves, would be suitable only for low-speed roads such as those in mountainous terrain. A sufficient length of tangent between the curves should usually be provided to allow removal of the superelevation from the first curve and attainment of adverse superelevation for the second curve. (See discussion of Attainment of Superelevation, Section 7-17.) Generally a separation distance of at least 100 m (328 ft) is required to allow proper transition from one curve to another. Figure 7-8 provides variable definitions and basic equations for a reverse curve configured along two parallel tangent lines.

Numerous factors other than the minimum radius or maximum degree of curvature must be considered when designing a curve alignment for a highway. These additional factors can include superelevation, transitions, widening, connection to side streets, storm water drainage, and impacts on adjacent properties, to list just a few.

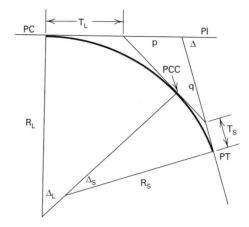

VARIABLES

R_L = Large curve radius *(1)*
R_S = Small curve radius *(1)*
Δ_L = Central angle of large radius curve
Δ_S = Central angle of small radius curve
Δ = Central angle or intersection angle
LT = Long tangent
ST = Short tangent

BASIC EQUATIONS FOR A COMPOUND CURVE (2-CENTERED)

$$\Delta = \Delta_L + \Delta_S$$

$$T_L = R_L \tan \frac{\Delta_L}{2}$$

$$T_L = R_S \tan \frac{\Delta_S}{2}$$

$$LT = T_L + p$$

$$ST = T_S + q$$

$$\frac{p}{\sin \Delta_S} = \frac{T_L + T_S}{\sin (180 - \Delta)} = \frac{q}{\sin \Delta_L}$$

NOTES: (1) Units for these variables can be expressed in either meters or feet.

FIGURE 7-7 Properties of a compound curve.

7-15 Superelevation of Curves

While traveling under open highway conditions along rural highways, most drivers gravitate toward a more or less uniform speed. In the event that the roadway is not designed properly, a vehicle must be driven at a reduced speed for safety as well as comfort of the occupants when moving from a tangent section to a curved section of the highway. This is due to the fact that a centrifugal force is acting on the vehicle that tends to cause an outward skidding away from the center of the curve.

As discussed in Section 7-7, most highways have a crowned pavement section that is provided to facilitate the flow of drainage away from the roadway surface. Along a tangent section of the horizontal alignment both sides of the pavement

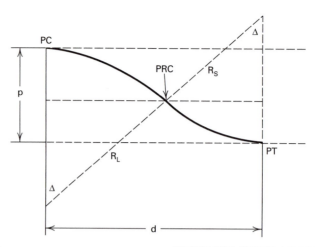

VARIABLES

R_L = Large curve radius *(1)*

R_S = Small curve radius *(1)*

PRC = Point of reverse curve

Δ = Central angle of each curve

p = Offset distance

d = Tangent distance

REVERSE CURVE EQUATIONS

$p = (R_L + R_S)(1 - \cos \Delta)$

$d = (R_L + R_S) \sin \Delta$

$\tan \dfrac{\Delta}{2} = \dfrac{p}{d}$

NOTES:
  (1) Units for these variables can be expressed in either meters or feet.
  (2) For low speed applications, such as in sub-division street design, due to the absence of a tangent between the two opposing curves.

FIGURE 7-8 Properties of a reverse curve.

crown slope away and downward from the centerline of the highway. This is commonly referred to as normal crown. It is apparent that when this type of pavement crown is carried along a curved section, the tendency to slip is retarded on the inside of the curve because of the banking effect of the crown. Conversely, the hazard of slipping is increased on the outside of the curve due to the outward sloping or negative crown of the pavement. In order to overcome this tendency to slip and to maintain average speeds, it is necessary to *superelevate* or bank both sides of the roadway cross section in a manner that helps to counteract the centrifugal force generated while traveling along a horizontal curve.

Analysis of the forces acting on the vehicle as it moves around a curve of constant radius indicates that the theoretical superelevation can be expressed in the following terms:

$$e + f = \frac{v^2}{gR} \qquad (7\text{-}2)$$

Metric System

$$e + f = \frac{v^2}{gR} = \frac{V^2}{127R} \qquad (7\text{-}3)$$

Traditional U.S. Units

$$e + f = \frac{v^2}{gR} = \frac{V^2}{15R} \qquad (7\text{-}4)$$

where   *where*
e = rate of superelevation (m per m) e = rate of superelevaton (ft per ft)
f = side-friction factor f = side-friction factor
v = velocity (meters/sec) v = velocity (ft/sec)
V = velocity (km/hr) V = velocity (mph)
R = radius of curvature (meters) R = radius of curvature (ft)
g = acceleration of gravity (9.8 m/sec^2) g = acceleration of gravity (32.2 ft/sec^2)

Research and experience have established limiting values for e and f. Use of the maximum e with safe f value in the formula permits determination of minimum curve radii for various design speeds. Present highway design practice suggests a maximum superelevation rate of 0.12 meter per meter (or ft per foot). Where snow and ice conditions prevail during winter driving, the maximum superelevation rate should not exceed 0.08 meter per meter (or ft per foot). Some states have adopted a maximum superelevation rate of 0.10; however, other rates may have application and can be utilized on some highway types when local conditions require special treatment.

The limiting value of the side friction factor, f, at which the tires begin to skid may be as high as 0.6 or higher. This upper limit of f is not the value used in equations 7-3 and 7-4. In design, engineers use only a portion of the side friction factor, accounting for the comfort and safety of the vast majority of drivers. Thus, in selecting maximum allowable side friction factors for design, engineers use an f-value corresponding to the point at which a driver begins to feel uncomfortable and react instinctively to avoid higher speed. Empirical studies have determined that such f-values range from about 0.17 at 30 km/hr (20 mph) to 0.09 at 120 km/hr (75 mph). Values within this general range may be noted in Table 7-4.

EXAMPLE 7-1

Calculation of Superelevation Rates Calculate the superelevation rates for a roadway with a design speed of 100 km/hr that has a wide range of curve radii; i.e., R = 1750, 875, 585, 440, 350, and 295 m. (These values correspond to degrees of curve, D = 1, 2, 3, 4, 5, and 6.) Use e_{max} = 0.10. Compare the results with those obtained from Figure 7-9.

From Table 7-4, use a maximum side friction value of 0.12.

By equation 7-3, the following values of e are calculated:

R(m)	Computed value of e	Recommended Design e, Fig. 7-9
1750	-0.075	$+0.032$
875	-0.030	$+0.060$
585	$+0.015$	$+0.078$
440	$+0.059$	$+0.095$
350	$+0.105$	Exceeds e_{max}
295	$+0.147$	Exceeds e_{max}

Discussion

For the two sharpest curves, the combination of the maximum superelevation rate and the maximum side friction factor is insufficient to offset the centrifugal force. These two curves are too sharp for the given design speed and maximum superelevation rate and would be unsuitable for the stated conditions.

TABLE 7-4 Minimum Radius for Limiting Values of *e* and *f*, Rural Highways and High-Speed Urban Streets

Design Speed (km/h)	Maximum e	Maximum f	Total (e + f)	Calculated Radius (meters)	Rounded Radius (meters)
30	0.04	0.17	0.21	33.7	35
40	0.04	0.17	0.21	60.0	60
50	0.04	0.16	0.20	98.4	100
60	0.04	0.15	0.19	149.2	150
70	0.04	0.14	0.18	214.3	215
80	0.04	0.14	0.18	280.0	280
90	0.04	0.13	0.17	375.2	375
100	0.04	0.12	0.16	492.1	490
110	0.04	0.11	0.15	635.2	635
120	0.04	0.09	0.13	872.2	870
30	0.06	0.17	0.23	30.8	30
40	0.06	0.17	0.23	54.8	50
50	0.06	0.16	0.22	89.5	90
60	0.06	0.15	0.21	135.0	135
70	0.06	0.14	0.20	192.9	195
80	0.06	0.14	0.20	252.0	250
90	0.06	0.13	0.19	335.7	335
100	0.06	0.12	0.18	437.4	435
110	0.06	0.11	0.17	560.4	560
120	0.06	0.09	0.15	755.9	755
30	0.08	0.17	0.25	28.3	30
40	0.08	0.17	0.25	50.4	50
50	0.08	0.16	0.24	82.0	80
60	0.08	0.15	0.23	123.2	125
70	0.08	0.14	0.22	175.4	175
80	0.08	0.14	0.22	229.1	230
90	0.08	0.13	0.21	303.7	305
100	0.08	0.12	0.20	393.7	395
110	0.08	0.11	0.19	501.5	500
120	0.08	0.09	0.17	667.0	665
30	0.10	0.17	0.27	26.2	25
40	0.10	0.17	0.27	46.7	45
50	0.10	0.16	0.26	75.7	75
60	0.10	0.15	0.25	113.4	115
70	0.10	0.14	0.24	160.8	160
80	0.10	0.14	0.24	210.0	210
90	0.10	0.13	0.23	277.3	275
100	0.10	0.12	0.22	357.9	360
110	0.10	0.11	0.21	453.7	455
120	0.10	0.09	0.19	596.8	595

Source: From *A Guide for Metric Conversion*, copyright 1993. American Association of State Highway and Transportation Officials, Washington, DC. Used by permission.

At the other extreme, negative values of e were computed. Along these curves, all of the centrifugal force could be offset without exceeding the recommended f_{max} value of 0.12, even with zero superelevation. AASHTO (*1*) favors a distribution of superelevation that provides a logical relation between the side friction factor and the applied superelevation rate and recommends a slight but positive amount of superelevation for the two flattest curves.

AASHTO (*1*) provides graphs and tables showing recommended design values of superelevation for a wide range of values of e_{max}, design speed, and radius of curve.

From Eqs. 7-3 and 7-4, the minimum radius or maximum safe degree of curvature for a given design speed can be determined from the rate of superelevation and side-friction factor. The minimum safe radius R can be calculated for both the metric and traditional U.S. unit systems as follows:

Metric System	**Traditional U.S. Units**
$$R = \frac{V^2}{127(e + f)}$$	$$R = \frac{V^2}{15(e + f)}$$

Using the definition of the degree of curvature D in terms of the radius R, as defined in the traditional U.S. units, a useful relationship for determining the maximum safe degree of curvature D can be expressed in terms of e and f for a given design speed as follows:

$$D = \frac{5729.6}{R} \quad \text{or} \quad D = \frac{85,900(e + f)}{V^2} \tag{7-5}$$

The relationship between superelevation and degree of curvature is illustrated in Figure 7-9 and Table 7-4 for selected metric design speeds. An equivalent table, presented in traditional U.S. units, is included in the appendix.

7-16 Spirals or Transition Curves

Transition curves provide a gradual change from the tangent section to the circular curve and vice versa. For most curves, drivers can follow a transition path within the limits of a normal lane width, and a spiral transition in the alignment is not necessary. However, along high-speed roadways with sharp curvature, transition curves may be needed to prevent drivers from encroaching into adjoining lanes.

A curve known as the Euler spiral or clothoid is commonly used in highway design. The radius of the spiral varies from infinity at the tangent end to the radius of the circular arc at the end of the spiral. The radius of the spiral at any point is inversely proportional to the distance from its beginning point.

The minimum length of the transition curve is given as (*3*)

Metric System	**Traditional U.S. Units**
$$L_S = 0.035 \frac{V^3}{R} \tag{7-6}$$	$$L_S = 1.6 \frac{V^3}{R} \tag{7-7}$$

where
L_S = length of the transition (m)
V = velocity (km/hr)
R = radius of curvature (meters)

where
L_S = length of the transition (ft)
V = velocity (mph)
R = radius of curvature (ft)

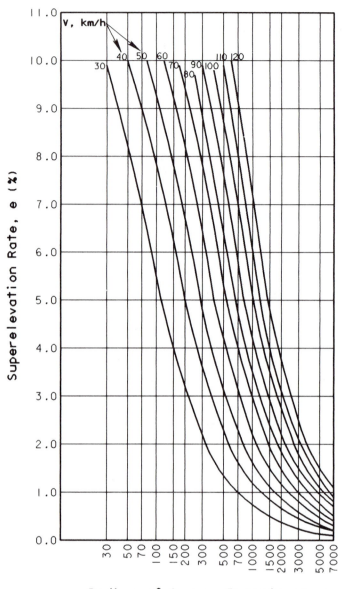

Radius of Curve, R (meters)

$e_{max} = 10.0\%$

FIGURE 7-9 Design superelevation rates, $e_{max} = 0.10$. (Source: *A Policy on Geometric Design of Highways and Streets*, copyright 1994, American Association of State Highway and Transportation Officials, Washington, DC. Used by permission.)

These equations allow for a rate of increase of centripetal acceleration of about 2 ft/s³, which is considered to be satisfactory for highway design.

When a transition curve is used in combination with a superelevated section, the superelevation should be attained within the limits of the alignment transition. Reference 3 gives more detailed information on the deisgn of transition curves for highways.

7-17 Attainment of Superelevation

The transition from a tangent, normal crown section to a curved superelevation section must be accomplished without any appreciable reduction in speed and in such a manner as to ensure safety and comfort to the occupants of the traveling vehicle.

In order to effect this change, the normal crown road section will have to be tilted or banked as a whole to provide the superelevation cross section required for a given design speed. This tilting usually is accomplished by rotating the roadway section about the centerline axis. The effect of this rotation is to lower the inside edge of pavement and, at the same time, to raise the outside edge without changing the centerline grade. Another method is to rotate about the inner edge of the pavement as an axis so that the inner edge retains its normal grade but the centerline grade is varied, or rotation may be likewise about the outside edge. Figure 7-10 provides schematics that depict all three methods for attaining superelevation. Rotation about the centerline is used by a majority of state highway departments, but for flat grades too much sag is created in the ditch grades by this method. On grades below 2 percent, rotation about the inside edge is preferred. Regardless of which method is utilized, care should be exercised to provide for drainage in ditch sections and along adjacent gutters in superelevated areas.

The roadway on full superelevated sections should be a uniform inclined section perpendicular to the direction of travel. When a crowned surface is rotated to the desired superelevation, the change from a crowned section to a uniformly inclined section would be accomplished gradually at a consistent rate along a length measured along the centerline. This may be effected by first transitioning the roadway section from the centerline to the outside edge of a level section, referred to as removing the crown. Second, raise the outside edge to an amount equal to one-half the desired superelevation, and at the same time change the inner edge to a uniform section, creating a continuous inclined plane for the whole roadway section. Third, continue rotation about the centerline until the desired superelevation is obtained.

The distance required for accomplishing the transition from a normal to a superelevated section, commonly referred to as the transition runoff, is a function of the design speed and the rate of superelevation. Recommended minimum lengths of superelevation runoff for two-lane pavements, based primarily on appearance and comfort, are shown in Table 7-5.

AASHTO (*1*) recommends that the superelevation runoff lengths for wider pavements be computed as follows:

- 3-lane pavements—1.2 times length for 2-lane highway
- 4-lane undivided pavements—1.5 times length for 2-lane highway
- 6-lane undivided pavements—2.0 times length for 2-lane highway

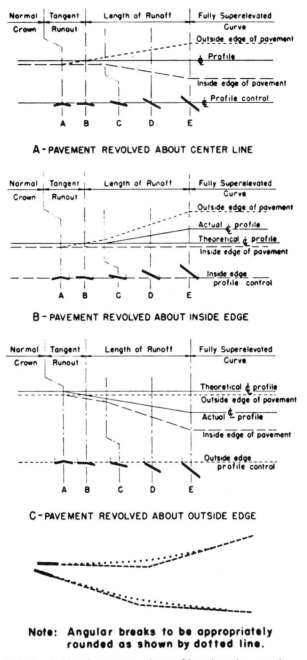

A - PAVEMENT REVOLVED ABOUT CENTER LINE

B - PAVEMENT REVOLVED ABOUT INSIDE EDGE

C - PAVEMENT REVOLVED ABOUT OUTSIDE EDGE

Note: Angular breaks to be appropriately rounded as shown by dotted line.

FIGURE 7-10 Diagrammatic profiles showing methods of attaining superelevation. (Courtesy American Association of State Highway and Transportation Officials.)

Superelevation is usually started on the tangent at some distance before the curve starts, and full superelevation is generally reached beyond the point of curvature (*PC*) of the curve. In curves with transitions, the superelevation can be attained within the limits of the spiral. In curves with a large radius or small degree of curvature where no horizontal transition is used, between 60 and 80

TABLE 7-5 Length Required for Superelevation Runoff, Two-Lane Pavements

Superelevation Rate in %	*L—Length of Runoff (m) for Design Speed (km/h) of:*									
	30	*40*	*50*	*60*	*70*	*80*	*90*	*100*	*110*	*120*
	3.6 m lanes									
2	20	25	30	35	40	50	55	60	65	70
4	20	25	30	35	40	50	55	60	65	70
6	30	35	35	40	40	50	55	60	65	70
8	40	45	45	50	55	60	60	65	70	75
10	50	55	55	60	65	75	75	80	85	90
12	60	65	65	75	80	90	90	95	105	110
	3.0 m lanes									
2	20	25	30	35	40	50	55	60	65	70
4	20	25	30	35	40	50	55	60	65	70
6	25	30	30	35	40	50	55	60	65	70
8	35	35	40	40	45	50	55	60	65	70
10	40	40	45	50	55	60	65	70	75	75
12	50	55	55	60	65	75	75	80	85	90

Source: From *A Policy on Geometric Design of Highways and Streets 1994,* copyright 1993. American Association of State Highway and Transportation Officials, Washington, DC. Used by permission.

percent of the superelevation runoff is put into the tangent. In order to obtain smooth profiles for the pavement edges, it is recommended that the breaks at cross sections be replaced by smooth curves. See inset included in Figure 7-10.

7-18 Widening of Curves

Extra width of pavement may be necessary on curves. As a vehicle turns, the rear wheels follow the front wheels on a shorter radius, and this has the effect of increasing the width of the vehicle in relation to the lane width of the roadway. Studies of drivers traversing curves have shown that there is a tendency to drive a curved path longer than the actual curve, shifting the vehicle laterally to the right on right-turning curves and to the left on left-turning curves. Thus, on right-turning curves the vehicle shifts toward the inside edge of the pavement, creating a need for additional pavement width. The amount of widening needed varies with the width of the pavement on tangent, the design speed, and the curve radius or degree of curvature.

Present roadway design practice requires no widening when the radius is greater than 195 m (9° curvature) on a 2-lane pavement 7.2 m (24 ft) wide. For smaller radius curves or greater curvatures and/or narrower pavements, widening is normally from 0.6 to 1.2 m (2 to 4 ft), depending on design speed and pavement width. It is suggested that no pavement be widened less than 0.6 m (2 ft).

7-19 Vertical Alignment

The vertical alignment of the roadway and its effect on the safe, economical operation of the motor vehicle constitute one of the most important features of a highway design. The vertical alignment, which consists of a series of straight profile lines connected by vertical parabolic curves, is known as the *profile grade line.* When the profile grade line is increasing from a level or flat alignment, this condition is referred to as a "plus grade," and when the grade is decreasing from a level alignment, the grade is termed a "minus grade." In analyzing grade and grade controls, the designer usually studies the effect of change on the centerline profile of the roadway.

In the establishment of a grade, an ideal situation is one in which the cut is balanced against the fill without a great deal of borrow or an excess of cut material to be wasted. All earthwork hauls should be moved in a downhill direction if possible and within a relatively short distance from the origin, due to the expense of moving large quantities of soil. Ideal grades have long distances between points of intersection, with long curves between grade tangents to provide smooth riding qualities and good visibility. The grade should follow the general terrain and rise or fall in the direction of the existing drainage. In rock cuts and in flat, low-lying or swampy areas, it is necessary to maintain higher grades with respect to the existing ground line. Future possible construction and the presence of grade separations or bridge structures can also act as control criteria for the design of a vertical alignment.

7-20 Grades and Grade Control

Changes of grade from plus to minus should be placed in cuts, and changes from a minus grade to a plus grade should be placed in fills. This will generally give a good design, and many times it will avoid the appearance of building hills and producing depressions contrary to the general existing contours of the land. Other considerations for determining the grade line may be of more importance than the balancing of cuts and fills.

Urban projects usually require a more detailed study of the controls and a finer adjustment of elevations than do rural projects. It is often best to adjust the grade to meet existing conditions to allow reuse of existing features such as pavement and to avoid the additional expense of doing otherwise.

In the analysis of grade and grade control, one of the most important considerations is the effect of grades on the operating costs of the motor vehicle. An increase in gasoline consumption, a reduction in speed, and an increase in emissions and noise are apparent when grades are increased. An economical approach would be to balance the added cost of grade reduction against the added annual costs and impacts of vehicle operation without grade reduction. An accurate solution to the problem depends on the knowledge of traffic volume and type, which can be obtained only by means of a traffic survey.

While maximum grades vary a great deal in various states, AASHTO recommendations make maximum grades dependent on design speed and the surrounding topography (*1*). Present design practice limits grades to 5 percent for a design speed of 110 km/hr (70 mph). For a design speed of 50 km/hr (30 mph), maximum grades typically range from 7 to 12 percent, depending on the roadway classification and the surrounding topography (*1*).

Wherever long sustained grades are used, the designer should not substantially exceed the critical grade without provision of climbing lanes for slower-moving vehicles. Critical grades vary from 500 m (1700 ft) for a 3 percent grade to 150 m (500 ft) for an 8 percent grade. Figure 7-2, which was discussed in Section 7-3, presents the relationship between critical grade, length, and truck slow-down speed.

Long sustained grades should be less than the maximum grade used on any particular section of a highway. It is often preferred to break the long-sustained uniform grade by placing steeper grades at the bottom and reducing the grades near the top of the ascent. Dips in the profile grade, which have the tendency to hide vehicles from view, should be avoided.

Minimum grades are governed by drainage conditions. Level grades may be used in fill sections in rural areas when crowned pavements and sloping shoulders can take care of the pavement surface drainage. However, it is preferred that the profile grade be designed to have a minimum grade of at least 0.3 percent under most conditions in order to secure adequate drainage.

7-21 Vertical Curves

The parabolic curve is used almost exclusively in connecting profile grade tangents. The primary reasons for the use of this type of curve in vertical highway alignments is the convenient manner in which the vertical offsets can be computed and the smooth transitions created from tangent to curve and then back to tangent. When a vertical curve connects a positive grade with a negative grade, it is referred to as a "crest curve." Likewise, when a vertical curve connects a negative grade with a positive grade, it is termed a "sag curve." Various configurations of crest and sag curves are illustrated in Figure 7-11. Variable definitions and basic equations for a typical vertical curve are presented in Figure 7-12. The sign conventions for g_1 and g_2 allow the use of the same formulas in the calculation of offsets and elevations for a sag curve also. Furthermore, the variable relationships and corresponding formulas are applicable for both the metric and traditional U.S. unit systems.

Offsets and spot elevations along vertical curves may be computed from the formulas provided in Figure 7-12. It is usually necessary to calculate elevations at every even 20-m (50-ft) station, while some special paving operations may require elevations at 10-m (25-ft) intervals or less. It is often necessary to compute other critical points on the vertical curve in order to ensure proper drainage, clearance, or connections to side streets. The high point or low point of a parabolic curve is seldom located at a point vertically above or below the vertex of the

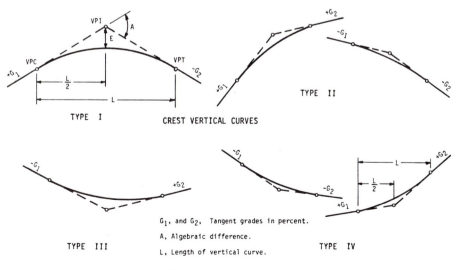

FIGURE 7-11 Types of crest and sag vertical curves. (Courtesy American Association of State Highway and Transportation Officials.)

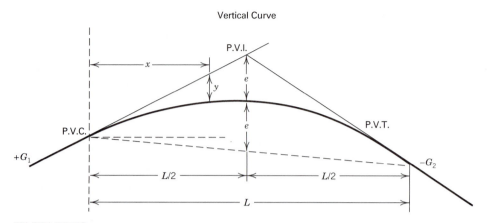

VARIABLES

VPI = Vertical point of intersection *(1)* E_{PI} = Elevation of VPI *(2)*
VPC = Vertical point of curvature *(1)* E_{PC} = Elevation of VPC *(2)*
VPT = Vertical point of tangency *(1)* E_{PT} = Elevation of VPT *(2)*
G_1 = Grade of initial tangent g_1 = Grade of initial tangent in percent
G_2 = Grade of final tangent g_2 = Grade of final tangent in percent
L = Length of vertical curve *(3)*
A = Algebraic difference in grade between g_1 and g_2
K = Vertical curve length coefficient as determined for stopping sight distance *(4)*
x = Horizontal distance to point on curve, measured from VPC *(3)*
E_x = Elevation of point on curve located at distance x from VPC *(2)*
x_m = Location of min/max point on curve, measured from VPC *(3)*
E_m = Elevation of min/max point on curve at distance x_m from VPC *(2)*
e = External distance = middle ordinate
y = Offset of curve from initial grade line

VERTICAL CURVE EQUATIONS

$$A = g_2 - g_1$$

$$K = \frac{L}{A}$$

$$e = \frac{(G_1 - G_2)L}{8} = \frac{AL}{800} = \frac{A^2 K}{800}$$

For high (low) point on curve,

$$x_m = \frac{g_1 L}{g_2 - g_1} = \frac{g_1 L}{A}$$

For any point p on curve,

$$y = \frac{(G_2 - G_1)x^2}{2L} = \frac{A x^2}{200L} = \frac{x^2}{200K}$$

$$E_x = E_{PC} + G_1 x + \frac{(G_2 - G_1)x^2}{2L}$$

similarly,

$$E_x = E_{PC} + \frac{g_1 x}{100} + \frac{x^2}{200K}$$

NOTES:
(1) Centerline stations expressed in either meters or feet.
(2) Elevations can be expressed in either meters or feet.
(3) Units for these variables can be expressed in either meters or feet.
(4) Dimensionless variable related to stopping sight distance, with different values that are utilized for either the metric or traditional US unit systems.

FIGURE 7-12 Properties of a typical vertical curve.

intersecting tangent grades, and it can fall on either side of this point depending on the configuration of the initial and final tangents. A method for determining the station and elevation of the high or low point of a vertical curve is included in Figure 7-12.

The discussion and formulas presented in this section apply only for a symmetrical curve, that is, one in which the tangents are of equal length. The unequal tangent or unsymmetrical vertical curve is a compound vertical parabolic curve. In general, this curve treatment is warranted only where a symmetrical curve cannot meet imposed alignment conditions, such as vertical clearance requirements. For a demonstration of vertical curve calculations, consider the following numerical example for a crest vertical curve.

EXAMPLE 7-1   **Calculation of Elevations Along a Crest Vertical Curve**   A plus 3.0 percent grade intersects a minus 2.0 percent grade at station $4+350$ and at an elevation of 190.500 m. Given that a 250-m length of curve is utilized, determine the station and elevation of the PC and PT. Calculate elevations at every 20-m station and locate the station and elevation of the high point of the curve. A sketch of the given conditions is shown below.

$g_1 = +3.0\%$ PI Sta $= 4+350$ m
$g_2 = -2.0\%$ PI EL $= 190.500$ m

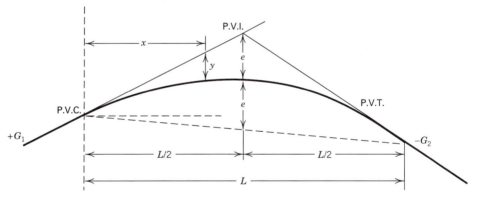

Vertical Curve

$G_1 = +0.03$ m per m $L = 250$ m
$G_2 = -0.02$ m per m $L/2 = 125$ m

Station locations for the PC and PT are

$$\text{PC Sta} = \text{PI Sta} - L/2 \quad = 4+350 - 125 = 4+225$$
$$\text{PT Sta} = \text{PC Sta} + L \quad = 4+225 + 250 = 4+475$$

Elevations for the PC and PT are

$$E_{PC} = E_{PI} - G_1 (L/2) = 190.500 - 0.03(125) = 186.750 \text{ m}$$
$$E_{PT} = E_{PI} - G_2 (L/2) = 190.500 - 0.02(125) = 188.000 \text{ m}$$

Location of high point can be calculated as follows:

$$x_m = \frac{g_1 \, L}{g_2 - g_1} = \frac{3.0(250)}{5.0} = 150 \text{ m}$$

High point Sta = PC Sta + 150 m = 4+225 + 150 = 4+375

Elevation of high point can be calculated as follows:

$$E_x = E_{PC} + G_1 \, x_m + \frac{(G_2 - G_1)x_m^2}{2L}$$

$$E_x = 186.750 + 0.03(150) + \frac{(-0.05)150^2}{2(250)} = 189.000 \text{ m}$$

Calculations for point elevations at even 20-m stations along the vertical curve can be conveniently tabulated as follows;

Station	x (meters)	Elevation on initial tangent $(E_{PC} + G_1 \, x)$	y	Final elevation on curve (Elev on tan − y)
4+240	15	187.200	−0.023	187.178
4+260	35	187.800	−0.123	187.678
4+280	55	188.400	−0.303	188.098
4+300	75	189.000	−0.563	188.438
4+320	95	189.600	−0.903	188.698
4+340	115	190.200	−1.323	188.878
4+360	135	190.800	−1.823	188.978
4+380	155	191.400	−2.403	188.998
4+400	175	192.000	−3.063	188.938
4+420	195	192.600	−3.803	188.798
4+440	215	193.200	−4.623	188.578
4+460	235	193.800	−5.523	188.278
4+475	250	194.250	−6.250	188.000

SIGHT DISTANCE

Safe highways must be designed to give drivers a sufficient distance of clear vision ahead so that they can avoid hitting unexpected obstacles and can pass slower vehicles without danger.

Sight distance is the length of highway visible ahead to the driver of a vehicle. When this distance is not long enough to permit passing of a slower vehicle, the controlling criterion is to provide stopping sight distance. The stopping distance is the minimum distance required to stop a vehicle traveling near the design speed before it reaches a stationary object in the vehicle's path. This stationary object may be another vehicle or some other object within the roadway. When the sight distance is long enough to enable a vehicle to overtake and pass another vehicle on a two-lane highway without interference from opposing traffic, this distance is referred to as "passing sight distance."

7-22 Minimum Stopping Sight Distance

Sight distance at every point should be as long as possible but never less than the minimum stopping distance. The minimum stopping sight distance is based on the sum of two distances: (1) the distance traveled from the time the object is sighted to the instant the brakes are applied and (2) the distance required for stopping the vehicle after the brakes are applied. The first of these two distances is dependent on the speed of the vehicle and the perception time and brake-reaction time of the driver. The second distance depends on the speed of the vehicle; the condition of brakes, tires, and roadway surface; and the alignment and grade of the highway.

Perception-Reaction Distance

As indicated in Chapter 5, a certain amount of time is required for a driver to perceive and react and for the brakes to be actuated. For purposes of design, a combined perception-reaction and brake-reaction time of 2.5 sec is recommended. The corresponding distance traveled during this time is 2.5 times the speed in meters per second (or feet per second).

Braking Distance

The approximate braking distance of a vehicle on a level highway is determined by the following relationship, expressed in both the metric and traditional U.S. unit systems:

$$d = \frac{v^2}{2fg} \tag{7-6}$$

where d = braking distance (meters or feet)
v = vehicle velocity when the brakes are applied (meters/sec or ft/sec)
g = acceleration of gravity
f = coefficient of friction between tires and roadway

Metric System

$$d = \frac{v^2}{2fg} = \frac{V^2}{254f} \tag{7-7}$$

Traditional U.S. Units

$$d = \frac{v^2}{2fg} = \frac{V^2}{30f} \tag{7-8}$$

where
d = braking distance (meters)
v = velocity (meters/sec)
V = velocity (km/hr)
g = acceleration of gravity (9.8 m/sec^2)

where
d = braking distance (ft)
v = velocity (ft/sec)
V = velocity (mph)
g = acceleration of gravity (32.2 ft/sec^2)

It is assumed that the friction force is uniform throughout the braking period. This is not strictly true; it varies as some power of the velocity. Other physical factors affecting the coefficient of friction at the condition and pressure of the tires, type and condition of the surface, and climate conditions such as rain, snow, and ice. Friction factors for skidding are assumed to vary from 0.60 at 50 km/hr (30 mph) to 0.55 at 110 km/hr (70 mph) for dry pavements. For wet pavements these values are much lower. Recommended minimum stopping sight distances in metric units are shown in Table 7-6, and a similar table for traditional U.S. units is provided in the appendix.

TABLE 7-6 Stopping Sight Distance on Wet Pavements

Design Speed (km/h)	Assumed Speed for Condition (km/h)	Brake Reaction Time (sec)	Brake Reaction Distance (m)	Coefficient of Friction f	Braking Distance on Level (m)	Stopping Sight Distance Computed (m)	Stopping Sight Distance Rounded for Design (m)
30	30–30	2.5	20.8–20.8	0.40	8.8–8.8	29.6–29.6	30–30
40	40–40	2.5	27.8–27.8	0.38	16.6–16.6	44.4–44.4	50–50
50	47–50	2.5	32.6–34.7	0.35	24.8–28.1	57.4–62.8	60–70
60	55–60	2.5	38.2–41.7	0.33	36.1–42.9	74.3–84.6	80–90
70	63–70	2.5	43.7–48.6	0.31	50.4–62.2	94.1–110.8	110–120
80	70–80	2.5	48.6–55.5	0.30	64.2–83.9	112.8–139.4	120–140
90	77–90	2.5	53.5–62.5	0.30	77.7–106.2	131.2–168.7	140–170
100	85–100	2.5	59.0–69.4	0.29	98.0–135.6	157.0–205.0	160–210
110	91–110	2.5	63.2–76.4	0.28	116.3–170.0	179.5–246.4	180–250
120	98–120	2.5	68.0–83.3	0.28	134.9–202.3	202.9–285.6	210–290

Source: From *A Guide for Metric Conversion*, copyright 1993. American Association of State Highway and Transportation Officials, Washington, DC. Used by permission.

Effect of Grade on Stopping Distance When a highway is on a grade, the formulas for braking distance are modified to

Metric System	**Traditional U.S. Units**

$$d = \frac{V^2}{254(f \pm G)} \quad (7\text{-}8)$$

$$d = \frac{V^2}{30(f \pm G)} \quad (7\text{-}9)$$

in which G is the percent of grade divided by 100. The safe stopping distances on upgrades are shorter and on downgrades are longer horizontal stopping distances. Where an unusual combination of steep grades and high speed occurs, the minimum sight distance should be adjusted to provide for this factor.

7-23 Measuring Stopping Sight Distance

The sight distance available over a crest vertical curve depends on the fundamental characteristics of the curve, namely the algebraic difference in grades and the length of curve. The basic relationships are

$$\text{when } S < L, \ L = \frac{AS^2}{100(\sqrt{2h_1} + \sqrt{2h_2})^2} \quad (7\text{-}10)$$

$$\text{when } S > L, \ L = 2S - \frac{100(\sqrt{2h_1} + \sqrt{2h_2})^2}{A} \quad (7\text{-}11)$$

where S = sight distance available over crest vertical curve (ft)
L = length of vertical curve (ft)
A = algebraic difference in grades (percent)
h_1 = height of eye of average driver (ft)
h_2 = height of object sighted (ft)

For purposes of design, it is assumed that the height of eye of the average driver $h_1 = 1.07$ m (3.5 ft) and the height of a stationary object $h_2 = 0.15$ m (0.5 ft). With these assumed values and the required stopping sight distance from Table 7-6, equations 7-10 and 7-11 can be used to calculate the minimum length of vertical curve for a given algebraic difference in grades and design speed.

Design Controls for Length of Vertical Curve

AASHTO (*1*) has published design controls for vertical curve length, expressed in terms of a rate of vertical curvature, K, meters (or feet) per percent of A, the algebraic difference in grades. Table 7-7 shows such controls for crest vertical curves. For example, for a design speed (and assumed travel speed) of 100 km/hr, a maximum rate of vertical curvature, $K = 105$ meters per percent of grade change is indicated. With a plus grade of, say, 2.5 percent intersecting a minus 1.5 percent, the algebraic difference in grades would be 4.0 percent and the minimum length of vertical curve would be $4 \times 105 = 420$ m. Note that the resulting curve length based on the criteria given in Table 7-7 is a minimum length. It may be desirable to use a longer curve length to balance the earthwork or to provide more favorable operating conditions or a more aesthetically pleasing design.

AASHTO (*1*) has also published design controls for the minimum length of sag vertical curves. Limits on the minimum length of vertical curve in sag curves are based on headlight sight distance, rider comfort, drainage control, and for general appearance of the vertical alignment. Recommended maximum K-values for sag vertical curves are given in Table 7-8.

7-24 Minimum Passing Sight Distance

The majority of U.S. highways carry two lanes of traffic moving in opposite directions. In order to pass slower-moving vehicles, it is necessary to use the lane

TABLE 7-7 Design Controls for Crest Vertical Curves Based on Stopping Sight Distance

Design Speed (km/hr)	Assumed Speed for Condition (km/hr)	Coefficient of Friction f	Stopping Sight Distance Rounded for Design (m)	Rate of Vertical Curvature, K (length [m] per % of A) Computed[a]	Rounded for Design
30	30–30	0.40	30–30	2.17–2.17	3–3
40	40–40	0.38	50–50	4.88–4.88	5–5
50	47–50	0.35	60–70	8.16–9.76	9–10
60	55–60	0.33	80–90	13.66–17.72	14–18
70	63–70	0.31	100–120	21.92–30.39	22–31
80	70–80	0.30	120–140	31.49–48.10	32–49
90	77–90	0.30	140–170	42.61–70.44	43–71
100	85–100	0.29	160–210	61.01–104.02	62–105
110	91–110	0.28	180–250	79.75–150.28	80–151
120	98–120	0.28	210–290	101.90–201.90	102–202

[a]Using computed values of stopping sight distance.

Source: From *A Guide for Metric Conversion,* copyright 1993. American Association of State Highway and Transportation Officials, Washington, DC. Used by permission.

TABLE 7-8 Design Controls for Sag Vertical Curves Based on Stopping Sight Distance

Design Speed (km/h)	Assumed Speed for Condition (km/h)	Coefficient of Friction f	Stopping Sight Distance Rounded for Design (m)	Computed[a]	Rounded for Design
30	30–30	0.40	30–30	3.88–3.88	4–4
40	40–40	0.38	50–50	7.11–7.11	8–8
50	47–50	0.35	60–70	10.20–11.54	11–12
60	55–60	0.33	80–90	14.45–17.12	15–18
70	63–70	0.31	110–120	19.62–24.08	20–25
80	70–80	0.30	120–140	24.62–31.86	25–32
90	77–90	0.30	140–170	29.62–39.95	30–40
100	85–100	0.29	160–210	36.71–50.06	37–51
110	91–110	0.28	180–250	42.95–61.68	43–62
120	98–120	0.28	210–290	49.47–72.72	50–73

[a]Using computed values of stopping sight distance.

Source: From *A Guide for Metric Conversion,* copyright 1993. American Association of State Highway and Transportation Officials, Washington, DC. Used by permission.

of opposing traffic. If passing is to be accomplished safely, the vehicle driver must be able to see enough of the highway ahead in the opposing traffic lane to have sufficient time to pass and then return to the right traffic lane without cutting off the passed vehicle and before meeting the oncoming traffic. The total distance required for completing this maneuver is the passing sight distance.

Empirical research conducted over a period of many years has established minimum sight distances required to provide safe passing maneuvers. The details of how this research was conducted and the assumptions that underlie the design controls for vertical alignment to allow passing are given elsewhere in the literature and will not be repeated here. Recommended minimum sight distances to allow safe passing maneuvers are given in Table 7-9.

7-25 Measuring Passing Sight Distance

Equations 7-10 and 7-11 can be used to measure available sight distance for passing maneuvers. In this case, the height of eye of the average driver, h_1, is assumed to 1.07 m (3.5 ft) as before. The height of object, h_2, for passing maneuvers is assumed to be 1.30 m (4.25 ft). With these assumed values and the required passing sight distances shown in Table 7-9, the equations can be used to calculate the minimum length of vertical curve for given algebraic difference in grades and design speed. Such curve lengths would allow safe passing along two-lane, two-way roads.

Design Controls for Length of Vertical Curve (Passing Allowed)

If passing is to be allowed along two-lane, two-way roads, much longer crest vertical curve lengths are required. This is reflected in greater values of K, the length of vertical curve per percent change in A. Values of K that will provide for minimum passing sight distance are shown in Table 7-9.

TABLE 7-9 Design Controls for Crest Vertical Curves Based on Passing Sight Distance

Design Speed (km/hr)	Minimum Passing Sight Distance for Design (m)	Rate of Vertical Curvature, K[a], Rounded for Design (length [m] per percent of A)
30	217	50
40	285	90
50	345	130
60	407	180
70	482	250
80	541	310
90	605	390
100	670	480
110	728	570
120	792	670

[a]Computed from rounded values of passing sight distance.

Source: From *A Policy of Geometric Design of Highways and Streets 1994,* copyright 1994. American Association of State Highway and Transportation Officials, Washington, DC. Used by permission.

7-26 Design Standards

Because of functional differences, a wide array of design standards must be used for the many types of facilities comprising a highway system. Design standards vary widely for different functional classes. For example, freeways are designed predominantly for traffic movement, and freeway standards are characterized by high design speeds, wide lanes, and straight horizontal and vertical alignments. At the other end of the spectrum, standards for subdivision streets reflect the emphasis on land access function. Such standards, exemplified by Tables 7-10 and 7-11, are based on design speeds of 30 to 55 km/hr (20 to 35 mph). In fact, local streets should be designed to discourage excessive speeds through the use of curvilinear alignment and discontinuities in the street system (*4*).

Within a functional class, design standards may vary with the type of terrain, anticipated traffic to be served, and whether the highway is to be in an urban or rural area. For example, see Table 7-12.

COMPUTER APPLICATIONS

The use of computers and computer networks is continuing to revolutionize the field of highway design. Digital data formats have allowed engineers to automate many of the design tasks, translating design criteria through the design process, directly reflecting the information in the project constructing drawings. This is accomplished through an integrated design environment that links design activities, such as horizontal alignment, cross sections, profiles, and quantities, with the final production of drawings that are created through the use of multifaceted software programs. An example flow chart of a typical software design package is presented in Figure 7-13.

One of the primary benefits of using a computer design program is to enhance

TABLE 7-10 Local Street Design Guidelines

Terrain classification →	Level			Rolling			Hilly		
Development density →	Low	Medium	High	Low	Medium	High	Low	Medium	High
Right-of-way width (ft)[a]	50	60	60	50	60	60	50	60	60
Pavement width (ft)	22–27	28–34	36	22–27	28–34	36	28	28–34	36
Type of curb (V = vertical face, R = roll-type, 0 = none)	0/R	V	V	V	V	V	V	V	V
Sidewalks and bicycle paths (ft)	0	4-6	4-6	0	4-6	4-6	0	4-6	4-6
Sidewalk distance from curb face (ft)	—	6	6	—	6	6	—	6	6
Minimum sight distance (ft)	←	200	→	←	150	→	←	110	→
Maximum grade (%)	←	4	→	←	8	→	←	15	→
Maximum cul-de-sac length (ft)	1000	700	700	1000	700	700	1000	700	700
Minimum cul-de-sac radius (right-of-way) (ft)	←	50	→	←	50	→	←	50	→
Design speed (mph)[b]	←	30	→	←	25	→	←	20	→
Minimum centerline radius of curves (ft)	←	250	→	←	175	→	←	110	→
Minimum tangent between reverse curves (ft)	←	50	→	←	50	→	←	50	→

[a]1 ft = 0.3048 m.

[b]1 mph = 1.6093 km/hr.

Source: Recommended Guidelines for Subdivision Streets: A Recommended Practice, Institute of Transportation Engineers, Washington, DC (1984).

TABLE 7-11 Collector Street Design Guidelines

Terrain classification	Level			Rolling			Hilly		
Development density	Low	Medium	High	Low	Medium	High	Low	Medium	High
Right-of-way width (ft)[a]	70	70	70	70	70	70	70	70	70
Pavement width (ft)	36	36	40	36	36	40	36	36	40
Type of curb (v = vertical face)	V	V	V	V	V	V	V	V	V
Sidewalk width (ft)	←	4-6	→	←	4-6	→	←	4-6	→
Sidewalk distance from curb face (ft)	←	10	→	←	10	→	←	10	→
Minimum sight distance (ft)	←	250	→	←	200	→	←	150	→
Maximum grade	←	4	→	←	8	→	←	12	→
Minimum spacing along major traffic route (ft)	←	1300	→	←	1300	→	←	1300	→
Design speed (mph)[b]	←	35	→	←	30	→	←	25	→
Minimum centerline radius (ft)[c]	←	350	→	←	250	→	←	175	→
Minimum tangent between reverse curves (ft)	←	100	→	←	100	→	←	100	→

[a]1 ft = 0.3048 m.
[b]1 mph = 1.6093 km/hr.
[c]Assumes superelevation.

Source: Recommended Guidelines for Subdivision Streets: A Recommended Practice, Institute of Transportation Engineers, Washington, DC (1984).

TABLE 7-12 Minimum Design Speeds for Rural Roads in North Carolina

Traffic Volume	Type of Terrain, km/hr (mph)		
	Level	*Rolling*	*Mountainous*
Current ADT < 50	64 (40)	48 (30)	32 (20)
Current ADT 50–250	64 (40)	48 (30)	32 (20)
Current ADT 250–400	80 (50)	64 (40)	32 (20)
Current ADT 400–750 and			
DHV 100–200	80 (50)	64 (40)	48 (30)
DHV 200–400	80 (50)	64 (40)	48 (30)
DHV > 400	80 (50)	64 (40)	48 (30)

Source: Design Manual, Vol. I, *Roadway,* North Carolina Department of Transportation, Raleigh.

the ability of an engineer to conduct numerous design iterations for the purpose of improving and refining the design without expending a large amount of time or effort. Another valuable feature is the ability to view the resulting effect of the design modification on the construction plans without the need to conduct the numerous intermediate steps that have been associated with the more traditional manual design methods of the past. It is important to note that the design engineer must have a firm understanding of all the criteria, standards, and design methods necessary to design a safe and efficient highway as described in the preceding sections of this chapter in order to utilize the computer programs correctly.

A full discussion of computer applications in highway design is beyond the scope of this text. The many companies that develop, distribute, and support computer aided design and drafting (CADD) software programs have numerous manuals describing programming commands and techniques for efficiently using their respective software products. The focus of the following paragraphs will be to present an overview of the most important features that this computer technology has contributed to an enhancement of the highway design process.

7-27 Digital Terrain Modeling

The utilization of this feature within an automated computer design process involves creating a digital database for the project limits and in areas immediately surrounding the proposed contruction. Survey and topographic information describing the project area can be collected through conventional field survey techniques or through controlled aerial photography. These spatially oriented data are located within the context of a three-dimensional grid, which is referenced in the traditional variables x, y, and z. Coordinates for x and y represent the horizontal location of the data points, and the variable z serves to provide an elevation of each point, typically measured with respect to sea level.

Field-collected survey points at consistently spaced intervals, perpendicular to the centerline at approximately 20 m (50 ft) extending to the outer limits of the project area, and at all surface-evident breaks in the terrain, such as ditches and ridge lines, are located in a digital format with respect to x, y, and z dimensions. A data collector is utilized that allows easy transfer of survey information into a consistent format that can be read by a DOS-based desktop computer. The field-collected information is utilized by one of the numerous computer aided design and drafting (CADD) programs, and an analytical method referred to as trian-

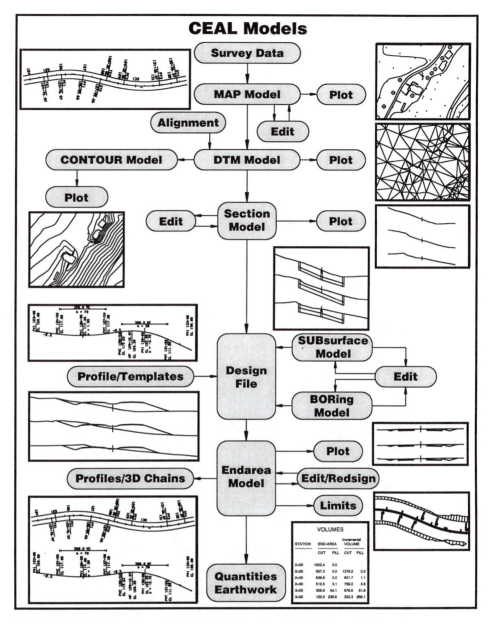

FIGURE 7-13 Design flow chart of a computer-aided design and drawing preparation software program entitled *Civil Engineering Automation Library (CEAL)*. (Courtesy CLM/Systems, Inc.)

gulated irregular network (TIN) is created through an interpolation between the various data points that were collected within the project limits. The results of the TIN is a digital terrain model (DTM) for the project area, which serves to create a computer surface model that represents the project topography. The results of this modeling procedure can be used to efficiently evaluate construction requirements of a new highway design.

Once a similar three-dimensional model of the proposed highway improvements is designed, these models, which are commonly referred to as surfaces, can be merged and further utilized in the design process. This merging of data allows determination of earthwork quantities and a number of other volume-based calculations, such as amount of rock excavation, limits of de-mucking in low-lying areas, and asphalt pavement quantities, given that the initial database can support such evaluations. An example of a menu-driven screen showing a view of merged surfaces between existing and proposed features is presented in Figure 7-14.

7-28 Coordinate Geometry

Another major feature of a computer aided design and drafting (CADD) system is the ability to calculate alignments, bearings, lengths, and curves within the context of a referenced horizontal grid system. The most commonly used grid system for highway design purposes is the state plane coordinate system. Each state in the United States typically has its own plane coordinate system or systems depending on the size of the state's area, which is utilized to reference property ownership and control monuments and other important topographical features on a statewide basis. Orientation of grid coordinates is commonly referenced in north and east directional units, where point coordinates increase in value or magnitude as one moves toward each of the respective coordinate directions.

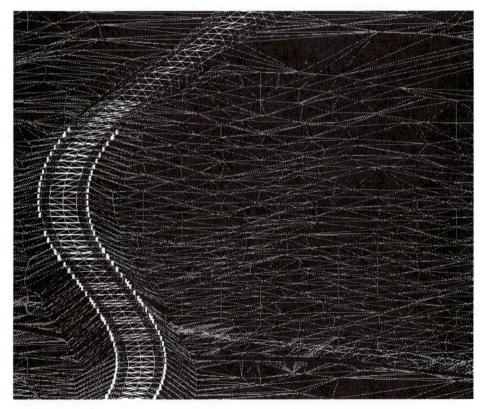

FIGURE 7-14 Computer screen example of a menu-driven software program showing a digital terrain model (DTM) and triangulated irregular network (TIN) created in *Intergraph Microstation*. (Courtesy Intergraph Corporation.)

Once the existing database is located with respect to the local state plane coordinate system, a wide variety of alignment and location calculations can be performed, allowing determination of a position for required improvements in relation to existing project features. Bearing-to-bearing intersections; concentric offsets for curb, sidewalk, and right-of-way locations; curve-to-curve intersections; and station distance and perpendicular offsets are but a few of the commonly utilized features that all coordinate geometry programs can calculate. The ease of conducting quick and efficient design iterations is a significant benefit derived from the use of this program and is a valuable tool for refining and optimizing the design of a proposed highway improvement project.

7-29 Plan Preparation

Database information and intermediate design calculations conducted during the creation of a DTM and coordinate geometry location of the highway improvements are all utilized to generate plan sheets, which when correctly processed should require only a minimum amount of drafting work. A small amount of drafting time is typically required to ensure consistent formats, standard symbols, and other agency-specific requirements that need to be incorporated into the plan set. Scripted command routines, which are tailored specifically for agency standards, can be developed and utilized to reduce the amount of effort required for drafting and can be used to ensure consistency in the plan preparation process. Uniformity is a significant aspect of the highway design process that cannot be overlooked. Large transportation agencies typically have numerous highway projects all being constructed by different general contractors, and drawing consistency is often related to minimized construction problems and other associated conflicts.

As previously metioned, the most frequently utilized programs are menu driven for ease in use and are structured with the intention of integrating a wide variety of database information and hardware components. A typical dual screen computer workstation is shown in Figure 7-15. The ability to customize screen configuration and information is a valuable feature that allows the designer to efficiently move through the various levels of a related database structure. Figure 7-16 shows a view of other hardware-associated components necessary to link a design workstation with equipment that allows data entry, provide file storage, produce tabular output data, and create drawing plots. Finally, Figure 7-17 provides a systemwide organization of computer components necessary for use within a network environment that is typically utilized by state transportation agencies and other large highway design groups.

PLANNING AND DESIGN OF BICYCLE FACILITIES

Through public participation and the effective work of numerous special interest groups, most state and local transportation agencies have adopted and are implementing policies that promote the use of bicycling as a viable mode of travel within the transportation system. During the 1960s, 1970s, and 1980s vast differences in opinion were evident about which type and what configuration of bike facilities is optimal to create a safe and functional environment for bicycle riders. Currently, the focus of providing safe and useful bicycle facilities emphasizes an integration of two basic infrastructure approaches: on-street improvements in-

FIGURE 7-15 An *Intergraph Microstation* computer-aided design and drafting computer work station. (Courtesy Intergraph Corporation.)

tended to better accommodate bicycles into the overall stream of traffic and multiuse paths that provide a good environment for a variety of path users, including cyclists.

Due to a shift in national transportation objectives that was spearheaded by the Intermodal Surface Transportation Efficiency Act (ISTEA) enacted by the U.S. Congress in 1991, a much greater emphasis has been placed on reducing levels of mobile source air emissions and decreasing other environmental impacts

FIGURE 7-16 One combination of hardware components that supports CLM/Systems software. (Courtesy CLM/Sustems, Inc.)

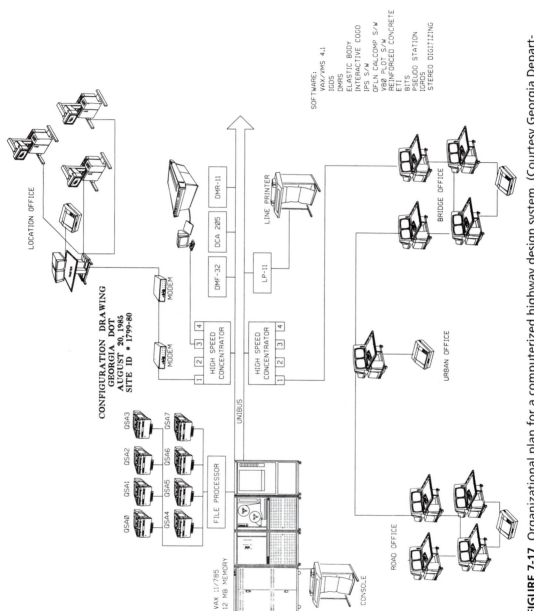

FIGURE 7-17 Organizational plan for a computerized highway design system. (Courtesy Georgia Department of Transportation.)

related to motor vehicle traffic that adversely affect the quality of life in urban areas. Within the overall strategy of accomplishing these broad-based goals, bicycle facilities and other related transportation improvements have been given increased importance. The change in emphasis for transportation agencies has established a national policy for bicycles that is twofold in nature: to provide transportation improvements that will accommodate the current levels of bicycling and to provide new facilities that will serve to increase the current levels of bicycle use.

Planning and design innovations for accomplishing these two bicycling objectives have been developed and adopted at various levels of the national, state, and local government agencies. Fundamental criteria standards and engineering methods used in the planning and design of bicycle facilities will be briefly presented in the following paragraphs. It is important to reiterate that a large degree of variation still exists among jurisdictions in the recommended practices for conducting bicycle planning and facility design.

7-30 Bicycle Facility Planning

Bicycle planning procedures are similar to those for other modes of travel. An effective process should be conducted in conjunction with the planning efforts for other modes of transportation and should include the following activities:

1. Establish overall bicycle program goals, objectives, and suitable measures of effectiveness.
2. Collect information on the characteristics of bicycle use, bicycle travel corridors, and the identification of bicycle traffic generators.
3. Create a detailed inventory of existing conditions.
4. Establish and adopt planning principles, design standards, and appropriate design.
5. Forecast demand for bicycle facilities.
6. Analyze and integrate facility improvements; final selection of type, location, and priority should be determined based on an evaluation of the factors listed in Table 7-13 (5).
7. Develop a comprehensive bicycle network plan to be used as the basis for all facility design and construction, created through a firm understanding of funding sources and funding procedures.

TABLE 7-13 Factors Affecting Selection of a Bicycle Facility

1. Barriers	10. Pavement surface quality
2. Accidents	11. Truck and bus traffic
3. Directness	12. On-street motor vehicle parking
4. Access	13. Traffic volumes and speeds
5. Attractiveness	14. Cost/funding
6. Security	15. Local laws
7. Delays	16. Bridges
8. Use conflicts	17. Intersection conditions
9. Maintenance	

Source: From *A Guide for the Development of Bicycle Facilities,* copyright 1991. American Association of State Highway and Transportation Officials, Washington, DC. Used by permission.

Design Bicyclist

Before an effective planning process for bicycle facilities can be established, it is of utmost importance that the differences in bicyclists' riding capabilities and their purposes for cycling be evaluated and understood. Within the context of the national objective of developing facilities that will accommodate current levels and promote an increase in bicycling, three classification groups of cyclists have been identified (6). These classification groups for bicyclists are described in Ref. 6 as

Group A—Advanced Bicyclists: These are experienced riders who can operate under most traffic conditions. They comprise the majority of current users of collector and arterial streets and are best served by the following: .

- direct access to designations usually via the existing street and highway system
- the opportunity to operate at maximum speed with minimum delays
- sufficient operating space on the roadway or shoulder to reduce the need for either bicyclists or motor vehicle operators to change position when passing

Group B—Basic Bicyclists: These are casual or new adult and teenage riders who are less confident of their ability in traffic without special provisions for bicycles. Some will develop greater skills and progress to the advanced level, but there will always be millions of basic bicyclists. They prefer

- comfortable access to destinations, preferably by a direct route, using either low-speed, low traffic-volume streets or designated bicycle facilities
- well-defined separation of bicycles and motor vehicles on arterial and collector streets (bike lanes or shoulders) or separate bike paths

Group C—Children: These are preteen riders whose roadway use is initially monitored by parents. Eventually they are accorded independent access to the system. They and their parents prefer the following:

- access to key designations surrounding residential areas, including schools, recreation facilities, shopping, or other residential areas
- residential streets with low motor vehicle speed limits and volumes
- well-defined separation of bicycles and motor vehicles on arterial and collector streets or separate bike paths

For the purpose of planning and design, the three classification groups are converted into a two-tiered approach for evaluation of bicycle facility needs. Bicycle facilities for Group A are focused on providing on-street improvements, which are in most cases beneficial to all roadway users. Likewise, facilities for Group B/C emphasize special improvements such as multiuse paths and other facilities that are in general separate from significant streets of the existing roadway network. The best approach is to integrate both approaches in order to establish an areawide network of bicycle improvements that will be capable of increasing bicycle use.

Forecasting Demand

A forecast of bicycle demand is necessary to determine where bicycle facilities are required, to evaluate the optimal facility type, and to establish the use of preestablished design criteria. Unfortunately, there are no generally accepted methods for this purpose and little experience in forecasting demand, especially latent or potential demand. Forecasts of demand can be based on special surveys (e.g., origin-destination studies), experience with other nearby facilities, and experiences in other cities or countries. Exercise caution when basing demand estimates on experience at other locations, being sure to allow for differences in social attitudes and physical conditions. Significant factors that have historically been responsible for large bicycling demands are

1. good temperate weather conditions
2. favorable and generally level surrounding terrain
3. various demographic characteristics (e.g., percent of young adults, university population, auto ownership, etc.)
4. use and availability of other transportation modes

Development of a Bicycle Facility Plan

A comprehensive bicycle facility plan should be integrated into the overall transportation master plan of the entire region typically prepared by the area's Metropolitan Planning Organization (MPO). Furthermore, the plan should specifically address the needs of both the A Group and the B/C Group of bicyclists. It is also important to create efficient connections with other travel modes, such as public transit. Policies that allow bicycles to be transported on public transit rail and/or buses can be a critical consideration in an overall improvement plan and can have a significant effect on bicycle use.

Other related bicycle amenities, such as parking, lockers, and security, may need to be incorporated into a comprehensive bicycle plan to ensure that use of the constructed linear facilities such as paths, lanes, and routes is enhanced. For the sake of brevity, the discussion of this text as related to bicycle facility design will focus on the linear components such as on-street and multiuse path improvements.

7-31 Design of Bicycle Facilities

As described in Ref. 6, there are five basic types of facility improvements that can be used to accommodate bicyclists:

- Shared lane: shared motor vehicle/bicycle use of a standard width travel lane
- Wide outside lane: an outside travel lane with a width of at least 4.2 m (14 ft)
- Bike lane: a portion of the roadway designated by striping, signing, and/or pavement markings for preferential or exclusive use of bicycles
- Shoulder: a paved portion of the roadway to the right of the edge stripe designed to serve bicyclists
- Separate bike path: a facility physically separated from the roadway and intended for bicycle use

The first four categories are considered on-street improvements, and a wide variety of design considerations are required to provide a safe environment for cycling within the overall traffic stream. An extensive list of bicycle planning factors that would also carry over into the facility design phase is provided in

Table 7-13. Traffic conditions are a major consideration when evaluating the suitability of a highway for accommodating bicycle traffic. In order to evaluate the facility needs of A and B/C bicyclists, three ranges of motor vehicle volumes have been established to assess risk or exposure of the cyclist to traffic (6). These traffic ranges are specified in terms of annual average daily traffic (AADT) volumes and include the following categories; under 2,000 AADT; 2,000 to 10,000 AADT; and over 10,000 AADT. Other design considerations specifically related to creating a suitable bicycling environment along existing roadways include drainage grates, railroad crossings, and traffic control devices.

The separate path category of bicycle facilities has traditionally had operational and maintenance problems, which in most cases are the result of oversights in the design and construction process. One of the most important considerations is that a bike or multiuse path is, in essence, a mini-roadway and should be designed and constructed as such. A horizontal and vertical alignment must be calculated for the desired design speed. This alignment should be staked in the field for construction, just as would be provided for a similar major highway project. Subgrade, subbase, pavement courses, and shoulder grading should be provided in a similar fashion to that for a roadway construction project.

Specific design requirements of facility design elements will be presented in the following sections. However, it should be noted that many bicycle facilities in the past have not been developed using this type of pragmatic approach, and this has resulted in substandard facilities that cannot be adequately maintained and have not accommodated continuous operation of a safe and efficient facility for use by bicycle riders and other trail users.

Width of Bicycle Facilities

The minimum width for a bike lane is 1.2 m (4 ft). It is important to evaluate the edge conditions of the roadway to determine if a lane width greater than the minimum width is desirable. When on-street parking is present, the minimum bike lane width increases to 1.5 m (5 ft). Additional widths are preferable when substantial truck traffic is present or when motor vehicle speeds exceed 55 km/h (35 mph). Both cases are depicted in Figure 7-18. To illustrate the variation in design criteria at the state level, the Oregon Department of Transportation provides a standard bike lane width of 1.8 m (6 ft) and provides these lanes along roadway routes that accommodate 50 bikes or more per day (7). Based on the previous description of A and B/C categories for bicyclists, Table 7-14 provides additional criteria for selecting a recommended bicycle facility treatment, given specified conditions (6).

In the design of a separate bike path, the width of pavement and corresponding operating width are primary considerations. The minimum recommended for a 2-direction bike path is 2.4 m (8 ft). A width of 3 m (10 ft) is desirable when heavier bicycle volumes are expected, pedestrian use is anticipated, and/or maintenance vehicles will travel along the path. In other instances when heavy mixed used traffic is likely, steep grades exist, and/or larger maintenance vehicles require access, a path width of 3.7 m (12 ft) is desirable. Figure 7-19 depicts a typical bike path. In addition to the pavement width, the need to provide an adequate graded shoulder and vertical/lateral clear distances is another important consideration. A minimum shoulder width of 0.6 m (2 ft) should be included in the typical section and must be graded to drain water away from the pavement surface. Figure 7-20 provides other clearance requirements for various path types, along

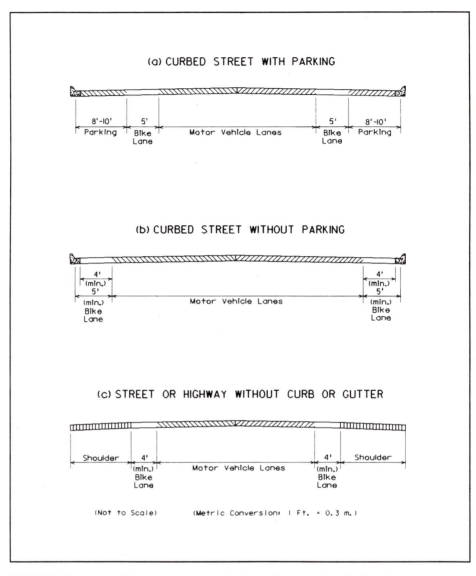

FIGURE 7-18 Typical bicycle lane cross sections. (Source: *Guide for the Development of Bicycle Facilities,* copyright 1991, American Association of State Highway and Transportation Officials, Washington, DC. Used by permission.)

with vegetation maintenance requirements, which are important due to the fact that many paths are routed through wooded areas (*8*).

Bikeway Design Speed

Previous speed studies of bicycle riders indicate that most cyclists travel within a range of 11 to 24 km/hr (7 to 15 mph), with an average of 16 to 18 km/hr (10 to 11 mph). The speed of a cyclist is dependent on several factors that should be carefully considered prior to the selection of a design speed for a bicycle facility. Factors affecting bicycle design speed include the type of bicycle, condition of the bicycle, purpose of trip, direction and speed of the wind, and condition of the

TABLE 7-14 Recommended On-Street Bike Facility Treatment for Group B/C Bicyclists, Along an Urban Roadway Section With No Parking

Average Motor Vehicle Operating Speed	Average Annual Daily Traffic (AADT) Volume											
	Less than 2,000				2,000–10,000				Over 10,000			
	Adequate Sight Distance		Inadequate Sight Distance		Adequate Sight Distance		Inadequate Sight Distance		Adequate Sight Distance		Inadequate Sight Distance	
		Truck, Bus, RV				Truck, Bus, RV				Truck, Bus, RV		
Less than 30 mph	wc 14	wc 14	wc 14	wc 14	wc 14	wc 14	wc 14	wc 14	bl 5	bl 5	bl 5	bl 5
30–40 mph	bl 5	bl 5	bl 5	bl 5	bl 5	bl 6	bl 6	bl 5	bl 5	bl 6	bl 6	bl 5
41–50 mph	bl 5	bl 5	bl 5	bl 5	bl 6	bl 6	bl 6	bl 6	bl 6	bl 6	bl 6	bl 6
Over 50 mph	bl 6	bl 6	bl 6	bl 6	bl 6	bl 6	bl 6	bl 6	bl 6	bl 6	bl 6	bl 6

1 mi/h = 1.61 km/h

wc = wide curb lane*   sh = shoulder   sl = shared lane   bl = bike lane*   na = not applicable

*WC numbers represent "usable widths" of outer lanes, measured from lane stripe to edge of gutter pan, rather than to face of curb. If no gutter pan is provided, add 1 ft (0.3 m) minimum for shy distance from face of curb. BL numbers indicate minimum width from the curb face. The bike lane stripe should lie at least 4 ft (1.2 m) from the edge of the gutter pan, unless the gutter pan is built with adequate width to serve as a bike lane by itself.

Source: Selecting Roadway Design Treatments to Accommodate Bicycles, Federal Highway Administration, U.S. Department of Transportation, Washington, DC (1994)

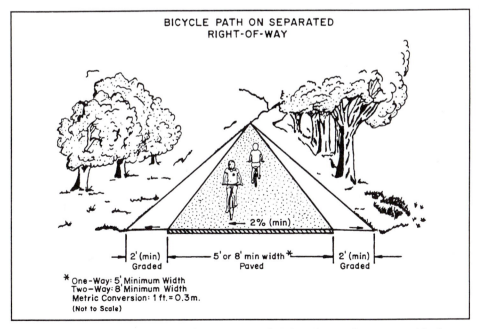

FIGURE 7-19 Bicycle path located on separated right-of-way. (Source: *Guide for the Development of Bicycle Facilities,* copyright 1991, American Association of State Highway and Transportation Officials, Washington, DC. Used by permission.)

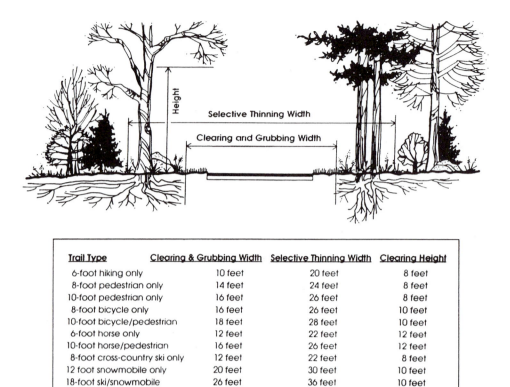

Trail Type	Clearing & Grubbing Width	Selective Thinning Width	Clearing Height
6-foot hiking only	10 feet	20 feet	8 feet
8-foot pedestrian only	14 feet	24 feet	8 feet
10-foot pedestrian only	16 feet	26 feet	8 feet
8-foot bicycle only	16 feet	26 feet	10 feet
10-foot bicycle/pedestrian	18 feet	28 feet	10 feet
6-foot horse only	12 feet	22 feet	12 feet
10-foot horse/pedestrian	16 feet	26 feet	12 feet
8-foot cross-country ski only	12 feet	22 feet	8 feet
12 foot snowmobile only	20 feet	30 feet	10 feet
18-foot ski/snowmobile	26 feet	36 feet	10 feet

FIGURE 7-20 Bicycle path vertical and lateral clearance dimensions. (Source: Federal Highway Administration, Conflicts on Multiple-Use Trails: Synthesis of Literature and State of the Practice, U.S. Department of Transportation, Washington, DC [1994].)

bicyclist. Good design methodology supports the assumption that the bicycle path should be designed to accommodate the faster bicyclists and a minimum design speed of 32 km/hr (20 mph) is recommended for most cases. Since higher speeds are likely to occur on downgrades, a design speed of 48 km/hr (30 mph) should be used when this type of grade or surrounding terrain condition is present.

Curvature and Superelevation

Using the same formula and variables described for Eq. 7-2 in Section 7-16, a numerical relationship for speed curvature and superelevation is utilized to develop important cross sectional parameters for the design of a bike path. The superelevation rate for most bike path designs will vary from a minimum of 2 percent to a maximum of approximately 5 percent. Table 7-15 presents design speeds, friction factors and minimum radius that should be used in the design of a bike path.

Profile Grades

The maximum grade that a bicyclist will be able to negotiate depends on the capability of the individual cyclist and the length of grade, as well as the condition of the bicycle and road surface, weather conditions, and other related factors.

TABLE 7-15 Minimum Radii for Paved Bicycle Paths

Minimum radius based on $e = 2$ percent.

Design Speed, V km/hr (mph)	Friction Factor, f	Minimum Radius, R meters (feet)
30 (20)	0.27	29.0 (95)
40 (25)	0.25	47.5 (155)
50 (30)	0.22	76.2 (250)
60 (35)	0.19	118.9 (390)
65 (40)	0.17	172.2 (565)

Source: From *A Guide for the Development of Bicycle Facilities,* copyright 1991. American Association of State Highway and Transportation Officials, Washington, DC. Used by permission.

Grades that are greater than 5 percent are undesirable. However, where the grade must exceed 5 percent, the length of the grade should be not more than 150 m (500 ft), and a higher design speed as well as additional width should be utilized in steeper grade sections. For unpaved paths with a wearing surface of crushed stone aggregate, grades steeper than 3 percent should not be used.

Sight Distance

Sight distance criteria for bicycle facility design are similar to the conditions previously described for motor vehicle highways. Minimum vertical curve lengths may be determined using an eye height for the cyclist $h_1 = 1370$ mm (4.5 ft) and object height $h_2 = 150$ mm (0.33 ft). As described in Section 7-22, minimum stopping distance is the sum of the perception-reaction distance and the braking distance. Additional evaluations should be included in the design to ensure that adequate sight distance is provided around curves and at intersections with other bike paths or with motor vehicle roadways.

Bikeway Capacity

Limited research has been performed to establish useful criteria for evaluation of the capacity of bikeway and bike path facilities. Available evidence suggests that "once basic operating space requirements are met, the bikeway would have ample capacity for most all situations" (*9*). Empirical studies indicate that a single-direction bike lane with an effective width of 1.2 m (4 ft) has a capacity of 1,275 bicycles per hour. Likewise, a 2-direction bike path with an effective width of 2.4 m (8 ft) has a capacity of approximately 1,900 bicycles per hour (*9*).

When mixed use of paths with pedestrians, joggers, roller bladers, and so on is experienced and conflicts between path users increase, the level of operation and capacity for bicyclists is adversely affected. Various approaches have been developed to improve safety for all path users. A greater width is the single most effective method of improving the safety and operation of a multiuse path. Other approaches of signing, striping, and channelizing the path specifically for the various travel modes that utilize the facility have also proven to be effective treatments for reducing undesirable conflicts (*10*).

PLANNING AND DESIGN FOR PEDESTRIANS

Each year approximately 7,000 pedestrians are killed after being struck by motor vehicles, and about 70,000 are injured. Many pedestrians who do not have access to an automobile or public transit, such as the young, the elderly, and the handicapped, face special difficulties in moving about streets and sidewalks congested with motor vehicles and other pedestrians. Increasing traffic volumes and pedestrian accident losses have been accompanied by a growing public awareness of the need for safe and convenient pedestrian facilities. In the following paragraphs, standards and criteria for the planning and design of pedstrian facilities will be briefly described.

7-32 Sidewalks

The need for sidewalks and width of sidewalks depend primarily on the type and density of land development and the volume of pedestrian and vehicular traffic. In rural areas, sidewalks are seldom provided since pedestrian traffic is usually very light. Along residential subdivision streets, 1.5-m (5-ft) wide sidewalks are recommended where the development density exceeds 2 dwelling units per acre (4). Sidewalks should generally be provided along arterial streets that are not provided with shoulders, even though pedestrian traffic is light (1). A minimum width of 1.8 m (6 ft) is recommended for commercial areas and major school routes.

Sidewalk widths in central business districts and other high-density areas may be determined objectively on the basis of capacity analyses. The quality and rate of pedestrian flow is dependent on the amount of space available per pedestrian. According to Pushkarev and Zupan (11), pedestrian flow can be grouped into seven categories ranging from "open flow" to "jammed." The seven groups are listed in Table 7-16 along with corresponding rates of flow and the amount of space per person.

Fruin (12) described six levels of pedestrian flows analogous to those for

TABLE 7-16 Characteristics of Pedestrian Flow in a Homogeneous Stream

Type of Flow	Flow Rate per/min/m (per/min/ft)	Space per Person m² (ft²)	Description of Flow
Open	< 1.6 (< 0.5)	> 50 (> 530)	Complete freedom to select speed and direction of movement
Unimpeded	1.6–6.6 (0.5–2.0)	50–12 (530–130)	Frequent indirect interaction with others
Impeded	6.6–19.7 (2.0–6.0)	12–4 (130–40)	Constant indirect interaction with others
Constrained	19.7–32.8 (6.0–10.0)	4–2 (40–24)	Crossing and passing movements are possible but with interference and likely conflicts
Crowded	32.8–45.9 (10.0–14.0)	2–1.5 (24–16)	Probability of conflicts fairly high; passing is difficult
Congested	45.9–59.0 (14.0–18.0)	1.5–1 (16–11)	Frequent body contacts; difficult to walk at normal pace
Jammed	59.0–82.0 (18.0–25.0)	1–0.2 (11–2)	Only shuffling movement is possible

Source: Boris S. Pushkarev with Jeffrey M. Zupan, *Urban Space for Pedestrians,* MIT Press, Cambridge, MA (1975).

TABLE 7-17 Level of Service Description for Walkaways

Level of Service	Flow Rate per/min/m (per/min/ft)	Space per Person m² (ft²)	Description of Flow
A	23 or less (7 or less)	3.3 (35)	Pedestrians freely select walking speed and are able to bypass slower pedestrians and avoid crossing conflicts
B	23–33 (7–10)	2.3–3.3 (25–35)	Where reverse direction or crossing movements exist, minor conflicts occur, slightly lowering mean speeds and volumes
C	33–49 (10–15)	1.4–2.3 (15–25)	Freedom to select individual walking speed and freely pass others is restricted
D	49–66 (15–20)	0.9–1.4 (10–15)	The majority of persons have normal walking speeds restricted and reduced; reverse flow and crossing movements are severely restricted
E	66–82 (20–25)	0.5–0.9 (5–10)	Virtually all persons have normal walking speeds restricted requiring frequent adjusting of gait; at lower end of range, forward progress made only by shuffling
F	Variable up to 82 (Variable up to 25)	0.5 or less (5 or less)	All pedestrian walking speeds are extremely restricted; frequently, unavoidable contact with others; forward progress made only by shuffling

Source: John J. Fruin, *Pedestrian Planning and Design,* Metropolitan Association of Urban Designers and Environment Planners, Inc., New York (1971).

highway flows as described in Chapter 6. These are described in Table 7-17 and illustrated in Figure 7-21.

In estimating pedestrian flows along sidewalks, the effective width should be reduced by an additional 0.6 m (2 ft) or more to account for the constricting effects of mailboxes, fire hydrants, and other street furniture.

EXAMPLE 7-2 **Sidewalk Flow Evaluation** A new office building is expected to add 1,000 pedestrians to a 5.5-m (18-ft) sidewalk during the peak 10-minute period. The sidewalk already has a flow of 1,600 pedestrians during the peak period. Because of parking meters, light standards, and other obstructions, subtract 0.9 m (3 ft) from the actual sidewalk width to obtain the effective width. Characterize the quality of flow using Tables 7-15 and 7-16.

The pedestrian volume P per width of sidewalk per minute is

Metric System

$$P = \frac{2,600 \text{ pedestrians}}{10 \text{ min} \times 4.6 \text{ m}} = 56.5$$

Traditional U.S. Units

$$P = \frac{2,600 \text{ pedestrians}}{10 \text{ min} \times 15 \text{ ft}} = 17.3$$

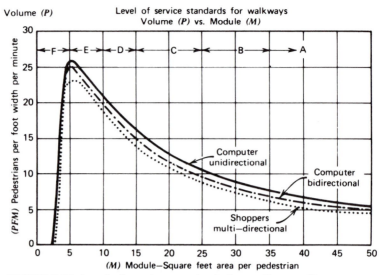

FIGURE 7-21 Level of service standards for walkways. (Source: John J. Fruin, *Pedestrian Planning and Design,* Metropolitan Association of Urban Designers and Environmental Planners, Inc., New York, 1971).

According to Table 7-16, the sidewalk flow would be termed "congested." By Table 7-17 and Figure 7-21, a level of service D would be expected.

The numbers of pedestrians can be determined by manual counts, analysis of time-lapse photography made from a stationary aerial vantage point, and studies of a series of timed, sequentially overlapping aerial photographs taken from a helicopter or light aircraft.

The forecasting of pedestrian traffic generated by proposed buildings is a complex process, the details of which are beyond the scope of this text. Generally,

TABLE 17-18 Daily and Peak-Period Trip Generation Rates for Various Building Uses

Building Use	In and Out Trips per 93 m² (1000 ft²) of Floor Space		
	Daily	Peak 15 min	Time of Peak Period
Cafeteria	492[a]	22.36	12:00 noon
Department store	252	15.64	12:45 P.M.
Supermarket	285	11.97	5:00 P.M.
Restaurant	173	10.20	1:15 P.M.
Office, headquarters	14.2	0.98	8:45 A.M.
Office, headquarters	13.2	0.97	5:00 P.M.
Office, mixed use	17.3	0.89	4:45 P.M.
Residence	8.3	0.25	5:45 P.M.

[a]10-hr rate only.

Source: Boris S. Pushkarev with Jeffrey M. Zupan, *Urban Space for Pedstrians,* MIT Press, Cambridge, MA (1975).

for a given building, estimates of such traffic can be made by reference to empirical trip generation rates. See Table 7-18. In aggregating pedestrian volumes for a given location, it is necessary to account for variations in peaking patterns (which depend on the type of building use) and in trip lengths (which depend on trip purpose). For further information on forecasting pedestrian traffic, refer to information included in Refs. 11 and 12.

Generally, sidewalks should be located well away from the curb face. Such a location provides increased safety, facilitates driveway gradient design, and provides an area for street furniture and temporary storage of snow. Many agencies place the sidewalks 0.3 m (1 ft) from the right-of-way line (*4*).

7-33 Provisions for Elderly and Handicapped Pedestrians

Special considerations must be given to elderly and handicapped pedestrians. Curb heights should generally not exceed 175 mm (7 in), and at locations where significant wheelchair traffic is expected, ramps should be provided. Recommended ramp designs are shown in Figure 7-22. The suggested 1:12 ramp slope should also be satisfactory for bicycle traffic.

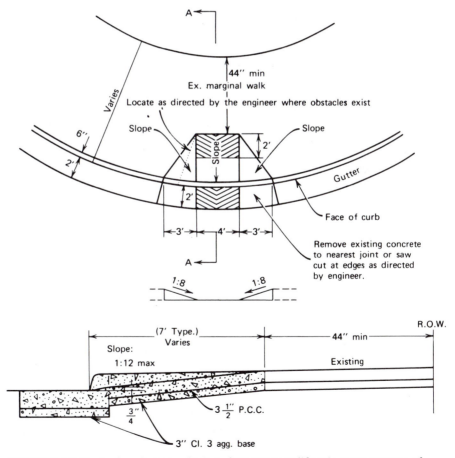

FIGURE 7-22 Typical curb ramp design. (Courtesy California Department of Rehabilitation.)

Sidewalk pavements should be free of holes, bumps, and other irregularities, and the surface should be skid-resistant. Gratings and manhole covers, which tend to become slippery and create tripping hazards, should be removed from crosswalks and other walkway areas wherever feasible. Special barriers and surface textures that can be felt are required in most areas to act as a guide for visually impaired pedestrians.

PROBLEMS

7-1 Given a horizontal curve with a radius of 410 m, a Δ angle of 32 degrees, and the P.I. station of 1 + 120.744, compute the curve data and the station of the P.T. Compute the deflection angles at even 20-m stations.

7-2 Given Δ = 35 degrees, D = 4.25 degrees, and the P.I. station 75 + 28.10, compute the curve data and the station of the P.T. using conventional U.S. units. Compute the deflection angles at even 100-ft stations.

7-3 What superelevation rate would you recommend for a roadway with a design speed of 100 km/hr and a radius of curvature of 500 m? Assume f = 0.11.

7-4 What superelevation rate would you recommend for a roadway with a design speed of 75 mph and a degree of curve of 4 degrees? Assume f = 0.11.

7-5 A 1-km long racetrack is to be designed with turns 250 m in length at each end. Determine the superelevation rate you would recommend for a design speed of 140 km/hr, assuming f = 0.20.

7-6 A 1-mile long racetrack is to be designed with ¼-mile turns at each end. Determine the superelevation rate for a design speed of 70 mph, assuming f = 0.20.

7-7 A plus 5.1 percent grade intersects a minus 2.8 percent grade at station 68 + 70 at an elevation of 327.50 ft. Calculate the centerline elevations for every even 100-ft station for a 500-ft vertical curve.

7-8 A plus 2.8 percent grade intersects a minus 5.1 percent grade at station 2 + 087.224 at an elevation of 190.280 m. Calculate the centerline elevations at every 20-m station for a 150-m vertical curve.

7-9 A vertical parabolic curve is to be used under a grade-separation structure. The curve is 250 m long and the minus grade of 3.8 percent intersects the plus grade of 5.4 percent at station of 0 + 120.424. Calculate the low point of the curve if the intersection at station 0 + 120.424 is at elevation 230.085.

7-10 A vertical parabolic curve is to be used as a sag curve, and it is desired to know the location and elevation of the low point. The curve is 750 ft long, and the minus grade of 3.6 percent intersects the plus grade of 6.2 percent at station 240 + 80.00. Calculate the low point of the curve if the intersection at station 240 + 80.00 is at elevation 426.84.

7-11 Calculate the stopping sight distance over the crest of a 450-m vertical curve with a plus grade of 5.6 percent and a minus grade of 3.2 percent.

7-12 Calculate the stopping sight distance over the crest of a 1500-ft vertical curve with a plus grade of 4.4 percent and a minus grade of 2.3 percent.

7-13 Calculate the passing sight distance over the crest described in Problem 7-11.

7-14 Calculate the passing sight distance over the crest described in Problem 7-12.

7-15 Given a horizontal curve with a 410-m radius, estimate the minimum length of spiral necessary for a smooth transition from tangent alignment to the circular curve. The design speed is 90 km/hr.

7-16 Given a horizontal curve with a 1360-ft radius, estimate the minimum length of spiral necessary for a smooth transition from tangent alignment to the circular curve.

7-17 The superelevation rate is 0.084. Prepare a plot showing the ordinates required for the transition from the tangent section to the full superelevated section, assuming that the pavement is 7.3-m wide.

7-18 A total of 1,000 pedestrians use a sidewalk during the peak 10-min period. If the effective width is 2.6 m, characterize the quality of flow using Tables 7-16 and 7-17.

REFERENCES

1. *A Policy on Geometric Design of Highways and Streets,* American Association of State Highway and Transportation Officials, Washington, DC (1990 and 1994, metric edition).

2. *Highway Capacity Manual,* Transportation Research Board Special Report No. 209 (1994).

3. Barnett, Joseph, *Transition Curves for Highways,* U.S. Government Printing Office, Washington, DC (1940).

4. *Recommended Guidelines for Subdivision Streets,* Institute of Transportation Engineers, Washington, DC (1984).

5. *Guide for the Development of Bicycle Facilities,* American Association of State Highway and Transportation Officials, Washington, DC (1991).

6. Federal Highway Administration, *Selecting Roadway Design Treatments to Accommodate Bicycles,* U.S. Department of Transportation, Washington, DC (1994).

7. *1992 Oregon Bicycle Plan,* Oregon Department of Transportation, Salem, OR (1992).

8. Federal Highway Administration, *Current Planning Guidelines and Design Standards Being Used by State and Local Agencies for Bicycle and Pedestrian Facilities,* Case Study No. 24, National Bicycling and Walking Study, U.S. Department of Transportation, Washington, DC (1992).

9. *Bikeways—State of the Art, 1974,* prepared for the Federal Highway Administration by DeLeuw Cather and Company, San Francisco (1974).

10. Federal Highway Administration, *Conflicts on Multiple-Use Trails: Synthesis of Literature and State of the Practice,* U.S. Department of Transportation, Washington, DC (1994).

11. Pushkarev, Boris S., with Jeffrey M. Zupan, *Urban Space for Pedestrians,* MIT Press, Cambridge, MA (1975).

12. Fruin, John J., *Pedestrian Planning and Design,* Metropolitan Association of Urban Designers and Environmental Planners, Inc., New York (1971).

CHAPTER 8

ROADSIDE DESIGN

In Chapter 6 we described the nature and extent of the highway accident problem. Here we will examine methods and procedures that highway and traffic engineers may use to deal with the problem by providing safer roadsides.

It is estimated that approximately 60 percent of fatal highway crashes involve only one vehicle, and in about 70 percent of these crashes, the vehicle leaves the roadway and either overturns or collides with a fixed object (1). Thus, a great opportunity exists to lessen accident losses by providing "forgiving" roadsides for those motorists who leave the roadway.

It is appropriate and helpful that engineers make every reasonable effort to prevent crashes from occurring by providing suitable horizontal and vertical align-ment, adequate lane widths and shoulders, appropriate signing and signaling, and so forth, as described elsewhere. They must acknowledge, however, that motorists will continue to run off the road for a variety of reasons. Many run off because of inattentiveness, excessive speed, and other driver errors. Others are driving under the influence of alcohol or other drugs. Some drivers leave the roadway deliberately to avoid a collision with another vehicle or an object on the roadway. Many other roadway factors (e.g., poor alignment, slippery pavement, inadequate drainage, etc.) and vehicular factors (e.g., worn tires, faulty brakes) may cause motorists to leave the roadway.

Once a vehicle leaves the roadway, the probability of an accident depends on the speed and trajectory of the vehicle and what lies in its path. If an accident does occur, its severity is dependent upon several factors, including the use of restraint systems by vehicle occupants, the type of vehicle, and the nature of the roadside environment. Of these factors, the highway engineer generally has a significant measure of control over only one—the roadside environment. *(1)*

8-1 Clear Roadside Recovery Areas

During the 1960s, it became apparent that many motorists were running off high-speed highways and colliding with trees, poles, and other fixed objects along the roadside. Many of these objects were sign supports, bridge piers, culverts, ditches, and other design features of the roadway. Highway engineers began to appreciate the concept of "forgiving roadsides," which would decrease the number and lessen the severity of off-road crashes.

Highway research focused on roadside hazards suggested that a clear and unobstructed roadside area extending about 9 m (30 ft) from the traveled way would allow approximately 85 percent of the vehicles leaving the roadway to avoid a collision. Figure 8-1, which was based on studies (*2, 3*) of 600 fatal off-

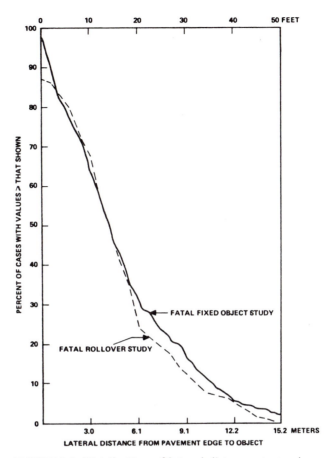

FIGURE 8-1 Distribution of lateral distances to crash point for studies of fatal fixed-object crashes and rollover crashes. (SOURCE: Paul H. Wright, and Paul Zador. Study of Fatal Rollover Crashes in Georgia. *Transportation Research Record 819*, 1981).

road crashes in Georgia, shows that only about 15 percent of the vehicles in those crashes came to rest farther than 9 m (30 ft) from the traveled way.

It was clear that the provision of clear roadside recovery areas along all of the nation's 3.9 million miles of streets and highways would be a massive and expensive undertaking. In view of the magnitude of this problem, Wright and Robertson (2) developed priorities for roadside hazard modification and recommended that top priority be given to roadsides in the vicinity of curves sharper than 6° or those with radius less than 290 m (955 ft). Their research indicated that the greatest needs for improvement exist on nonlocal roads and along the outside of horizontal curves, particularly those accompanied by downhill grades greater than 2 percent. More recent studies (4, 5) examining the relationship between accident rates and horizontal curvature have shown a breakpoint between 4 and 5° and further, that accident rates increase almost linearly for curves greater than 5°. Interestingly, Lamm and co-workers (6) have concluded that for curves less than or equal to 5°, consistency in horizontal alignment exists, and the horizontal alignment does not create inconsistencies in vehicle operating speeds. They classify curves less than or equal to 5° as good design, those between 5 and 10° as fair design, and those greater than 10° as poor design. These studies indicate that special attention should be given to roadsides in the vicinity of horizontal curves with curvatures sharper than 5° or those with radius less than 350 m (1148 ft).

In the early 1970s, highway agencies began to try to provide a traversable and unobstructed roadside extending 9 m (30 ft) from the driving lane, particularly on high-speed, heavily-traveled highways. It soon became apparent that in some circumstances, the 9-m (30-ft) border was insufficient; in other situations, a border that wide was not needed.

AASHTO (1) recommends clear zone dimensions based on traffic volumes and speeds and on roadside geometry. Table 8-1 gives recommended clear zone distances for safe roadside design. It is important to recognize that these distances "suggest only the approximate center of a range to be considered and not a

TABLE 8-1 Clear Zone Distances in Meters from Edge of Driving Lane

Design Speed	Design ADT	Fill Slopes			Cut Slopes		
		6:1 or flatter	5:1 to 4:1	3:1	3:1	4:1 to 5:1	6:1 or flatter
64 km/hr or less	Under 750	2.1–3.0	2.1–3.0	[b]	2.1–3.0	2.1–3.0	2.1–3.0
	750–1500	3.0–3.7	3.7–4.3	[b]	3.0–3.7	3.0–3.7	3.0–3.7
	1500–1600	3.7–4.3	4.3–4.9	[b]	3.7–4.3	3.7–4.3	3.7–4.3
	Over 6000	4.3–4.9	4.9–5.5	[b]	4.3–4.9	4.3–4.9	4.3–4.9
72–80 km/hr	Under 750	3.0–3.7	3.7–4.3	[b]	2.4–3.0	2.4–3.0	3.0–3.7
	750–1500	3.7–4.3	4.9–6.1	[b]	3.0–3.7	3.7–4.3	4.3–4.9
	1500–6000	4.9–5.5	6.1–7.9	[b]	3.7–4.3	4.3–4.9	4.9–5.5
	Over 6000	5.5–6.1	7.3–8.5	[b]	4.3–4.9	5.5–6.1	6.1–6.7
88 km/hr	Under 750	3.7–4.3	4.3–5.5	[b]	2.4–3.0	3.0–3.7	3.0–3.7
	750–1500	4.9–5.5	6.1–7.3	[b]	3.0–3.7	4.3–4.9	4.9–5.5
	1500–6000	6.1–6.7	7.3–9.1	[b]	4.3–4.9	4.9–5.5	6.1–6.7
	Over 6000	6.7–7.3	7.9–9.8[a]	[b]	4.9–5.5	6.1–6.7	6.7–7.3
96 km/hr	Under 750	4.9–5.5	6.1–7.3	[b]	3.0–3.7	3.7–4.3	4.3–4.9
	750–1500	6.1–7.3	7.9–9.8[a]	[b]	3.7–4.3	4.9–5.5	6.1–6.7
	1500–6000	7.9–9.1	9.8–12.2[a]	[b]	4.3–5.5	5.5–6.7	7.3–7.9
	Over 6000	9.1–9.8[a]	11.0–13.4[a]	[b]	6.1–6.7	7.3–7.9	7.9–8.5
104–112 km/hr	Under 750	5.5–6.1	6.1–7.9	[b]	3.0–3.7	4.3–4.9	4.3–4.9
	750–1500	7.3–7.9	8.5–11.0[a]	[b]	3.7–4.9	5.5–6.1	6.1–6.7
	1500–6000	8.5–9.8[a]	10.4–12.8[a]	[b]	4.9–6.1	6.7–7.3	7.9–8.5
	Over 6000	9.1–10.4[a]	11.6–14.0[a]	[b]	6.7–7.3	7.9–9.1	8.5–9.1

[a]Where a site specific investigation indicates a high probability of continuing accidents, or such occurrences are indicated by accident history, the designer may provide clear zone distances greater than 9 m as indicated. Clear zones may be limited to 9 m for practicality and to provide a consistent roadway template if previous experience with similar projects or designs indicates satisfactory performance.
[b]Because recovery is less likely on the unshielded, traversable 3:1 slopes, fixed objects should not be present in the vicinity of the toe of these slopes. Recovery of high speed vehicles that encroach beyond the edge of shoulder may be expected to occur beyond the toe of slope. Determination of the width of the recovery area at the toe of slope should take into consideration right of way availability, environmental concerns, economic factors, safety needs, and accident histories. Also, the distance between the edge of the travel lane and the beginning of the 3:1 slope should influence the recovery lane provided at the toe of slope. While the application may be limited by several factors, the fill slope parameters that may enter into determining a maximum desirable recovery area are illustrated in Figure 8.2.

Source: From Roadside Design Guide, copyright 1989. American Association of State Highway and Transportation Officials, Washington, DC. Used by permission.

precise distance to be held as absolute" (*1*). In selecting an appropriate clear zone width, the designer should consider site-specific conditions, as well as economic and environmental factors. AASHTO also suggests that the dimensions given in Table 8-1 may need to be modified for horizontal curvature in situations where accident histories indicate a need, or "where a site-specific investigation shows a definitive accident potential which could be significantly lessened by increasing the clear zone width and such increases are cost effective." For such circumstances, the suggested clear zone values in Table 8-1 should be multiplied by the curve correction factors given in Table 8-2. These correction factors should be applied only to the outside of curves.

8-2 Roadside Topography

When a motorist leaves the roadway, the vehicle's lateral encroachment and trajectory is affected by the geometry of the roadside. Most roadsides are not flat. Usually, the motorist encounters an embankment, a cut slope, or a ditch.

The slopes of embankments may be parallel to the flow of traffic or at an angle to the flow of traffic.

Embankments Parallel to the Flow of Traffic

AASHTO (*1*) defines three categories of slopes of embankments parallel to the flow of traffic:

TABLE 8-2 Horizontal Curve Adjustments

| Degree of Curve | K_{cz} (Curve Correction Factor) | | | | | | |
| | Design Speed, mph | | | | | | |
	40	45	50	55	60	65	70
2.0	1.08	1.10	1.12	1.15	1.19	1.22	1.27
2.5	1.10	1.12	1.15	1.19	1.23	1.28	1.33
3.0	1.11	1.15	1.18	1.23	1.28	1.33	1.40
3.5	1.13	1.17	1.22	1.26	1.32	1.39	1.46
4.0	1.15	1.19	1.25	1.30	1.37	1.44	
4.5	1.17	1.22	1.28	1.34	1.41	1.49	
5.0	1.19	1.24	1.31	1.37	1.46		
6.0	1.23	1.29	1.36	1.45	1.54		
7.0	1.26	1.34	1.42	1.52			
8.0	1.30	1.38	1.48				
9.0	1.34	1.43	1.53				
10.0	1.37	1.47					
15.0	1.54						

$$CZ_c = (L_c)(K_{cz})$$

where CZ_c = clear zone on outside of curvature, m (ft)

L_c = clear zone distance, m (ft); see Table 8.1

Clear zone correction factor is applied to outside of curves only. Curves flatter than 2° do not require an adjusted clear zone.

Source: From *Roadside Design Guide,* copyright 1989. American Association of State Highway and Transportation Officials, Washington, DC. Used by permission.

1. *Recoverable slopes.* Motorists encroaching on recoverable embankment slopes can generally stop their vehicles or slow them enough to return safely to the roadway. Slopes 4 to 1 or flatter are generally considered to be recoverable.

2. *Nonrecoverable slopes.* These slopes are traversable, but motorists who encroach on them will be unable to stop or to return to the roadway safely. Slopes between 3 to 1 and 4 to 1 are considered to be nonrecoverable. Because most errant vehicles will not stop on such slopes, the recovery area must extend to the toe of the slope, and a clear runout area at the base with a slope of 6 to 1 or flatter is desirable.

3. *Critical slopes.* Critical slopes are those on which a vehicle is likely to overturn. Slopes generally steeper than 3 to 1 fall into this category.

An example of a parallel embankment slope design is shown as Figure 8-2. This design, known as a "barn roof" section, provides a relatively flat area adjacent to the roadway, bordered by a steeper foreslope. It is more economical to construct than a recoverable slope that extends all the way to the original ground line and is considered to be safer than providing a constant steeper slope from the edge of the shoulder.

Embankments at an Angle to the Flow of Traffic

Motorists who leave the roadway may be confronted with cross slopes or embankments at an angle to the flow of traffic. Such obstacles include driveways, intersecting side roads, and median crossovers. Such slopes are of more concern to errant motorists because they are usually struck head-on by out-of-control vehicles. Along high-speed roadways, such cross slopes should be no steeper than 6 to 1 and preferably 10 to 1 or flatter. Embankment cross slopes steeper than 6 to 1 may be appropriate in urban areas or for low-speed facilities. Obstacles of this type often occur at side streets and driveways where ditches flow into drain-

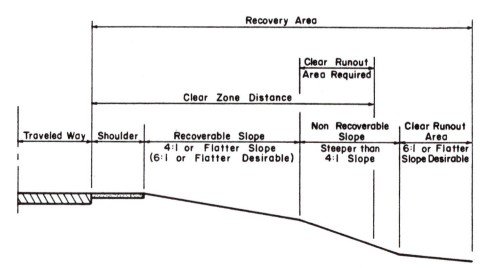

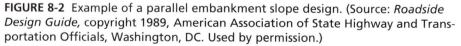

FIGURE 8-2 Example of a parallel embankment slope design. (Source: *Roadside Design Guide,* copyright 1989, American Association of State Highway and Transportation Officials, Washington, DC. Used by permission.)

age structures. In such instances, consideration should be given to an alternative drainage design in which water flows into a grated drop inlet or else is routed to a safer location.

8-3 Ditches and Drainage Structures

Ditches and drainage structures are integral and necessary elements of the highway cross-section. These elements must, however, be thoughtfully and properly designed, constructed, and maintained with consideration given to their effect on a safe roadside environment.

Ditches

Generally speaking, the side slopes of ditches should be as flat as possible consistent with drainage requirements. Preferably, the bottoms of ditches should be wide and rounded so as to be traversable by out-of-control vehicles. If deep, narrow side ditches must be used, they may need to be shielded by roadside barriers. Recommended side slopes of ditches are shown in Figures 8-3 and 8-4. Slope combinations that fall within the shaded regions of the figures are considered to be traversable.

Drainage Structures

It is also important that curbs, cross-drainage structures, parallel drainage culverts, and drop inlets be designed with proper consideration for safety. To enhance roadside safety, the following guidelines apply (*1*):

1. Eliminate all non-essential drainage structures.
2. Design, relocate, or modify drainage structures so they are traversable and not otherwise hazardous to an out-of-control vehicle.
3. Shield hazardous drainage features that cannot be redesigned or relocated with a suitable traffic barrier.

Curbs

The use of curbs along high speed roadways should be avoided. It is common for vehicles that are sliding sideways or slipping to trip and overturn upon impact with curbs. Curbs placed in front of roadside or median barriers can cause a vehicle to mount or cross over the barrier. If a curb must be used along high speed highways where a barrier is present, it is preferable to place it flush with the barrier or behind it.

Cross-Drains

Cross-drainage structures, which are designed to carry water underneath the roadway embankment, are varied in size and design. A description of such structures is given in Chapter 11. Care should be taken to ensure that cross-drains do not become a hazard, either as a protruding fixed object or as an opening into which a vehicle can drop.

   If a slope is traversable, the preferred safety treatment for a cross-drain is to extend it or shorten it to intercept the embankment slope and design the inlet or outlet end of the culvert to match the slope of the embankment. For single round culverts 0.9 m (30 in.) in diameter, no other treatment is needed. Larger

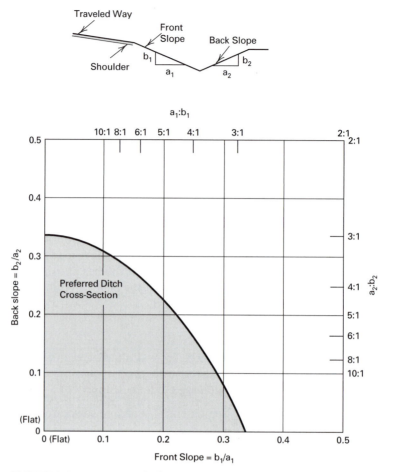

FIGURE 8-3 Recommended cross sections for ditches with abrupt slope changes. This figure is applicable to all **V**-ditches, rounded ditches with a bottom width less than 2.4 m (8 ft), and trapezoidal ditches with bottom widths less than 1.2 m (4 ft). (Source: *Roadside Design Guide*, copyright 1989, American Association of State Highway and Transportation Officials, Washington, DC. Used by permission.)

culverts can be made traversable by passenger cars by using bar grates or pipes to reduce the clear opening width. A recommended design is shown as Figure 8-5.

Under circumstances in which it is difficult to make the ends of culverts traversable, it may be desirable to extend the culvert ends beyond the clear recovery zone. In other instances, such as where the cost of extending large drainage structures is prohibitive, the preferred treatment may be to shield the ends of the drainage structure with appropriate traffic barriers.

Parallel Drainage Structures

Parallel drainage structures are placed generally parallel to the flow of traffic. These structures are commonly found under driveways, median crossovers, in-

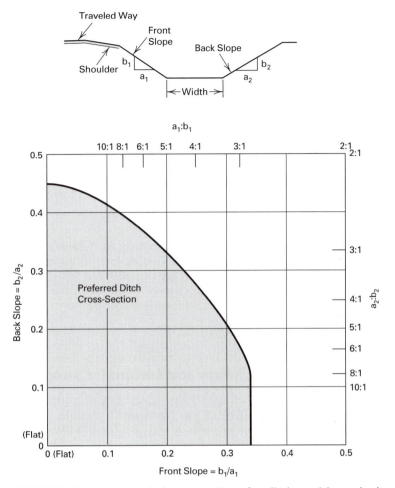

FIGURE 8-4 Recommended cross sections for ditches with gradual slope changes. The figure is applicable to rounded ditches with a bottom width of 2.4 m (8 ft) or more and trapezoidal ditches with bottom widths equal to or greater than 1.2 m (4 ft). (Source: *Roadside Design Guide,* copyright 1989, American Association of State Highway and Transportation Officials, Washington, DC. Used by permission.)

tersecting side roads, and other traveled ways generally at a right angle to the main street or highway. Such structures can be particularly hazardous because they can be struck head-on by out-of-control vehicles.

Safety guidelines for dealing with parallel drainage are similar to those for cross-drains. It is recommended, however, that in order to make such structures traversable, designers make the slopes of intersecting roadways as flat as possible but no more than 6 to 1 for areas likely to receive a high-speed impact. The ends of vulnerable drainage structures should be outfitted with sloped grates consisting of pipes or bars placed at a right angle to the main flow of traffic. Research has shown that pipes or bars constructed on 600 mm (24-in.) centers will usually function satisfactorily. Figure 8-6 shows a recommended end treatment for parallel drainage structures.

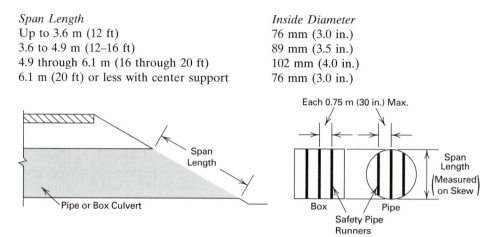

Span Length	Inside Diameter
Up to 3.6 m (12 ft)	76 mm (3.0 in.)
3.6 to 4.9 m (12–16 ft)	89 mm (3.5 in.)
4.9 through 6.1 m (16 through 20 ft)	102 mm (4.0 in.)
6.1 m (20 ft) or less with center support	76 mm (3.0 in.)

FIGURE 8-5 Recommended design criteria for safety treatment of cross-drains. The figure shows recommended pipe runner sizes for various span length for cross-drainage structures. The safety pipe runners are Schedule 40 pipes placed on centers of 0.75 m (30 in.) or less. (Source: *Roadside Design Guide,* copyright 1989, American Association of State Highway and Transportation Officials, Washington, DC. Used by permission.)

8-4 Breakaway Sign and Luminaire Supports

We have emphasized earlier the desirability of providing traversable areas along roadsides that are free of any obstacles that might be struck. This option is not always possible. For example, highway signs and lighting must be placed near the roadway if they are to serve their intended purpose.

Whenever possible, large signs such as those used on expressways should be placed on overhead grade separation structures or behind existing roadside barriers. Similarly, in certain circumstances, luminaires may be placed atop retaining walls, behind barriers, or otherwise shielded from encroaching vehicles. Where signs and luminaires must be located in areas exposed to errant vehicles, breakaway hardware should be used in most cases.

The term "breakaway support" refers to all types of sign, luminaire, and traffic signal supports that are designed to break or yield when struck by a vehicle (*1*). Most highway agencies have standard designs for structural breakaway supports for such highway elements (*7*).

Breakaway sign supports should not be placed in drainage ditches where erosion or freezing could interfere with the proper operation of the yielding support. Care must also be taken to ensure that the support is impacted at the bumper height, about 500 mm (20 in.) above the ground, and that vehicles not be snagged by protruding support foundations. Breakaway supports may not be suitable in urban locations where there are large concentrations of pedestrians and bicyclists who could be struck by sign hardware after a crash (*1*).

A variety of breakaway support designs have been proposed and evaluated on the basis of full-scale crash tests. Large wooden sign supports can be suitably modified by drilling holes through the center of the post parallel to the face of the sign. Such a modification may greatly lessen the force of impact without significantly reducing the resistance to forces caused by winds.

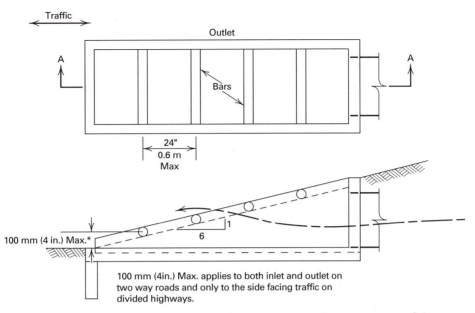

FIGURE 8-6 Recommended inlet and outlet design for safety treatment of drains parallel to traffic. (Source: *Roadside Design Guide,* copyright 1989, American Association of State Highway and Transportation Officials, Washington, DC. Used by permission.)

Two designs that have given satisfactory service for metal sign supports are shown in Figure 8-7. The small post slip-base design shown in Figure 8-7*a* is used for single post mountings for small signs and route markers. When struck, the connecting bolts shear and the entire assembly rotates over the moving vehicle. The hinged joint and slip-base support shown in Figure 8-7*b* is used for larger signs. A vehicle striking the post just above the slip base causes it to rotate about the hinged joint and clear the vehicle as it passes (*8*).

Several types of breakaway luminaire supports are commonly used including slip base, frangible coupling (couplers), and those with a frangible base (cast aluminum transformer base) (*1*).

Like sign supports, breakaway luminaire supports are designed to release in shear when impacted at the typical bumper height. As long as the negative side slopes between the roadway and the luminaire support are no steeper than 6 to 1, vehicles can be expected to strike the support at a suitable height (*1*).

Generally speaking, available breakaway hardware is suitable for luminaire poles not higher than about 17 m (55 ft). It is also recommended that in order to mitigate the consequences of a pole falling on a vehicle, the mass of the luminare support should not exceed about 450 kg (1,000 lb).

LONGITUDINAL BARRIERS

Longitudinal barriers are used to prevent vehicles from leaving the roadway and crashing into roadside obstacles, overturning, or crossing into the path of vehicles traveling in the opposite direction. In the following paragraphs, we discuss three

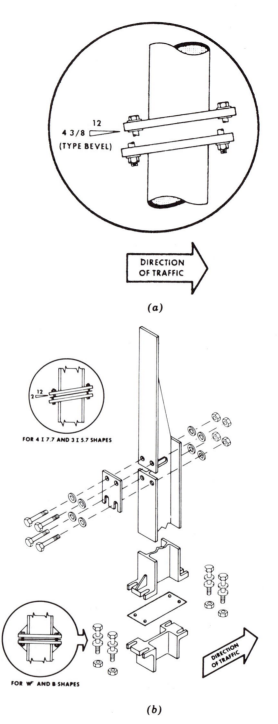

FIGURE 8-7 Breakaway sign supports. (Courtesy Federal Highway Administration.)

main classes of longitudinal barriers: roadside barriers, median barriers, and bridge railings.

Before they are placed into service, longitudinal barriers and related energy-absorbing terminals and transitions are usually tested and evaluated by full-scale vehicle crash tests. Highway engineering professionals have developed recommended procedures for the safety performance evaluation of these and other highway features (*9, 10*). These crash tests are performed according to exacting standards in which vehicle type and mass, angle of impact, speed, and other test parameters are observed and carefully controlled.

8-5 Roadside Barriers

Roadside barriers are used to shield motorists from hazards located along each side of a roadway. They are intended to redirect errant vehicles away from a hazard rather than to protect or prevent damage to a sign or other highway appurtenance. Their primary purpose is to prevent a vehicle from leaving the roadway and striking a fixed object or traversing a terrain feature that is considered more hazardous than the barrier itself. Because they may also constitute a fixed object hazard, roadside barriers should be used with care and discretion.

Various types of roadside barrier systems are in use at the present time including flexible, semirigid, and rigid systems. Flexible and semirigid barriers are commonly referred to as guardrails. The most important of these are the **W**-beam guardrail, the cable guardrail, and the box beam guardrail. These may be installed on wood, steel, or concrete posts. Some agencies have successfully used the so-called weak post system. In this system, the posts collapse when struck, and the rail deflects and absorbs the energy of impact. When strong posts are used, the rail should be blocked out or supported away from the post to minimize snagging of the vehicle and lessening the possibility of a vehicle vaulting over the barrier. Design features of some of these guardrail systems are given in Table 8-3. Concrete and stone masonry walls are examples of rigid barrier systems.

Roadside barriers are most commonly used to shield motorists from embankments and roadside obstacles. They are occasionally used to protect bystanders and pedestrians, especially at locations that have a history of run-off-the-road accidents.

As Figure 8-8 illustrates, the height and slope of embankments are the primary factors that determine whether guardrails should be used along embankments. Note that, generally speaking, barriers are not warranted on embankments with side slopes flatter than 3 to 1. Similar objective criteria are not available as warrants for guardrails to shield motorists from fixed obstacles or to protect bystanders from errant vehicles.

Lateral Placement of Barriers

It is important that roadside barriers be placed properly in relation to the edge of pavement and the hazard being shielded. Guidelines for proper placement of barriers include the following:

1. The barrier should be placed as far away from the traveled way as conditions permit.
2. The barrier should be placed far enough from the edge of pavement so as not to cause a motorist to reduce speed or change the vehicle position on

TABLE 8-3 Design Features of Guardrails That Have Performed Satisfactorily

System	Cable	W Beam (Steel Weak Post)	Box Beam	Blocked-Out W Beam (Wood Post)
Deflection	12 ft.[a]	8 ft	4 ft 2 ft	2 ft
Post spacing	16 ft	12 ft 6 in. nominal	6 ft 4 ft	6 ft 3 in.
Post	S3 × 5.7	S2 × 5.7		8 × 8-in. Douglas fir
Beam	Three $\frac{3}{4}$-in.-diam. steel cables	Steel **W** section	6 × 6 × 0.180-in. steel tube	Steel **W** section
Offset brackets			L5 × 3½ × ¼ in.-steel angle 4½ in. lg	8 × 8 × 14-in. Douglas fir blocks
Mounting	$\frac{5}{16}$-in.-diam. steel hook bolts	$\frac{5}{16}$-in.-diam. steel bolt	$\frac{3}{8}$-in.-diam. steel bolt (beam to angle)	$\frac{5}{8}$-in. carriage bolts
Footing	¼-in.- steel plate welded to post	¼-in. steel plate welded to post	¼-in. steel plate welded to post	None
Developed by	New York	New York	New York	California
Remarks	Revised 1971	Revised 1971	Increase height of rail from 30 to 35 in. on the outside of super-elevated curve. Revised 1971	Southern yellow pine is acceptable alternative to Douglas fir.

[a]1 ft = 0.3048 m.

Source: A Handbook of Highway Safety Design and Operating Practices, Federal Highway Administration, Washington, DC (1973 and 1978 Editions).

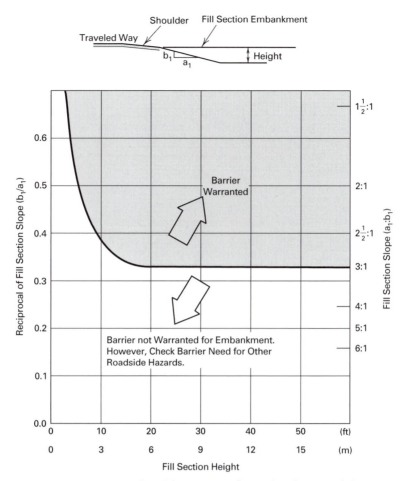

FIGURE 8-8 Comparative risk warrants for embankments. (Source: *Roadside Design Guide,* copyright 1989, American Association of State Highway and Transportation Officials, Washington, DC. Used by permission.)

the roadway. This distance, called the "shy line," varies with design speed. Recommended shy line distances are given in Table 8-4.

3. On embankments, guardrails should be placed no closer than 0.6 m (2 ft) from the outer edge of the shoulder to provide adequate post support.

4. The lateral placement of guardrails should take into account its tendency to deflect upon impact. Where the guardrail is shielding a rigid object, the barrier-to-hazard distance should be not less than the dynamic deflection of the guardrail.

Length of Barrier

The length of the barrier should be sufficient to prevent vehicle access behind the protective system. To provide adequate shielding along one-way facilities, the minimum required length, exclusive of specially designed terminal or end sections is

TABLE 8-4 Recommended Shy Line Offset Values

Design Speed, km/hr (mph)	Shy Line Offset Values, m (ft)
129 (80)	3.7 (12.0)
112 (70)	3.0 (10.0)
96 (60)	2.4 (8.0)
80 (50)	2.0 (6.5)
64 (40)	1.5 (5.0)
48 (30)	1.1 (3.5)

Source: Adapted from *Roadside Design Guide.* Copyright 1989. American Association of State Highway and Transportation Officials, Washington, DC. Used by permission.

$$X = \frac{L_H - L_2}{(L_H)/(L_R)} \tag{8-1}$$

where L_H = lateral distance from pavement edge to hazard
L_2 = lateral distance from pavement edge to barrier
L_R = encroachment distance parallel to the pavement edge

These relationships are illustrated in Figure 8-9.

Equation (8-1) applies to an installation in which the barrier is placed paralleled to the edge of pavement. Reference 1 gives an equation to calculate the length of barrier when its end is flared away from the edge of pavement.

End Treatments and Transitions

Special care should be exercised when designing and installing barrier ends and transitions. Unless properly designed, exposed guardrail ends can "spear" a vehicle and injure or kill its occupants. One approach to avoiding the spearing hazard is to flare guardrail ends away from the pavement and anchor them in an existing or false cut. Another method of treating the approach end of the guardrail is to design it to collapse and absorb the energy of the crash. Figure 8-10 shows a breakaway cable terminal that is designed to collapse upon impact.

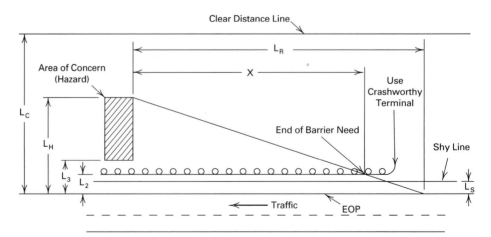

FIGURE 8-9 Guardrail barrier layout relationships. (SOURCE: Adapted from Ref. 1.)

FIGURE 8-10 Photograph of a breakaway cable terminal.

AASHTO's *Roadside Design Guide* (*1*) describes 10 barrier end treatments, including several experimental and proprietary devices that have been developed as a solution to the end-spearing problem.

Where guardrails transition to bridge parapets or other type of barrier, connections should be carefully designed. Unless special precautions are taken, there will be a tendency for the guardrail to deflect excessively and "pocket" the vehicle, directing it into the end of the barrier. Some concepts of barrier transition design are shown in Figure 8-11.

Median Barriers

Median barriers are used primarily to prevent vehicles from crossing the median and encroaching into opposing lanes. Although cross-median crashes are rare events, they are usually high-speed crashes that tend to be severe in terms of vehicular damages and personal injuries.*

Engineers now believe that freeways and expressways with average daily traffic volumes greater than 20,000 and median widths narrower than 9 m (30 ft) require the installation of median barriers. Adverse accident experience may indicate a need for barriers when the average daily traffic is less than 20,000 and the median width exceeds 9 m (30 ft).

Like roadside barriers, a number of median barriers have been developed and tested by full-scale crash tests. Design features of median barrier systems that

---

*During the late 1960s and early 1970s, when median barriers were not commonly provided along Atlanta's expressways, cross-median crashes in that city accounted for only about 4 percent of the expressways accidents but approximately half of the expressway fatalities (*12*).

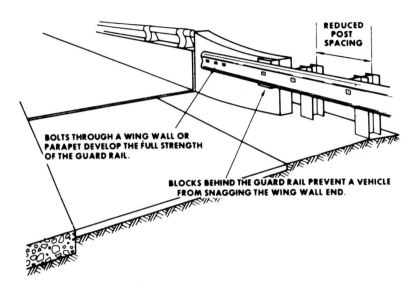

REDUCED
POST
SPACING

BOLTS THROUGH A WING WALL OR
PARAPET DEVELOP THE FULL STRENGTH
OF THE GUARD RAIL.

BLOCKS BEHIND THE GUARD RAIL PREVENT A VEHICLE
FROM SNAGGING THE WING WALL END.

FIGURE 8-11 Barrier transition design. (Courtesy Federal Highway Administration.)

have performed satisfactorily are given in Table 8-5. Barrier systems perform best when a level surface is provided in front of the barrier. Curbs, dikes, sloped shoulders, and stepped medians can cause a vehicle to vault over the barrier or strike it and overturn (*13*).

In narrow medians, concrete barriers such as those shown in Figure 8-12 are recommended. Such barriers are often used at retaining walls and along the face of rock cuts. If such a barrier is placed on an existing paved median, 25-mm-(1-in.) diameter dowels 200 m (8 in.) long should be placed on 1.2-m (4-ft) center-to-center spacing as shown in the figure.

Several other barrier systems have been considered including an experimental self-restoring rail designed to deflect upwards and back during contact and to return to its pre-impact position after a vehicle is redirected (*1*). Earth berms have also been used successfully to mitigate hazards located in medians such as bridge piers.

As with roadside barriers, the ends of median barriers require special attention. Untreated barrier ends can result in the beam penetrating the passenger compartment or cause intolerable decelerative forces in the crash. To mitigate these problems, metal barrier ends are often flared away from the road and anchored in the backslope. Rigid barrier ends may be shielded by earth berms or provided with energy-absorbing crash cushions.

8-7 Bridge Railings

Bridge railings are special types of longitudinal barriers designed to prevent vehicles from running off the edge of bridges or culverts. They differ from roadside barriers in that they are connected to the structure and are usually designed to have little or no deflection. The most appropriate type of bridge railing for a given location depends on the design speed, the traffic volume, and the percentage of heavy trucks and buses in the traffic stream. A large number of full-scale crash tests have shown the performance capabilities of a variety of bridge railings.

TABLE 8-5 Design Features of Median Barriers That Have Performed Satisfactorily

System	MB1 Cable	MB2 "W" Beam	MB3 Box Beam	MB4W Blocked-Out "W" Beam (Wood Post)
Deflection	11 ft[a]	7 ft	4 ft	2 ft
Post spacing	8 ft, 0 in.	12 ft, 6 in. nominal	6 ft, 0 in.	6 ft, 3 in.
Post	H2-1/4 × 4.1	S3 × 5.7	S3 × 5.7	8 × 8 in. Douglas fir
Beam	Two 3/4 in. diam. steel cables	Two steel "W" sections	8 × 6 × 1/4 in. steel tube	Two steel "W" sections Two C6 × 8.2 steel sections (rub rails).
Offset brackets				Two 8 × 8 × 14 in. Douglas fir blocks
Mountings	1/2 in. diam. steel "U" bolts	5/16 in. diam. bolts	Steel paddles	5/8 in. carriage bolts
Footings	Details vary with application	1/4 in. steel plate welded to post	1/4 in. steel plate welded to post	None
Developed by	California	New York	New York	California

[a]1 ft = 0.3048 m.

Source: A Handbook of Highway Safety Design and Operating Practice. Federal Highway Administration, Washington, DC, revised 1973 and 1978.

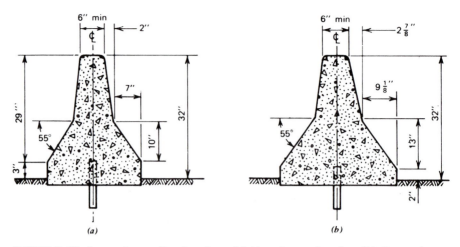

FIGURE 8-12 Concrete median barriers. (*a*) New Jersey barrier. (*b*) General Motors barrier. (Courtesy Transportation Research Board.)

The results of these tests, as well as field experience, costs, and aesthetics, usually enter into the selection process.

A large number of bridge railing systems are in common use. One of the simplest systems is a "thrie beam" rail mounted approximately 800 mm (32 in.) above the deck on wood or steel posts at 2.5 m (8 ft, 4 in.) centers. This is the only non-rigid bridge railing that has been successfully crash tested, and it is intended for low-volume secondary roads.

A system known as the Texas Type T6 has a rail made of two 12-gauge **W**-beam guardrails welded together to form a tubular shape. This strong rail is secured to a moderately weak steel **I**-beam posts spaced on 1.9-m (6.25-ft) centers. It has a standard mounting height of 685 mm (27 in.).

Another system known as the Oklahoma modified TR-1 bridge railing consists of a reinforced concrete post and beam system. The beam is 330 mm by 432 mm (1 ft 1 in by 1 ft 5 in) and is supported by 178 mm (7 in.) thick, 1.5 m (5 ft) long posts with 1.5 m (5 ft) openings between adjacent posts. The beam is offset 114 mm (4.5 in.) toward the roadway to minimize any potential for vehicle snagging.

Several varieties of reinforced concrete railings with a "safety shape" on its front face are designed to contain and redirect impacting vehicles. The shape of its front face is practically identical to that of a side of a concrete median barrier (Fig. 8-12). These railing are commonly 0.8 m (2 ft 8 in.) in height. In order to redirect heavy trucks, some agencies add a post-mounted steel railing atop the concrete barrier to provide a total height of about 1270 mm (50 in.).

Additional information on the selection and design of bridge railings may be found in Refs. 1 and 11.

8-8 Crash Cushions

Crash cushions are used to decelerate errant vehicles to a stop, greatly reducing the severity of head-on impact with a fixed object by spreading the energy of impact over time and space. Most crash cushions are also designed to redirect a vehicle away from a hazard for angled impacts. They may be seen in front of retaining walls, bridge piers, and abutments and in off-ramp gore areas. A variety of crash cushions have been proposed and tested, including rows of barrels, entrapment nets, and arrays of containers filled with sand or water. Most crash cushions are proprietary systems that have been carefully designed and tested by their manufacturers.

The design of crash cushions usually employs one of two concepts of mechanics: the kinetic energy principle or the conservation of momentum principle. In the first, it is assumed that the kinetic energy of a moving vehicle is converted to work by crushing some plastically deformable or crushable material. Hydraulic energy-absorbing systems also fall in this category. These types of crash cushions require a rigid support to resist the vehicle impact force that crushes the energy absorbing material. Example 8-1 illustrates the design of a crash cushion by the kinetic energy principle.

EXAMPLE 8-1 **Crash Cushion Design By Kinetic Energy Principle** A crash cushion is to be placed at an elevated expressway gore to safely decelerate a 2040-kg car traveling at a speed of 96 km/hr (26.7 m/sec). Two-hundred liter steel barrels with a 178-mm-diameter hole in the center of each end will serve as the basic element. Laboratory studies have indicated that a dynamic force of 40 kN is required to crush one barrel from its original 0.6 m diameter to approximately 0.15 m. The dynamic energy consumption is therefore $e_d = 0.45$ m $\times$ 40 kN = 18 kJ.

Determine the number and arrangement of barrels that will stop the car and the average deceleration level.

Solution The kinetic energy is

$$K.E. = 1/2 \text{ mass} \times \text{velocity}^2 = (2040 \times 26.7^2)/2$$
$$K.E. = 727{,}148 \text{ J}$$

The number of barrels required is

$$N_b = 727.148/18 = 40.4 \text{ barrels, say 41 barrels}$$

Suppose the barrels are arranged as shown in Figure 8-13. The barrier stopping force is a stepped function corresponding to the number of barrels in a row, as the figure illustrates.

Vehicle penetration into the crash cushion is determined by equating the kinetic energy to the area under the force-penetration curve, which represents the amount of work performed by the crushing of the barrels. The penetration of the single, double, and triple rows of barrels is $0.45 + 3 \times 0.45 + 6 \times 0.45 = 4.5$ m. Let X_1 be the penetration of the barrels arranged in four rows and X_2 the total penetration. The kinetic energy is

$$K.E. = 727.148 = (40 \times 0.45) + (80 \times 1.35) + (120 \times 2.7) + 160X_1$$
$$X_1 = 1.7 \text{ m}$$
$$X_2 = 1.7 + 4.5 = 6.2 \text{ m}$$

The average deceleration level is

$$G_a = \frac{v_0^2}{2 X_2^2} = \frac{(26.7)^2}{2(6.2)} = 57.5 \text{ m/sec}^2 \text{ or } 5.8g$$

A similar example in conventional U.S. units is given in Appendix B.

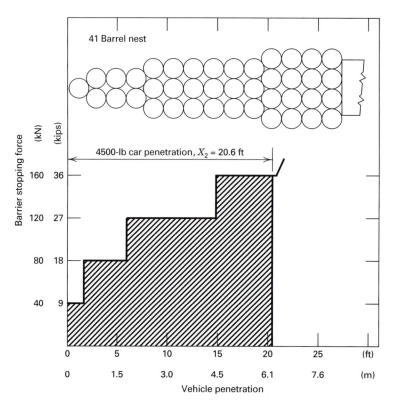

FIGURE 8-13 Analysis of crash cushion system, Example 8-1. (Courtesy National Cooperative Highway Research Program, Transportation Research Board.)

The conservation of momentum principle assumes that the momentum of a moving vehicle is transferred to an expendable mass of material located in the vehicle's path. This type of system is referred to as an inertial barrier. This type of barrier needs no rigid backup since the kinetic energy of the vehicle is being transferred to the other mass rather than being absorbed.

Consider an inertial barrel comprising sand-filled plastic barrels configured as shown in Figure 8-14. According to the principle of conservation of momentum,

$$W_v V_0 = W_v V_1 + W_1 V_1 \tag{8-2}$$

where W_v = mass of vehicle
V_0 = original impact velocity
W_1 = mass of sand in first barrel(s)
V_1 = velocity after first impact

and

$$V_1 = V_0 W_v/(W_v + W_1) \tag{8-3}$$

The general equation is

$$V_n = V_{n-1} W_v/(W_v + W_n) \tag{8-4}$$

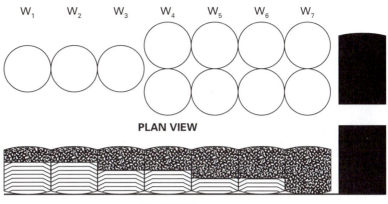

PLAN VIEW

ELEVATION

FIGURE 8-14 Configuration of example sand-filled plastic barrels.

EXAMPLE 8-2 **Crash Cushion Design By Conservation of Momentum** A 2040-kg car impacts a sand-filled inertial barrier head on at a speed of 72 km/hr (20 m/sec). The system, which is shown below, was designed for a speed of 80 km/hr. Determine the speed after impact with each of the rows of barrels and the average rate of deceleration. The barrels are 0.91 m (3 ft) in diameter, and the length of the system is 6.4 m (21 ft). The numbers shown on the sketch show the masses of each barrel of sand in kilograms.

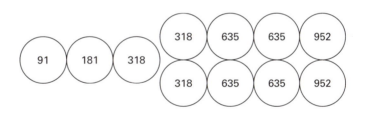

Solution By equation (8-4),

$$V_1 = (72)(2040)/(2040 + 91) = 68.9 \text{ km/hr}$$
$$V_2 = (68.9)(2040)/(2040 + 181) = 63.3 \text{ km/hr}$$
$$V_3 = (63.3)(2040)/(2040 + 318) = 54.8 \text{ km/hr}$$
$$V_4 = (54.8)(2040)/(2040 + 636) = 41.8 \text{ km/hr}$$
$$V_5 = (41.8)(2040)/(2040 + 1270) = 25.8 \text{ km/hr}$$
$$V_6 = (25.8)(2040)/(2040 + 1270) = 15.9 \text{ km/hr}$$
$$V_7 = (13.3)(2040)/(2040 + 1904) = 8.2 \text{ km/hr}$$

Note that, theoretically, the vehicle cannot be stopped completely by this concept; however, it is usually adequate to show that the computed vehicle velocity decreases to about 16 km/hr (10 mph). The remaining energy is dissipated as the vehicle "bulldozes" through the modules (*1*).

The average deceleration is

$$G_a = \frac{V_0^2 - V_7^2}{2\,D} = \frac{(20)^2 - (2.3)^2}{2(6.4)} = 30.8 \text{ m/sec}^2$$

$$G_a = 30.8/9.8 = 3.1\ g$$

The selection of an appropriate crash cushion depends not only on its dynamic performance characteristics but also on the space available for the cushion and the installation, maintenance, and damage repair costs.

Additional information on the location, selection, and maintenance of various types of highway traffic barriers is given in Ref. 13.

8-9 Conclusion

In this chapter we have only been able to briefly outline some of the more important aspects of roadside design. The reader is referred to the references listed at the end of this chapter for a more complete treatment of this important area of study.

REFERENCES

1. *Roadside Design Guide.* American Association of State Highway and Transportation Officials, Washington, DC (1989).

2. Wright, Paul H., and Robertson, Leon S. Studies of Roadside Hazards for Projecting Fatal Crash Sites. *Transportation Research Record 609,* Transportation Research Board, Washington, DC (1976).

3. Wright, Paul H. and Paul Zador. Study of Fatal Rollover Crashes in Georgia. *Transportation Research Record 819,* Transportation Research Board, Washington, DC (1981).

4. Krammes, Raymond A. *Design Speed and Operating Speed in Rural Highway Alignment Design.* Paper No. 940996, Prepared for Presentation at the 73rd Annual Meeting of the Transportation Research Board, Washington, DC (1994).

5. Anderson, Ingrid B. and Krammes Raymond A. *Speed Reduction as a Surrogate for Accident Experience at Horizontal Curve on Rural Highways.* Paper No. 940994, Prepared for Presentation at the 73rd Annual Meeting of the Transportation Research Board, Washington, DC (1994).

6. Lamm, Ruediger, Choueiri, Elias M., Hayward, John C., and Paluri Anand. Possible Design Procedure to Promote Design Consistency in Highway Geometric Design on Two-Lane Rural Roads. *Transportation Research Record 1195,* Transportation Research Board, Washington, DC (1988).

7. *Standard Specifications for Structural Supports for Highway Signs, Luminaires, and Traffic Signals.* American Association of State Highway and Transportation Officials, Washington, DC (1986).

8. *Handbook of Highway Safety Design and Operating Practices.* U.S. Department of Transportation, Federal Highway Administration, Washington, DC (1978).

9. *Recommended Procedures for the Safety Performance Evaluation of Highway Safety Appurtenances.* National Cooperative Highway Research Program Report 230, Transportation Research Board, Washington, DC (1980).

10. *Recommended Procedures for the Safety Performance Evaluation of Highway Features.* National Cooperative Highway Research Report 350, Transportation Research Board, Washington, DC (1993).

11. *Standard Specifications for Highway Bridges,* 11th Ed. American Association of State Highway and Transportation Officials, Washington, DC (1992).

12. Wright, Paul H., and Arrillaga, Bert. Simulation of Cross-Median Crashes. *Highway Research Record 388,* Highway Research Board, Washington, DC (1972).

13. Miche, J. D., and Bronstad, M. E. *Location, Selection, and Maintenance of Highway Traffic Barriers.* National Cooperative Highway Research Program Report No. 118, Transportation Research Board, Washington, DC (1971).

PROBLEMS

8-1 Compute the minimum length of guardrail required to shield a hazard located 25 ft from the pavement edge of a one-way facility. The guardrail is to be located 12 ft from the pavement edge. Assume that vehicles will encroach at an angle of 11° to the pavement edge. If the roadway has two-way operation, what minimum length of guardrail will be required?

8-2 Compute the minimum length in meters of guardrail required to shield a hazard located 8 m from the edge of a one-way facility. The guardrail is to be located 3.5 m beyond the pavement edge. The encroachment angle is 15°.

8-3 Given the crash cushion described in Example 8-1, how far will a 2500-lb car traveling 55 mph at impact penetrate the crash cushion? Calculate the average deceleration rate in *g*'s.

8-4 Given the crash cushion described in Example 8-1, which a 1200-kg car impacts at a speed of 80 km/hr: Estimate the penetration in meters and the average deceleration in meters per second squared and in *g*'s.

8-5 Given the crash cushion described in Example 8-2, which a 3500-lb car impacts at a speed of 60 mph: Compute the speed of the vehicle after impact with each of the rows of barrels and the average rate of deceleration in feet per second squared and in *g*'s.

8-6 Given the crash cushion described in Example 8-2, which a 1585-kg car impacts at a speed of 80 km/hr: Compute the speed of the vehicle after impact with each of the rows of barrels and the average rate of deceleration in meters per second squared and in *g*'s.

CHAPTER 9

INTERSECTIONS, INTERCHANGES, TERMINALS

In Chapter 7, we presented standards and procedures for the design of highway links, the facilities that serve "line-haul" transportation. In this chapter, we will discuss the design of facilities to accommodate trips at nodes and termini. We will discuss the design of (1) intersections and interchanges that provide for change in travel direction and (2) parking facilities and terminals that provide for interaction with passengers and the storage of vehicles at the termini.

INTERSECTIONS AT GRADE, GRADE SEPARATIONS, AND INTERCHANGES

9-1 Intersections at Grade

Most highways intersect at grade, and the intersection area should be designed to provide adequately for turning and crossing movements, with due consideration given to sight distance, signs, grades, and alignment.

Simple intersections at grade consist of a junction of three, four, or more road approaches. A junction of three approaches is indicated as a "branch," **T**, or **Y**. A branch may be defined as an offshoot of a main-traveled highway, and it usually has a small deflection angle. A **T** intersection is one in which two approaches intersect to form a continuous highway and the third approach intersects at, or nearly at, right angles. A **Y** intersection is one in which three approaches intersect at nearly equal angles. In addition to the above types of intersection, the flared intersection may be used. This consists of additional pavement width or additional traffic lanes at the intersection area. A few examples of these types of intersections are given in Figure 9-1.

The design of the edge of the pavement for a simple intersection should provide sufficient clearance between the vehicle and the other traffic lanes. It is frequently assumed that all turning movements at intersections are accomplished at speeds of less than 32 km/hr (20 mph), and the design is based on the physical characteristics of the assumed design vehicle.

The design of the curb line at intersections depends on the composition of the traffic, the angle of intersection, the right-of-way available, the width and number of lanes on the intersecting streets, and the needs of pedestrians. To facilitate traffic movements, the curb line should closely conform to the turning path of the largest class of vehicles that, in significant numbers, are to turn at the intersection. Where traffic predominantly consists of passenger vehicles, arcs of circles with radii of 4.6 to 7.6 m (15–25 ft) may be sufficient. Whenever feasible, corner

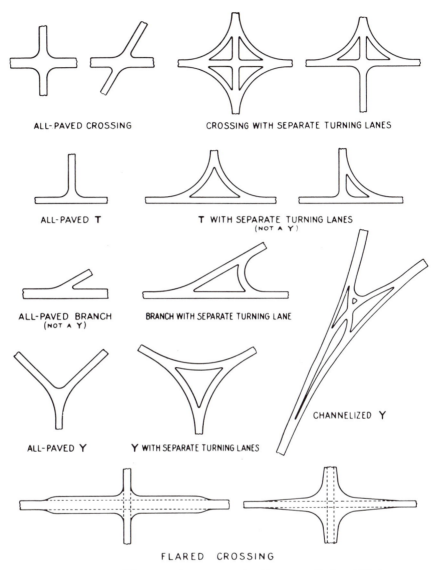

ALL-PAVED CROSSING   CROSSING WITH SEPARATE TURNING LANES

ALL-PAVED T   T WITH SEPARATE TURNING LANES
(NOT A Y)

ALL-PAVED BRANCH   BRANCH WITH SEPARATE TURNING LANE
(NOT A Y)

CHANNELIZED Y

ALL-PAVED Y   Y WITH SEPARATE TURNING LANES

FLARED CROSSING

FIGURE 9-1 Types of intersections at grade. (Courtesy Federal Highway Administration.)

radii of 7.6 to 9.1 m (25 and 30 ft) should be provided, respectively, for minor and major cross streets on new construction and reconstruction (*1*). Tables 9-1 and 9-2 give recommended minimum edge of pavement designs at intersections. It will be noted that three-centered compound curves are recommended, especially for sharp turns and for the larger classes of vehicles. Alternatively, simple curves combined with spirals or tapers to conform to the paths of vehicles may be used.

9-2 Islands and Channels

Traffic islands or pavement markings may be used at intersections to facilitate the safe and orderly flow of vehicles and pedestrians. Properly designed chan-

TABLE 9-1 Minimum Edge of Pavement Designs for Turns at Intersections—Simple Curves and Tapers

Angle of Turn (degrees)	Design Vehicle	Simple Curve Radius (m)	Simple Curve Radius with Taper		
			Radius (m)	Offset (m)	Taper (m:m)
60	P	12	—	—	—
	SU	18	—	—	—
	WB-12	28	—	—	—
	WB-15	45	29	1.0	15:1
	WB-19	50	43	1.2	15:1
	WB-20	60	43	1.3	15:1
	WB-29	46	29	0.8	15:1
	WB-35	—	54	1.3	20:1
75	P	11	8	0.6	10:1
	SU	17	14	0.6	10:1
	WB-12	—	18	0.6	15:1
	WB-15	—	20	1.0	15:1
	WB-19	—	43	1.2	20:1
	WB-20	—	43	1.3	20:1
	WB-29	—	26	1.0	15:1
	WB-35	—	42	1.7	20:1
90	P	9	6	0.8	10:1
	SU	15	12	0.6	10:1
	WB-12	—	14	1.2	10:1
	WB-15	—	18	1.2	15:1
	WB-19	—	36	1.2	30:1
	WB-20	—	37	1.3	30:1
	WB-29	—	25	0.8	15:1
	WB-35	—	35	0.9	15:1
105	P	—	6	0.8	8:1
	SU	—	11	1.0	10:1
	WB-12	—	12	1.2	10:1
	WB-15	—	17	1.2	15:1
	WB-19	—	35	1.0	30:1
	WB-20	—	35	1.0	30:1
	WB-29	—	22	1.0	15:1
	WB-35	—	28	2.8	15:1
120	P	—	6	0.6	10:1
	SU	—	9	1.0	10:1
	WB-12	—	11	1.5	8:1
	WB-15	—	14	1.2	15:1
	WB-19	—	30	1.5	25:1
	WB-20	—	31	1.6	25:1
	WB-29	—	20	1.1	15:1
	WB-35	—	26	2.8	15:1

Source: From *A Policy on Geometric Design of Highways and Streets,* copyright 1994. American Association of State Highway and Transportation Officials, Washington, DC. Used by permission.

TABLE 9-2 Minimum Edge of Pavement Designs for Turns at Intersections—Three Centered Compound Curves

Angle of Turn (degrees)	Design Vehicle	3-Centered Compound Curve Radii (m)	Symmetric Offset (m)	3-Centered Compound Curve Radii (m)	Asymmetric Offset (m)
60	P	—	—	—	—
	SU	—	—	—	—
	WB-12	—	—	—	—
	WB-15	60-23-60	1.7	60-23-84	0.6-2.0
	WB-19	120-30-120	4.5	34-30-67	3.0-3.7
	WB-20	122-30-122	2.4	76-38-183	0.3-1.8
	WB-29	76-24-76	1.4	61-24-91	0.6-1.7
	WB-35	198-46-198	1.7	61-43-183	0.5-2.4
75	P	30-8-30	0.6	—	—
	SU	36-14-36	0.6	—	—
	WB-12	36-14-36	1.5	36-14-60	0.6-2.0
	WB-15	45-15-45	2.0	45-15-69	0.6-3.0
	WB-19	134-23-134	4.5	43-30-165	1.5-3.6
	WB-20	128-23-128	3.0	61-24-183	0.3-3.0
	WB-29	76-24-76	1.4	30-24-91	0.5-1.5
	WB-35	213-38-213	2.0	46-34-168	0.5-3.5
90	P	30-6-30	0.8	—	—
	SU	36-12-36	0.6	—	—
	WB-12	36-12-36	1.5	36-12-60	0.6-2.0
	WB-15	55-18-55	2.0	36-12-60	0.6-3.0
	WB-19	120-21-120	3.0	48-21-110	2.0-3.0
	WB-20	134-20-134	3.0	61-21-183	0.3-3.4
	WB-29	76-21-76	1.4	61-21-91	0.3-1.5
	WB-35	213-34-213	2.0	30-29-168	0.6-3.5
105	P	30-6-30	0.8	—	—
	SU	30-11-30	1.0	—	—
	WB-12	30-11-30	1.5	30-17-60	0.6-2.5
	WB-15	55-14-55	2.5	45-12-64	0.6-3.0
	WB-19	160-15-160	4.5	110-23-180	1.2-3.2
	WB-20	152-15-152	4.0	61-20-183	0.3-3.4
	WB-29	76-18-76	1.5	30-18-91	0.5-1.8
	WB-35	213-29-213	2.4	46-24-152	0.9-4.6
120	P	30-6-30	0.6	—	—
	SU	30-9-30	1.0	—	—
	WB-12	36-9-36	2.0	30-9-55	0.6-2.7
	WB-15	55-12-55	2.6	45-11-67	0.6-3.6
	WB-19	160-21-160	3.0	24-17-160	5.2-7.3
	WB-20	168-14-168	4.6	61-18-183	0.6-3.8
	WB-29	76-18-76	1.5	30-18-91	0.5-1.8
	WB-35	213-26-213	2.7	46-21-152	2.0-5.3

Source: From *A Policy on Geometric Design of Highways and Streets,* copyright 1994. American Association of State Highway and Transportation Officials, Washington, DC. Used by permission.

nelization systems increase intersection capacity and decrease conflicts and accidents.

Islands in an intersection serve one or more of the following purposes (*1*):

1. Separation of conflicts.
2. Control of angle of conflict.
3. Reduction of excessive pavement areas.
4. Regulation of traffic flow in the intersection area.
5. Arrangements to favor a predominant turning movement.
6. Protection of pedestrians.
7. Protection and storage of turning and crossing vehicles.
8. Location of traffic control devices.

Islands are generally grouped into three major classes: directional or channelized, divisional, and refuge. General types and shapes of islands are shown in Figure 9-2.

Directional islands are designed primarily to guide the motorist through the intersection by indicating the intended route. Where spacious area exists at an intersection and leaves much to the discretion of the driver, islands may be used to channel the motorist into the desired lane by placing a channeling island in the little-used portion of the intersection.

The placing of directional islands should be such that the proper course of travel is immediately evident and easy to follow. A complicated system of islands where the desired course of travel is not immediately evident may result in confusion of the motorist and may be of more hindrance than help in maintaining

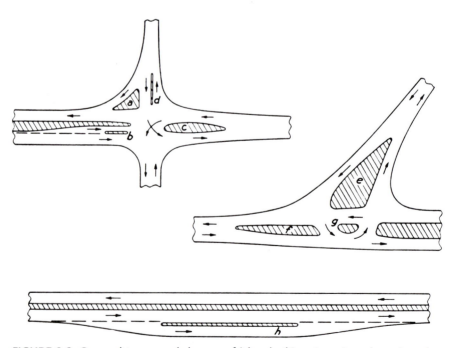

FIGURE 9-2 General types and shapes of islands. (Courtesy American Association of State Highway and Transportation Officials.)

a steady traffic flow. Islands should be so placed that crossing streams of traffic will pass at approximately right angles and merging streams of traffic will converge at flat angles. By use of such angles there will be less hindrance to traffic on the thoroughfare, and the possibility of accidents in the intersection will be decreased.

Divisional islands are most frequently used on undivided highways approaching intersections. They serve to alert the driver to the intersection and regulate the flow of traffic into and out of the intersection. Their use is particularly advantageous for controlling left-turning traffic at skewed intersections.

A refuge island is located at or near crosswalks to aid and protect pedestrians crossing the roadway. Refuge islands are most generally used on wide streets in urban areas for loading and unloading of transit riders.

Islands should be large enough to command attention. AASHTO (*1*) recommends that islands have an area of at least 7 m² (75 ft²) and preferably 9.2 m² (100 ft²). This means that the sides of triangular islands should be not less than about 3.6 m (12 ft) but preferably 4.6 m (15 ft) after rounding of corners. It is recommended that elongated or divisional islands be at least 1.2 m (4 ft) wide and 6 to 7.6 m (20–25 ft) long (*1*).

Islands should be visible to drivers for a sufficient distance to allow appropriate response. Generally speaking, islands should not be used on or near the crest of vertical curves or where sight distance is restricted due to horizontal curvature.

> For median curbed islands, the face of the curb at the approach island nose should be offset at least 0.6 m (2 ft), and preferably 1.2 m (4 ft) from the normal (median) edge of pavement. The median curbed island should then be gradually widened to its full width. For other curbed islands, the total nose offset should be 1.2 to 1.8 m (4–6 ft) from the normal edge of through pavement, and 0.6 to 0.9 m (2–3 ft) from the pavement edge of a turning roadway. (*1*)

Steep-sided island curbs approximately 175 mm (7 in.) in height are recommended in congested areas to provide a greater degree of protection for pedestrians. Mountable curbs are appropriate in outlying areas where pedestrian volumes are low. Painted flush islands may be used in sparsely developed areas where approach speeds are relatively high, pedestrian traffic is light, and signs, signals, or lighting standards are not needed on the island area.

9-3 Rotary Intersections

A rotary intersection is one in which all traffic merges into and emerges from a one-way road around a central island. Advantages of a rotary intersection include the following:

1. Traffic moves continuously from all legs at reduced speed at periods of low volume.
2. Because crossing movements are eliminated, accidents are likely to be less serious.
3. Where more than four intersection legs are involved, the design layout may be simplified. In such a case, a channelized layout may become extremely complicated.
4. The cost of an at-grade intersection may be considerably less than that of grade separation structures.

Disadvantages include the following:

1. A rotary can handle no more traffic than an adequately designed channelized layout.
2. It has been found that unsatisfactory functioning occurs when two or more legs approach design capacity.
3. The area involved to satisfy proper geometric design of a rotary is extensive. This requirement for large flat areas may render the cost prohibitive in urban areas or areas of difficult topography.
4. Channelization will often prove more acceptable if large pedestrian traffic is expected.
5. Most rotaries are designed to function at low speeds. If high speeds are anticipated the large lengths of required weaving sections may cause prohibitive land costs.
6. For proper functioning, it has been found that adequate lighting must be provided and access controlled on the intersecting legs. Failing these provisions, the intersection may become a serious safety hazard.

Because of these disadvantages, highway engineers in the United States seldom recommend the use of rotary intersections for new construction.

9-4 Grade Separations and Interchanges

Intersections at grade can be eliminated by the use of grade-separation structures that permit the cross flow of traffic at different levels without interruption. The advantage of such separation is the freedom from cross interference with resultant saving of time and increase in safety for traffic movements.

Grade separations and interchanges may be warranted (1) as part of an express highway system designed to carry volumes of traffic, (2) to eliminate bottlenecks, (3) to prevent accidents, (4) where the topography is such that other types of design are not feasible, (5) where the volumes to be catered for would require the design of an intersection at grade of unreasonable size, and (6) where the road user benefit of reducing delays at an at-grade intersection exceeds the cost of the improvement (*1*).

An interchange is a grade separation in which vehicles moving in one direction of flow may transfer by the use of connecting roadways. These connecting roadways at interchanges are called ramps.

Many types and forms of interchanges and ramp layouts are used in the United States. These general forms may be classified into four main types:

1. **T** and **Y** interchanges.
2. Diamond interchanges.
3. Partial and full cloverleafs.
4. Directional interchanges.

T and Y Interchanges

Figure 9-3 shows typical layouts of interchanges at three-legged junctions. The geometry of the interchange can be altered to favor certain movements by the provision of large turning radii and to suit the topography of the site. The trumpet interchange has been found suitable for orthogonal or skewed intersections.

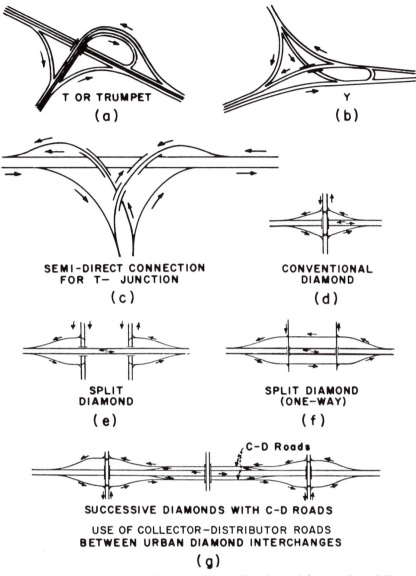

FIGURE 9-3 Highway interchanges. (Partially adapted from Adaptability of Interchanges to Interstate Highways, *Transactions of the American Society of Civil Engineers,* Vol. 124, p. 558.)

Figure 9-3*a* favors the left turn on the freeway by the provision of a semidirect connecting ramp. Figure 9.3*c* indicates an intersection where all turning movements are facilitated in this way.

Diamond Interchanges

The diamond interchange is adaptable to both urban and rural use. The major flow is grade-separated, with turning movements to and from the minor flow achieved by diverging and merging movements with through traffic on the minor flow. Only the minor flow directions have intersections at grade. In rural areas, this is generally acceptable owing to the light traffic on the minor flow. In urban areas, the at-grade intersections generally will require signalized control to prevent serious interference of ramp traffic and the crossing arterial street. The design of the intersection should be such that the signalization required does not impair the capacity of the arterial street. To achieve this, widening of the arterial may be necessary in the area of the interchange. Care must also be taken in the design of the ramps so that traffic waiting to leave the ramps will not back up into through lanes of the major flow.

One disadvantage of the diamond interchange is the possibility of illegal wrong-way turns, which can cause severe accidents. If the geometry of the intersection may lead to these turns, the designer can use channelization devices and additional signing and pavement marking. Wrong-way movements are, in general, precluded by the use of cloverleaf designs.

Figure 9-3*d* shows the conventional diamond interchange. Increased capacity of the minor flows can be attained by means of the arrangement shown in Figure 9-3*e* or Figure 9-3*f.* The arrangement shown in Figure 9-3*g* is suitable where two diamond ramps are in close proximity. Weaving movements which would, in this case, inhibit the flows of the major route, are transferred to the parallel collector-distributor roads.

Single Point Urban Interchanges (2)   A new type of diamond interchange has emerged recently called the single-point urban interchange (SPUI). With the SPUI, all through and left turn movements converge into a single, signalized intersection area instead of two separate intersection areas normally found with a diamond interchange (Fig. 9-4). The advantage of this design feature is that all intersecting movements can be served by a single signal with, at most, one stop required to pass through the interchange. Proponents of the SPUI claim that it provides improved flows, safer operation, and reduced right-of-way needs compared with other alternative configurations.

The SPUI is an unusual design requiring crossroad drivers to rely heavily on guide signing, pavement markings, and lane use signing in order to travel safely through the intersection area. Designers of the SPUI will need to exercise special care to ensure that the intersection functions as it should. Merritt makes these suggestions:

1. A design speed of 32 to 64 km/hr (20–40 mph) for the left-turning roadways.
2. Long, constant radii of left-turning roadways ranging from about 52 to 122 m (170–400 ft).
3. A minimum of 1.8-m (6-ft) clearance between the outside edge lines of opposing left-turning movements and 3-m (10-ft) vehicle body clearance.
4. Increased pavement width on turning roadways to accommodate the off-tracking characteristics of large trucks.
5. An exclusive, actuated pedestrian signal phase to accommodate pedestrian movements, or directions to pedestrians to the nearest intersection where they can safely cross the crossroad.
6. Control of access on the crossroad within the functional boundary of the intersections; no driveways or entrance approaches should be permitted in the SPUI vicinity.

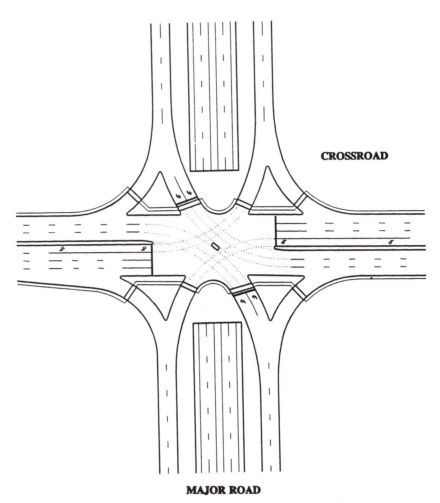

CROSSROAD

MAJOR ROAD

FIGURE 9-4 An overpass single point urban interchange. (Courtesy Transportation Research Board.)

7. Lane markings at a high level of visibility to enhance the delineation of left-turn movements and provide the appropriate level of positive guidance.

Partial and Full Cloverleafs

The partial cloverleafs shown in Figure 9-5 are sometimes adopted in place of the diamond interchange. Traffic can leave the major flow either before or after the grade-separation structure, depending on the quadrant layout. The intersections at grade for the minor road are present as for the diamond interchange, but the probability of illegal turning movements can be reduced. By the provision of two on-ramps for each direction of the major route was in Figure 9.5c, left-turn traffic on the minor route can be eliminated.

The more conventional arrangement of the full cloverleaf, which can be adapted to nonorthogonal layouts, eliminates at-grade crossings of all traffic streams for both major and minor roads. The ramps may be one-way, two-way,

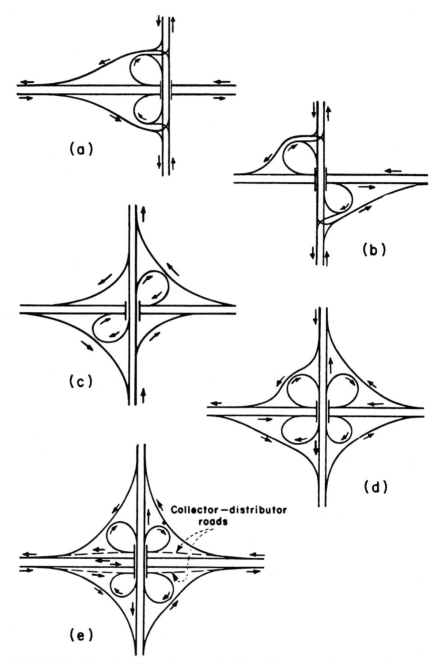

FIGURE 9-5 Cloverleaf interchanges. (Adapted from *A Policy on Geometric Design of Rural Highways—1965,* American Association of State Highway and Transportation Officials.)

two-way separated, or two-way unseparated roads. Although all crossing movements are eliminated, the cloverleaf design has some disadvantages: (1) the layout requires large land areas and (2) decelerating traffic wishing to leave the through lanes must weave with accelerating traffic entering the through lanes. Figure 9-5e is a layout using collector-distributor roads to overcome this second disadvantage. The use of collector-distributor roads also has certain operational and signing advantages. An exiting driver makes only one decision (i.e., to transfer to the cross road) while traveling on the high-speed, high-volume facility. The second decision—for example, to proceed north or to proceed south—is made off the high-speed facility (*3*).

Figure 9-6 shows a full cloverleaf interchange in California.

Directional Interchanges

Directional interchanges are used whenever one freeway joins or intersects another freeway. The outstanding design characteristic of this type of interchange is the use of a high design speed throughout, with curved ramps and roadways of large radius. The land requirements for a directional interchange are therefore very large. In cases where volumes for certain turning movements are small, design speeds for these movements are reduced and the turnoff is effected within a loop. Figure 9-7 shows a four-level directional interchange.

Recently, experience with directional interchanges has revealed operational problems associated with left-hand entrance and exit ramps. Most drivers expect to exit freeways to the right and to enter from the right. When those expectations are violated, confusion, erratic maneuvers, and accidents sometime result.

Consider the alternative designs shown in Figure 9-8. Design A is a directional interchange in which a driver wishing to go right bears right and a driver wishing to go left bears left. A northbound driver must make two decisions as he ap-

FIGURE 9-6 A full cloverleaf interchange in California.
(Courtesy Federal Highway Administration.)

FIGURE 9-7 A four-level directional interchange. (Courtesy Federal Highway Administration.)

proaches the first exit: (1) that he is going to leave the freeway at this exit, and (2) whether his destination is to the right (east) or left (west). Confronted with such interchange configuration, a driver may be in the right-hand lane, decide that he must go west to reach his destination, and be faced with a need to cross several lanes of high-speed traffic in order to exit to the left.

Design B (Figure 9-8) is a preferred single exit configuration on which a northbound driver makes decisions one at a time. The driver exits to the right and then decides whether to proceed east or west while on the lower speed, less congested exit ramp.

9-5 Design of Ramps and Turning Roadways

A "turning roadway" is a connecting roadway for traffic turning between two legs of an intersection. The term "ramp" is used to refer to all types of turning roadways that connect two or more legs at an interchange (*1*). The principles for the design of open highways are generally applicable to the design of ramps and turning roadways. There are, of course, differences. The design speeds for turning roadways are considerably lower than for the open highway but should be as high as practicable.

Research has shown that drivers will operate on turning roadways at higher speeds in relation to curvature than on the open highway. Thus, larger side-friction factors are used for design. AASHTO (*1*) recommends friction factors of 0.32, 0.23, and 0.18 for design speeds of 24, 40, and 56 km/hr (15, 25, and 35 mph), respectively. On the other hand, limitations in space preclude the devel-

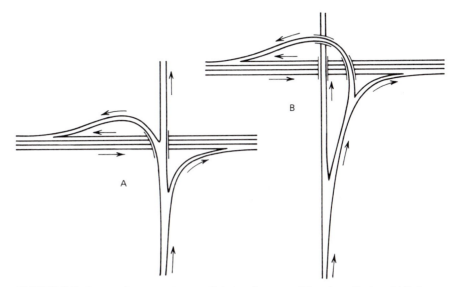

FIGURE 9-8 One-exit versus two-exit interchanges. (Courtesy Federal Highway Administration.)

opment of superelevation rates on intersection curves as large as those on the open highway. With appropriate selection of values for a friction factor and a maximum rate of superelevation, the minimum radius of curvature can be calculated by Eq. 7-2.

Compound and transition curves are recommended for turning roadways to more closely fit the natural paths of vehicles. Where compound curves are used, it is preferred that the ratio of the flatter radius to the sharper radius not exceed 1.75. Where that ratio exceeds 2, a spiral transition should be placed between the two curves (*1*).

Gradients for turning roadways should be as flat as feasible and generally should not exceed 6 percent. Stopping sight distance should be provided for turning roadways.

The recommended widths of turning roadways depend on the radius of curvature of the inner edge of the pavement, the class of design vehicle, and the type of operation (one-way, two-way, etc.). The recommended widths vary from 3.6 to 15 m (12–50 ft), depending on the conditions.

9-6 Railroad Grade Intersections

In the design of a highway that intersects a railroad at grade, consideration must be given to approach grades, sight distance, drainage, volume of vehicular traffic, and the frequency of regular train movements at the particular intersection. The particular type of surfacing and kind of construction at railroad crossings at grade will depend on the class of railroad and kind of roadway improvement.

All railroad intersections at grade require proper advance warning signs. At crossings on heavily traveled highways where conditions justify, automatic devices should be installed. Recommended standards for railroad-highway grade-crossing protection have been adopted by the Association of American Railroads.

The use of grade separations at railroad crossings is recommended at all main-line railroads that consist of two or more tracks and all single-line tracks when regular train movements consist of six or more trains per day. Other considerations for separating railroad and highway traffic are the elements of delay and safety.

Railroad grade-separation structures may consist of an overpass on which the highway is carried over the railroad or an underpass that carries the highway under the railroad. The selection of the type of structure will depend in large part on the topographic conditions and a consideration of initial cost. Drainage problems at underpasses can be serious. Pumping of surface and subsurface water may have to be carried on a large part of the time, and failure of power facilities sometimes causes flood conditions at the underpass with the resultant stoppage of traffic.

DESIGN OF PARKING FACILITIES

Widespread parking problems exist in business districts, on college campuses, at airports, and in industrial areas and other highly developed areas. Shortages of parking spaces in close proximity to major traffic generators have worsened as automobile registrations have increased and more and more people have opted to travel by automobile. Parking problems have been aggravated in many cities as curb parking has been eliminated to improve flow.

The procedures for conducting parking surveys are described in Chapter 3. In the following paragraphs, we will discuss principles for the location, layout, and design of off-street and on-street parking facilities.

9-7 Location of Off-Street Parking Facilities

Perhaps the most important aspect of parking facility planning and design is the choice of a site. Experience has shown that an improperly located lot or garage is likely to fail or have limited use, even if located only a few blocks away from the center of parking demand.

Most parkers are reluctant to walk even short distances from the parking location to their ultimate destination. The maximum walking distance a parker will tolerate depends on trip purpose, city size, and the cost of parking. The tolerance of longer walking distance also correlates with parking duration. Long-term parkers such as workers and those attending sporting and other special events may be willing to walk a maximum of 600 m (2000 ft), especially if the rates are attractive (*4*). At airports, where travelers are carrying baggage, maximum walking distances from the parking space to the curb front generally should not exceed 300 m (1000 ft). Parking for short-term parkers such as at convenience stores, banks, and fast foot establishments should be located within about 30 m (100 ft) of the building entrance (*5*). In general terms, the smaller the city and the higher the parking cost, the less distance a parker is willing to walk.

Preferably, parking lots and garages should be located on or near major arterials; garages should have access to two or more streets. It is also desirable that both lots and garages be accessed by right-hand turning movements. Approach routes to the parking facility should be capable of handling forecasted traffic at an acceptable level of service. Experience suggests that parking facilities exceeding about 1,200 spaces may require extensive traffic access provisions and/or nearby freeway access (*5*).

Safe and convenient pedestrian access to parking facilities should be provided. To alleviate parkers' fear for their personal security, pedestrian walkways should be well lighted, suitably marked, and free of blind corners. Most pedestrians prefer overhead bridges to tunnels. To the extent feasible, pedestrian conflicts with vehicular traffic should be avoided.

In summary, parkers prefer parking facilities near their destination that can be accessed with ease, used without fear for personal safety, and charge low or no fees.

9-8 Layout of Parking Lots and Garages

Preferably, a parking lot or garage should be rectangular with cars parked on both sides of access aisles. Parking sites should be wide enough to provide two or more bays approximately 18 m (60 ft) wide. Relatively long aisles, preferably 76 m (250 ft) or longer, funtion well. Aisles running with the long dimension of the lot or garage provide a better search pattern and yield more spaces per unit area.

Ninety degree parking is generally used for two-way traffic. Right-angled parking tends to require slightly less area per parked car than other configurations.

Parking stalls oriented at angles of 45 to 75° to the aisles are often used with one-way circulation. Where such an arrangement is used to serve a building along the end of a parking lot, drivers must circulate at least once next to the building during entry or exit, causing more vehicle-pedestrian conflict than the two-way 90° arrangement.

If space permits, it is desirable to provide pedestrian sidewalks between adjacent lines of parking cars. When sidewalks are provided, it is usually necessary to install wheel stops to prevent vehicle encroachment. For 90° pull-in parking, the wheel stop setback from the edge of the curb should be about 0.75 m (2.5 ft). For back-in parking, a setback of at least 1.2 m (4.0 ft) is recommended (*4*).

Stall widths of 2.6 to 2.7 m (8.5–9.0 ft) are commonly used for parking facilities in the United States. Somewhat wider stalls are preferred in high turnover spaces such as supermarket lots and other areas where packages are customarily placed in cars (*4*). Stall depths of 5.4 to 5.6 m (18.0 to 18.5 ft) are commonly used.

The downscaling of the size of U.S. automobiles and the need to minimize the amount of parking space allocated to each car have resulted in two design strategies to accommodate the smaller cars more efficiently. These strategies have involved either a general reduction in the standard dimensions of automobile stalls or the provision of spaces designed for the exclusive use of small cars.

Where it is feasible to provide separate parking spaces for the exclusive use of small cars, a stall dimension of 2.3 by 4.6 m (7.5 by 15.0 ft) is typically used.

The most common and preferred layout pattern is the bumper-to-bumper interlocked configuration illustrated by Figure 9-9. The herringbone or nested interlock pattern is sometimes used with 45° parking. In that configuration, the bumper of one car faces the fender of another car, necessitating the installation of wheel stops and increasing the probability of vehicular damage. Parking layout dimensions for 9 by 18.5 ft stalls arranged at various angles are given by the table that accompanies Figure 9.9. The best parking layout for a given site, usually determined by a process of trial, will depend primarily on these characteristics:

- the size and shape of the available area.
- type of facility (attendant, self-park).

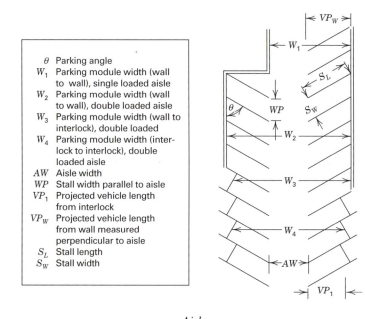

θ Parking angle
W_1 Parking module width (wall to wall), single loaded aisle
W_2 Parking module width (wall to wall), double loaded aisle
W_3 Parking module width (wall to interlock), double loaded
W_4 Parking module width (interlock to interlock), double loaded aisle
AW Aisle width
WP Stall width parallel to aisle
VP_I Projected vehicle length from interlock
VP_W Projected vehicle length from wall measured perpendicular to aisle
S_L Stall length
S_W Stall width

Parking Angle and Projected Vehicle Length		Stall Widths		Aisle Widths	Module Widths				Clearances	
		S_W	WP	AW	W_1	W_2	W_3	W_4	c	l
	Large-car Vehicle 1956 mm by 5461 mm (77″ by 215″)									
90°	VP_W − 5.61 (18.42)	2.59	2.59	7.33	12.94	18.56	18.56	18.56	0.46	0.61
	VP_I − 5.61 (18.42)	(8.50)	(8.50)	(24.04)	(42.46)	(60.88)	(60.88)	(60.88)	(1.5)	(2.0)
75°	VP_W − 5.93 (19.45)	2.59	2.68	6.45	12.38	18.31	18.06	17.80	0.46	0.61
	VP_I − 5.68 (18.62)	(8.50)	(8.80)	(21.17)	(40.62)	(60.07)	(59.24)	(58.41)	(1.5)	(2.0)
60°	VP_W − 5.84 (19.16)	2.59	2.99	4.29	10.13	15.97	15.48	14.99	0.46	0.61
	VP_I − 5.35 (17.55)	(8.50)	(9.82)	(14.09)	(33.25)	(52.41)	(50.80)	(49.19)	(1.5)	(2.0)
45°	VP_W − 5.25 (17.21)	2.59	3.66	3.35	8.60	13.84	13.15	12.46	0.46	0.61
	VP_I − 4.55 (14.94)	(8.50)	(12.02)	(11.0)	(28.21)	(45.42)	(43.15)	(40.88)	(1.5)	(2.0)
	Small-car Vehicle 1676 mm by 4445 mm (66″ by 175″)[a]									
90°	VP_W − 4.60 (15.08)	2.29	2.29	6.79	11.38	15.98	15.98	15.98	0.46	0.61
	VP_I − 4.60 (15.08)	(7.50)	(7.50)	(22.27)	(37.35)	(52.43)	(52.43)	(52.43)	(1.5)	(2.0)

[a]Small car spaces normally are considered only for 90-degree layouts.

FIGURE 9-9 Parking stall layout elements expressed in meters (feet). (Adapted from *Parking*, Eno Foundation for Transportation, 1990.)

- type of parker (especially long-term versus short-term).
- type of operation (pull-in, back-in, one-way, two-way, etc.).

9-9 Parking Garage Design Criteria

Special design criteria for parking garages, taken largely from Ref. 5, are given in the following paragraphs.

Single entrances and exits, with multiple lanes, are preferable to several openings. Entrances and exits should be located away from street intersections to prevent congestion on the street. Lane widths of 3.6 to 4.3 m (12–14 ft) are typically used for entrances and exits. Lanes should be tapered to a width of 2.7 to 3.0 m (9–10 ft) for approaches to ticket dispensers and cashiers' booths.

When self-parking is used, there is little need for reservoir space at the entrance. Studies indicate that the capacity of an entrance lane without a ticket dispenser ranges up to 700 or more vehicles per hour. With an automatic ticket dispenser, the capacity is approximately 400 vehicles per hour. One entrance lane for every 300 to 500 spaces is typically provided.

Discharge capacities for each gate-controlled exit lane range between 150 and 300 vehicles per hour. One exit lane should be provided for every 200 to 250 spaces, depending on average parking duration.

If attendant parking is used, a storage reservoir will be needed. The reservoir space required will depend on the passenger unloading time and the time required for parking each vehicle. Table 9-3 gives the number of reservoir spaces required for various vehicle arrival and storage rates. Use of the recommended values should result in overloading less than 1 percent of the time.

If possible, clear span construction should be provided with column spacing equal to the unit parking depth (module width). A clear ceiling height of 2.1 m (7 ft) and preferably 2.3 m (7.5 ft) should be provided.

Vehicular access between floors can be provided by sloped floors or ramps. For self-park facilities, the floor slopes should not exceed 3 to 4 percent. Floor slopes up to 10 percent can be used for attendant park facilities. Ramp slopes should preferably not exceed 10 percent. One-way straight ramps should be at least 2.7 m (9 ft) wide. Curved ramps are usually at least 3.6 to 4.0 m (12–13 ft) in width. A minimum radius of curvature of 9.1 m (30 ft) is recommended, measured at the face of the outer curb of the inside lane.

Counterclockwise circulation is preferred for parking garages. When helical ramps are used, the down ramp should be placed inside and the up ramp outside.

9-10 Curb Parking

Curb parking tends to seriously impede traffic flow and contribute to conflicts and accidents. Most highway and traffic engineers therefore recommend that curb parking be prohibited along major streets. It should also be prohibited adjacent to fire plugs, within pedestrian crosswalks and bus stops, and in the vicinity of intersections, alleys, and driveways. Where permitted, curb parking should be regulated to minimize its effect on accidents and congestion and to ensure that available parking spaces are used appropriately and efficiently.

TABLE 9-3 Reservoir Spaces Required for Various Vehicle Arrival and Storage Rates

	Average Number of Cars Arriving During Peak Hour					
Rate of Storage	*20*	*60*	*100*	*140*	*180*	*220*
0.9 × rate of arrival	14	26	35	43	50	56
0.95 × rate of arrival	13	23	30	36	42	47
1.00 × rate of arrival	12	20	25	29	33	36
1.05 × rate of arrival	11	17	20	22	23	24
1.10 × rate of arrival	10	13	15	15	15	15

Source: Adapted from *Traffic Design of Parking Garages.* The Eno Foundation, Inc., Landsdowne, VA (1957).

Properly marked stalls tend to make it easier for drivers to park, lessening the likelihood of congestion and crashes. One study (*4*) found a 43 percent reduction in the average parking time after suitably marked parking stalls were painted. Stalls are especially recommended for locations where there is great demand for curb parking spaces and where parking meters are used.

Three basic types of curb stalls are used: (1) end stalls, (2) interior stalls, and (3) paired parking stalls. End stalls are situated adjacent to intersections, alleys, driveways, and other retricted areas. End stalls are typically 6.1 m (20 ft) in length. Interior stalls are usually 6.7 m (22 ft) long, providing approximately 1 .4 m (4.5 ft) between adjacent cars for maneuvering. Paired parking consists of pairs of contiguous 5.5-m (18-ft) stalls separated by a 2.4-m (8-ft) open space that may be used for maneuvering. Paired parking is frequently used in conjunction with double parking meters, that is, two meters installed on a single post.

Curb parking stalls are designated by white lines extending out from the curb a distance, typically, of 2.1 m (7 ft). The end row of curb parking spaces is usually marked with an inverted **L**-shaped line, and interior stalls are designated by a **T**-shaped line.

Curb parking stalls should not be placed closer than 6.1 m (20 ft) to the nearest sidewalk edge at nonsignalized intersections. At signalized intersections, a clearance to the nearest sidewalk edge of 15.2 m (50 ft) and preferably 30 m (100 ft) should be observed. Parking stalls should not be placed closer than 4.6 m (15 ft) from fire hydrants and driveways.

TRUCK TERMINAL PLANNING AND DESIGN

The planning and design of facilities to accommodate the loading, unloading, and parking of trucks can have a significant impact on the operational efficiency of the street and highway system. Although truck loading docks and berths are usually located off the rights-of-way of public roadways, inadequacies in such facilities will cause bottlenecks and congestion on nearby streets and highways. Furthermore, trucks making pickups and deliveries for commercial establishments often make use of street and alley space for maneuvering into truck loading docks and berths. It is apparent that, while public highway agencies do not usually become involved in the planning and design of truck loading and unloading facilities, it is in the public interest that such facilities be properly planned and designed.

9-11 Truck Terminal Planning

Truck terminal planning involves the determination of the overall size and general layout of facilities required to handle the forecast quantity of freight. This includes estimating the number of loading berths (doors) as well as the amount of space required for the handling and temporary storage of the freight. To make reliable estimates of the number of loading berths, it is necessary to make a thorough analysis of the truck movements and cargo-handling procedures. The design year workload must also be forecast, taking into account daily and seasonal variations in the rate of truck arrivals and the effect of loading and unloading rates and procedures. Detailed procedures for the planning of shipper-motor carrier facilities, which are beyond the scope of this text, are given in Ref. 7.

TABLE 9-4 Minimum Recommended Apron Lengths and Dock Approach Lengths for Truck Terminals

Overall Length of Tractor-Trailer		Apron Length		Dock Approach Length	
(m)	*(ft)*	*(m)*	*(ft)*	*(m)*	*(ft)*
12.2	40	13.1	43	25.3	83
13.7	45	14.9	49	28.7	94
15.2	50	17.4	57	32.6	107
16.8	55	18.9	62	35.7	117
18.2	60	21.0	69	39.3	129
19.8	65	22.9	75	42.7	140

Source: Adapted from *Shipper-Motor Carrier Dock Planning Manual.* The Operations Council, The American Trucking Associations, Washington, DC (1970).

9-12 Truck Terminal Design Criteria

To allow proper clearances on each side of 2.6-m- (8.5-ft-) wide vehicles, a minimum width of 3.4 m (11 ft), and preferably 3.6 m (12 ft), is recommended for each spot (7). The dock approach length is the length of the largest tractor-trailer combination using the dock plus the apron length necessary to maneuver the vehicle in and out of the parking spot. As a rule, the dock approach should be at least twice the length of the longest tractor-trailer combination. Minimum recommended apron and dock approach lengths are given in Table 9-4. If 3.4-m (11-ft) widths are used for dock spots, the apron lengths shown in the table should be increased about 0.6 m (2 ft).

The apron dock approach area should be nearly level. Although most trucks are designed to negotiate a 15 percent grade, the startup grade for pulling away from a dock should not exceed 3 percent (6).

The dock height should be as close as possible to the bed height of the vehicles using the dock. Generally, trailer bed heights vary from about 1220 to 1320 mm (48–52 in.), whereas pickup and delivery vehicle bed heights range from about 1120 to 1220 mm (44–48 in.). In order to ensure that vehicle doors can be opened and closed, it is best to select a dock height that is too low rather than too high.

Vertical clearances of at least 4.6 m (15 ft) should be provided to accommodate the 4.1-m- (13.5-ft-) high trailers.

PROBLEMS

9-1 Sketch a right-angled **T** intersection using metric units and WB-15 trucks. Show all dimensions of the curb line. Assume that both streets will have two-way operation.

9-2 Work Problem 9-1 in conventional U.S. units.

9-3 Sketch a right-angled **T** intersection in metric units that is to accommodate passenger cars, buses, and WB-19 trucks. The minor or intersecting road is one way, allowing right turns only. Show all dimensions of the curb line.

9-4 Work Problem 9-3 using conventional U.S. units.

9-5 How many large cars could be placed in a commercial parking lot 74 m wide (street frontage) and 115 m deep? Assume two driveways, one-way circulation, double-loaded aisles, and a 75° parking angle.

9-6 How many large cars could be placed in a commercial parking lot 376 ft wide (street frontage) and 332 ft deep? Assume two driveways, one-way circulation, double-loaded aisles, and a 75° parking angle.

9-7 Prepare a report on how you would provide for handicapped motorists in the parking lot described in Problem 9-5. Comply with applicable federal, state, or local regulations.

9-8 Sixty cars arrive at a parking garage during the peak hour. If cars can be stored at a maximum rate of 66 per hour, estimate the number of required reservoir spaces.

9-9 Prepare a report on the experience in your city on the location and lengths of bus stops. Indicate whether bus stops are located on the near side or far side of intersections or at midblock. Discuss the advantages and disadvantages of the system and indicate steps that could be taken to alleviate any traffic or safety probems related to bus movements.

REFERENCES

1. *A Policy on Geometric Design of Highways and Streets.* American Association of State Highway and Transportation Officials, Washington, DC (1994).

2. Merritt, David R. Geometric Design Features of Single-Point Interchanges. *Transportation Research Record 1385,* Transportation Research Board, Washington, DC (1993).

3. *Dynamic Design for Safety Seminar Notes.* Prepared by the Institute of Traffic Engineers for the Federal Highway Administration, Arlington, VA (1975).

4. *Parking Principles.* The Transportation Research Board Special Report No. 125, Washington, DC (1971).

5. Weant, Robert A., and Levinson, Herbert S. *Parking.* Eno Foundation for Transportation, Westport, CT (1990).

6. *Traffic Design of Parking Garages.* Eno Foundation for Transportation, Lansdowne, VA (1957).

7. *Shipper-Motor Carrier Dock Planning Manual.* The Operations Council, The American Trucking Associations, Washington, DC (1970).

CHAPTER 10

HIGHWAY MASS TRANSIT FACILITIES

Highway engineers have long recognized that the facilities they plan and design must accommodate a wide range of vehicle types and sizes. Events of the 1970s, especially increasing urban congestion and scarcity of fuel, heightened the awareness of engineers that steps could be taken to enable the highway system to be used more efficiently. Deliberate actions could be taken to encourage more extensive use of streets and highways by buses, vans, and other high-occupancy vehicles (HOVs).* The highway design would not only accommodate mass transit vehicles but also facilitate the movement of people rather than vehicles and give priority to HOV transportation.

In this chapter, we describe the planning and design of special facilities for buses and other high-occupancy vehicles and examine some traffic control measures and institutional programs designed to encourage more efficient use of streets and highways, especially in heavily congested urban areas.

10-1 Bus Stops on Streets

Bus stops that are properly located, adequately designed, and effectively enforced can improve bus service and expedite traffic flow (*2*). Preferably, the frequency of stops should not exceed 5 or 6 per kilometer (8 to 10 per mile). Practically speaking, this means that stops should be provided every block where city blocks are greater than about 150 m (500 ft) long and every other block where blocks are shorter.

Bus stops may be located on the near side or far side of intersections or at midblock. The choice of location will depend on the geometric design of the intersection, the proximity of large generators of passengers, whether turning movements are permitted at the intersection, and a number of other factors.

Some of the advantages to *far-side* bus stops include

1. They make the curb lane available for traffic, increasing the intersection capacity.
2. They reduce conflicts between stopped buses and right-turning vehicles.

---

*The definition of HOV varies. Commonly, it includes all vehicles with two or more occupants; however, some jurisdictions have restricted HOV lanes to vehicles with three or more occupants (*1*).

3. They lessen the problem of buses blocking the view of drivers approaching the intersection.

4. They reduce pedestrian accidents (*3*).

On the other hand, *near-side* bus stops are preferred where transit flows are heavy but traffic and parking problems are not extensive (*2*). Near-side stops are usually near a crosswalk, making it easier for passengers to board the buses. Also, with near-side stops there is less interference with traffic turning into the bus route from a side street. Furthermore, bus drivers tend to prefer near-side stops because it is easier for the buses to reenter the traffic stream, especially where curb parking is allowed.

Midblock bus stops are sometimes provided in downtown areas where several bus routes converge and require long boarding areas. At midblock areas, buses create less serious sight restrictions to drivers and pedestrians. In addition, passengers are able to assemble at less congested areas of the sidewalk.

Sufficient space should be provided at bus stops to permit bus drivers to easily pull in and out of the traffic stream and to park parallel to the curb with the center door not more than 0.3 m (1 ft) from the curb (*4*). Recommended lengths of curb bus loading zones are given in Table 10-1.

Figure 10-1 shows recommended bus stop design standards for stops at far-side, near-side, and midblock locations. These dimensions are based on an assumed 12.2 m (40-ft) bus length and should be adjusted for buses of different length. Where two or more buses are expected to use a bus stop simultaneously, 14 m (45 ft) should be added to the lengths shown for each additional bus.

10-2 Bus Turnouts on Streets

Several cities have used bus turnouts by relocating the curb and flaring the street width. Bus turnouts should be considered along heavily traveled high-speed arterial streets, especially when bus dwell times are relatively long. Reference 2 suggests that bus turnouts are warranted whenever the following conditions exist:

1. Right-of-way width is sufficient to permit construction of the turnout without impinging on needed sidewalk space.
2. Peak-hour traffic in the curb lane is at least 500 vehicles per hour.
3. Curb parking is prohibited, at least during periods of peak traffic flow.
4. Average bus dwell time exceeds 10 sec per stop.
5. Bus volumes are insufficient to justify an exclusive bus lane, but bus vol-

TABLE 10-1 Recommended Curb Bus Loading Zone Lengths for One 12.2 m (40 ft) Bus[a]

Type of Loading Zone	Length, meters (ft)
A. Far-side stop	24–30 (80–100)
B. Far-side stop after right turn	40–46 (130–150)[b]
C. Near-side stop	27–32 (90–105)
D. Midblock stop	40–49 (130–160)

[a]Add 13.7 m (45 ft) to lengths shown for each additional bus expected to use bus stop simultaneously.

[b]Lengths are measured from near curb line of intersecting street.

Source: Adapted from *Bus Use of Highways,* NCHRP Report 155, Transportation Research Board, Washington, DC (1975).

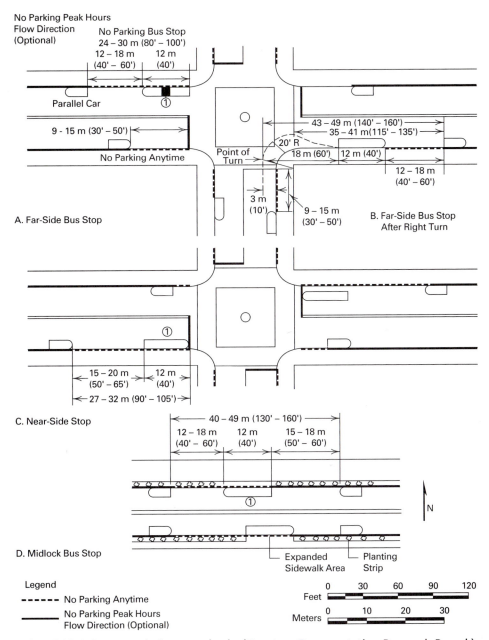

FIGURE 10-1 Bus stop design standards. (Courtesy Transportation Research Board.)

umes should be at least 100 per day and consist of 10 to 15 buses transporting 400 to 600 passengers during the peak hour.

Figure 10-2 shows typical layouts for bus turnouts at far-side, near-side, and midblock locations. Figure 10-3 gives recommended dimensions for bus bays to accommodate one standard 40-ft bus.

The capacity of bus turnouts depends on the peak-hour bus volume and the average dwell time. Figure 10-4 provides some guidelines for determining the

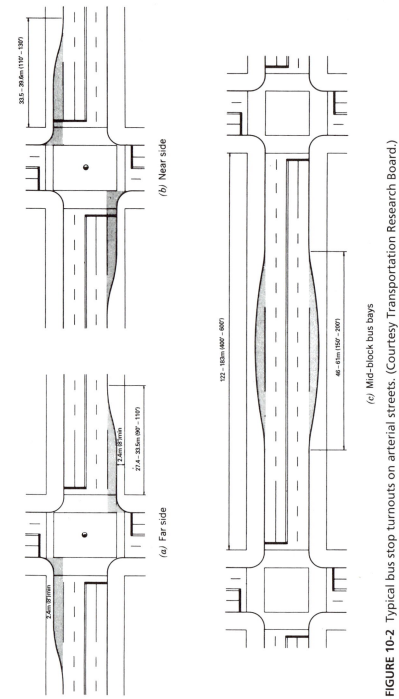

2.4m (8')min

27.4 – 33.5m (90' – 110')

2.4m (8')min

(a) Far side

33.5 – 39.6m (110' – 130')

(b) Near side

122 – 183m (400' – 600')

46 – 61m (150' – 200')

(c) Mid-block bus bays

FIGURE 10-2 Typical bus stop turnouts on arterial streets. (Courtesy Transportation Research Board.)

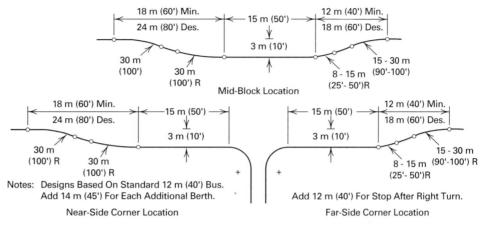

FIGURE 10-3 Recommended dimensions for bus bays to accommodate one standard 12-m (40-ft) bus. (Courtesy Transportation Research Board.)

number of buses that will have to be accommodated at each turnout. The curves assume random (Poisson) arrivals and are based on a confidence level of 95 percent. The criteria are also applicable for determining the lengths of bus stops.

10-3 Bus Stops on Freeways

In certain instances, special bus stops are needed at interchanges and street crossings to provide the public with the benefit of freeway bus operations. Criteria for the location of bus stops at freeway interchanges include (1) the need for a transit transfer point, (2) the density of population in the area, (3) transit generators within less than one block of the freeway route, and (4) forecasts of future usage of expressway by transit vehicles.

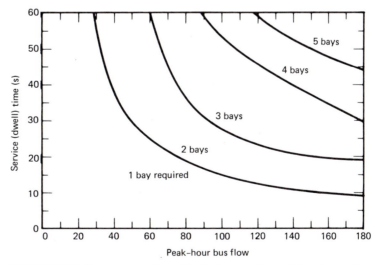

FIGURE 10-4 Recommended criteria for number of bus bays at bus stops and turnouts.

In planning and locating freeway bus stops, there is a need to maximize bus patron convenience but to minimize delays to buses and to other vehicles in the traffic stream. The spacing of the stops should allow the buses to operate at or near the speed of the traffic on the highway. Based on this criterion, AASHTO (*3*) suggests a spacing of 3.2 km (2 miles) or more.

Bus stops along freeways may be placed at the street level or the freeway level. Street-level stops are appropriate in or near downtown areas, and either street- or freeway-level stops are suitable for suburban and rural areas. Bus stops at the cross-street level are more convenient for entering and leaving passengers but cause delays for through passengers.

Street-level bus stops are not well suited to cloverleaf or directional interchanges, and AASHTO (*3*) recommends that they not be provided at such interchanges. At diamond interchanges, the preferred location for bus stops is adjacent to on-ramps. There, the bus stop may consist of a widened shoulder area, or it may be on a separate roadway.

Freeway-level bus stops should be isolated from the through freeway lanes. Bus stops generally should be located on a separate roadway at least 6.1 m (20 ft) wide to permit buses to pass a standing or stalled bus (*4*). Acceleration and deceleration lanes at least 30.5 m (100 ft) in length should be provided (*2*). It is usually necessary to provide fencing between the bus stop area and through lanes to deny pedestrians access to the freeway lanes.

Bus stops should be at least 24 m (80 ft) in length to allow simultaneous loading or unloading by two buses. The pedestrian loading platform should be at least 1.5 m (5 ft) and preferably 2.4 m (8 ft) in width.

Because bus stops along freeways are usually located at intersecting streets or highways, stairs or ramps for pedestrians are required. Efforts should be made to minimize the total height of stairways, and landings should be provided about every 2 m (6 or 8 ft) (*3*). The gradients on pedestrian ramps should not exceed 8.33 percent.

Benches and shelters are recommended, and telephones and other amenities may also be warranted.

Figure 10-5*a* shows a layout for a freeway-level bus stop where the street crosses over the freeway without an interchange. Figure 10-5*b* illustrates two layout plans for an undercrossing street without an interchange. In the upper portion of the figure, the turnout to the bus stop is ramped to the level of the frontage road, making it unnecessary for passengers to climb ramps or stairs. In the lower part of Figure 10-5*b*, stairs are shown leading from the bus stop to the undercrossing street. The layout also features a pedestrian walkway built over the freeway.

10-4 Reserved Bus Lanes

In central business districts and other areas that experience heavy concentrations of bus traffic, lanes may be reserved for exclusive bus use or shared by buses, taxis, and right-turning vehicles. Reserved bus lanes may operate in the same direction as or counter to automobile traffic flow. (Lanes of the latter type are referred to as "contraflow lanes.") Reserved bus lanes are commonly located along curbs but in certain instances may be located in street medians.

Reserved bus lanes may be delineated by painted lines and markings or by some physical separation such as curbs, flexible stanchions, or traffic islands.

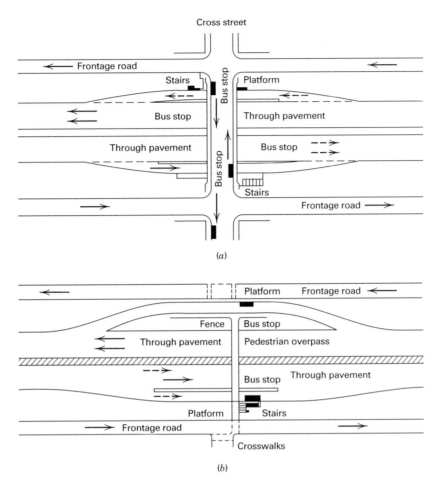

FIGURE 10-5 Layout of a freeway-level bus stop. (*a*) Street crossing over freeway. (*b*) Street crossing under freeway. (Courtesy American Association of State Highway and Transportation Officials.)

Normal flow curb lanes are usually separated from other lanes by a solid white line 200 to 300 mm (8 to 12 in.) wide. Normal flow median lanes are separated from other lanes by double lines 150 mm (6 in.) wide: white lines facing traffic in the same direction, and yellow lines facing traffic in the opposite direction. Contraflow bus lanes are separated from traffic in the opposite direction by double 150 mm (6-in.) yellow lines, a median divider, or some other physical separation (*2*).

Large BUS ONLY or BUS LANE messages should be painted on the pavement at least once each block and twice where the block lengths exceed 90 m (300 ft). Special signs that conform to the *Manual on Uniform Traffic Control Devices for Streets and Highways* (*5*) should be installed to facilitate the control of traffic.

Reserved curb lanes are warranted under the following conditions (*2*):

1. Curb parking should be prohibited whenever the bus lane is in effect.
2. Where the bus lane preempts a lane used by automobiles, there should be

sufficient capacity in the remaining lanes or parallel streets to accommodate displaced traffic.

3. Two lanes, and preferably three lanes, should remain for other traffic in the same direction (normal flow reserved lanes) or in the opposite direction (contraflow lanes).

4. Essential services and access to adjacent property should be maintained. It may be possible to provide access to buildings from the rear or during off-peak periods.

5. The number of peak-hour bus passengers transported in the reserve lane should equal or nearly equal the average number of passengers carried by cars in the adjacent lanes.

6. The bus traffic should be at or above the following levels:

	Normal Direction	*Contraflow*
One-way peak-hour traffic	30–40 buses	40–60 buses
One-way peak-hour movement of people	1200–1600 people	1600–2400 people
Daily use of reserved lanes	300 buses	400 buses
Busiest 20-min flow rate	60 buses/hr	60 buses/hr

Three-meter (ten-foot) lane widths may be used for reserved bus lanes used for normal direction of flow. Wider lanes, preferably measuring 3.6 m (12 ft), are desired for contraflow lanes. Typical curb bus lane configurations are shown in Figure 10-6, and typical configurations for contraflow lanes are shown in Figure 10-7.

Median lanes are warranted under the following conditions (*2*):

1. The bus lanes are to replace street railway operations in the center of the street, establishing the precedent for center-of-the-street loading.

2. The street should have a wide median that can be paved for buses without harmful environmental effect.

3. The street should be wide enough to provide at least two lanes on each side of the median. (The curb lane may be used for parking.)

4. The number of bus passengers should equal or exceed the average number of passengers carried by automobile in the adjacent lanes throughout a

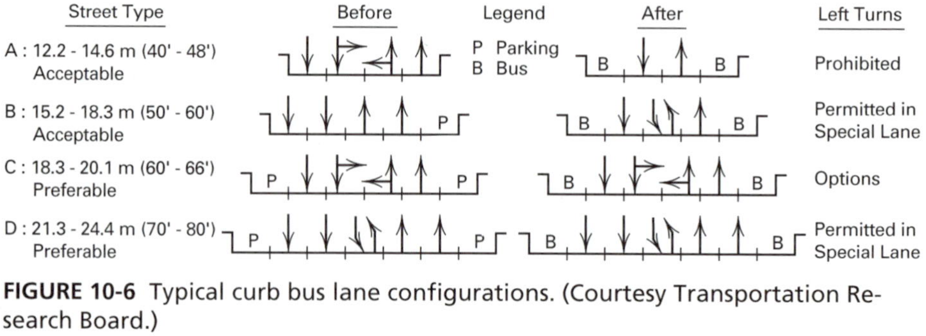

FIGURE 10-6 Typical curb bus lane configurations. (Courtesy Transportation Research Board.)

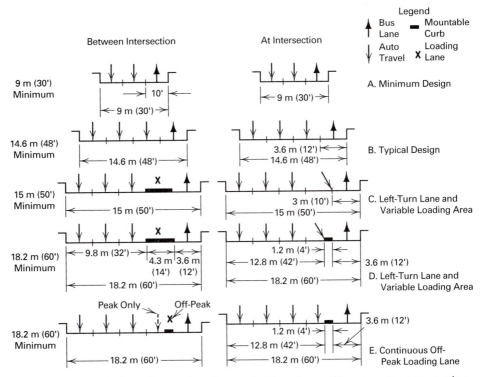

FIGURE 10-7 Typical configurations for contraflow lanes. (Courtesy Transportation Research Board.)

12-hr base period or, alternatively, the periods the lanes are in effect. Where peak-hour median lanes are provided, buses should transport 1.25 times the number of passengers carried by cars in the adjacent same-direction lanes.

5. The bus traffic should equal or exceed the following amounts:

One-way peak-hour traffic	60–90 buses/lane
One-way peak-hour movement of people	2400–3600 people/lane
Daily use of median lane	600 buses/lane

6. Conflicting left-turning movements should be prohibited or channeled into lanes outside the median.

The effectiveness of reserved bus lanes has not been definitely established, and some consider the success of such measures to be rather limited (*3*). It is difficult, for example, to reserve the curb lane of arterial streets for the exclusive use of buses, because other vehicles making right turns must make use of this lane. There are problems of enforcement as well as concerns about safety and increased traffic on the remaining lane street space. Nevertheless, the evidence that is available suggests that, when properly conceived and implemented, reserved bus lanes can result in significant improvements in bus operations that are not outweighed by increases in accidents, nonbus congestion, and detrimental impacts on abutting property.

In a study of contraflow bus lanes in Chicago (*6*), it was found that there was a significant improvement in bus operations after the installation of the lanes. Average travel speeds of the buses increased 22 percent, resulting in an annual cost saving of approximately $400,000. The study revealed that during peak periods, the contraflow lanes accommodated more people in one lane than were moved on the remaining through lanes and in only one-tenth the number of vehicles.

10-5 Busways

Busways are specially designed roadways for exclusive or predominant use by buses. They are used (*2*)

- to provide line-haul express service to the city center.
- to provide feeder service to rail transit lines.
- to provide short bypasses of areas of serious congestion.

Busways may be constructed in separate rights-of-way, along the side of a freeway, or in a freeway median. They may (1) be completely grade-separated and provide levels of service comparable to those provided by freeways or rail rapid transit or (2) incorporate some at-grade intersections and provide a level of service comparable to that on arterial streets or light rail transit lines. These two classes of busways have been designated as class A and class B, respectively (*2*).

Typically, busways have two-lane, two-way operation. The cross section consists of two lanes, each 3.4 to 4.0 m (11 to 13 ft) in width, plus space to accommodate breakdowns. A single breakdown lane may be constructed between the two traffic lanes, or breakdown space may be provided as shoulders along the outside of each lane. Typical sections for these two design approaches are shown as Figure 10-8.

Busways may operate with conventional right-hand operations (*normal flow*) or with the buses keeping to the left of the centerline (*contraflow*). Contraflow operation permits the use of common island station platforms between the two

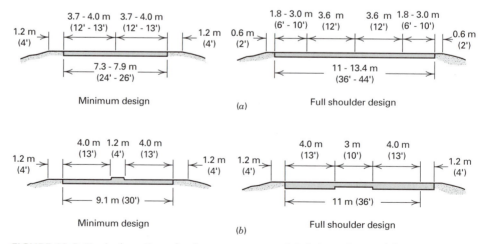

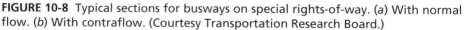

FIGURE 10-8 Typical sections for busways on special rights-of-way. (*a*) With normal flow. (*b*) With contraflow. (Courtesy Transportation Research Board.)

lanes, providing efficiencies in station security, supervision, and maintenance and simplifying access to the platforms. However, contraflow operation is not recommended whenever trucks and/or HOVs might use the bus lanes.

Suggested busway design criteria are given in Table 10-2. The pavement widths shown in the table are for tangent sections. Pavements on two-lane busways should be widened 0.5 to 1.0 m (1.5 to 3.0 ft), depending on normal pavement width, design speed, and sharpness of curve. The ramp widths shown are for one-lane, one-way operation with no provision for passing.

Busway design criteria for horizontal and vertical curvature generally conform to AASHTO highway standards. The table values have been modified somewhat to allow for the special design and operational characteristics of buses and to allow for the later conversion of busways to rail facilities, where appropriate.

Figure 10-9 illustrates a busway in a depressed freeway median with normal flow.

10-6 Other Measures to Encourage Bus and High-Occupancy Vehicle Use

We have described several initiatives that can be taken to foster the use and improve the flow of buses and high-occupancy vehicles on streets and highways, Some of the initiatives described are physical facilities; others are operational changes. There are other measures that will positively affect mass transit and HOV use. Such measures include routine traffic engineering techniques that improve traffic flow generally, such as planning and effectuation of one-way street systems and implementation of progressive traffic signal systems along arterial streets. In addition, there are measures designed to give priority to bus and HOV movements. Some of these priority measures are described in the following sections of this chapter.

Ramp Metering with HOV Bypass Lanes

A number of traffic engineering agencies use ramp metering to control traffic flow on freeways. Special traffic signals on entrance ramps permit only those vehicles to enter the freeway that can be accommodated without unduly disrupting main-line flow. Vehicles are required to wait at the ramp signal, artificially creating queues. A few agencies have used metered ramps to provide preferential treatment for high-occupancy vehicles. They use bypass ramps such as the one illustrated in Figure 10-10 to allow buses and possibly HOVs to move ahead of vehicles waiting in the queue.

The effectiveness of bypass lanes in inducing greater use of carpools, vanpools, and buses has not been well established. A study of Highway 50 bypass lanes in Sacramento, California, revealed a tendency for automobile occupancy to increase where the carpool bypass lane was implemented; however, the researchers reported that at least some of these changes probably resulted from a shift in route of existing carpools rather than the formation of new carpools (7).

Bus Bypass of Queues at Congestion Points

Special bypass facilities may also be suitable for relieving chronic congestion on freeways at bottlenecks that cannot reasonably be remedied by more conventional methods. Such facilities may be (1) short sections of busways built on separate rights-of-way or (2) special bus lanes that are segregated from other

TABLE 10-2 Suggested Busway Design Criteria

Item	Class A Busway	Class B Busway
Loads	H-20-S-16-44	H-20-S-16-44
Design speed (km/hr)		
Desirable	112	80
Minimum	80	48
Lane width (m)		
With paved shoulders	3.6[a]	3.4–3.6[a]
Without paved shoulders	4.0[a]	3.6[a]
Paved shoulder width[b] (m)	2.4–3.0	1.8–2.4
Total paved width (m)		
Normal flow	7.9–13.4	7.3–12.2
Special flow	9.1–11.0	
Contraflow	9.1–11.0	
Minimum viaduct width[c] (m)	8.5	8.5
Minimum tunnel width[d] (m)	9.5	9.5
Minimum vertical clearance (m)		
Desirable	4.4–5.5[e]	4.4[e]
Absolute minimum		3.8
Min. lat. dist. to fixed obstructions[f] (m)		
Left	1.1	0.6
Right	1.9	1.0
Maximum superelevation (%)	8.0[g]	8.0[g]
Min. radius of horiz. curves (m)		
112 km/hr	488	488
94 km/hr	350	350
80 km/hr	229	229
64 km/hr	137	137
48 km/hr	76	76
Absolute min. radius[h] (m)		
Conv. to conventional rail	76	76
Convertible to light rail	31	31
Nonconvertible	9.1	9.1
Maximum gradients (%)		
Desirable		
Convertible to rail	3–4	3–4
Other	5	6
Ramps, up	6	7
Ramps, down	7	8
Absolute		
Main line	8	8
Ramps	10	10
Ramps		
Design speed (km/hr)	48–56	24–40
Lane width (m)		
With paved shoulders	3.6	3.6
Without paved shoulders	4.3	4.0
Paved shoulder width (m)	2.4	2.4
Total paved width (m)	4.3–6.7	4.0–6.1

[a]Increase lane width 0.3 m when nonmountable-type curbs are used adjacent to travel lane.
[b]Applies only to normal flow busways.
[c]Curb to curb; excludes pedestrian walks and width required by curbs.
[d]Inside envelope.
[e]Varies according to requirements of the selected rail system.
[f]Distance from edge of traveled lane to vertical face of a noncontinuous obstruction, such as a bridge pier.
[g]May be reduced to 6.0% in regions where roadway icing is a consideration.
[h]Inner lane edge.

Source: Bus Use of Highways: Planning and Design Guidelines (2). Transportation Research Board, Washington, DC (1975).

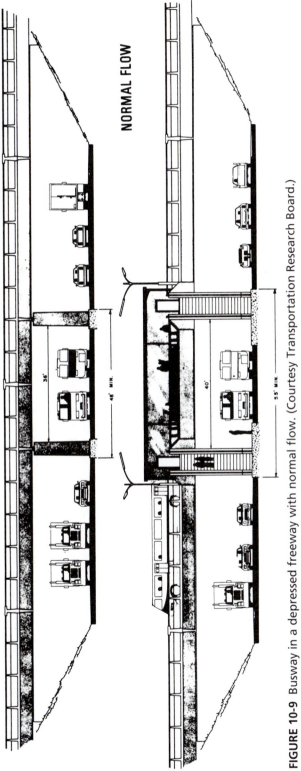

NORMAL FLOW

FIGURE 10-9 Busway in a depressed freeway with normal flow. (Courtesy Transportation Research Board.)

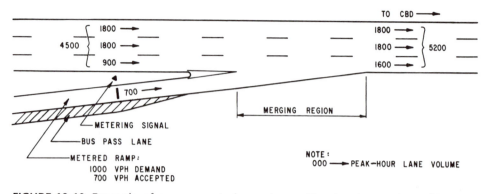

FIGURE 10-10 Example of a ramp metering system with a bus bypass lane. (Courtesy Transportation Research Board.)

lanes by signs and markings on some physical separation. Provision of bus bypass facilities may be appropriate at bridges, toll plazas, and bottlenecks caused by highway construction.

Bus bypass facilities at congestion points are generally warranted where more than 60 design-year peak-hour buses are forecast and it is expected that each bus will save at least 5 min (*2*). At toll plazas, such facilities may be warranted if buses can avoid an aggregate delay of 30 to 50 min/hr. For example, Ref. 2 recommends a warrant of 15 buses/hr, each saving 2 or 3 min, or 10 buses/hr, each saving 5 min.

It is important that special bus bypass facilities be properly signed and marked and that adequate sight distance be provided at the beginning and at the end of the bypass. Such facilities should not reduce the highway capacity through the bottleneck.

Special Traffic Control Measures

Traffic engineering agencies can give priority to bus movements by actions designed to facilitate bus-turning movements at intersections. One action is to allow buses to make left turns that are prohibited to other vehicles (with signs such as NO LEFT TURN, BUSES EXCEPTED). The objective may be to facilitate bus routing or to remove buses from a heavily traveled arterial.

At signalized intersections, a special bus phase may be incorporated in the signal timing plan to allow buses to turn across heavily traveled traffic streams. Signalization of turning movements is considered to be justified wherever buses are required to cross more than two traffic lanes, each accommodating 500 or more vehicles during the peak hour (*2*).

Priority treatment for buses in mixed flow can also be achieved by signal preemption. This normally involves extending the green phase along the main street in the direction of bus movements. The goal is to improve bus flows by reducing the number of stops due to cross-street traffic. Signal preemption can be achieved by (1) passive systems that are based on signal coordination and improved flow for all the main street traffic or (2) active systems that incorporate special bus detection equipment and signal controllers to adjust the signal phase to give priority to bus movements.

The preemption of traffic signals must take into account its effect on the total signal network. The objective is to reduce total person delay without unduly limiting side-street movements by other vehicles and pedestrians. To justify preemption, there should be at least 10 to 15 buses transporting 400 to 600 people during the peak hour and a daily volume of at least 100 buses (*2*).

Signal preemption can result in significant improvements in bus flows. Bus delays at traffic signals have been reported to constitute 10 to 20 percent of overall bus trip times and to cause nearly half of all delays (*8*).

10-7 Park-and-Ride Facilities

Park-and-ride facilities encourage automobile users to park in outlying areas and complete their journey by express bus or rail rapid transit. Such facilities may be located 2 to 48 km (1 to 30 miles) from a city's central business district (CBD). Change-of-mode lots served by buses are usually closer to the CBD than are lots served by rail.

The layout of park-and-ride facilities should minimize pedestrian travel and give priority to interchanging transit passengers. Facilities should be provided for feeder bus loading and unloading, passenger car loading and unloading (kiss-and-ride), and the short-term and long-term parking. Preferably, the movement of buses and cars should be separated; however, at stations with fewer than 12 to 15 terminating buses per peak hour, buses may share parking area roadways with automobile traffic (*2*).

Kiss-and-ride facilities should allow for passengers to be dropped off and picked up close to the station entrance and should incorporate short-term parking for 20 to 60 spaces.

The layout and design of the automobile parking lot for park-and-ride facilities conform to the standards for off-street lots generally, which are described in Chapter 9.

A sketch of a typical park-and-ride facility is shown in Figure 10-11.

10-8 Paratransit

Thus far, we have focused primarily on facilities and programs designed to improve the movement of buses and other high-occupancy vehicles. In this section, we examine more closely the other HOVs and approaches that can be taken to encourage their use.

The term "paratransit" is used to describe a broad range of transportation services falling between conventional fixed-route bus service and the private individually occupied automobile (*8*). Paratransit is a low-capacity service that includes the following types of transportation:

1. *Demand-responsive transportation* refers to the family of transportation that usually involves minibuses or vans that are directed from a central dispatching office. Like taxis, they provide on-demand, door-to-door, flexible-route service. Unlike most taxis, ride sharing is accomplished as a matter of operating policy and does not require the passengers' consent. The terms *dial-a-ride* and *shared taxi* have also been used to describe this type of transportation. Demand-responsive transportation is a rapidly growing but very small component of mass transportation. A large part of its market has been groups having special mobility needs: the handicapped, the elderly, and, to some extent, the economically disadvantaged. Demand-

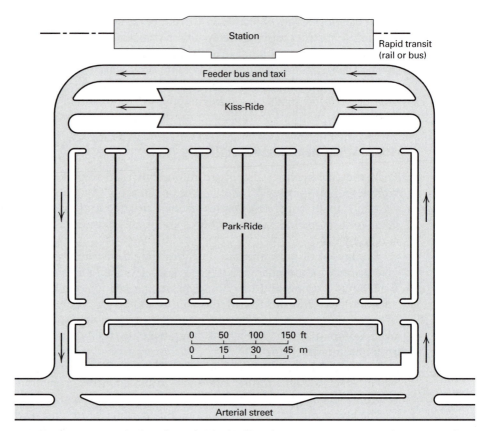

FIGURE 10-11 A typical park-and-ride facility. (Courtesy Transportation Research Board.)

responsive transportation has also been used for commuters in low-density areas to provide feeder service to conventional fixed-route transit.

2. *Vanpools and carpools* constitute a high-occupancy mode of paratransit that supplements conventional transit. It primarily serves commuters, and its major contribution is to help ameliorate highway congestion that occurs during the morning and afternoon rush hours. The vehicles used in vanpools and carpools are usually owned by individuals, and the service is arranged by subscription. Some employers encourage vanpooling and carpooling by administering a ride-matching program, by providing priority parking, and by permitting slight adjustments in work hours.

10-9 Alternative Work Schedules

Another approach to relieving peak-hour congestion on streets and highways is to manage the travel by (1) shifting demand to less congested times when surplus capacity is available and (2) reducing the need to travel (*9*). This can be accomplished by means of alternative work schedules.

Three basic types of alternative work schedules may be used (*9*):

1. Staggered work hours, by which groups of employees are assigned different starting times, for example, at 15-min intervals over a 2-hr period.

2. Flexible work hours, where individual workers are allowed to have some control over their own working hours but all employees are required to be present during a core time.

3. Compressed work weeks, by which workers work fewer than 5 days per week but work more hours per day, for example, 4 days per week, 10 hr per day.

At large places of employment, alternative work schedules can significantly reduce the number of employees arriving and departing during the peak period. Research has shown that staggered and flexible work hours result in lower travel times and improved load factors on transit. Reductions in travel time ranging from 2.5 to 8 min have been reported for commuters who participate in flexible work hours programs (9). Significant but somewhat smaller benefits accrue to all commuters who use the affected highways.

REFERENCES

1. Roark, John J. *Enforcement of Priority Treatment for Buses on Urban Streets.* Synthesis of Transit Practice, National Cooperative Transit Research and Development Program, Transportation Research Board, Washington, DC (1982).

2. Levinson, Herbert S., Adams, Crosby L., and Hoey, William F. *Bus Use of Highways: Planning and Design Guidelines.* National Highway Cooperative Highway Research Program Report 155, Transportation Research Board, Washington, DC (1975).

3. *A Policy on Geometric Design of Highways and Streets.* American Association of State Highway and Transportation Officials, Washington, DC (1994).

4. *Bus Stops for Freeway Operations.* ITE Recommended Practice, Institute of Transportation Engineers, Washington, DC (1971).

5. *Manual on Uniform Traffic Control Devices for Streets and Highways.* Federal Highway Administration, Washington, DC (1988).

6. LaPlante, John, and Harrington, Tim. Contraflow Bus Lanes in Chicago: Safety and Traffic Impacts. *Transportation Research Record 957,* Transportation Research Board, Washington, DC (1984).

7. Rogers, Christy A. Effects of Ramp Metering with HOV Bypass Lanes on Vehicle Occupancy. *Transportation Research Record 1021,* Transportation Research Board, Washington, DC (1985).

8. Alschuler, David M. Labor Protection, Labor Standards, and the Future of Paratransit. *Special Report No. 186* Transportation Research Board, Washington, DC (1979).

9. *Alternative Work Schedules: Impacts on Transportation.* Synthesis of Highway Practice 73, National Cooperative Highway Research Program, Transportation Research Board, Washington, DC (1980).

CHAPTER 11

DRAINAGE AND DRAINAGE STRUCTURES

One of the most important considerations in locating and designing rural highways and city streets is providing adequate drainage. Adequate and economic drainage is absolutely essential for the protection of the investment made in a highway structure and for safeguarding the lives of the persons who use it.

The flow of surface water with which the highway engineer is concerned generally results from precipitation in the form of rain or snow or melting ice. A portion of the surface water infiltrates the soil, while the remainder stays on the surface of the ground and must be carried on, beside, beneath, or away from the traveled way. Artificial drainage resulting from irrigation, street cleaning, and similar operations may also be of consequence in some cases. In certain instances the control of underground water (ground water) may be important, as in the case of an underground flow encountered in a highway cut or in a location where the water table lies close to the surface of the ground.

Measures taken to control the flow of surface water are generally termed "surface drainage," whereas those dealing with ground water in its various forms are called "subsurface drainage" or, more simply, "subdrainage."

In the following pages, we will discuss some of the fundamental concepts of highway and street drainage. Surface drainage in essentially rural areas is discussed in considerable detail; accompanying this is a discussion of measures for the prevention of erosion of shoulders, side slopes, and side ditches. Considerable space is devoted to the location, design, and construction of culverts. Attention is briefly given to highway bridges. Material is also presented relative to subdrainage, and the chapter concludes with a brief discussion of drainage in municipal areas.

SURFACE DRAINAGE

The portions of the highway structure that provide for surface drainage in rural locations include the roadway crown, shoulder and side slopes, longitudinal ditches (channels), culverts, and bridges. Divided highways in rural areas also have inlets and storm drains (underground pipes) in the median strip to handle a portion of the surface flow.

11-1 Pavement and Shoulder Cross Slopes

Consistent with other design objectives, highway designers should ensure that precipitation is removed from the pavement as expeditiously as possible. As was indicated in Chapter 7, roadway surfaces are normally crowned (or sloped as in the case of a superelevated section) to facilitate the removal of surface water from the wearing surface. The recommended crown or cross slope tends to be a compromise between the needs of vehicular traffic and those for drainage. Pavement cross slopes should be steep enough to ensure expeditious drainage, but not so steep as to cause driver annoyance, discomfort, or hazard. Recommended ranges of pavement cross slopes are given in Table 11-1. The amount of cross slope varies with the type of surface, being generally small for relatively impervious surfaces such as portland cement concrete and large for previous surfaces such as gravel or earth.

Shoulders are normally sloped to drain away from the pavement surface. Precipitation that occurs on the shoulder area largely flows to the side ditches or the median swale, as does that which falls on the roadway proper. As Table 11-1 indicates, recommended shoulder cross slopes vary from about 2 to 6 percent depending on the type of surface and whether or not curbs are provided (*1*).

11-2 Side Slopes and Side Ditches

Open side ditches are generally provided in cut sections in highway locations in rural areas to provide for surface drainage. Side ditches may also be constructed along embankment sections when needed to supplement natural drainage channels. Both flat-bottomed and **V**-section ditches are used, with preference being given to the former type, with slope changes in the ditch section being rounded to improve appearance and prevent erosion. In either case, side slopes are made as flat as possible consistent with drainage requirements and limiting widths of right-of-ways. Deep, narrow side ditches are to be avoided whenever possible because of the increased hazards presented by such construction. Where they must be used, adequate provision should be made for safeguarding traffic through use of traffic barriers. Warrants for traffic barriers as well as recommended front slopes and back slopes for ditches are discussed further in Chapter 8.

Water flows in side ditches in a direction that is generally parallel to the roadway centerline. Grades used in open ditches may also be roughly the same as those used on the highway centerline; on the other hand, flat roadway grades and steeper ditch grades in the same location are very frequently used. In very flat country, ditch grades as low as 0.1 or 0.2 percent may be used, while in rolling

TABLE 11-1 Recommended Ranges of Cross Slopes for Pavements and Shoulders

Roadway Element	Range in Rate of Cross Slope (%)
High-type surface	
Two lanes	1.5–2.0
Three or more lanes in each direction	1.5 minimum; increase 0.5–1.0% per lane; 4.0 maximum
Intermediate surface	1.5–3.0
Low-type surface	2.0–6.0
Urban arterials	1.5–3.0; increase 1.0% per lane
Shoulders	
Bituminous or concrete	2.0–6.0
With curbs	≥4.0

Source: Drainage of Highway Pavements. Highway Engineering Circular No. 12, Federal Highway Administration, Washington, DC (March 1984).

or mountainous terrain the maximum grade may be dictated only by the necessity for preventing erosion.

Side ditches provide open channels for the removal of surface water from within the limits of the highway right-of-way. In certain circumstances, areas adjacent to the right-of-way may also contribute to the flow. The water must be carried to an outlet, in the form of either a natural or an artificial drainage channel. The ditch must be hydraulically capable of handling the anticipated flow of surface water in such fashion that the roadway structure is not endangered or the safety of the motorist threatened. The hydraulic design of side ditches and other open channels is treated more fully later in this chapter.

11-3 Prevention of Erosion of Shoulders and Side Slopes

The flow of surface water adjacent to highways is frequently accompanied by detrimental soil erosion that may result in the destruction of productive soils, the creation of areas of unsightly appearance, and the clogging of ditches and drainage structures. Erosion may also endanger the stability of side slopes in embankment and cut sections. A number of construction measures designed to prevent or minimize erosion are discussed in the paragraphs immediately following.

Any decision that is made regarding the installation of any of the special features discussed hereafter with the objective of preventing erosion on shoulders and side slopes must be based largely on judgment and knowledge of local conditions. Many factors enter into such a decision, including such items as intensity and duration of rainfall, soil types and condition, magnitude of slopes, climatic conditions, and many others.

Curbs, Gutters, and Flumes

Curbs and gutters are sometimes used at the edge of the traveled way on parkways and rural highways to prevent the water that flows from the crowned surface from spilling over and eroding the shoulders and side slopes. On urban highways, curbs and gutters are more commonly used, providing a measure of protection to pedestrians as well as facilitating drainage. Curbs may also be placed on the outer edge of surfaced shoulders when water threatens to erode the embankment slope. At intervals, water that collects along such curbs is channeled into a spillway, paved ditch, or flume and carried into the side ditch or pipe drain. The spacing of the inlets depends on the design discharge, the geometric configuration of the curb and gutter section, and the extent to which water is permitted to spread on the traveled way. More information on the hydraulic design of such facilities is given later in this chapter and in Ref. 1.

Turf Culture

One of the easiest and most effective ways to reduce erosion on side slopes and in side ditches is through the cultivation and development of a firm turf. Every effort is usually made in areas subject to erosion to encourage the growth of native grasses on exposed slopes. Formation of a firm turf may be accomplished by seeding, sprigging, or sodding the slope with suitable native grasses. In recent years, widespread use has been made of mats and linings of various kinds to encourage turf formation. It is possible in most cases to establish adequate growths of grass that serve greatly to reduce the erosive effects of a flowing stream or sheet of water in areas of moderate rainfall or on moderate slopes.

Intercepting Ditches

Shallow ditches may be placed at the top of outside cut slopes to intercept surface water and thus prevent erosion. Such ditches, which are usually quite shallow and narrow, are termed "intercepting ditches." Water conveyed in such a ditch is collected at suitable intervals along the top of the slope and discharged down the slope in paved or otherwise protected spillways into the longitudinal ditch, or spread out over the adjacent land.

Slope Protection Under Severe Erosion Conditions

Highways are often subjected to very severe erosive action in which the measures thus far discussed provide only a partial solution of the problem. One example of this situation might be a highway location in a mountainous region subject to heavy rain and snowfalls. In such areas, a large portion of the location will probably consist of side-hill sections with the side slopes being made as steep as possible in the interests of economy. Slopes of 1.5 to 1, 1 to 1, or even steeper are not unusual and are, of course, very susceptible to erosion. The slopes may be protected, for example, by the use of riprap or hand-placed rock.

Still another situation in which a highway embankment may be subjected to very severe erosion is that existing when the location parallels a large or rapidly flowing stream. Such a location may involve a change in alignment of the stream, with resultant scour of the embankment slope. The slope may be protected by a riprap, a paved revetment, steel sheet or timber piling, cribbing, or any one of a number of other protective devices.

11-4 Culverts, Bridges, and Storm Drains

Culverts and bridges constitute the "cross-drainage" system of a highway in a rural location through which water flowing in natural streams or collected on the high side of the right-of-way is transmitted from one side of the highway to the other. Additional information on culverts is contained in Sections 11-16 through 11-19.

Where space is restricted, in median swales, in urban areas, or where the natural slope of the ground is unsuited for drainage by open channels, storm drains are provided for the disposal of surface water. Storm drains and appurtenant structures are discussed when consideration is given to the drainage of city streets; see Sections 11-35 through 11-38.

DESIGN OF SURFACE DRAINAGE SYSTEMS

The design of surface drainage systems for a highway may be divided into three major phases: an estimate of the quantity of water that may be expected to reach any element of the system; the hydraulic design of each element of the system; and the comparison of alternative systems, alternative materials, and other variables in order to select the most economical system that can be devised. In the third phase, attention must be given to selecting the system that has the lowest annual cost when all variables are taken into consideration.

11-5 Hydrologic Approaches and Concepts

A variety of aproaches have been used to estimate the quantity of runoff for drainage design. When a drainage structure is to handle the flow of an existing stream, as in the case of some culverts and most bridges, the flow used for hydraulic design may be based on available records for that stream. For such "gaged" sites, statistical analyses can be performed on the recorded stream flow to provide an estimated peak design flow for a given "return period."

The term "return period" refers to the estimated frequency of rare events such as floods. Selection of the frequency of occurrence of the design storm is largely a matter of experience and judgment, although department policy may establish the interval to be used for a given situation. The return period is a statistical matter. For example, if the system is designed for a return period of 25 years, the statistical assumption is that the system will accommodate the most severe storm to occur once in 25 years. It is apparent that selection of a return period of 100 years instead of 25 would mean designing for a more severe storm and, in general, a more costly system. Conversely, if the frequency is 10 years, the intensity of the design storm will be less and in most cases a less costly drainage system will result, although economic losses from use of the short time might offset the savings in construction costs.

On the Interstate System, the Federal Highway Administration requires that all drainage facilities other than bridges and culverts be designed for storms with a frequency at least as great as 10 years, except that a 50-year frequency is used for underpasses or other depressed roadways where ponded water can be removed only by the storm drainage system.

Generalized flood-frequency studies have been published for a number of states. Some of these studies have been made by the U.S. Geological Survey in cooperation with the various state highway departments, and studies of this type are continuing. Flood-frequency studies of this type are based on statistical analysis of streamflow records. Usually, they indicate the mean annual flood as a function of the size of the drainage area for each hydrologic region within the state. Factors for computing floods of any frequency (within limits) in terms of the mean annual flood are given on a separate graph. The regional flood curves usually cover drainage areas of about 100 to 2000 square miles. Separate flood curves may be given for the larger rivers, which cross regional boundaries and seldom conform to the trend given by the smaller rivers.

If suitable streamflow records are not available, useful information may be gained from observations of existing structures and the natural stream. Drainage installations above and below the proposed location may be studied and a design based on those that have given satisfactory service on other portions of the same stream. Lacking this information, an examination of the natural channel may be made, including the evidences left by flood crests that have occurred in the past, and an estimate made of the quantity of water that has been carried by the stream during flood periods.

Stream gage data for particular regions have also been used to develop statistical regression equations for most areas of the country (*2*). These equations normally require basic watershed parameters such as the drainage area and the average slope of the stream. With such data, it is possible to estimate peak design flows for ungaged sites within the hydrologic region.

Special studies have focused on urban watersheds where growth and development have contributed to flooding and unexpected failures of highway drainage facilities. Regression models recommended for urban runoff include, in general order of significance, equivalent rural discharge, a "basin development factor," the contributing drainage area, slope, rainfall intensity, storage, and the amount of impervious area (*3*).

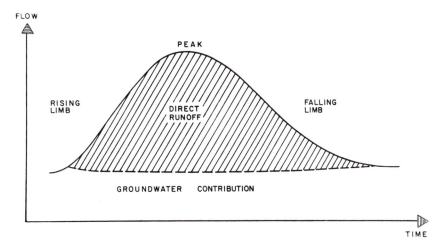

FIGURE 11-1 A flood hydrograph. (Courtesy Federal Highway Administration.)

Flood Hydrographs

A flood wave passing a point along a stream follows a common pattern known as a flood hydrograph (Fig. 11-1). The flow increases to a maximum, then recedes. As has been suggested, engineers have traditionally designed highway drainage facilities to accommodate the peak flow. Using the peak flow does not take into consideration the attenuating effect of upstream storage. If this storage is taken into account, the required size of the hydraulic structure may be significantly reduced. (*2*)

In urban areas, it may be especially desirable or necessary to make storage an integral part of the drainage design. Storage has the effect of broadening and flattening the flood hydrograph and decreasing the potential for downstream flooding. Additional information on storage routing procedures is given in Refs. 2 and 4.

11-6 Rainfall Intensity

The estimation of peak runoff for drainage design is accomplished by consideration of severe storms that occur at intervals and during which the intensity of rainfall and runoff of surface water are far greater than at other periods. Rainfall intensity during the design storm is a function of occurrence, duration, and intensity.

As Figure 11-2 illustrates, the intensity of rainfall for a particular return period varies greatly with the duration of rainfall. The average rate for a short time—such as 5 min—is much greater than for a longer period, such as 1 hr. In the design of many highway drainage systems, the duration chosen corresponds to the time of concentration. (See Section 11-8).

An accurate estimate of the probable intensity, frequency, and duration of rainfall in a particular location can be made only if sufficient data have been collected over a period of time. If such information is available, standard curves may be developed to express rainfall-intensity relationships with an accuracy sufficient for drainage problems. A number of state highway departments have developed such curves for use in specific areas. Rainfall intensity-duration data

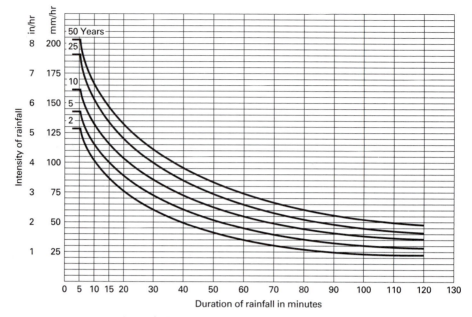

FIGURE 11-2 Typical rainfall intensity-duration curves. (Courtesy Federal Aviation Administration.)

have been published by the National Weather Service for various sections of the United States. One such publication is a rainfall-frequency atlas (5), which contains maps of rainfall frequency for durations of 30 min and 1, 2, 3, 6, 12, and 24 hr for periods of 1, 2, 5, 10, 25, 50, and 100 years.

11-7 Surface Runoff

When rain falls on a pervious surface, part of it passes into the soil and the remainder disappears over a period of time by either evaporation or runoff or both. In the design of highway drainage systems, the amount of water lost by evaporation is negligible; thus, drainage must be provided for all rainfall that does not infiltrate the soil or is not stored temporarily in surface depressions within the drainage area.

The rate at which water infiltrates the soil is dependent on the following factors: type and gradation of the soil; soil covers; moisture content of the soil; amount of organic material in the soil; temperature of the air, soil, and water; and presence or absence of impervious layers near the surface. Rates of infiltration on bare soil are less than on turfed surfaces. Frozen soil is impervious, and rain infiltrates very little until the frozen layer thaws. The rate of infiltration is assumed to be constant during any specific design storm.

The rate at which runoff occurs depends on the nature, degree of saturation, and slope of the surface. The rate of runoff is greater on smooth surfaces and initially slower where vegetation is present. On pavements and compacted surfaces runoff occurs at a high rate that varies with the slope and character of the surface at each point. For use in connection with the design of a drainage system for a particular area, these variables are considered and a coefficient of runoff selected. Values of the coefficient of runoff for use in one method of estimating

TABLE 11-2 Coefficients of Runoff to be Used in the Rational Formula

Type of Drainage Area	Coefficients of Runoff, C
Rural Areas[a]	
Concrete or asphalt pavement	0.8–0.9
Asphalt macadam pavement	0.6–0.8
Gravel roadways or shoulders	0.4–0.6
Bare earth	0.2–0.9
Steep grassed areas (2:1)	0.5–0.7
Turf meadows	0.1–0.4
Forested areas	0.1–0.3
Cultivated fields	0.2–0.4
Urban Areas	
Flat residential, with about 30% of area impervious	0.40
Flat residential, with about 60% of area impervious	0.55
Moderately steep residential, with about 50% of area impervious	0.65
Moderately steep built-up area, with about 70% of area impervious	0.80
Flat commercial, with about 90% of area impervious	0.80

[a]For flat slopes or permeable soil, use the lower values. For steep slopes or impermeable soil, use the higher values.

the quantity of flow (the rational method) are given in Table 11-2. If the drainage area being considered is composed of several types of surfaces, then a coefficient may be chosen for each surface and the coefficient for the entire area computed as a weighted average of the individual areas.

11-8 Time of Concentration

After selecting the design storm frequency, computations are made to determine the duration of the rainfall that produces the maximum rate of runoff. The duration of rainfall required to produce the maximum rate of runoff is known as the "time of concentration." The time of concentration usually consists of the time of overland flow plus the time of flow in the drainage system. The time of overland flow is the time required for a particle of water to flow from the most remote point in any section of the drainage area being considered to the point where it enters the drainage system. To this must be added the time of flow in the drainage system, from the intake to the point being considered. The time of overland flow varies with the slope, type of surface, length, and other factors. A number of empirical studies have been published on the time of overland flow to the defined channels, for example, Ref. 6. When the particular drainage area consists of several types of surfaces, the time of overland flow must be determined by adding together the respective times computed for flow over the lengths of the various surfaces from the most remote point to the inlet. Estimates of time of flow in the drainage system can be made from observed or computed velocities of flow.

11-9 The Rational Method

One of the most common methods of estimating runoff from a drainage area is the rational method. Its popularity is due to the fact that it combines engineering judgment with calculations made from analysis, measurement, or estimation. The method is based on the direct relationship between rainfall and runoff. In conventional U.S. units, it is expressed by

$$Q = CIA \tag{11-1}$$

where Q = runoff (ft³/sec)
   C = a coefficient representing the ratio of runoff to rainfall. Typical values of C are given in Table 11-2.
   I = intensity of rainfall (in./hr for the estimated time of concentration)
   A = drainage area in acres. The area may be determined from field surveys, topographic maps, or aerial photographs.

In metric units, the equation becomes

$$Q \text{ (m}^3\text{/sec)} = 0.0028 \, CIA \tag{11-2}$$

where I is expressed in millimeters per hour and *A* in hectares.

In view of the preceding discussion of the many variables involved in the rainfall-runoff relationship, shortcomings of this method are apparent. Application of the method should be confined to relatively small drainage areas (up to 200 acres, according to the Federal Highway Administration).

11-10 Computer Software in Hydrology

The Federal Highway Administration has sponsored the development of an integrated system of seven hydrology and hydraulic computer programs in flood computation and drainage design known as HYDRAIN (7). The software requires a minimum equipment component of MS DOS 3.0 or better, IBM XT, 640K RAM, hard disk, and monochrome monitor. The HYDRAIN software is available through FHWA distribution agents including McTrans Center, University of Florida, and PCTrans, University of Kansas.

One of the programs included in the integrated package is a hydrology program known as HYDRO. This system is a computer-based subset of Hydraulic Engineering Circular No. 19 (4) and includes some areas of Hydraulic Engineering Circular No. 12 (1).

Among other of the many capabilities of HYDRO, users are able to

1. estimate a time of concentration and intensity of rainfall for any site in the continental United States based on location and basin characteristics.
2. create a duration versus rainfall intensity curve for any user provided frequency and location in the continental United States are given.
3. determine a peak flow using the rational method.
4. determine a peak flow using regression equations developed by state and federal agencies.

DESIGN OF SIDE DITCHES AND OTHER OPEN CHANNELS

11-11 The Manning Formula

With the quantity of water expected to reach any given point in the drainage system known, the design of side ditches, gutters, stream channels, and similar facilities is based on established principles of flow in open channels. The principles also apply to flow in conduits with a free water surface. Most commonly used for design is Manning's formula, which applies to conditions of steady flow in a uniform channel and has the following form:

$$V = \frac{R^{2/3}S^{1/2}}{n} \tag{11-3}$$

where V = mean velocity (m/sec)

R = hydraulic radius (m); this is equal to the area of the cross section of flow (m²) divided by the wetted perimeter (m)

S = slope of the channel (m/m)

n = Manning's roughness coefficient. Typical values of n are given in Table 11-3.

In conventional U.S. units, in which V is expressed in ft/sec and R in ft, a unit conversion factor of 1.486 is added to the numerator of the Manning formula.

Also applicable is the equation of continuity:

$$Q = VA \tag{11-4}$$

where Q = discharge (m³/sec)

A = area of the flow cross section (m²)

TABLE 11-3 Value of Manning's Roughness Coefficient

Type of Channel or Structure	*Values of* n
Open Channels for Type of Lining Shown	
Smooth concrete	0.013
Rough concrete	0.022
Riprap	0.03–0.04
Asphalt, smooth texture	0.013
Good stand, any grass—depth of flow more than 6 in.	0.09–0.30
Good stand, any grass—depth of flow less than 6 in.	0.07–0.20
Earth, uniform section, clean	0.016
Earth, fairly uniform section, no vegetation	0.022
Channels not maintained, dense weeds	0.08
Natural Stream Channels	
(Surface Width at Flood Stage is 100 ft)	
Fairly regular section	
Some grass and weeds, little or no brush	0.030–0.035
Dense growth of weeds, depth of flow materially greater than weed height	0.035–0.05
Some weeds, light brush on banks	0.035–0.05
Some weeds, heavy brush on banks	0.05–0.07
Some weeds, dense willows on banks	0.06–0.08
For trees within channel, with branches submerged at high stage, increase all above values by	0.01–0.02
Irregular sections, with pools, slight channel meander; *increase* values given above about	0.01–0.02
Culverts	
Concrete pipe and boxes	0.012
Corrugated metal	
Unpaved	0.024–0.027
Paved 25%	0.021–0.026
Fully paved	0.012

In conventional U.S. units, discharge, Q, is expressed in ft³/sec, and cross-sectional area, A, in ft².

A family of charts has been published by the Federal Highway Administration for the solution of Manning's equation for various common channel cross sections (*8, 9*). One of these charts is shown in Figure 11-3.

11-12 Types of Open Channel Flow

At this point it is necessary to introduce the concepts of the "theory of critical flow" as it relates to the flow of water in open channels and culverts. The theory of critical flow gives rise to the definition of "critical depth," which is the depth where the flow changes from "tranquil" to "rapid" or "shooting." For example, water flowing down a relatively flat slope in an open channel is in tranquil flow, while that tumbling down a steep slope is rapid.

In the same channel, with the same quantity of flow, the flow can be changed from tranquil to rapid by an increase in slope. In such a case, the depth of flow decreases from that existing in the section of tranquil flow to a lower value in the section of rapid flow. The decrease in depth begins at some point ahead of the crest of the steep slope and continues gradually over some distance. In the section of rapid flow, since the quantity of flow is the same, the dimensions of the channel the same, and the depth less, it follows from the equation of continuity that the velocity is greater. Critical depth is defined as the depth corresponding to the change from tranquil to rapid flow.

Critical velocity and critical slope are the velocity and slope that correspond to uniform flow at critical depth. Critical depth is independent of channel slope and roughness, but the critical slope is a function of the slope of the channel.

Tranquil flow exists when the normal depth of water in an open channel is greater than the critical depth; conversely, when the depth is less than critical, the flow is rapid.

Rapid flow is difficult to control, because abrupt changes in alignment or cross section produce waves that travel downstream, alternating from side to side and sometimes causing the water to overtop the sides of the channel. Changes in channel slope, shape, or roughness cannot be reflected upstream except for very short distances; this condition is called "upstream control." This type of flow is common in steep flumes and mountain streams.

Tranquil flow is relatively easy to control. Changes in channel shape, slope, or roughness affect the stream for some distance upstream (downstream control). This kind of flow often occurs in streams in plains and valley regions where slopes are relatively flat.

Critical depth is important in analysis because it is always a hydraulic control. The flow must pass through critical depth in going from one type of flow to the other.

11-13 Design Procedure

Hydraulic design procedures are difficult to simplify because of the wide variety of choices presented to the designer in a typical case and the various assumptions that must be made. Design is usually based on an assumption of uniform flow because the error involved is relatively slight in most cases.

Basically, the design of a highway drainage channel is done in two parts. The first part involves the selection of a channel section that will carry the given discharge on the available slope. The second part is the determination of the

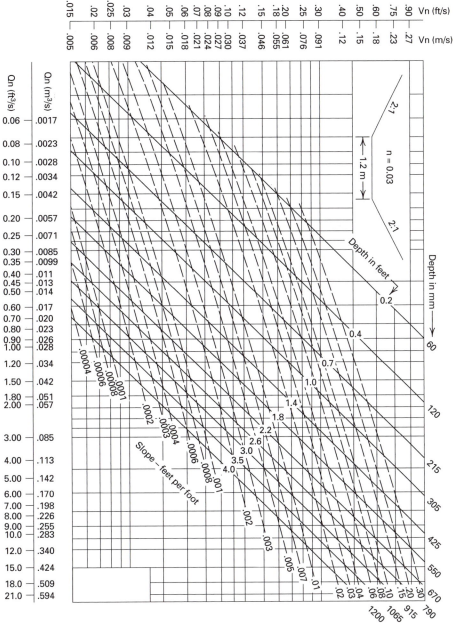

FIGURE 11-3 Graphical solution of Manning's equation for one trapezoidal channel section with 2-1 side slopes and bottom width 1.2 m (4 ft). (Adapted from Reference 9.)

protection required (if any) to prevent erosion of the drainage channel (see Section 11-14).

Use of charts like the one in Figure 11-3 gives a direct solution of the Manning equation for uniform flow in trapezoidal channels with 2 to 1 side slopes and with fixed bottom width. Depths and velocities shown in the chart apply accurately only to channels in which uniform flow at normal depth has been established by sufficient length of uniform channel on a consistent slope when the flow is not affected by backwater.

Figure 11-3 applies to a channel with any roughness coefficient, *n*. To use the figure, compute the quantity, *Q* times *n*, and enter the graph on the abscissa with the product $Q \cdot n$. Project a vertical line upward to the intersection with the slope line. There, by interpolation, read the depth. From the intersection, project a horizontal line to the left ordinate and read the value for the product $V \cdot n$. Compute the velocity by dividing this product by the roughness coefficient, *n*.

EXAMPLE 11-1   **Use of Figure 11-3** Determine the depth and velocity of flow in a trapezoidal channel (n = 0.02) with 2 to 1 side slopes and a 1.2-m bottom width discharging 4.25 m³/sec on a 2 percent slope.

   Solution Use the $Q \cdot n$ metric scale: $Q \cdot n$ = 4.25 (0.02) = 0.085 m³/sec. From 0.085 on the $Q \cdot n$ scale, project a line upward to the intersection with the line S_0 = 0.02. At this point, d_n = 550 mm, and from the left ordinate, by interpolation, the product $V \cdot n$ = 0.070. V = (0.070/0.02) = 3.5 m/sec.

   In conventional U.S. units, the product $Q \cdot n$ = 3.00, d_n = 1.8 ft, and $V \cdot n$ = 0.23. V = (0.23/0.02) = 11.5 ft/sec.

The following equation allows the calculation of critical depth of flow in non-rectangular channels:

$$\frac{Q^2}{g} = \frac{A^3}{b} \qquad (11\text{-}5)$$

where Q = flow in m³/sec (ft³/sec)
        A = the cross-sectional area in m² (ft²)
        b = width of the channel at the surface in m (ft)
        g = 9.8 m/sec² (in conventional U.S. units, 32.2 ft/sec²)

EXAMPLE 11-2   **Calculation of Critical Depth** A flow of 28 m³/sec occurs in a trapezoidal channel with a base width of 3 m and 2 to 1 side slopes. Calculate the critical depth.

   Solution The cross-sectional area is

$$A = 3\,d_c + 2\,d_c^2$$

and

$$b = 3 + 4\,d_c$$

By equation (11-5):

$$\frac{(28)^2}{9.8} = \frac{(3\,d_c + 2\,d_c^2)^3}{(3 + 4\,d_c)}$$

This problem is solved by trial using Equation (11-5). The critical depth, d_c, is equal to 1.5 m.

Open channel graphs are also available which allow users to determine directly critical depth, critical velocity, and critical slope with input of flow for specified channel shape and dimensions (9).

11-14 Prevention of Erosion in Drainage Channels

Erosion of highway drainage channels may create unsightly and hazardous ditches, cause pollution of nearby lakes and streams, and increase maintenance costs of drainage structures because of sedimentation. To prevent erosion, it may be necessary to provide protective linings in the bottom and along the sides of drainage channels. There are two general classes of protective linings: rigid linings and flexible linings. Common types of rigidly lined channels are those paved with portland cement concrete or soil-cement. Examples of flexible linings are rock riprap and vegetation.

Rigid channel linings are more expensive than flexible linings and, being smoother, may cause objectionable high velocities at the end of the lining. On the other hand, rigid linings are capable of preventing erosion under severe service conditions, along steep slopes, and in restricted areas where steep sidewall slopes are required.

Flexible linings are generally unsuitable for severe service conditions, for example, where the depth or velocity of flow is great; however, such linings are relatively inexpensive, easy to maintain, and aesthetically pleasing.

Temporary Linings

With increasing attention being paid to the environmental impact of highway construction, temporary linings are recommended to protect seeded channels until vegetation has been established. Examples of such linings include jute yarn net, fiberglass roving, knitted plastic netting interwoven with paper strips, and mats made from curled wood or synthetic fibers.

Design Approaches

Since rigid linings are resistant to erosion, the maximum depth of flow that can be accommodated in paved channels generally depends only on the required freeboard. The Federal Highway Administration (*10*) recommends a vertical freeboard height of 150 mm (0.5 ft) for small drainage channels. For large channels, wave height may have to be considered when establishing freeboard requirements.

Traditionally, the type of lining provided in drainage channels was determined on the basis of maximum allowable velocities. Currently, the preferred design method is based on the concept of maximum permissible shear stress or tractive force (*10*). The overall approach to the design of channel linings involves comparing a *computed* shear stress on a channel to empirically derived *permissible* shear stresses for various lining materials.

The computed shear stress is a function of flow depth; it is usually computed for the maximum depth of uniform flow as determined by Manning's equation. Experience has shown that the shear stress is maximum around bends where currents and eddies cause higher shear stresses on the sides and bottom compared to stresses along straight sections. Reference 10 provides a dimensionless adjustment factor to account for this effect.

The design flow rates used for channel lining design are usually based on a 5- or 10-year return period. For design purposes, a trapezoidal or triangular channel shape is assumed.

Channel slope is one of the major parameters affecting the computed shear stress, and it is usually fixed and the same as the road profile. Generally, most flexible lining materials are suitable for protecting channels with gradients up to 10 percent. Riprap and wire-enclosed riprap are more suitable for channel gradients steeper than 10 percent.

Reference 10 provides a table and two figures to give permissible shear stress values for a variety of channel linings.

11-15 Stream Enclosures

Side ditches paralleling highway locations in rural areas are sometimes eliminated by enclosing the stream in a pipe drain. Conditions that would make this desirable include the elimination of a deep, narrow side ditch made necessary by a narrow right-of-way, the desired widening of an existing pavement built in a narrow right-of-way, and the elimination of erosion. This method has also been employed for the somewhat unusual purpose of enclosing a surface stream so that the highway could be built in the middle of a narrow valley rather than on a side-hill location.

Stream enclosures of the type that have been described are, generally speaking, storm sewers and are designed in a fashion similar to that used in the design of municipal drainage systems. Appurtenant structures such as inlets, catch basins, and manholes are also normally included in the design.

DESIGN OF CULVERTS

A culvert is usually, although not always, differentiated from a bridge by virtue of the fact that the top of the culvert does not form part of the traveled roadway. More frequently, culverts are differentiated from bridges on the basis of span length. On an arbitrary basis, structures having a span of 20 ft or less will be called culverts, whereas those having spans of more than 20 ft will be called bridges. This line of division is by no means standard, and span lengths of 8 to 20 ft are employed by various organizations as limiting culvert lengths. Culverts also differ from bridges in that they are usually designed to flow full under certain conditions, while bridges are designed to pass floating debris or vessels.

Culverts are to be found in three general locations: at the bottom of depressions where no natural watercourse exists; where natural streams intersect the roadway; and at locations required for passing surface drainage carried in side ditches beneath roads and driveways to adjacent property.

11-16 Principles of Culvert Location

The majority of culverts are installed in natural watercourses that cross the roadway, either at right angles or on a skew. In addition to selecting the proper location or "station number" for the culvert crossing with respect to the centerline of the road, the alignment and grade of the culvert are of importance.

The location of the centerline of the culvert on the centerline of the road may be determined by inspection of the plans or in the field. This location will generally be on the centerline of an existing watercourse or at the bottom of a depression if no natural watercourse exists.

The alignment of the culvert should generally conform to the alignment of the natural stream, and the culvert should, if possible, cross the roadway at right angles in the interests of economy. Skew culverts, located at an angle to the centerline of the road, are needed in many instances. The selection of the natural direction of the stream is somewhat difficult in some areas, where the stream bed is not in a fixed position but shifts with the passage of time. In such a case, judgment must be exercised in selecting the most desirable location for the culvert, and some channel improvements may be necessary to ensure the proper functioning of the culvert after it is built. Where meandering streams are encountered, the water should be carried beneath the roadway at the earliest opportunity. Any changes that are necessary in the direction of the culvert itself should be effected gradually so that excessive head losses and consequent reduction in flow may be avoided.

Similarly, the grade of the culvert should generally conform to the existing grade of the stream. If the grade is reduced through the culvert, the velocity may be reduced, sediment carried in the water will be deposited at the mouth or in the length of the culvert, and the capacity of the structure will thus be further reduced. Culvert grades that are greater than those existing in the natural channel may result in higher velocities through the culvert and at the outlet end. Undesirably high velocities at the outlet will result in scour or erosion of the channel beyond the culvert and may make it necessary to install elaborate and costly protective devices. Changes in grade within the length of the culvert should also be avoided.

11-17 Hydraulic Design of Culverts

Earlier in this chapter, we discussed concepts and design procedures related to estimation of the quantity of runoff from a drainage basin. In this section, we will briefly discuss principles and techniques for the hydraulic design of culverts. For a more complete treatment of the subject, the reader is advised to consult Ref. 2.

The purpose of hydraulic design is to provide a drainage facility or system that will adequately and economically provide for the estimated flow throughout the design life without unreasonable risks to the roadway structure or nearby property.

Hydraulic design of culverts involves the following general procedure:

1. Obtain all site data and plot a roadway cross section at the culvert site, including a profile of the stream channel.
2. Establish the culvert invert elevations at the inlet and outlet and determine the culvert length and slope.
3. Determine the allowable headwater depth and the probable depth of tailwater during the design flood.
4. Select a type and size of culvert and the design features of its appurtenances that will accommodate the design flow under the established conditions. The design of the culvert inlet, discussed in Section 11-18, is especially important to the overall hydraulic efficiency of the structure.
5. Examine the need for energy dissipators, and, where needed, provide appropriate protective devices to prevent destructive channel erosion. (See Section 11-19.)

Whenever a constriction such as a culvert is placed in a natural open channel, there is an increase in the depth of water just upstream of the constriction. The allowable level of the headwater upstream of the culvert entrance is generally the principal control on the culvert size and inlet geometry. The allowable headwater depth depends on the topography and the nature of land use in the culvert vicinity. In establishing the allowable headwater depth, the designer should consider possible harmful effects that flooding may cause, such as damages to the pavement, interruptions to traffic, and inundation of nearby property.

Types of a Culvert Flow

The type of flow occurring in a culvert depends on the total energy available between the inlet and outlet. The available energy consists primarily of the potential energy or the difference in the headwater and tailwater elevations. (The velocity in the entrance pool is usually small under ponded conditions, and the velocity head or kinetic energy can be assumed to be zero.) The flow that occurs naturally is that which will completely expend all of the available energy. Energy is thus expended at entrances, in friction, in velocity head, and in depth.

The flow characteristics and capacity of a culvert are determined by the location of the *control section*. A control section in a culvert is similar to a control valve in a pipeline. The control section may be envisioned as the section of the culvert that operates at maximum flow; the other parts of the system have a greater capacity than is actually used.

Laboratory tests and field studies have shown that highway culverts operate with two major types of control: *inlet control* and *outlet control*. Examples of flow with inlet control and outlet control are shown, respectively, by Figures 11-4 and 11-5.

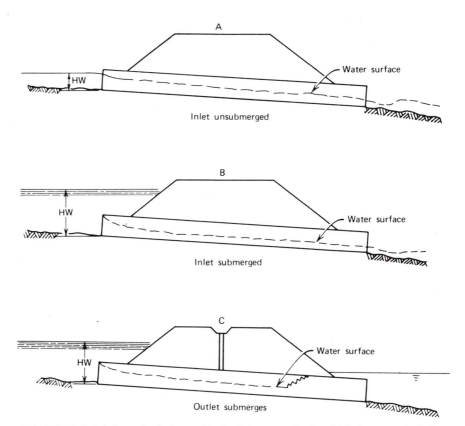

FIGURE 11-4 Inlet controls for culverts. (Courtesy Federal Highway Administration.)

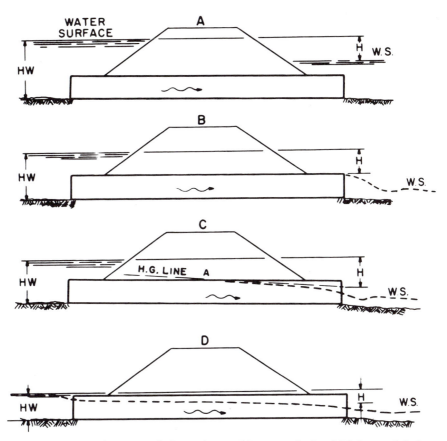

FIGURE 11-5 Outlet controls for culverts. (Courtesy Federal Highway Administration.)

Culverts Flowing with Inlet Control

Under inlet control, the discharge capacity of a culvert depends primarily on the depth of headwater at the entrance and the entrance geometry (barrel shape, cross-sectional area, and type of inlet edge). Inlet control commonly occurs when the slope of the culvert is steep and the outlet is not submerged.

On the basis of experimental work, analytical relationships have been developed for culverts with inlet control. These relationships are complex, however, and for a given entrance shape and flood condition (i.e., submerged, unsubmerged) are applicable only within a specified range of discharge factors.* For design expedience, the Federal Highway Administration has published a series of nomographs and design charts (2). Examples of available nomographs and charts are shown as Figures 11-6 and 11-7.

Culverts Flowing with Outlet Control

Maximum flow in a culvert operating with outlet control depends on the depth of headwater and entrance geometry and the additional considerations of the elevation of the tailwater at the outlet and the slope, roughness, and length of

---

*Discharge factor $= Q/D^{5/2}$, where Q is flow and D is culvert diameter.

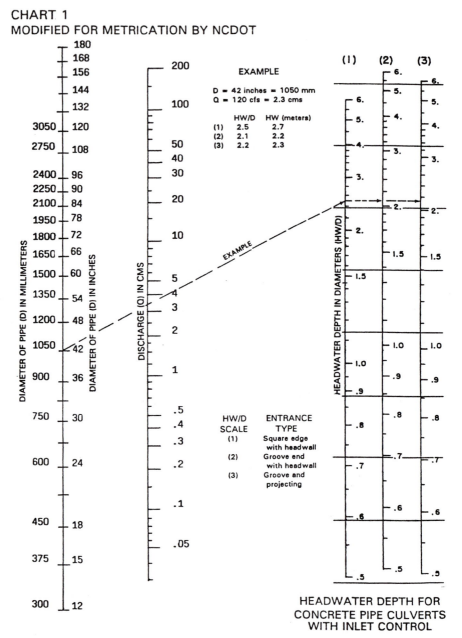

CHART 1
MODIFIED FOR METRICATION BY NCDOT

EXAMPLE

D = 42 inches = 1050 mm
Q = 120 cfs = 2.3 cms

	HW/D	HW (meters)
(1)	2.5	2.7
(2)	2.1	2.2
(3)	2.2	2.3

HW/D SCALE | ENTRANCE TYPE
(1) | Square edge with headwall
(2) | Groove end with headwall
(3) | Groove and projecting

HEADWATER DEPTH FOR
CONCRETE PIPE CULVERTS
WITH INLET CONTROL

FIGURE 11-6 Nomograph for determining headwater depth for concrete pipe culverts with inlet control. (Courtesy Federal Highway Administration and North Carolina Department of Transportation.)

the culvert. This type of flow most frequently occurs on flat slopes, especially where downstream conditions cause the tailwater depth to be greater than the critical depth.

For culverts flowing full, the difference in head between the upstream and downstream water surface H is equal to the velocity head plus the energy lost at the entrance and in the culvert:

$$H = [1 + K_e + (19.6 \, n^2 L)/(R^{4/3})] \, (V^2/2g) \qquad (11\text{-}5)$$

CHART 2
MODIFIED FOR METRICATION BY NCDOT

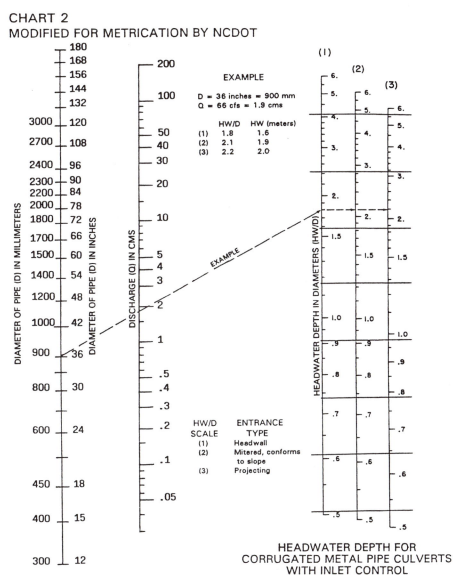

EXAMPLE

D = 36 inches = 900 mm
Q = 66 cfs = 1.9 cms

	HW/D	HW (meters)
(1)	1.8	1.6
(2)	2.1	1.9
(3)	2.2	2.0

HW/D SCALE	ENTRANCE TYPE
(1)	Headwall
(2)	Mitered, conforms to slope
(3)	Projecting

HEADWATER DEPTH FOR
CORRUGATED METAL PIPE CULVERTS
WITH INLET CONTROL

FIGURE 11-7 Nomograph for determining headwater depth for corrugated metal pipe culverts with inlet control. (Courtesy Federal Highway Administration and North Carolina Department of Transportation.)

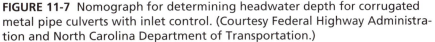

where　H = difference in elevation between the headwater and tailwater surfaces, illustrated in Figure 11-5a, or between the headwater surface and the crown of the culvert at the outlet, as shown in Figure 11-5b.

K_e = entrance loss coefficient; see Table 11-4
n = Manning's roughness coefficient; see Table 11-3
L = length of culvert (m)
R = hydraulic radius (m)
V = velocity (m/sec)
g = gravity constant (9.8 m/sec²)

TABLE 11-4 Entrance Loss Coefficients K_e; Outlet Control; Full or Partly Full

Type of Structure and Design of Entrance	Coefficient K_e
Pipe—concrete	
Projecting from fill, socket end (groove-end)	0.2
Projecting from fill, sq. cut end	0.5
Headwall or headwall and wingwalls	
Socket end of pipe (groove-end)	0.2
Square-edge	0.5
Rounded (radius $= \frac{1}{12}D$)	0.2
Mitered to conform to fill slope	0.7
End section conforming to fill slope	0.5
Beveled edges, 33.7° or 45° bevels	0.2
Pipe or pipe-arch—corrugated metal	
Projecting from fill (no headwall)	0.9
Headwall or headwall and wingwalls square-edge	0.5
Mitered to conform to fill slope, paved or unpaved slope	0.7
End section conforming to fill slope	0.5
Beveled edges, 33.7° or 45° bevels	0.2
Box—reinforced concrete	
Headwall parallel to embankment (no wingwalls)	
Square-edged on three edges	0.5
Rounded on three edges to radius of $\frac{1}{12}$ barrel dimension, or beveled edges on three sides	0.2
Wingwalls at 30° to 75° to barrel	
Square-edged at crown	0.4
Crown edge rounded to radius of $\frac{1}{12}$ barrel dimension, or beveled top edge	0.2
Wingwall at 10° to 25° to barrel	
Square-edged at crown	0.5
Wingwalls parallel (extension of sides)	
Square-edged at crown	0.7
Side or slope tapered inlet—all three culvert types	0.2

Source: Hydraulic Design of Improved Inlets for Culverts. Federal Highway Administration, Washington, DC (1972).

Where the critical depth falls below the crown of the culvert at the outlet, as shown in Figures 11-5c and 11-5d, the headwater can be determined analytically only by tedious and time-consuming backwater computations (2). Such computations can generally be avoided by using the design charts published by the Federal Highway Administration (2). From charts such as these, the headwater depths for both inlet and outlet control may be determined for practically all combinations of culvert size, material entrance geometry, and discharge.

To understand the use of Figures 11-6 and 11-7, the reader should study the examples on the nomographs.

Performance Curves

A culvert at a specific location may operate in a variety of ways, depending on the headwater; culvert diameter, area, and shape; and other physical conditions (e.g., slope, entrance shape, culvert roughness, etc.). Generally speaking, with low headwater, culverts tend to have entrance control and operate like a weir; with high headwater, they tend to operate with outlet control and as an orifice.

It is sometimes desirable to develop culvert performance curves that show flow rates versus headwater depth or elevation. A performance curve can be developed by the following steps (*2*):

1. Select a range of flow rates and determine the corresponding headwater elevations. These flow rates should fall above and below the design discharge and cover the entire flow range of interest. Both inlet and outlet control headwaters should be calculated.

2. Combine the inlet and outlet control performance curves to define a single performance curve for the outlet.

3. Reference 2 also describes a method for determining headwater elevations and flow rates when flooding results in roadway overtopping.

Figure 11-8 shows a hypothetical culvert performance curve with roadway overtopping.

11-18 Improved Culvert Inlet Design

With inlet control, flow in the culvert barrel is very shallow and the potential capacity of the barrel is generally wasted. Because the barrel is usually the most expensive component of the structure, flow under inlet control tends to be uneconomic. Surveys of culvert design practices by highway agencies indicate that millions of dollars could be saved each year by using improved inlet design concepts. An article by Normann (*11*) in *Civil Engineering,* reproduced in part in the following paragraphs, describes fundamental concepts of improved inlet design. For more detailed information on this important subject, the reader should refer to the FHWA publication *Hydraulic Design of Highway Culverts* (*2*).

In 1967, culvert research sponsored by the FHWA was completed by the National Bureau of Standards. Based on this research, a design manual was formulated, in which the best improved inlet designs were selected and compiled, based on several factors

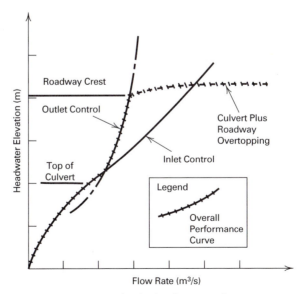

FIGURE 11-8 Hypothetical culvert performance curve with roadway overtopping. (Courtesy Federal Highway Administration.)

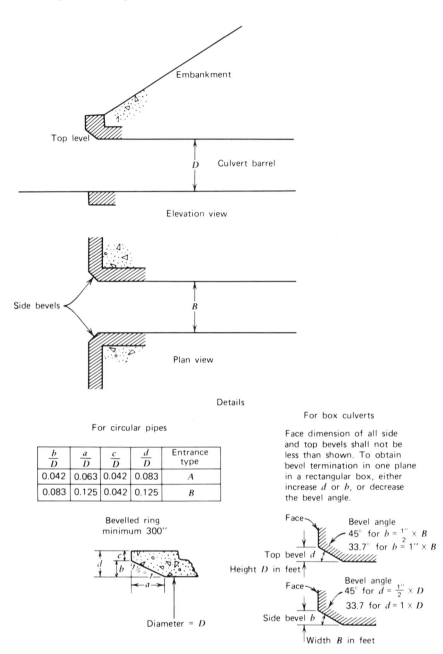

FIGURE 11-9 Bevel-edged inlets.

For circular pipes

$\dfrac{b}{D}$	$\dfrac{a}{D}$	$\dfrac{c}{D}$	$\dfrac{d}{D}$	Entrance type
0.042	0.063	0.042	0.083	A
0.083	0.125	0.042	0.125	B

including hydraulic efficiency, easy of construction and maintenance, and debris passage capability.

Three basic improved inlet designs are presented: bevel edged inlets [Figure 11-9], side-tapered inlets [Figure 11-10], and slope-tapered inlets [Figure 11-11]. These inlets improve hydraulic performance in two basic ways: (1) by reducing the flow contraction at the culvert inlet and more nearly filling the barrel and (2) by lowering the inlet control section and thus increasing the effective head exerted at the control section for a given headwater pool elevation.

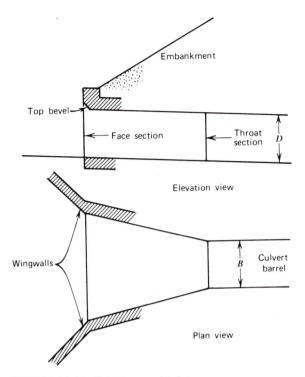

FIGURE 11-10 Side-tapered inlets.

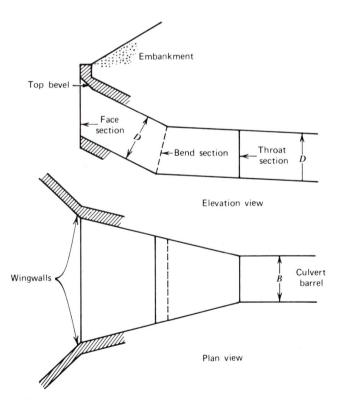

FIGURE 11-11 Slope-tapered inlets.

Bevel-edged inlet

The bevel-edged inlet shown in Figure 11-9 improves hydraulic performance by reducing the inlet contraction. This is the least sophisticated inlet improvement, and also the least expensive. As a minimum, this degree of improvement should be used on all culverts, both in inlet and outlet control. For concrete pipe culverts, the groove end, left intact and facing upstream, will serve essentially the same purpose as the bevel-edged inlet.

Side-tapered inlet

The side-tapered inlet in Figure 11-10 increases hydraulic efficiency by further reducing the contraction at the inlet control section, which should be the throat section at or near the design discharge. The face section is designed large enough so as to not restrict the flow. The roof and floor of the inlet are straight line extensions of the culvert roof and floor, and the tapered side walls meet the barrel walls at a smaller angle than the beveled edges (9.5 to 14° vs 33 to 35°). Also, the throat section is somewhat lower than the face, thus concentrating more head on the control section for a given headwater elevation.

Slope-tapered inlet

The slope-tapered inlet [in Figure 11-11] provides an efficient control section at the throat, similar to that provided by the side-tapered inlet, and further increases the concentration of head on the throat. In this design, the face section remains near the stream bed elevation and the throat is lowered by incorporating a fall within the inlet structure. This fall in the inlet reduces the slope of the barrel and increases the required excavation. The larger the fall required to pass the design discharge, the larger will be the inlet structure.

The slope-tapered inlet incorporates both methods of increasing hydraulic performance: reducing the entrance contraction and lowering the control section. Of course, any of the other inlets, conventional bevel-edged or side-tapered, may be lowered by constructing a sump upstream of the entrance, thus obtaining increased head on the inlet control section.

In addition to their use on new installations, improved inlets, especially bevel-edged and side-tapered inlets may be added to existing barrels to increase hydraulic performance if the existing inlet is operating in inlet control. Many times, this will preclude the construction of a new barrel when the existing culvert is undersized.

Similar improvements may be applied to pipe culverts. As an example of the resulting economy, savings of $1500 per culvert, or a total of $30,000 were recently reported on an interstate highway project in Georgia from the use of precast side-tapered inlets for pipe culverts.

New culvert design procedure

"Hydraulic Charts for the Selection of Highway Culverts," (U.S. Department of Transportation, Hydraulic Engineering Circular No. 5*) is probably the most widely used and accepted culvert design manual in the world. In the conventional design procedure, both inlet and outlet flow control are evaluated. The headwater depth (depth of upstream pool above the inlet invert) is computed for the design discharge assuming that (1) inlet control governs, and (2) that outlet control governs. Then, the highest of the two required headwater depths defines the control type and the culvert is categorized as being in either "inlet control" or "outlet control." Assumptions are simplified to expedite the computations, and by comparing headwaters, the difficult task of defining the actual flow profile through the culvert barrel is avoided. However, in this conven-

---

*This circular has been replaced by Hydraulic Design Series No. 5 (2).

tional design method, no attempt is made to modify the unbalanced flow condition that may exist. The inlet control performance curve represents the actual flow capacity of the culvert, while the more favorable outlet performance control curve is the culvert barrel potential. Thus, the full capacity of the barrel is not attained and an uneconomic situation exists.

11-19 Energy Dissipators

In the vicinity of culverts, high velocities, abrupt changes in direction of flow, and the presence of waves and eddies may cause severe erosion or "scour" to occur. A common problem at culvert outlets, scour may also occur in the vicinity of culvert inlets and at other locations in roadside channels. Unless preventive measures are taken, scour may cause unsightly and hazardous holes in the channel, damage the embankment, or even cause a culvert structure to fail. It is possible that such problems can be overcome by the use of protective channel listings as discussed in Section 11-14.

Where severe erosion problems are likely to occur, the Federal Highway Administration (*12*) recommends that consideration be given to the employment of energy dissipators. Such devices provide an effective means of controlling velocities and smoothing the flow.

Following are examples of energy-dissipating devices:

1. Drop structures that change the channel slope from steep to mild combined with special stilling basins to dissipate the kinetic energy.
2. Blocks, sills, or other roughness elements installed in the channel to increase the resistance or force and stabilize a hydraulic jump.
3. Stilling wells that dissipate energy by forcing the flow to travel directly upward to reach the downstream channel.

The design of energy-dissipating devices is a complex process that is beyond the scope of this book. The subject is adequately treated in Ref. 12.

11-20 Computer Software for Hydraulic Design

The HYDRAIN (*7*) integrated system of computer programs (mentioned in Section 11-10) includes the following software to aid in hydraulic design of ditches, channels, and culverts:

1. HYCLV analyzes and designs circular, rectangular, elliptical, and other culvert shapes. It ranks entrance edge conditions from most efficient to least efficient. It makes it possible to compare hydraulic characteristics of different culvert types for a given site. With the software, it is possible to compute exit velocities and to develop hydraulic performance curves of both individual culverts and an entire system.
2. HY8 is a BASIC program that allows the user to investigate the hydraulic performance of a culvert system, including the actual hydraulic structure as well as hydrologic inputs, storage and storage considerations, and energy dissipation devices and strategies. This software automates the methods presented in Hydraulic Design Series No. 5 (*2*), Hydraulic Engineering Circular No. 14 (*12*), and Hydraulic Engineering Circular No. 19 (*4*).
3. HYCHL provides guidance for the design of stable roadside channels and makes it possible to analyze the performance capabilities of various chan-

nel linings. This software also assists in the design of irregular channel riprap linings.

11-21 Culvert Types and Materials

Materials most commonly used in the construction of culverts are reinforced concrete and corrugated metal. Less frequently, culverts are made from timber, cast-iron pipe, vitrified-clay pipe, and, occasionally, stone masonry.

Concrete Culverts

Reinforced-concrete pipe intended for use in culverts is made in diameters of 300 to 3600 mm (12–108 in.) and in various lengths, the usual length being 1.2 to 2.4 m (4–8 ft). Standard specifications establish five classes of pipe with strengths increasing from class I through class V (*13*). The specifications show cross-sectional areas of reinforcing steels and concrete strengths for three series of wall thicknesses. Reinforcement may be either circular or elliptical. For special applications, reinforced-concrete culvert pipe may be manufactured with a cross section other than circular—elliptical and "arch" shapes being in quite common use. Concrete culvert pipes have tongue-and-groove or bell-and-spigot joints; the joints are sealed during construction with portland cement mortar, rubber gaskets, or other materials. A greater or lesser amount of care is required in the preparation of the foundation on which the pipes are to be laid. This preparation or "bedding" may vary from simple shaping of the bottom of a trench, or of the ground on which the pipe is laid, to embedment of the pipe in a concrete cradle, depending on foundation conditions, loads on the pipe, and other factors. Pipe culverts are most frequently constructed in what is termed the "projection condition." That is, the culvert is constructed on the surface of the ground, in the open, and the fill is built around it. In such cases, only a nominal amount of attention need be given to bedding of the pipe in normal soils and normal heights of fill.

Concrete box culverts are constructed in place with square or rectangular cross sections; single box culverts vary in size from 0.6 to 3.6 m (2–12 ft) square, depending on the required area of waterway opening. Most state highway departments use standard designs for various sizes of box culverts; perhaps the most commonly used sizes of concrete box culverts are in the range of 1.2 to 2.4 m (4–8 ft) square. Rectangular cross sections are used where it is desired to reduce the height of the culvert to provide adequate cover between the top of the culvert and the roadway surface. The use of box culverts has declined in recent years, largely because of the time required for their construction.

Both concrete pipe and concrete box culverts are built with more than one opening where additional waterway area is required and when it is desired to avoid the use of excessively large single pipes or boxes. Such installations are called "multiple culverts" and may, for example, be "double" or "triple" concrete pipe or concrete box culverts.

Concrete arches are sometimes used in place of concrete box culverts, although difficulties attendant on their proper design and construction have somewhat restricted their use. Concrete arch bridges are used more frequently.

Corrugated Metal Culverts (Steel)

Corrugated steel is used in various forms in the construction of culverts for use in highway drainage.

Corrugated metal (galvanized steel) pipe is made in diameters of 200 to 2440 mm (8–96 in.) and in lengths of 6 to 12 m (20–40 ft). Various thicknesses of metal are used, generally from 16 to 8 gage. The corrugations that are formed in the sheet metal are 68 mm (2⅔ in.) from crest to crest and 13 mm (½ in.) deep. Standard pipe is manufactured by bending the corrugated sheet metal into a circular shape and riveting the longitudinal joint. Helically corrugated pipe has a folded seam rather than a riveted longitudinal joint. In the field, lengths of corrugated metal pipe may be joined by a pipe sleeve or by a connecting band that is several corrugations in length—angle irons are riveted to the two ends of the band and connected by bolts. The band is slipped over the ends of the pipes to be connected and the bolts drawn tight to form the connection.

The maximum desirable diameter of standard corrugated metal pipe is 2.4 m (8 ft). This fact led to the development of a method of construction using heavier, curved, corrugated metal plates that are bolted together to form circular pipes or arches. The plates are curved, corrugated, and galvanized at the factory, shipped to the field site, and there bolted together to form the desired structure. The plates are heavier than normal corrugated metal pipe, being available in gages designated as 1, 3, 5, 7, 8, 10, and 12. They are made in various widths and in lengths of 1.8 and 2.4 m (6 and 8 ft). Pipes up to 6.4 m (21 ft) in diameter have been fabricated by this method, whereas arches of almost any desired combination or rise and span are possible; maximum standard size is a rise of 4 m (13 ft 2 in.) and a span of 6.3 m (20 ft 7 in.). Development of the "compression ring" method has allowed rational design of multiple-plate structures to fit any combination of circumstances.

In many culvert installations headroom is limited, and a circular pipe that has sufficient hydraulic capacity is not suitable. In such cases, corrugated metal "pipe arches" may be used. Pipe arches made of standard corrugated metal are available in sizes varying from a span of 460 mm (18 in.) and a rise of 280 mm (11 in.) to a span of 1830 mm (72 in.) and a rise of 1120 mm (44 in.). Figure 11-13 shows a standard drawing for a corrugated metal pipe arch with a flared end section. A typical example of a location where this type of culvert may be used to advantage lies in the culvert opening required to pass water flowing in a side ditch beneath an approach road to a rural highway.

In the interest of increased durability, corrugated metal pipe is sometimes furnished with the invert of the pipe covered with a thick bituminous mixture that completely fills the corrugations in this section of the pipe. Such pipe is called "paved-invert" pipe. The remainder of the pipe may also be coated with a bituminous material.

Miscellaneous Culvert Types

Other materials that may be used in the construction of culverts include vitrified-clay pipe, cast-iron pipe, and timber. In addition, masonry arch culverts are still constructed in some localities where suitable stone is cheaply available, although the use of this type of construction has declined in recent years. Each of these types of culvert has advantages and may be used in areas where the material is economically available and for special-purpose structures.

11-22 Selection of Culvert Type

The type of culvert selected for use in a given location is dependent on the hydraulic requirements and the strength required to sustain the weight of a fill or moving wheel loads. After these items have been established, the selection is then largely a matter of economics. Consideration must be given to durability and to the cost of the completed structure, including such items as first cost of manufactured units and costs of transportation and installation. Maintenance costs should also be considered in any overall comparison of the cost of different culvert types.

11-23 Culvert Headwalls and Endwalls

Headwalls and endwalls are provided on culverts principally to protect the sides of the embankment against erosion. Some authorities refer to the upstream wall as a "headwall" and the downstream one as an "endwall." In the following discussion, the term headwall will be used generically.

In addition to their function in erosion control, headwalls may serve to prevent disjointing of section pipe culverts and to retain the fill. Materials most commonly used for culvert headwalls are concrete, masonry (stone or rubble), and metal. Of these, concrete is most widely used because of its adaptability to all types of culverts and because it lends itself to pleasing architectural treatment.

In selecting the size and type of headwall to be used in a given case, matters of economy must again be given consideration. In additon, some weight must be given to aesthetic considerations, as the headwall is the principal portion of the average culvert structure which is visible to the traveler. Headwalls are not always necessary, of course, and their use should be avoided wherever it is feasible to do so because of their cost. Headwalls are an expensive portion of the average culvert installation; the headwall should be made as small as possible consistent with adequate design. Safety of traffic must also be considered in choosing the type of headwall to be used in a given case.

Many different types of headwall are used by different highway agencies; generally, each agency has developed standard designs that are used whenever possible. Several illustrations are given of typical headwall installations currently used by various organizations.

Figure 11-12 is a drawing that illustrates the details of standard endwalls used for multiple-pipe culverts by the Virginia Department of Highways. The flaring portions of this endwall are generally called "wing-walls." The endwall shown in Figure 11-13 is typical of the prefabricated metal end sections used in connection with small corrugated metal pipe culverts.

In the interest of safety, highway agencies should consider the placement of a traversible grate over the end of culverts if there is danger of vehicles colliding with the headwall. (See Figures 8-5 and 8-6.)

11-24 Culvert Intakes

In certain circumstances, for example, in easily eroded soils when high velocities occur, unusual precautions must be taken to prevent damage to the inlet of a culvert. Such precautions may consist only of paving ditches that carry water to the culvert entrance. In other cases, special culvert intakes may be provided.

As emphasized previously, consideration should also be given to the hydraulic properties of the culvert entrance. That is, the intake should be designed to minimize losses of head due to eddies and turbulent flow. In general, sharp corners or "breaks" in the culvert entrance should be avoided. In certain instal-

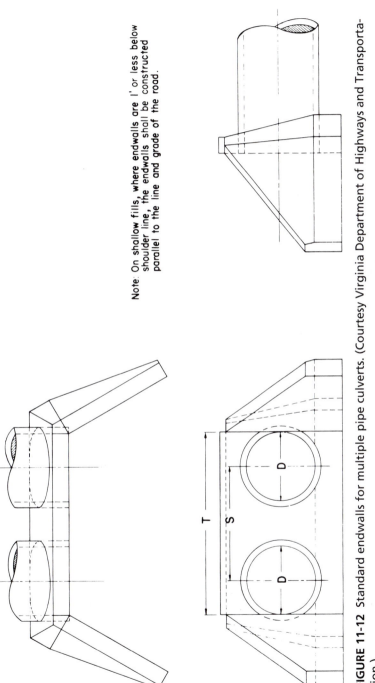

Note: On shallow fills, where endwalls are 1' or less below shoulder line, the endwalls shall be constructed parallel to the line and grade of the road.

FIGURE 11-12 Standard endwalls for multiple pipe culverts. (Courtesy Virginia Department of Highways and Transportation.)

313

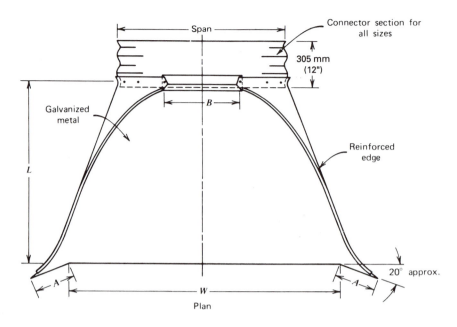

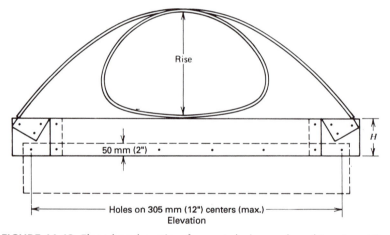

FIGURE 11-13 Flared-end section for metal pipe arches. (Courtesy Virginia Department of Highways and Transportation.)

lations, consideration must be given to prevention of clogging of the culvert entrance by drift and rubbish. Special auxiliary devices are sometimes constructed for this purpose (*14*).

HIGHWAY BRIDGES

As previously indicated, a bridge may be defined as a drainage structure that has a span of more than 6 m (20 ft). As a further distinguishing feature, bridge spans usually, although not always, rest on separate abutments, whereas culverts are

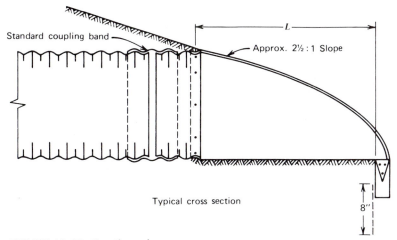

Standard coupling band

Approx. 2½ : 1 Slope

L

Typical cross section

8"

FIGURE 11-13 *Continued*

regarded as integral structures. Although the term "bridge" is usually associated with structures that are required to carry the roadbed over an established waterway, it may also be somewhat loosely applied to grade separation structures and elevated highways in urban areas (viaducts).

11-25 Bridge Location

In modern practice, bridges of relatively short span are located to conform with the general location of the highway, which has been previously determined. That is, the tentative location for the highway is established after an analysis of all the economic and engineering factors involved, and the bridge engineer is given the problem of providing an economical and adequate bridge design to conform to the roadway location. In some cases the location of a suitable stream crossing may be the most important single factor influencing the location of the highway in a given section; such is usually the case when long bridge spans are involved.

The ideal location for a bridge crossing is, of course, one in which the crossing is made at right angles to the centerline of the stream at the narrowest point, where the alignment of the approach pavement is straight, where the approach grade is slight, and where soil conditions are adequate for the installation of the most economical foundation for the span involved. This ideal combination of circumstances is encountered all too infrequently, except in structures of short span, and many bridges have been located on skew crossings, vertical curves, or with curving alignment. In such cases, considerations related to the general roadway location may still be regarded as controlling factors, and the required adjustments in the location of the bridge are made with these requirements in mind.

Many times, alternative locations of a proposed bridge may seem to offer somewhat similar advantages. A careful comparison must then be made of the several possible locations. The final decision should be based on a complete analysis, including factors related to traffic safety and operating conditions, fulfillment of the purpose of the road (e.g., the direct connection of population centers), and economics. Any complete analysis must include both the bridge and the approaches to it. A comparison of this type will generally result in the selection of one of the possible sites as the most desirable.

Once the general location is determined, the selected site must be subjected to careful scrutiny. This examination may be extremely detailed or somewhat cursory, depending largely on the size and importance of the contemplated structure. A complete survey of the bridge site may include an examination of the channel for some distance above and below the bridge crossing, a complete topographic map of the site, and an extensive soil survey of the area, including securing undisturbed soil samples where required and determining the required waterway opening and the requirements of navigation on large streams.

Although the preceding paragraphs have dealt largely with waterway crossings, similar factors govern the design of grade separation and the bridge portions of complex interchanges. Obviously, a grade separation structure designed to carry the roadway over an existing railroad presents the location engineer and designer with similar problems, as does the design of a highway overpass. Similarly, the general location of an elevated highway in an urban area is usually determined on the basis of maximum serviceability to traffic (and availability of right-of-way) so that the engineer must prepare a design suitable to conditions in a rather limited area.

11-26 Design of Waterway Opening

In many locations the natural stream channel is somewhat constricted by the bridge structure and roadway approaches. In the interests of economy, the roadway is frequently placed on an embankment on either side of the bridge span, the distance between abutments is reduced as much as possible, and piers may be placed in the stream channel. All these things serve in many cases to reduce severely the area through which the water must pass, particularly when the stream is at flood stage. Two results may immediately be noticed: during flood stage the velocity of the water through the bridge opening may be considerably increased, with resultant danger to the bridge structure through scour at abutments and piers, and the elevation of the water on the upstream side may be increased, with the result that the area subjected to flooding above the bridge site is increased and adjacent property is subjected to overflow beyond the limits of the normal floodplain. It thus seems axiomatic that the bridge must be designed to pass the flow occurring at flood stage without excessive velocity and without damage to property located above the bridge crossing. Estimating flood flows is best done by study of stream-gaging records but is sometimes based on observation of high water marks, the behavior of structures located on the same stream, and hydrologic computations.

Bridge openings are also normally designed to pass floating debris carried in the channel at normal and flood states. On navigable streams, requirements of navigation must be evaluated and provided for. Generally speaking, navigable streams are spanned by high-level crossings or movable bridges. Plans for bridges over navigable streams in the United States are subject to approval by the U.S. Corps of Engineers.

11-27 Bridge Clearances

Standards related to bridge clearances, both vertical and horizontal, are an important part of the design of highway bridges. Clearances recommended by AASHTO for bridges (and culverts, where applicable) are shown in Figure 11-14 (*15*). AASHTO specifies that the roadway width at bridges shall be equal to the full shoulder width of the approach section. Along curbed roadways, the full

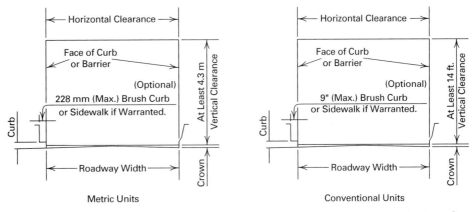

FIGURE 11-14 Clearance diagram for bridges. (Courtesy American Association of State Highway and Transportation Officials.)

width of the approach section should similarly be carried across the structure. For low-speed, low-volume roads, a minimum horizontal clearance of the width of the approach traveled way plus 2.4 m (8 ft) is recommended (*15*).

Provision of adequate clearances on new bridges is extremely important because literally thousands of existing bridges that are otherwise adequate have been rendered functionally obsolete by narrow roadways. The provision of adequate width in new construction is much easier and cheaper than modifications that must be made after the design width proves to be inadequate.

As Figure 11-14 indicates, the vertical clearance on highway bridges should be at least 4.3 m (14 ft) over the entire roadway. An allowance should be added for pavement resurfacing. Along the interstate system and other main highways, a 4.9-mm (16-ft) vertical clearance is usually provided.

11-28 Bridge Live Loads

Live loads used in the design of highway bridges are normally those established by AASHTO; two types of loadings are in use—H loadings and H-S loadings (*15*).

The H loadings consist of a two-axle truck or corresponding "lane loading," the latter being a certain uniformly distributed load and a concentrated load, which are equivalent in effect to the specified truck loading. Two H loadings are specified: H20-44 and H15-44. The meaning of these may be explained by examination of the H15-44 loading. The gross weight of this truck is 15 tons (30,000 lb), and this load is distributed 20 percent (6000 lb) on the front axle and 80 percent (24,000 lb) on the rear axle; the spacing between axles is 14 ft. The load has 10-ft clearance and load-lane width; the transverse spacing center-to-center of the wheels is 6 ft. The suffix "44" simply indicates that this standard, as explained, was adopted in 1944. The H20 loading is similarly specified.

The H-S loadings consist of a tractor truck with semitrailer, or the corresponding lane loading. Two H-S loadings are in common use: HS15-44 and HS20-44. These loadings are somewhat similar to the H loadings; the spacing between axles of the truck is fixed at 14 ft, while the axle spacing of the semitrailer is from 14 to 30 ft, this distance between varied to produce maximum stresses. The total load of the HS15-44 loading is 27 tons (54,000 lb), which is distributed as follows:

front axle of truck, 6,000 lb; rear axle of truck, 24,000 lb; and axle of semitrailer, 24,000 lb. Total load of the HS20-44 loading is 36 tons (72,000 lb), and it is similarly distributed.

The choice of load to be used in the design of a particular structure depends on the traffic expected to use the bridge during its anticipated life. Heavier loadings are generally used in the design of structures on the various primary systems, and lighter loads are used on other components of the highway system such as country roads. The HS20-44 loading is being used in the design of new bridges on the Interstate System, while many structures on the existing federal-aid system have been designed for H-15 loading. There is a definite trend toward the use of heavier loadings simply because the use of the heavier loads does not greatly increase the cost of the structure and seems to be consistent with the objectives of long-range planning.

11-29 Bridge Types

Under the stimulus of a greatly expanded highway construction program, engineers in the United States (and in most Western countries) have virtually revolutionized bridge design and construction methods in the past decade. The advances apply to short-, medium-, and long-span bridges.

For permanent bridges, the most commonly used materials are steel and concrete. Bridges of many different types are built with these materials, used singly or in combination. Timber may be used for temporary above-water construction, for the elements of a structure that lie below the waterline (particularly timber piles), or for short-span bridges located on secondary roads. A few short-span aluminum bridges have been built in the United States on an experimental basis.

The principal portions of a bridge may be said to be the "substructure" and the "superstructure." This division is used here simply for the convenience because in many bridges there is no clear dividing line between the two.

Common elements of the substructure are abutments (usually at the bridge ends) and piers (between the abutments). Piers and abutments often rest upon separately constructed foundations such as concrete spread footings or groups of bearing piles; these foundations are part of the substructure. Occasionally a bridge substructure comprises a series of pile bents in which the piles extend above the waterline and are topped by a pile cap that, in turn, supports the major structural elements of the superstructure. Such bents often are used in a repetitive fashion as part of a long, low, over-water crossing.

In recent years, the dividing lines between short-, medium-, and long-span bridges have blurred somewhat. Currently, spans of 20 to 100 ft are regarded as short by many designers, who have developed many standardized designs to handle these spans economically. Medium spans can range up to 400 ft in modern bridge practice, depending on the organization involved and the materials used. Long spans range up to 4000 ft or more, but a clear span above 1000 ft is comparatively rare.

Bridges may also be classed as "deck" or "through" types. In the deck type of bridge, the roadway is above the supporting structure; that is, the load-carrying elements of the superstructure are below the roadway. In the through type of bridge, the roadway passes between the elements of the superstructure, as in a through steel-truss bridge. Deck structures predominate; they have a clean appearance, provide the motorist with a better view of the surrounding area, and are easier to widen if future traffic requires it.

Thousands of short-span bridges have been built in the United States in recent years under the impetus of the accelerated federal-aid highway program and the increasing use of limited access.

Examples of short-span steel bridges include simple-span, wide-flange beam; simple-span, welded girder; continuous, wide-flange beam (the bridge is designed to act as a continuous structure over two or more supports); and continuous, welded girder. Typically, these steel bridges have reinforced-concrete decks designed for composite action with underlying steel members.

Examples of short-span concrete bridges include cast-in-place, reinforced concrete **T** beam (and slab); simple-span, prestressed, which incorporates precast, prestressed **I** girders or box girders topped by a cast-in-place deck; and cast-in-place box girder.

The designer of each medium- and long-span bridge tries to devise a structure that is best suited to the conditions encountered at that particular location. The result is an almost bewildering variety of structures that differ either in basic design principles or in design details.

General categories of steel bridges are briefly described in the following paragraphs:

Suspension bridges are used for very long spans or for shorter spans where intermediate piers cannot be built. An example is the Verrazano Narrows Bridge (Fig. 11-15), which was completed in 1964. The $305 million, 1298-m (4260-ft) structure spans the entrance to New York Harbor to join Staten Island and Brooklyn.

Girder bridges come in two basic varieties: plate and box girders.

Plate girders are used in the United States for medium spans. They generally are continuous structures with maximum depth of girder over the piers and minimum depth at midspan. The plate girders generally have an **I** cross section; they are arranged in lines that support stringers, floorbeams, and, generally, a cast-in-place concrete deck. The girders are shop-fabricated by welding; field connections generally are by high-strength bolts. Welded-steel box girder structures are generally similar to plate girder spans except for the configuration of the bridge cross section.

Rigid frames are used occasionally, most often for spans in the range of 23 to 30 m (75 to 100 ft) and for grade-separation structures.

Arch bridges are used for longer spans at locations where intermediate piers cannot be used and where good rock is available to withstand the thrusts at the arch abutments. A variation in the arch bridge is the tied arch, in which a horizontal tie that carries the roadway takes much of the horizontal thrust inherent in the arch form.

Truss bridges are built in many forms and in many locations for medium and long spans. Both deck and through trusses are built, with cantilever and continuous trusses being the most common.

Figure 11-16 illustrates the Hale Boggs Bridge, a modern cable-stayed bridge in Louisiana. Its 372-m (1222-ft) main span is the world's second longest for this type of bridge.

Concrete bridges come in nearly as great a variety as do steel bridges. Following are various types now in use and some examples of outstanding concrete structures:

FIGURE 11-15 The Verrazano Narrows Bridge, longest clear span in the world. (Courtesy American Institute of Steel Construction.)

Conventional reinforced concrete is used primarily for short spans, with a variety of innovations from the customary cast-in-place, slab-and-beam design.

Precast, prestressed concrete bridges are popular in the United States for short, simple spans; they generally consist of prestressed **I** beams or box girders that support a cast-in-place deck. One of the world's great water crossings, the 28.3-km (17.6-mile) Chesapeake Bay Bridge-Tunnel, contains over 19 km (12 miles) of repetitive spans that incorporate prestressed concrete cylinder piles, precast pile caps, and 22.8-m (75-ft)-long precast prestressed deck sections. Each span contains four parallel deck sections, each with a double **T** shape.

Cast-in-place prestressed concrete has been used for long-span bridges in Eu-

FIGURE 11-16 A modern cable-stayed bridge in Louisiana. (Courtesy State of Louisiana Department of Transportation and Development.)

rope and in a few locations in the United States. For example, a 201-m (660-ft)-long, four-span viaduct in Oakland, California, was prestressed (post-tensioned) with wires longitudinally, transversely, and vertically. In addition, it was stressed by a longtitudinal movement of the pier bases produced by jacking after the superstructure was completed.

Arches also are built in concrete. One of the world's longest concrete spans is an arch at Sydney, Australia. It is 304.8 m (1000 ft) in length and is made up of 512 huge, hollow concrete boxes, each 6.1 m (20 ft) deep.

11-30 Movable Bridges

The three basic types of movable bridges—bascule, vertical lift, and swing—are shown in Figure 11-17. They are used where a high-level bridge is uneconomical but provision must be made for navigation.

There is also a fourth type of movable bridge, the floating bridge. Three of these unusual structures are in the Puget Sound area of Washington State. The most recent of these incorporates a four-lane, 2.3-m (1.4-mile)-long floating section supported by 35 prestressed concrete pontoons, the largest of which is 110 m (360 ft) long, 18.2 m (60 ft) wide, and nearly 4 m (13 ft) deep. Each pontoon is secured by cables to precast concrete anchors on the lake bottom in about 60 m (200 ft) of water. A drawspan at the center provides for passage of ships.

HIGHWAY SUBDRAINAGE

As previously mentioned, the term "subdrainage" is related to the control of ground water encountered in highway locations. Subdrains are a necessary part of the complete drainage system for many highways in rural areas, and they

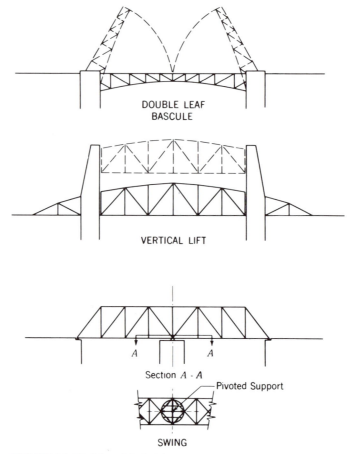

FIGURE 11-17 Movable bridges.

function along with adequate surface drainage facilities to prevent damage caused by water in its various forms.

A highway subdrain usually consists of a circular pipe laid at a suitable depth in a trench, which is then backfilled with porous, granular material. Materials principally used in subdrains include vitrified clay pipe, porous concrete pipe, and perforated corrugated metal pipe (both galvanized steel and aluminum). Clay pipe is usually laid with open joints, whereas concrete and perforated corrugated metal pipe are generally laid with sealed joints. Concrete pipe may also be laid with open joints. The size of drain is usually based on previous experience. Hydraulic design of subdrains is difficult but may be carried out on extensive projects. Pipes of 6- and 8-in. diameter are in common use, although pipes for subdrains are made as large as 24 in. The slope of the pipe should be sufficient to prevent the deposition or "settling out" of any solid material which may enter the pipe through the joints or perforations. Minimum recommended slope is 0.15 to 0.25 percent.

Subdrains are installed for a number of purposes, most of which may be included in the following classifications:

1. Control of seepage in cuts or sidehill locations; these installations are generally called "intercepting drains."

2. Lowering of ground-water table, as in swampy areas.
3. Base and shallow subgrade drainage.

11-31 Intercepting Drains

In rolling or mountainous terrain, cuts made during highway construction frequently expose flowing ground water or "seepage." Seepage that occurs through the cut slope may be a source of damage to the slope and to the roadway itself. Similarly, the seepage zone itself may not be invaded by the construction of a side-hill section, but the roadway and pavement structure may be located only slightly above the zone of flowing underground water and thus be subject to the detrimental effects of capillary action. In such cases the flow of underground water is intercepted by a subdrain located on the uphill side of the section so that the water is prevented from flowing beneath the pavement. An example of such an installation is shown in Figure 11-18. In this example, because there is an impervious soil layer located at a relatively shallow depth, the flow of seepage water is entirely cut off by the subdrain. In cases where the seepage zone is deeper, or where no impervious layer is found, the subdrain may simply be carried to sufficient depth to eliminate the effects of capillarity on the subgrade or base. Drains of the type that have been described are normally placed parallel to the centerline and are called "longitudinal drains." In some cases, seepage flow may occur in a direction parallel to the centerline. Transverse drains beneath the pavement may then be needed to intercept this longitudinal seepage.

The location of subdrains of this type before construction begins is somewhat difficult unless a very complete study is made of subsurface conditions. Information obtained from a soil survey may permit the approximate location of subdrains. Many organizations, however, prefer to provide necessary subdrainage on the basis of information gained during construction. Such a procedure demands the exercise of good engineering judgment in the field but should be more satisfactory than a decision made in the office on the basis of incomplete information regarding ground-water conditions. In any case, water collected in the subdrains must be carried to a suitable outlet.

11-32 Lowering of the Water Table

In many locations in flat terrain, the roadway may be built on a low embankment in the interests of economy and the base may be only 0.6 or 0.9 m (2 or 3 ft) above the water table. If the subgrade soil is subject to capillary action, water

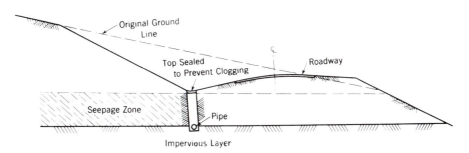

FIGURE 11-18 Intercepting subdrain.

will be drawn up into the subgrade and base with resultant loss of stability, possible frost damage, and similar detrimental effects. The answer to such a problem is simply to lower the water table a sufficient amount to prevent harmful capillary action. This is usually accomplished by the installation of parallel lines of subdrains at the edges of the roadway (or shoulders) at the proper elevation. In practice, the solution is frequently complicated by the difficulty of providing satisfactory outlets for the water collected in the subdrains. The depth required beneath the base varies with the type of soil encountered but is generally from 1 to 2 m (3 to 6 ft).

11-33 Base Drainage

Water that falls on the surface of flexible and rigid pavements may enter the base and subgrade through cracks in the surface, joints, and shoulders. If the base is relatively impervious, or if it lies above an impervious subgrade soil, this water may collect in the base and the upper portion of the subgrade. This effect may be noted even where water is not brought up into the base or subgrade from below by capillary action. Again, water thus trapped may cause weakening of the base material, and it is particularly serious in areas that are subject to frost action. Subdrains of the types that have been described may be used to remove water from the base, although experience has not always shown this type of installation to be successful.

11-34 Laying and Backfilling

Elements of the construction of a pipe subdrain of the type that has been described include the excavation of a suitable trench, preparation of the trench bottom, laying of the pipe, and backfilling with pervious material.

Excavation may be accomplished by either hand or machine methods. Trenches are usually quite shallow, excavated with vertical walls whenever possible, and lend themselves to construction through the use of a ditching machine in most localities. When the pipe is being laid in a pervious, water-bearing stratum, very little preparation of the trench bottom is necessary other than a nominal amount of shaping. In the case of an intercepting drain laid at the top of an impervious layer in order to ensure complete removal of the intercepted flow, the pipe is generally placed a short distance down into the impervious layer. If soft, unstable soils are encountered, sufficient granular material must be worked into the upper portion of the soil to ensure uniform support of the pipe.

Perforated circular pipe is laid with the perforations down, when located in pervious soils, so that there is less likelihood of clogging by the entrance of fine-grained soil into the pipe through the perforations. For the same reason, corrugated metal and concrete pipe is generally laid with sealed joints. Open joints in bell-and-spigot vitrified-clay pipe are protected from silting by covering them with tar paper or some similar materials.

Proper backfilling around the pipe and in the trench is of extreme importance if the drain is to function properly without an excessive amount of maintenance. The backfill material must be coarse enough to allow easy passage of water and fine enough to prevent intrusion of fine-grained soil into the pipe.

Criteria which have been developed for adequate filter soils are as follows:

$$\frac{D_{15\ \text{(filter)}}}{D_{85\ \text{(protected soil)}}} \leq 5$$

and

$$\frac{D_{15 \text{ (filter)}}}{D_{15 \text{ (protected soil)}}} \geq 5$$

where D_{15} is the grain diameter that is larger than 15 percent of the soil grains, that is, the 15 percent size on the grain-size distribution curve. (Similarly, for D_{85}.) For most fine-grained soils, standard concrete sand (AASHTO M6 fine aggregate for portland-cement concrete) forms a satisfactory filter material. When such sands contain a sufficient amount of fine gravel, and for pipes with small openings, they are safe against infiltration into the pipe. Otherwise, the relationship between the size of the filter material and the size of the perforations in the pipe is

$$\text{for slots } \frac{D_{85 \text{ (filter)}}}{\text{slot width}} > 1.2$$

$$\text{for circular openings } \frac{D_{85 \text{ (filter)}}}{\text{hole diameter}} > 1.0$$

In some states, material of this approximate gradation is used for the entire depth of the trench with the exception of the top 200 to 300 mm (8 to 12 in.), which may be filled with ordinary topsoil. In other states the upper portion of the trench is filled with a coarse material. Under certain circumstances it may be advisable to provide an impervious seal over the top of the trench to prevent washing of silt and other material down into the drain. Adequate compaction of the backfill is essential to prevent excessive subsidence of the ground over the subdrain and is especially important if the drain is to be beneath the roadway (as in the case of an interception drain designed to prevent longitudinal seepage) or beneath a shoulder.

DRAINAGE OF CITY STREETS

Precipitation that occurs on city streets and adjacent areas must be rapidly and economically removed before it becomes a hazard to traffic. The removal of surface water in municipal areas is accomplished by methods similar to those employed in the drainage of rural highways, except that the surface water is commonly carried to its eventual disposal point by means of underground pipe drains or "storm drains." The storm drains may be designed to provide for the flow of ground water as well as surface water, or additional subdrains may be required in certain areas. No further mention will be made of subdrainage in this discussion, as subdrainage installations in urban areas are very similar to those required in rural areas.

The surface-water drainage system in an average city may be considered to be composed of the following basic elements: pavement crown; curb and gutter; and the storm drains themselves, including such commonly used appurtenances as inlets, catch basins, and manholes.

11-35 Pavement Crowns, Curbs, and Gutters

Water falling on the pavement surface itself is, as in the case of rural highways, removed from the surface and concentrated in the gutters by the provision of an adequate crown; crown requirements previously discussed are applicable to city streets. The crown of any street should be made as small as possible, consistent with proper drainage, in the interest of appearance and safety.

The surface channel formed by the curb and gutter fulfills about the same drainage function as a side ditch on a rural highway, and it must be designed to adequately convey the runoff from the pavement and adjacent areas to a suitable collection point. Typical curb and gutter sections have been illustrated in Chapter 7. The longitudinal grade of the gutter must also be sufficient to facilitate rapid removal of the water; minimum grades are generally recommended as 0.2 or 0.3 percent. The grade of the gutter is usually the same as that of the pavement surface, although it may be different if so required for proper drainage. Hydraulically, gutters are designed as open channels of special shape.

Flow in the gutters generally continues uninterrupted to the intersection unless a low point or "sag" is reached. At such a point an inlet must be provided to carry the water into the storm sewer system. Inlets may also be required between intersections if long blocks are used.

The paved area formed by a street intersection must also be drained, and adjustment of the grades of city streets at intersections to provide for adequate drainage can be a very complex problem. Several analytical methods have been successfully used for determining the elevations required within the intersection to remove water from the intersection area and to adjust any differences in grades that may be present on the streets that approach and leave the intersection. The solution may also be accomplished in the field during the construction process without previous computation, especially where light street grades are involved.

11-36 Inlets

Inlets are generally provided at intersections to intercept the water flowing in the gutters before it can reach the pedestrian sidewalks. It is essential that enough inlets of sufficient size be provided to rapidly remove collected storm water. Improperly drained intersections are a constant source of annoyance and danger to both vehicular and pedestrian traffic.

The selection of the type of inlets to be used and their positioning in the area adjacent to the intersection are matters of individual choice, and considerable variation occurs in practice. As an example of the proper positioning of inlets at the intersection of two streets of normal width, the simple sketch of Figure 11-19 has been prepared.

On lightly traveled streets it may sometimes be desirable to provide for the flow of surface water across the intersection by means of a shallow paved gutter or "dip." Such a procedure is not generally recommended, but it may be necessary in the interest of economy.

Inlets of various designs are used by different organizations. The type most generally used on new construction may be called the "combined curb-and-gutter inlet." This type of inlet basically consists of a concrete or brick box with an opening in one side, near or at the bottom, into a circular pipe forming one of the laterals of the storm sewer system. Water enters the box through a metal grate placed over a horizontal opening in the gutter and through a relatively narrow opening in the face of the curb. This latter opening in the face of the curb may or may not be protected by a grating of some kind. Inlets of this type are either cast in place or composed of precast units. Inlets are also designed without

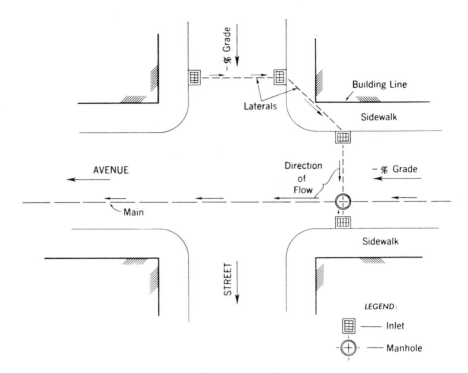

FIGURE 11-19 Drainage at an urban intersection.

the opening in the curb, the water flowing in only through a grated opening in the gutter. Some inlets have also been constructed with only curb openings.

While inlets of the types that have been described usually have only an outlet in the form of a circular pipe placed so that the invert of the pipe is flush (or nearly so) with the bottom of the box, they are also constructed with a pipe inlet (in addition to the curb and gutter openings) and a pipe outlet. In other words, the inlet may be a connecting link in the storm sewer system. Still another variation in design occurs in what may be called a "drop inlet." In a drop inlet, the water entering from the gutter must drop some considerable distance to the bottom of the box or may simply be conveyed directly into the storm sewer by means of a vertical pipe.

Inlets are susceptible to clogging with debris and ice, and they must be subjected to continual inspection and maintenance if they are to function properly. The proper hydraulic design of inlets to accommodate a given amount of water under given conditions is beyond the scope of this book. Recent research has lead to increased understanding of the hydraulic principles involved and the development of various design methods. Curb openings for inlets, on the average, are from 0.6 to 1.2 m (2–4 ft) in length and from 150 to 200 mm (6–8 in.) high.

11-37 Catch Basins

Catch basins are similar to inlets in their function and design. The principal difference between a catch basin and an inlet is that in a catch basin the outlet pipe (and inlet pipe if present) is placed some distance above the bottom of the chamber. The notion behind the installation of a catch basin is that debris flushed from the street into the drainage system is trapped in the bottom of the catch basin and so does not enter the storm sewer itself. This consideration is of less importance now than in former years because of the universal construction of permanent-type street surfaces, more efficient street-cleaning methods, and the general design of storm sewer systems to carry water at higher velocities of flow. This latter feature has made many sewer systems practically self-cleaning. Catch basins require good maintenance; if they do silt up and the debris is not removed, they function as inlets.

11-38 Manholes

Storm sewer systems are subject to partial or complete clogging, and facilities must be provided for cleaning them at regular intervals. In modern sewerage practice, manholes are generally placed at points where the sewer changes grade or direction, where junctions are made, and at intermediate points, usually at intervals of 90 to 150 m (300–500 ft). The opening provided must, of course, be large enough to permit a man to enter the manhole chamber and have room in which to work. They are usually about 1.2 m (4 ft) in diameter in their main portions and are carried to a depth sufficient to perform the purpose for which they are intended. The pipes generally enter and leave the manhole at the bottom of the chamber, and various arrangements are made for carrying the flow through the manhole with a minimum loss in head. Manholes are built of concrete masonry, concrete block, or brick, and cities generally employ a standard design of manhole to be used in all installations. Manhole covers (and frames) are generally of cast iron; the covers are circular in shape and about 0.6 m (2 ft) in diameter. When the entrance to the manhole occurs in the traveled way, as it usually does, special care must be taken in the design and placing of the cover relative to the finished street surface if traffic is not to be impeded by the presence of the manhole.

PROBLEMS

11-1 A culvert is to be designed for a 10-hectare flat, urban residential area. About 60 percent of the area is impervious. Assume that the rainfall intensity-duration chart shown as Figure 11-2 applies to this area. The time of concentration is 20 min. Using a 10–year return period, estimate the runoff (m³/sec) by the rational method.

11-2 Using the rational method, determine the expected runoff (ft³/sec) for a 300-acre drainage area located near Chicago, which has rainfall intensity of 2.5 in./hr if the return period is 10 years. The area is 10 percent bituminous pavement, 10 percent gravel, and the remaining 80 percent impervious soil with turf.

11-3 If the computer software program HYDRAIN is available to you, find the intensity-duration-frequency (IDF) curve for a 50-year storm at this location: Latitude 34 degrees 45 minutes, longitude 83 degrees 22 minutes.

11-4 A 1.2-m flat-bottomed channel with 2 to 1 side slopes is paved with smooth concrete and has a slope of 1.5 percent. The depth of flow is 790 mm. Compute the velocity of flow (m/sec). Check your answer by Figure 11-3.

11-5 Determine the depth and velocity of flow in a trapezoidal channel (n = 0.030) with 2 to 1 side slopes and 4-ft bottom width discharging 60 ft³/sec on a slope of 2 percent.

11-6 A 1.2-m flat-bottomed channel with 2 to 1 side slopes has a flow of 15 m^3/sec. Calculate the critical depth.

11-7 Determine an appropriate size in millimeters of circular concrete culvert that will accommodate a flow of 7 m^3/sec with an allowable headwater depth of 3 m. Assume the culvert has a groove end with a headwall and that the flow has inlet control.

11-8 Determine the appropriate size of circular corrugated metal culvert that will accommodate a flow of 1.5 m^3/sec with an allowable headwater depth of 2.7 m. Assume the culvert has a projecting entrance and that the flow has inlet control.

11-9 If the computer software HYDRAIN is available to you, solve problem 11-7 with the software component HYCLV.

11-10 Calculate the difference in elevation between the headwater and tailwater for a circular concrete culvert with a 45° beveled entrance that is flowing full. The culvert is 24 m long and has a diameter of 900 mm. The flow is 0.8 m^3/sec.

REFERENCES

1. *Drainage of Highway Pavements.* Highway Engineering Circular No. 12, Federal Highway Administration, Washington, DC (1984).

2. *Hydraulic Design of Highway Culverts.* Hydraulic Design Series No. 5, Federal Highway Administration, Washington, DC (1975).

3. Sauer, V. B., Thomas, W. O. Jr., Stricker, V. A., and Wilson, K. V. *Flood Characteristics of Urban Watersheds in the United States.* U.S. Geological Survey Water Supply Paper No. 2207, Prepared in Cooperation with U.S. Department of Transportation, Federal Highway Administration, Washington, DC (undated).

4. Masch, Frank D. *Hydrology.* Hydraulic Engineering Circular No. 19, Federal Highway Administration, Washington, DC.

5. *Rainfall-Frequency Atlas of the United States.* Technical Paper No. 40, U.S National Weather Service, Washington, DC (1961).

6. Kirpich, P. Z. Time of Concentration of Small Agricultural Watersheds. *Civil Engineering,* Vol. 10, No. 6 (1940).

7. Young, Kenneth, and Krolak, Joseph S. *HYDRAIN, Integrated Drainage Design Computer System: Version 5.0.* Developed by GKY and Associates, Inc., Springfield, VA, for the Office of Engineering and Highway Operations, Research and Development, Federal Highway Administration, McLean, VA (1994).

8. *Design of Roadside Drainage Channels.* Hydraulic Design Series No. 4, Federal Highway Administration, Washington, DC (1965).

9. *Design Charts for Open Channel Flow.* Hydraulic Design Series No. 3, Federal Highway Administration, Washington, DC (1973).

10. Chen, Y. H. and Cotton, G. K. *Design of Roadside Channels with Flexible Linings.* Hydraulic Engineering Circular No. 15, Federal Highway Administration, Washington, DC (1988).

11. Normann, Jerome M. Improved Design of Highway Culverts. *Civil Engineering* (March 1975).

12. *Hydraulic Design of Energy Dissipators for Culverts and Channels.* Hydraulic Engineering Circular No. 14, Federal Highway Administration, Washington, DC (1975).

13. *Standard Specifications for Transportation Materials,* 17th Ed. M170M, American Association of State Highway and Transportation Officials, Washington, DC (1994).

14. *Debris Control Structures.* Hydraulic Engineering Circular No. 9, Federal Highway Administration, Washington, DC (1971).

15. Standard Specifications for Highway Bridges, 15th Ed. American Association of State Highway and Transportation Officials, Washington, DC (1992).

CHAPTER 12

TRAFFIC ENGINEERING

12-1 Introduction

Before a highway is opened to traffic, careful thought must be given to conveying to drivers information concerning its proper use. Such information is given primarily by traffic-control devices.

Engineers must also recognize that despite their best efforts to promote traffic safety by well-designed traffic-control systems, vehicles will occasionally go out of control and possibly crash. Every effort must be made to prevent traffic accidents from occurring and to minimize the injuries and economic losses from those crashes that do occur.

In this chapter we will examine ways in which traffic engineers promote operational efficiency and safety through the use of traffic-control devices. This area encompasses signs, markings, signals, islands, work-zone traffic control, traffic control for school areas, traffic-control systems for railroad-highway grade crossings, and traffic control for bicycle facilities. The devices may be placed on, over, or adjacent to a street or highway and are intended to regulate, warn, or guide traffic.

Regulatory devices inform drivers of speed limits, one-way operation, turn prohibitions, right of way such as at a traffic signal, or the need to take some action such as stop. Warning devices indicate potentially hazardous curves, end of pavement, low clearance, railroad crossing, and so on. Guide devices include route markers, destination and distance signs, information signs such as at rest areas, and delineation of roadway by means of pavement markings, raised pavement markers, and delineators, which are post-mounted reflectors installed outside the edge of the shoulder.

TRAFFIC-CONTROL DEVICES AND SYSTEMS

In order to provide safe and efficient traffic flows, uniform standards have been developed for the use of all public street and highway facilities in the United States. The need for such standards became apparent in the late 1920s, and in 1935 a joint committee of the American Association of State Highway Officials and the National Conference on Street and Highway Safety developed and published the *Manual on Uniform Traffic Control Devices (MUTCD) (1)*. The committee is now called the National Committee on Uniform Traffic Control

Devices. The Federal Highway Administration has approved the *MUTCD* as the national standard for all streets and highways open to public travel, regardless of type or class or the governmental agency having jurisdiction. Also, the legislatures of many states have adopted the *MUTCD* by statute, making it a particular authoritative document. The manuals of the other states are not substantially different from the *MUTCD* and frequently call for more stringent requirements than the minimums expressed in the *MUTCD*.

The Federal Highway Administration also publishes a companion manual to the *MUTCD, Traffic Control Devices Handbook* (2). It is an operating guide. The *TCDH* augments the *MUTCD* by serving an interpretative function, chapter by chapter. It offers guidelines for implementing the standards and applications contained in the *MUTCD*. It does not merely repeat or paraphrase the *MUTCD*. Although the *TCDH* is not officially adopted by the statutes of any state, it is very authoritative because of its official status linking the *MUTCD* standards and warrants with the activities related to complying with these national uniform standards. Much of the material in this chapter has been abstracted from the 1988 edition of the *MUTCD* and the 1983 edition of the *TCDH*.

The application of traffic-control devices should be responsive to driver needs for information. This topic was introduced in Chapter 5, "Driver, Pedestrian, and Vehicle Characteristics," and includes crucial concepts such as driver expectancy. In the 1970s Alexander and Lunenfeld organized these concepts into a formal body of knowledge called Positive Guidance (3). The principles of Positive Guidance assist the traffic engineer to apply the traffic-control devices needed to warn of a hazard in such a way as to confront the driver with a single, simple decision at a time. Positive Guidance helps the engineer to look at the road from the driver's point of view, a skill that is abolutely essential.

Traffic-control devices include all signs, marking, and signals placed on or adjacent to a street or highway by public agencies in order to regulate, warn, or guide traffic. Any traffic control device, to be effective, should

- fulfill a need.
- command attention.
- convey a clear, simple meaning.
- command respect of road users.
- give adequate time for proper response.

Traffic-control devices should be reasonable and appropriate for the traffic requirements at the location used. The use of a traffic-control device at a location where it is not warranted tends to invite drivers to disregard the device and to have less respect for traffic-control devices in general.

Traffic-control devices should be properly designed. The size of the device, its shape, colors, contrast with the background, and lighting or reflectorization should draw attention. The sign, marking, or signal should simply and forthrightly convey a clear and simple message. The use of uniform devices following the *MUTCD* (1) simplifies the driver's task of recognizing and understanding the traffic-control messages and tends to increase the level of observance. In the interests of uniformity the National Committee has adopted the following color code for traffic-control devices:

Color	Meaning
Red	Stop or prohibition
Green	Indicated movements permitted, direction guidance
Blue	Motorist services guidance
Yellow	General warning
Black	Regulation
White	Regulation
Orange	Construction and maintenance warning
Brown	Public recreation and scenic guidance

Traffic-control devices should be placed and operated in a uniform, consistent manner. In this way, drivers can be expected to respond properly to the devices on the basis of previous exposure to similar traffic situations. A device should be placed as close as practicable to the driver's cone of vision and in reasonable proximity to the point, object, or situation to which it applies. For example, a STOP sign should be placed as near as possible to where the vehicle is to stop. The location and legibility of a STOP AHEAD sign, for example, should combine to give the driver traveling at normal speed sufficient time to make the proper response, such as to brake the vehicle to a stop.

Well-maintained devices that appear fresh and clean and are in good working order command the respect of vehicle operators and pedestrians. When no longer needed, traffic-control devices should be removed. Removal of a STOP sign at a multiway STOP location, however, is a special case requiring the attention of an experienced traffic engineer. Driver expectancies created by the multiway STOP signs will no longer be met once one of the signs is removed. The hazard of right-angle collision after removal can be reduced by installing special signs as shown in Figure 12-1 (*4*). This is an excellent application of the principles of Positive Guidance, particularly the concept of driver expectancy.

The *MUTCD* is not a "cookbook" that reduces the application of traffic-control devices to mere look-up. In fact, the decision whether to use a certain device at a particular location is based on engineering judgment, which in turn is to be based on an engineering study of the location. For example, suppose that an intersection currently under STOP-sign control is found through an engineering study to meet one or more of the *MUTCD* warrants for a traffic signal. This finding creates no legal requirement that a signal be installed there; the engineer is free to exercise discretion in the selection of intersections to be signalized. In fact, the engineer should positively exercise that discretion by formally prioritizing the candidate sites. Once the decision to signalize is made, the *MUTCD* controls many details of the installation (e.g., the height of the signal heads), but all of the important and interesting decisions such as phasing and timing are left to the judgment of the engineer. This fact makes traffic-operations engineering a challenging occupation, and an especially exciting and gratifying one, because the benefits of improved traffic control can usually be observed and measured soon after installation.

The selection and design of traffic-control devices should be an integral part of geometric design rather than a separate step. The potential for future operational problems can be significantly minimized if signing and marking are considered to be an intrinsic part of the geometric layout of the facility (*5*). For

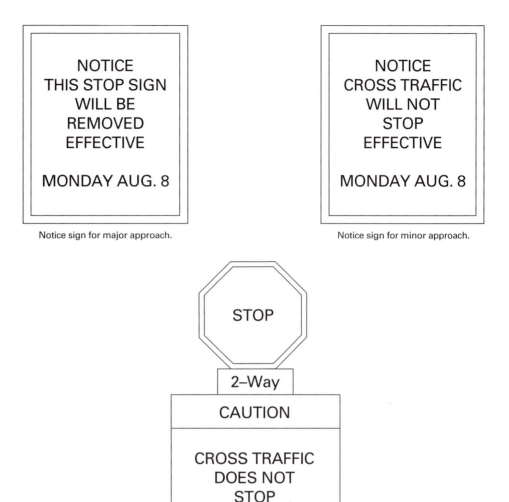

FIGURE 12-1 Signs needed when removing a STOP sign at a multiway STOP location. (Courtesy Transportation Research Board.)

example, a key test of the design of a complex interchange is a check to see that it is possible to locate guide signs appropriately.

Government agencies have a legal duty to maintain the roadway in a reasonably safe condition. There has been a fast-growing trend toward litigation and large awards in the areas of highway construction, operations, and maintenance. The *TCDH* includes a section explaining the basics of current liability laws as they pertain to the use of traffic-control devices. The *TCDH* points out that agencies could reduce their exposure to suit in this area by knowing the pertinent laws, keeping an inventory of devices, replacing them at the end of their effective life, and applying the state's own specifications and standards (*2*).

12-2 Traffic Signs

There are three functional classes of traffic signs: (1) regulatory, (2) warning, and (3) guide signs. Examples of these three classes are shown in Figure 12-2. The

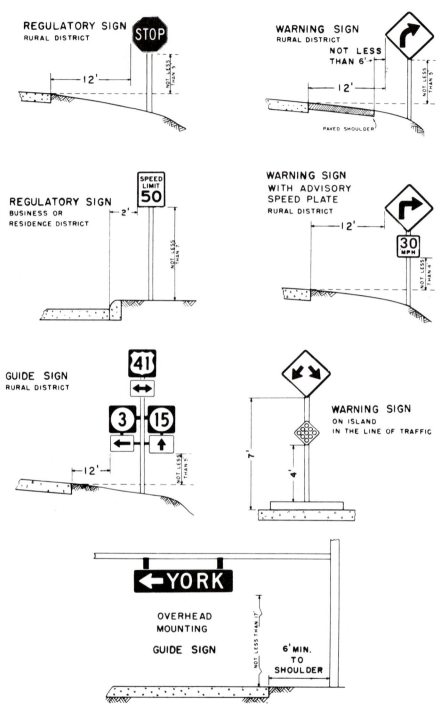

FIGURE 12-2 Typical highway signs. (Courtesy Federal Highway Administration.)

MUTCD (*1*) gives detailed specifications on color, size, shape, height, and placement of signs as well as warrants for their use. In many situations, however, the final decision as to whether to use a certain sign is left to the judgment of the engineer, guided by an engineering study that might, for example, indicate that a certain sign would be unnecessary at a certain location. There are special-use signs such as those with changeable messages (Fig. 12-3). Changeable-message systems provide drivers with real-time information on a current condition or problem, such as congestion resulting from a parade or other special event, a freeway incident such as spillage of a hazardous material, fog, or the operation of high-occupany vehicle lanes.

The *TCDH* (*2*) describes methods for fabricating, installing, and maintaining signs and gives examples of specifications, contract plan sheets, and work orders. To be more visible at night, signs are fabricated with retroreflective sheeting that reflects back to the driver a large percentage of the light from vehicle headlights. The technology of retroreflectivity is described in more detail in the next section.

The *MUTCD* cautions that care should be taken not to install too many signs. "A conservative use of regulatory and warning signs is recommended as these signs, if used to excess, tend to lose their effectiveness" (*1*). They should be used to fulfill a specific need, as warranted by facts and field studies. "Signs are essential where special regulations apply at specific places or at specific times only, or where hazards are not self-evident." (*1*). For example, exit ramps of a freeway commonly have design speeds lower than those of the freeway lanes. It may not be necessary to post signs for reduced speed limits on these ramp curves if the curves are readily visible and apparent to the approaching driver. Only if they are not self-evident would the additional signing be needed. Engineers need to avoid a proliferation of signs that could breed disrespect for all signs.

Regulatory signs give users notice of traffic laws or regulations. Such signs designate right of way (e.g., STOP, YIELD), indicate speed controls (e.g., SPEED

FIGURE 12-3 A changeable-message sign. (Courtesy Lite Vision, Inc.)

LIMIT 50, SPEED ZONE AHEAD), control movements (e.g., NO RIGHT TURN, KEEP RIGHT, ONE WAY), regulate parking (e.g., NO PARKING), control pedestrian movements (e.g., CROSS ONLY AT CROSS-WALKS), and regulate traffic in various other ways.

Warning signs direct attention to conditions on or adjacent to a street or highway that are potentially hazardous to traffic operations. Such signs require drivers to exercise caution, reduce speed, or make some maneuver in the interest of their own safety or that of other road users.

Typical locations or situations that may warrant the use of warning signs include

- intersections, entrances, and crossings.
- changes in horizontal alignment.
- grades.
- railroad crossings.
- narrow roadways.
- roadway surface conditions.
- advance warning of control devices.

Examples of warning signs are curve signs (showing curved arrow), STOP AHEAD signs, PAVEMENT ENDS signs, and advisory speed plates. The advisory signs supplement other types of warning signs, that is, an advisory speed is posted below the warning sign on the same post, as shown in one of the example drawings in Figure 12-2. Warning signs have a black legend and border placed on a yellow background. With few exceptions, warning signs have a diamond shape. A curve-warning sign, shown in the upper right of Figure 12-2, "may be used where engineering investigations of roadway, geometric, and operating conditions show the recommended speed on the curve to be greater than 48 km/hr (30 mph) and equal to or less than the speed limit . . ." (1). Figure 12-4 provides one method to determine the recommended speed (2). It is based on the equation

$$e + f = V^2/15R \qquad (12\text{-}1)$$

that was developed in Chapter 7 for determining rate of superelevation. The radius R and superelevation of the curve may be obtained from field measurement or from roadway plans and entered into the figure to arrive at the recommended speed V. Alternatively, a pendulum-like device called a "ball-bank indicator" can be used in a test car to determine the recommended speed without the need to know either the superelevation rate or the curve radius (2).

Guide signs indicate route designations, directions, distances, points of interest, and other geographic or cultural information. Two types of guide signs (route markers and a destination sign) are illustrated in Figure 12-2. Other examples of guide signs include JUNCTION signs, DETOUR signs, REST AREA signs, and service signs (FOOD, GAS, LODGING, etc.). Figure 12-5 shows a high-technology, internally illuminated guide sign. Such signs need only one to four 400-W metal halide bulbs to illuminate them. The bulbs are remotely located, up to 6 m (20 ft) from the sign, in boxes near support poles where they can be serviced without the need for catwalks, sign-face removal, or lane closures. The light is "piped" through tubes of optical lighting film to the sign box, where internal "feeder tubes" distribute it horizontally across the sign face.

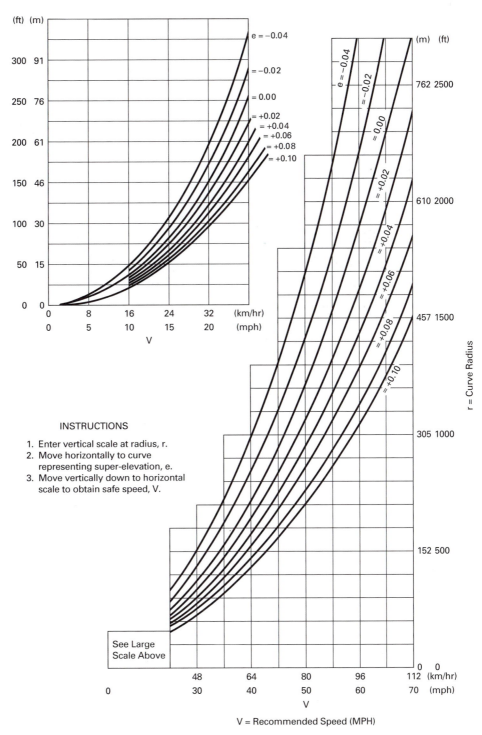

INSTRUCTIONS

1. Enter vertical scale at radius, r.
2. Move horizontally to curve representing super-elevation, e.
3. Move vertically down to horizontal scale to obtain safe speed, V.

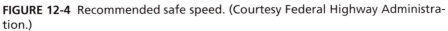

V = Recommended Speed (MPH)

FIGURE 12-4 Recommended safe speed. (Courtesy Federal Highway Administration.)

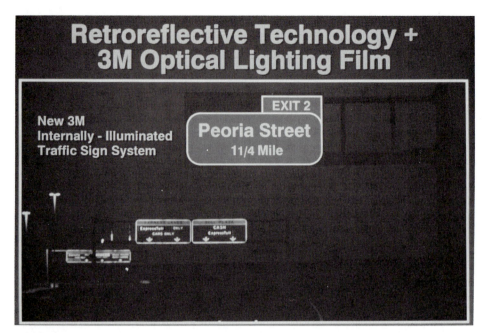

FIGURE 12-5 An internally lighted guide sign. (Courtesy 3M Corporation.)

12-3 Traffic Markings

Markings consist of paint (or more durable materials such as thermoplastics, epoxies, and polyesters) placed on the pavement, curb, or object to convey traffic regulations and warnings to drivers. Markings may be used alone or in combination with traffic signs or signals. Although markings are an effective and essential means of traffic control, neither paint nor the durable plastic pavement markings are effective on rainy nights, and they may be obliterated altogether by snow and ice.

Like the sheeting used for signs, most types of markings are retroreflective owing to the use of embedded, tiny glass beads or minute corner-cube retroreflectors. Figure 12-6 shows how round, transparent glass beads sprinkled onto a freshly painted stripe can refract and redirect light back toward the driver. Much brighter are the prismatic corner-cube mirrored surfaces of Figure 12-7, which gather and return light in a narrow beam. Each square centimeter of a marker has about 15 of these elements (2).

There are seven general classes of markings: (1) pavement markings, (2) curb markings, (3) raised pavement markings, (4) object markings, (5) delineators, (6)

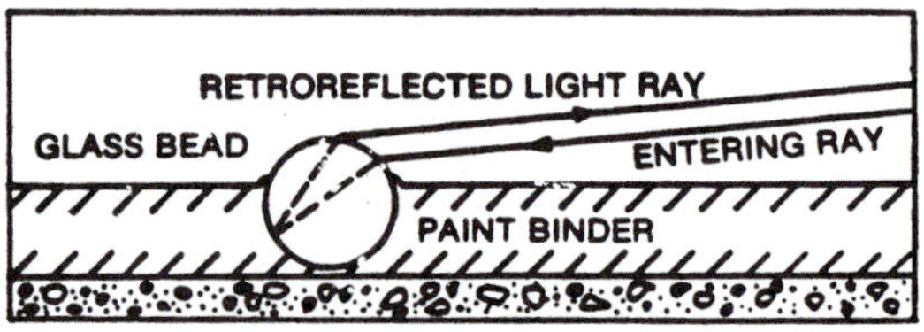

FIGURE 12-6 Glass-bead retroreflection. (Courtesy Federal Highway Administration.)

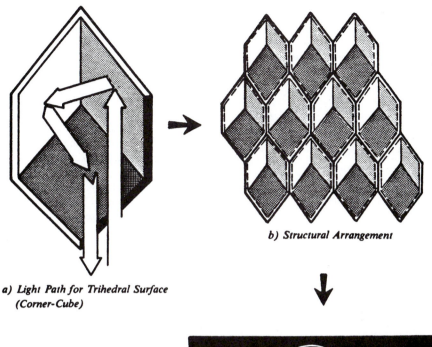

a) Light Path for Trihedral Surface
(Corner-Cube)

b) Structural Arrangement

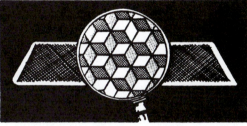

c) Magnified View of Precision Molded Corner-Cube
Reflex Elements

FIGURE 12-7 Corner-cube retroreflection. (Courtesy Federal Highway Administration.)

colored pavements, and (7) barricades and channelizing devices. The latter are discussed in the next section on traffic control in work zones.

The most common type of markings—pavement markings—includes longitudinal and transverse lines, words, and symbols. By choice of color, width, and type of longitudinal markings, the traffic engineer is able to convey a variety of messages to drivers, briefly stated as follows:

1. Broken lines are permissive in character, solid lines are restrictive, double lines indicate maximum restriction.
2. White lines delineate separation of traffic flows in the same direction; yellow lines delineate separation of flows in opposing directions.
3. The width of the line indicates degree of emphasis.

Transverse markings include pedestrian crosswalks, stop lines, and cross-hatched areas to discourage use of shoulders or to identify object hazards.

Typical applications of various classes of pavement markings are given in Table 12-1. Examples of pavement markings are illustrated in Figure 12-8.

TABLE 12-1 Typical Applications of Pavement Markings

Type	Typical Application
Longitudinal lines	
Broken white line	Lane line of multilane highway
Broken yellow line	Two-lane two-way roadway, overtaking and passing permitted
Solid white line	Pavement edge marking
Double solid white line	Channelizing line in advance of obstructions
Solid yellow line	Used with broken yellow line to indicate no-passing zone for traffic adjacent to solid line
Double broken yellow line	Edges of reversible traffic lanes
Dotted lines	Extension of lines through an intersection or interchange area
Transverse Markings	
Crosshatched shoulder markings	To discourage use of shoulders as a traffic lane
Pairs of solid white lines, 150 mm (6 in.) or more in width, spaced 1.8 m (6 ft) or more apart	Pedestrian crosswalks
Solid white lines, 300 to 600 mm (12 to 24 in.) in width	Stop lines to indicate where vehicles are required to stop

Curb markings are used for roadway delineation and for parking regulation. Color-coded markings may be used as a supplement to standard signs to indicate the type of parking regulation. When potentially hazardous objects such as bridge supports must be located within or adjacent to the roadway, they should be conspicuously marked. The *MUTCD (1)* specifies three alternative types of object-marker designs, including reflectors mounted in a prescribed way on a rectangular or diamond-shaped panel and reflectorized all-yellow or yellow and black diagonally striped panels.

Delineators are used to aid night driving and are especially valuable where there is a change in horizontal alignment, where the alignment could be confusing, and where there is a transition in pavement width. Delineators normally consist of reflector elements with a minimum dimension of about 75 mm (3 in.) mounted on thin posts. White delineators are used along through roadways, and yellow delineators are placed alongside interchange ramps, speed-change lanes, and median crossovers. A high-technology lighted guidance tube, shown in Figure 12-9, mounts on such things as barriers and guard rails, to delineate hazardous sections of roadway. The 100-mm- (4-in.-) diameter polycarbonate plastic tube appears luminous to drivers to guide them through curves or bad weather.

Colored pavements are sometimes used to supplement other traffic control devices. Red pavements are specified for approaches to a STOP sign that is in continuous use; yellow pavements for medians separating opposing traffic flows; and white for shoulders, crosswalks, and channelizing islands.

12-4 Traffic Control in Work Zones

The *MUTCD (1)* and the *TCDH (2)* offer extensive guidance for traffic controls for street and highway construction, maintenance, utility and emergency opera-

a—Typical pavement marking with offset lane lines continued through the intersection and optional crosswalk lines and stop limit lines.

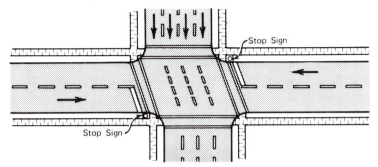

b—Typical pavement marking with optional double turn lane lines, pavement messages, crosswalk lines, and stop limit lines.

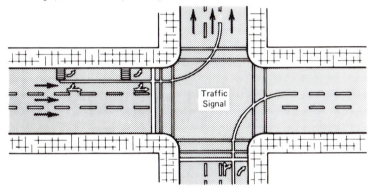

c—Typical pavement marking with optional turn lane lines, crosswalk lines, and stop limit lines.

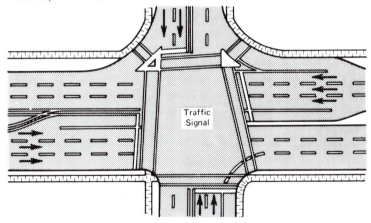

FIGURE 12-8 Typical pavement-marking applications. (Courtesy Federal Highway Administration.)

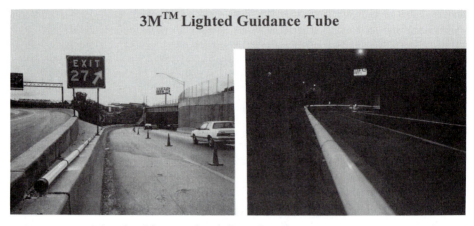

FIGURE 12-9 Lighted guidance tube delineation. (Courtesy 3M Corporation.)

tions such as the management of crashes, stalled vehicles, and so on. "Motorists should be guided in a clear and positive manner while approaching and traversing construction and maintenance work areas" (*2*). Work-zone traffic control encompasses signs, signals, lighting devices, markings, barricades, channelizing, and hand-signaling devices such as flags. Figure 12-10 illustrates channelizing devices and high-level warning devices (*1*).

The *TCDH* (*2*) defines a "traffic-control zone" as the distance between the first advance warning sign and the point beyond the work area where traffic is no longer affected. The five parts of a traffic-control zone are shown in Figure 12-11. Note that there is to be a buffer space, which is an open or unoccupied area between the transition and the work area, to provide a margin of safety for both traffic and workers.

Figure 12-12 illustates the use of traffic-control devices on a multilane highway where two of the four lanes in one direction must be closed. The figure includes formulas for calculating the taper length L. Strict adherence to these equations is important, as a taper put in "by eye" will usually be much shorter than required and can cause a crash due to the inability of the approaching driver to move laterally in time.

It was stated earlier that all traffic-control devices need to command respect. An important ingredient of respect is credibility. Warning signs should meet an important need by preparing the driver to detect and recognize a hazard that actually exists.

> When the hazard is present, the value of the warning is reinforced. However, when the hazard is not there or not apparent to the driver, the credibility of the warning or regulation is reduced. . . . (*3*)

Examples of unnecessary control include construction-zone signing with no work crew in sight and reduced-speed zoning that seems unnecessary. The *MUTCD* (*1*) points out that drivers will reduce their speed only if they see a need to do so.

AASHTO states that drivers "read the road" and the environment and rely more on that information than on traffic-control devices that may be indicating something different (*5*). The Federal Highway Administration agrees, explaining that

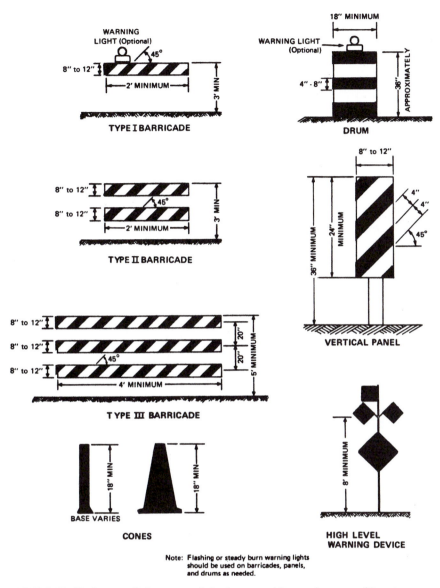

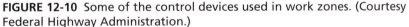

Note: Flashing or steady burn warning lights should be used on barricades, panels, and drums as needed.

FIGURE 12-10 Some of the control devices used in work zones. (Courtesy Federal Highway Administration.)

Guidance information comes primarily from the roadway and secondarily from traffic-control devices. When signs are in conflict with what the roadway indicates, the motorist is more likely to believe the road. (6)

It is common to see work-zone signs that have no relationship to the actual construction activity. "The traveling public soon beomes dubious of all traffic-control attempts, and may disregard or completely violate the traffic controls in the zone" (7). Experience with fog-warning systems has shown that if drivers are warned of fog and then find no fog they will ignore the message the next 10 to

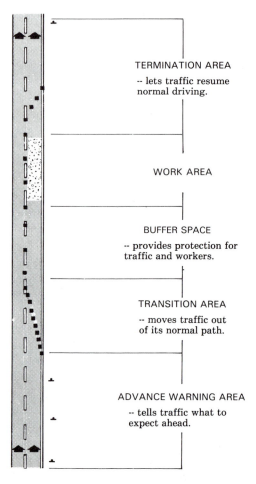

FIGURE 12-11 Areas in a traffic-control work zone. (Courtesy Federal Highway Administration.)

20 times they see it, until the credibility of the system has been slowly reestablished. "You can't cry wolf and get away with it." (8)

12-5 Traffic Signals

The most common types of signals are traffic-control signals (which include pedestrian signals), beacons, and lane-use control signals.

Traffic-Control Signals at Individual Intersections

Traffic-control signals are primarily used to control the movements of vehicular and pedestrian traffic at intersections. This section describes signals at individual intersections. Signals within one-half mile of each other can be coordinated into systems to give a "green-wave effect." Such systems can further reduce stops and delay and are discussed in the next major section, "Traffic-Control Signals for Arterial Streets."

The traffic-signal device (or head) is an assembly that contains one or more signal "faces." Each signal face, except in pedestrian signals, comprises three to

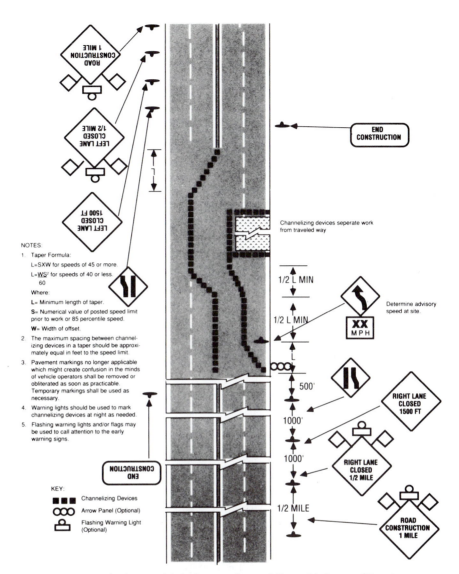

FIGURE 12-12 Closing multiple lanes of a multilane highway. (Courtesy Federal Highway Adminsitration.)

five optical units that display color-coded indications (red, yellow, green, green arrow, etc.) to the driver. A conventional optical unit consists of a lens, reflector, incandescent lamp, and lamp socket assembly. Light-emitted diode (LED) lamps are now being used to replace red incandescent lamps for several reasons. LEDS can effect a 6 to 1 reduction in electrical power consumption, do not require routine replacement, only rarely would need emergency replacement as a result of catastrophic failure, and can be expected to last 100,000 elapsed-time calendar hours, or 11.4 years, without replacement (*9*).

The optical units in each signal face are usually in a vertical line, arranged red, yellow, and green from top to bottom. The face may instead be mounted horizontally, with the red at the left, the yellow next, and the green at the viewers' right.

A steady circular green indication permits traffic to proceed in any direction that is lawful and practical. A green arrow allows vehicular movements that are completely protected from conflict with other movements. A steady circular yellow warns traffic of an impending change in the right-of-way assignment. The yellow may be followed by a relatively short red clearance interval, to clear the intersection before conflicting movements are released. A steady circular red indication is used to prohibit traffic from entering (*1*).

For through traffic, at least two signal faces must be provided on an approach to a signalized intersection, as one red signal could burn out or be blocked from view by a large vehicle ahead. To help ensure adequate visibility, the *MUTCD* specifies the location and height of signals and minimum visibility distance for various approach speeds.

The basic building-blocks of signal timing are signal "intervals," during which time there is no change in the indications for either vehicles or pedestrians. A green and the following yellow constitute a "phase," during which a certain movement or movements have the right of way. The time for one complete sequence of phases is the "cycle length." For example, Figure 12-13 shows one cycle of a simple two-phase, eight-interval sequence where a north-south street (Ø1) intersects an east-west street (Ø2) with no need for left-turn arrows; a new interval is seen to begin when the signals change from green and WALK to green and flashing DON'T WALK. (The symbol Ø stands for "phase.") The division or proportioning of the cycle time among the various intervals and phases is the "split" of the cycle.

Traffic signals are operated by cabinet-housed equipment that includes a con-

PHASE SEQUENCE CHART

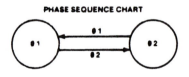

SIGNAL SEQUENCE CHART

Interval	1	2		3	4	5	6		7	8
Timer Status	WALK & MIN	PED CLR GRN	PASSAGE OR REST	YLW	RED CLR	WALK & MIN	PED CLR GRN	PASSAGE OR REST	YLW	RED CLR
Ø1 — G	①	▪ ▪ ▪								
Ø1 — Y				▬						
Ø1 — R					▬▬▬▬▬▬▬▬▬▬▬					
Ø1 — W	▬									
Ø1 — FDW		▬								
Ø1 — DW			▬▬▬▬▬▬▬▬▬▬▬							
Ø2 — G						▬▬	▪ ▪ ▪			
Ø2 — Y										
Ø2 — R	▬▬▬▬▬▬▬▬▬▬▬▬									▬
Ø2 — W						①				
Ø2 — FDW							▬			
Ø2 — DW	▬▬▬▬▬▬▬▬▬▬▬▬									

FIGURE 12-13 A simple signal-sequence chart. (Courtesy Peek Traffic.)

troller unit. This unit may be electromechanical, meaning that it has moving parts such as motors and gears, or it may be of solid-state electronic design. The controller unit times the duration of the signals and causes them to be turned on and off in accordance with plans selected by the traffic engineer.

Electromechanical controllers have a pretimed (fixed-time) operation, which means that there are no detectors to sense the minute-to-minute arrivals and adjust the lengths of the greens accordingly. The more modern solid-state types can be used to produce pretimed or actuated control, the latter including vehicle detectors (and perhaps pedestrian pushbuttons) on some or all of the approaches. Electromechanical controllers have many advantages, including ease of repair by electricians with ordinary skill, relatively high reliability against failure by show-ing green indications to conflicting movements, and resistance to damage by electrical transients such as lightning. They are still used in the United States, especially in the older and larger cities, and are common in some other countries. New installations in the United States, however, are usually solid state. These controllers are computers with enormous flexibility to acccept detector actuations and to adapt to widely ranging conditions of traffic flow, street geometry, and special situations such as diamond-interchange phasing and control of priority at railroad crossings. Figure 12-14 shows a modern controller known as Type 170. It was developed jointly by the states of California and New York. It is a general-purpose computer. The module labeled 412C contains the software that makes this Type 170 a traffic-signal controller. The module can be removed and replaced with another that would change the Type 170 into a ramp-metering controller (discussed in a later section), or even into a giant time-switch to turn irrigation sprinklers on and off in Southern California! Type 170 control equipment com-petes with other types designed to meet the standards of the National Electrical Manufacturers' Association (NEMA). NEMA-standard controllers, unlike the Type 170 models, have built-in software proprietary to each hardware manufac-

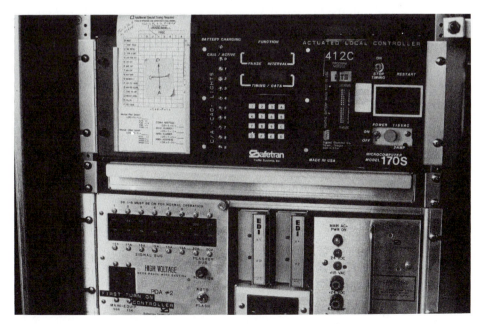

FIGURE 12-14 A Type 170 traffic-signal controller in its cabinet.

turer; however, a controller from one manufacturer can be interchanged with that from another for purposes of operating the signals at an individual intersection. A NEMA-standard controller is shown in its cabinet in the next section.

Pretimed Control at Individual Intersections Pretimed control uses no detectors to sense the minute-to-minute arrivals of traffic; therefore, it is seldom the best choice for individual signals uncoordinated with others because vehicle arrivals tend to fluctuate at these locations. It might be appropriate, however, at a high-rise for the elderly where the residents should not have to push pedestrian-call buttons, or at a remote location where it is costly in shop-personnel travel time to maintain loop detectors. Electromechanical pretimed controllers have a fixed phase-sequence, but there may be as many as three cycle lengths (plus flashing operation) that can be selected by time-of-day. Each cycle has a split that can be tailored to the average relative demands of the traffic usually present at that time of day on each approach. The solid-state pretimed controllers are more flexible in that they can vary the phase sequence; also, they have more choices of cycle length and split, normally selected by time of day.

Traffic-Actuated Control at Individual Intersections The minute-to-minute arrivals of vehicles on a signalized approach usually fluctuate greatly. Therefore, it is efficient to use traffic-actuated control that will bring the green only to an approach where there is waiting traffic and, upon serving those vehicles, will transfer the green promptly to others waiting on another approach. Traffic-responsive control can reduce stops, delay, fuel consumption, pollutant emission, vehicle operating cost, and the cost of motorist time, as compared to pretimed control.

The "eyes" of this design are detectors (sensors) installed in or over the pavement. Figure 12-15 shows a vehicle approaching an intersection and passing over a sensor loop sealed in a saw-slot several inches below the pavement surface. Sensor electronics in the curbside cabinet drive low-voltage alternating current through the loop at the resonant frequency, which is very high (in the FM radio range). The loop is the inductive element of this resonant circuit. A vehicle

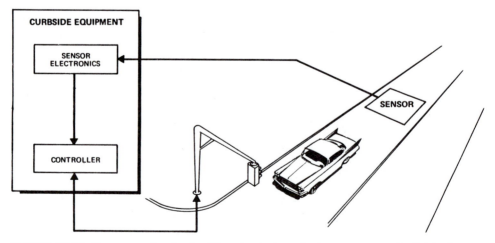

FIGURE 12-15 Components of a traffic-actuated control system at an individual intersection.

FIGURE 12-16 The loop of an inductive loop-detector.

entering the loop absorbs some of the energy and reduces the inductance slightly, causing a tiny increase in the resonant frequency. The increase actuates the detector's output relay, and a signal is sent to the controller unit that a vehicle has arrived on that approach. The controller either holds the green for the vehicle or else brings it at the earliest opportunity. Traffic moves and new actuations or "calls" are registered; thus the detector, controller, and signals form a closed control loop. Figure 12-16 shows a long loop being installed at the stop line of a signalized intersection. Other types of detectors are discussed in Section 12-6.

The basic technique in traffic-actuated control uses the following basic timing settings, all of which must be selected by the traffic engineer:

1. Minimum green time.
2. Passage time.
3. Maximum interval.

An intersection approach with simple detection is illustrated in Figure 12-17. The distance from the detector to the stop line influences the timing to be allocated to both the minimum green and the passage time.

The minimum green setting on the controller phase should be long enough to permit traffic stored between the stop line and the detector to start into motion and move into the intersection. If other traffic is waiting on a red signal, then the phase will be terminated at the end of the minimum green setting if no additional vehicles cross the detector. If a vehicle crosses the detector before the end of the minimum, then the green will be extended by one passage time, as shown in Figure 12-18, to provide sufficient green time to allow the vehicle to move from the detector to the stop line.

If the passage time is initiated and an additional vehicle is then detected, the unexpired portion of the passage time is canceled, and a new passage time is initiated. The process is repeated for additional vehicles crossing the detector until (1) no more vehicles are detected or (2) the green timing reaches a preset maximum. In either of these two cases, the timing of a yellow change interval is initiated, and the phase is terminated (provided, of course, other traffic is waiting on a red). If the last passage time does not completely time out (because of the maximum override), then the green will automatically be returned to that approach at the earliest opportunity to prevent trapping a vehicle at the stop line on a red signal that will not change until a new vehicle arrives and trips the detector.

This explanation makes it clear that the passage time setting serves two different functions: it is the extension of the green to "carry" the clearing vehicle from the detector to the stop line, and at the same time it is the maximum allowable time-gap between arriving vehicles, which allows them to hold the

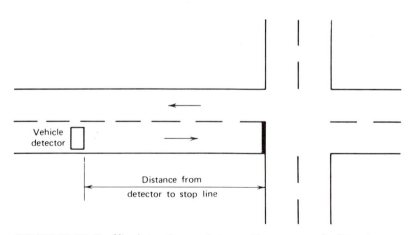

FIGURE 12-17 Traffic detection on intersection approach. (Courtesy Federal Highway Administration.)

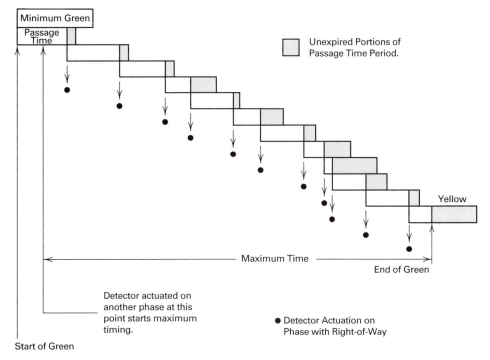

FIGURE 12-18 The timing of a basic actuated phase.

green against traffic waiting on the red (until the maximum time expires, of course). Because vehicles closely following one another will be about 2 sec apart, it follows that a gap of 3 or 4 sec is a reasonable setting of the passage time. A higher setting effectively cripples the gap-seeking logic of the controller. Short settings of minimum green and passage time usually produce snappy, efficient intersection operation.

There are several forms of traffic-actuated control, including semiactuated control and full-actuated control. In semiactuated control, detectors are placed on one or more, but not all, of the approaches to the intersection. The minor-street approaches are usually selected for detector placement. This type of operation provides for the following timing:

A preset minimum green time for the major street.

Interruption of major-street green only after the minimum green time has expired (with traffic detected to be waiting on the minor approach, of course).

Minor-street green controlled by the timing of the minimum green and passage-time extensions (up to the preset maximum). Unlike Figure 12-18, the maximum begins to time at the start of minor-street green, because the controller can only assume that traffic has immediately begun to wait on the major street.

Control will rest (dwell) in the major-street green in the absence of minor-street actuations.

In full-actuated control, detectors are placed on all approaches. Each green is an actuated phase that follows the logic of Figure 12-18. If consecutive vehicle actuations on a given phase occur at a shorter time interval than the timing of the passage time, then the green will remain on that phase up to a preset maximum (timed from the arrival of a waiting vehicle). If a passage time expires without a new actuation, that is, a gap greater than the passage time is detected, then the right of way will be given to the next phase for which there is demand. In the absence of waiting traffic, the green will rest in the phase that last received it; however, if one of the streets is more important than the other, the major-street phase can be set to "recall," meaning that it will continually "see" a waiting vehicle there and will bring the green to it at every opportunity.

This section has explained only the first principles of vehicle-actuated control at individual intersections. More-advanced concepts can be found in sources such as References 10 and 11.

Traffic-Control Signals for Arterial Streets

Along arterial streets it is often desirable to operate the signals as a system. By coordination of all the signals along a street, it is possible to provide progressive movement of traffic through several intersections without stopping. Reference 12 explains this process:

> The basic approach to arterial street signal control considers that vehicles traveling along the arterial street are released in platoons from a signal and thence travel in platoons to the next signal. Thus, it becomes desirable to establish a time relationship between the beginning of the arterial green at one intersection and the beginning of the arterial green at the next intersection so that traffic platoons may receive a green indication just as they approach a signalized intersection. This permits the continuous flow of traffic along an arterial street and aids in reducing delay.

> This traffic flow control concept for arterial streets can be presented graphically by a technique known as a time–space diagram, as shown in Figure 12-19. This figure introduces the need for several definitions, which are presented as follows:

> **Through-band.** The space between a pair of parallel speed lines that delineates a progressive movement on a time–space chart.

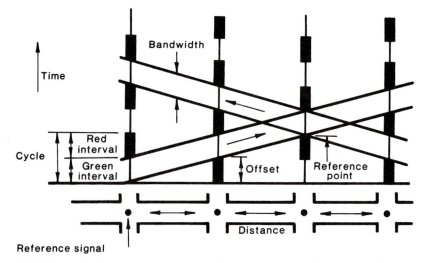

FIGURE 12-19 Time-space diagram. (Courtesy Federal Highway Administration.)

Band speed. The slope of the through band representing the progressive speed of traffic moving along the arterial.

Bandwidth. The width of the through band in seconds indicating the period of time available for traffic to flow within the band.

Offset. The time relationship expressed in seconds (or percentage of cycle length) determined by the difference between a defined interval portion of the coordinated phase green and a system reference point.

To establish a control system for an arterial street (or open network), it is necessary to develop a timing plan for all of the signals in the system. Such a timing plan consists of the following elements:

Cycle length. This must be the same for all signals in the system. Cycle length is usually established by examining traffic flow requirements to each intersection and computing the required length for each intersection. The largest cycle length required in the system is the one usually selected.

Cycle splits. A determination of the length of the various signal phases must be made for each individual intersection in the system. Phase lengths (splits) may vary from intersection to intersection.

Offset. A determination of an offset value must be made for each intersection in the system. The offset is usually established with reference to one master intersection in the system. (*12*)

Several different systems have been employed in providing for progressive movement, including

- a *simultaneous system,* in which several signals along a given street display the green indication at the same time. This system is usually employed where the block lengths are extremely short or where saturated conditions exist.

- a *single alternate system,* in which each signal shows an opposite indication to that of the next signal. Half-cycle offsets are used in this system that are alternated at each signal.

- a *double alternate system,* in which groups of two signals display the same indications simultaneously and adjacent groups display opposite indications.

- a *limited progressive system,* in which the signal controllers are equipped to operate with a single cycle length, a single cycle split, and a single timing schedule that permits continuous operation of platoons of vehicles along the street at the progression speed.

- a *flexible progressive system,* in which the common time cycle, splits, and offsets may be adjusted to conform to varying traffic conditions throughout the day or week.

Electromechanical controllers are coordinated by electrical cables that carry rudimentary commands asserted simply by turning on 120 V. More modern communications systems may use twisted copper-wire pairs, leased telephone lines, coaxial cable (perhaps for cable TV), fiber optic cables, radio, microwave, and even laser beams. Coordination does not require any interconnection at all, however. Developments in digital electronics have made it possible to keep signals synchronized one with another by means of precise clocks within the controller at each intersection. Time-based coordination can achieve enormous savings by eliminating the cost to install and operate an interconnect system. Because the timing plan can be changed only on the basis of time of day, time-based coordination is most satisfactory where traffic demands are stable and predictable.

Since the early 1980s, signal control for arterial streets in the United States has been obtained primarily through "closed-loop systems." Offered by most manufacturers of Type 170 and NEMA-standard equipment, closed-loop systems are so called because they not only allow commands to be sent from a master computer to the intersections, but also permit surveillance data on equipment operation to be returned from the field to a master or an operator at the traffic-control center (Fig. 12-20). Figure 12-15, above, becomes Figure 12-20, in which the double-headed dashed arrow indicates communications in both directions.

The control components of a typical closed-loop system are shown in Figure 12-21. The dial-up telephone line from a central personal computer (PC) to the field master is not expensive, and it allows the engineer at the traffic-control center, as well as the technicians at the signal shop, to monitor local-intersection operation and dispatch first-line maintenance personnel quickly to a location where equipment has malfunctioned. Timing data can be uploaded from the field and adjusted values (or commands) downloaded to the local controller without the need to drive to the intersection. Communications between the field master and the local controllers use a serial-transmission technique called "time-division multiplexing" (TDM), in which two or three dozen intersections can be coordinated by one or two twisted copper-wire pairs or a single fiber optic conductor. Figure 12-22 shows how a PC can display in real time the operation of the controller. These relatively inexpensive systems have revolutionized the degree of surveillance and control available to the traffic engineer for arterial signals.

Traffic Control Signal Warrants

A thorough investigation of traffic conditions and physical characteristics of the location should be made to determine the need for a traffic signal and to provide

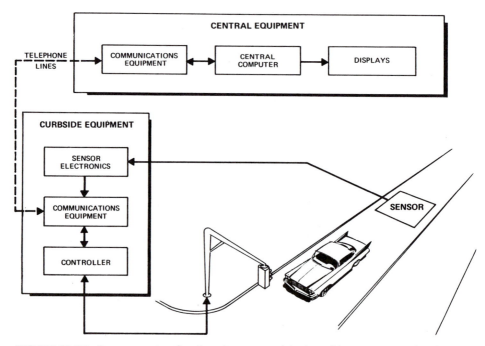

FIGURE 12-20 Components of a signal system with closed-loop communications.

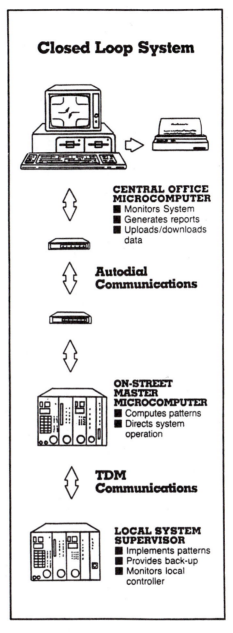

FIGURE 12-21 Components of a closed-loop system.

necessary data for the design and operation of the signal where it is warranted. *MUTCD* (*1*) lists 11 sets of conditions that warrant the installation of a traffic signal:

1. Where traffic volumes on intersecting streets exceed values specified in the *MUTCD* (*1*).
2. Where the traffic volume on a major street is so heavy that traffic on a minor intersecting street suffers excessive delay or hazard in entering or

FIGURE 12-22 A PC monitors the operation of a NEMA-standard controller in its cabinet.

crossing the major street. The *MUTCD* (*1*) gives traffic volumes on the major and minor street that satisfy this warrant.

3. Where vehicular volumes on a major street and pedestrian volumes crossing that street exceed specified levels.

4. Where inadequate gaps exist for schoolchildren to cross at established school crossings.

5. Where necessary to maintain proper grouping or platooning of vehicles and effectively regulate group speed. This is known as the "progressive movement" warrant.

6. Where the number of accidents correctable by traffic-signal control exceeds a specified value.

7. At a common intersection of two or more major routes within a traffic flow system. The purpose of this warrant is to encourage concentration and organization of traffic flow networks.

8. Where no single warrant is satisfied but where warrants 1 and 2 are satisfied to the extent of 80 percent or more of specified values.

9. Where in each of any four hours of an average day the hourly volumes on the major street (both approaches) and the minor street (one direction) exceed values specified in the *Manual* (*1*).

10. Where for one hour of the day minor street traffic suffers undue delay in entering or crossing the major street.

11. Where the peak hourly volumes on the major street (both approaches) and the minor street (one direction) exceed values specified in the *Manual* (*1*).

Reference 1 gives specific and quantitative descriptions of these warrants.

Beacons

Four types of beacons are recommended by the *MUTCD*: (1) hazard identification beacons, (2) speed limit sign beacons, (3) intersection control beacons, and (4) stop sign beacons. Hazard identification beacons are used in conjunction with appropriate warning or regulatory signs or markers to draw attention to obstructions in or immediately adjacent to the roadway or other potentially hazardous location or condition. Such beacons consist of one or more sections of a standard traffic signal head with a flashing circular yellow indication in each section.

Speed limit sign beacons are used to draw attention to fixed or variable speed limit signs. Such beacons commonly consist of two circular yellow indications that are aligned vertically and that are flashed alternately. The use of a flashing single large circular yellow lens section is also permitted.

Intersection control beacons are used to control traffic at intersections where traffic or physical conditions do not justify conventional traffic signals but high accident rates exist. Such beacons consist of one or more sections of a standard traffic signal head, having circular yellow or circular red indications in each face.

Stop sign beacons consist of one or two sections of a standard traffic signal head with a flashing circular red indication in each section. Stop sign beacons are mounted within 12 to 24 in. above the stop sign.

Lane-Use Control Signals

Lane-use control signals are normally used to permit or prohibit use of a traffic lane. They are most commonly used to control traffic on streets with reversible lanes. Examples of lane-use control signals include a steady downward green arrow, which indicates that a driver may use the lane over which the signal is located, and a steady red **X,** which means that the driver may not use the lane.

Several other types of signals are used but are not described here, including drawbridge signals, emergency traffic control signals, and train approach signals and gates.

12-6 Traffic Engineering Software

In recent years, several computer models have been developed to help traffic engineers understand and solve traffic problems. Although traffic engineering studies continue to be done by manual methods, computer approaches are being increasingly employed for traffic engineering analyses and design. Parsonson (*10*) has described the most common computer programs for signal timing optimization and traffic simulation. His description follows with but minor editorial changes.

Signal Timing Optimization Software

The following programs are the most widely used signal timing optimization software. Each has its own particular area of application and its own signal timing design philosophy.

SOAP The SOAP (*Signal Operations Analysis Package*) program develops fixed-time signal-timing plans for individual intersections. SOAP can develop timing plans for six design periods in a single run. It can also analyze 15-min volume data for up to 48 continuous time periods and determine which timing plan is best suited for each 15-min period. A data input manager is included with the program to facilitate data entry.

PASSER II and MAXBAND PASSER II (*Progression Analysis and Signal System Evaluation Routine*) and MAXBAND are known as bandwidth-optimization programs. They develop timing plans that maximize the through progression band along arterials of up to 20 intersections. Both programs work best in unsaturated traffic conditions where turning movements onto the arterial are relatively light. PASSER II and MAXBAND can also be used to develop arterial phase sequencing for input into a stop-and-delay optimization model such as TRANSYT-7F.

The latest version of PASSER II features enhanced program output, explicit treatment of permitted left turns, and a menu-driven, graphical input/output processor. The program also comes with a user-friendly input preprocessor.

TRANSYT-7F and SIGOP-III The *Traffic Signal Network Study Tool* (TRANSYT-7F) and the *Signal Timing Organization Program* (*SIGOP-III*) develop signal-timing plans for arterials or grid networks. The objective of both programs is to minimize stops and delay for the system as a whole, rather than maximizing arterial bandwidth.

TRANSYT-7F (Release 6, December 1988) was a cooperative effort by the Federal Highway Administration (FHWA), The University of Florida, and the California Department of Transportation (Caltrans). This version of the program features better treatment of actuated control, a bandwidth-constraint capability, and several other new features. The program is completely menu driven and comes with a data input manager and a number of other utility programs to assist in creating data files and displaying results.

Arterial Analysis Package The *Arterial Analysis Package* (AAP) allows the user to easily access PASSER II and TRANSYT-7F to perform a complete analysis and design of arterial signal timing. The package contains a user-friendly forms display program so that data can be entered interactively on a microcomputer. Through the AAP, the user can generate an input file for either of the two component programs to quickly evaluate various arterial signal-timing designs and strategies. The package also links to the "Wizard of the Helpful Intersection Control Hints" (WHICH), to facilitate detailed design and analysis of the individual intersections. The current program interfaces with TRANSYT-7F (Release 7), PASSER II (90), and WHICH.

The AAP software features pull-down menus, on-line help, and many automated features to make it easier to use TRANSYT-7F and PASSER together to perform a comprehensive signal-timing analysis and design.

Traffic Simulation Models

Simulation models allow the traffic engineer to evaluate a variety of proposed operational improvements before implementing the changes in the field. The models vary in their levels of detail and are classified as either microscopic (simulation of individual vehicles) or macroscopic (simulation of platoons of traffic). The models also apply to different types of facilities (e.g., signalized networks, freeways, or corridors).

TRAF-NETSIM TRAF-NETSIM is a microscopic simulation model that provides a detailed evaluation of proposed operational improvements in a signalized network. For example, TRAF-NETSIM can evaluate the effects of converting a street to one way, adding lanes or turn pockets, moving the location of a bus stop, or installing a new signal.

CORFLO The CORFLO model, formerly called TRAFLO, provides a macroscopic simulation of a corridor containing both signalized intersections and freeways. It also contains a traffic assignment model that can redistribute traffic flows in response to control or geometric changes in the corridor. For example, the model provides a powerful tool for analyzing alternative construction plans.

Signal-timing plans are typically developed using programs such as TRANSYT-7F or PASSER. This approach is known as first-generation control and is based on off-line signal timing plan generation with manual, time-of-day, or traffic-responsive system plan selection. Alternative control strategies for automatic generation of timing plans while the system is on-line and operating are known as second-generation control. These programs automatically collect detector data, perform calculations of signal timing, and immediately implement the plans on the system. These systems require the installation of numerous detectors in order to have the detailed traffic flow data required to calculate effective timing plans.

An alternative strategy, known as the 1.5 generation of control, is being developed in an effort to have a simpler, less expensive way of generating new signal-timing plans. It incorporates many of the automatic features of the second-generation programs in terms of the automatic linkage between traffic flow data and the timing programs and between the timing programs and the signal timing data base. In addition, the 1.5 generation of control provides for a manual review and adjustment of timing plans. This reduces the need for numerous detectors, because user intervention can compensate for the errors produced because of limited data during the manual review process.

12-7 Advanced Transportation Management Systems

The Intermodal Surface Transportation Efficiency Act of 1991 (ISTEA, pronounced "Ice Tea") launched the development of a National Intermodal Transportation System. Included was funding for Intelligent Vehicle-Highway Systems (IVHS), which is the application of advanced technologies in surveillance (sensing), communications, information processing, and control to increase the safety and efficiency of surface transportation. Sometimes called "Smart Cars and Smart Highways," IVHS has been divided into six interrelated system areas, each focused on technology or applications (*13*):

- **Technology-oriented systems:**
    Advanced Traffic Management Systems (ATMS)
    Advanced Traveler Information Systems (ATIS)
    Advanced Vehicle Control Systems (AVCS)
- **Applications-oriented systems:**
    Advanced Public Transportation Systems (APTS)
    Commercial Vehicle Operations (CVO)
    Advanced Rural Transportation Systems (ARTS)

The importance to IVHS of a comprehensive communications infrastructure is shown by Figure 12-23, which illustrates the IVHS vision of AT&T. In recognition of the broader potential of IVHS, it is often called Intelligent Transportation Systems (ITS).

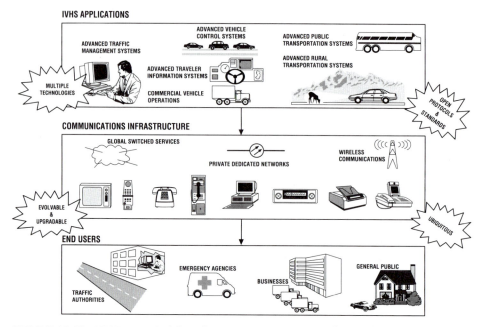

FIGURE 12-23 AT&T's IVHS vision. (Courtesy AT&T Network Systems.)

Components of ATMS

ATMS addresses technologies to monitor, control, and manage traffic on streets and highways. As described in Ref. 13, ATMS technologies include the following:

- Traffic control centers in major metropolitan areas to gather and report traffic information and to control traffic movement to enhance mobility and reduce congestion through such things as ramp, signal, and lane management or vehicle route diversion.
- Changeable message signs to provide current information on traffic conditions to highway users and suggest alternate routes.
- Priority control systems to provide safe travel for emergency vehicles when needed.
- Programmable, directional traffic signal control systems.
- Automated dispatch of tow, service, and emergency vehicles to accident sites.

ATMS Communications Systems

A key element of ATMS is the sharing of real-time (current) information among affected agencies and jurisdictions, permitting the collaboration needed to optimize traffic flow. The communications network required for an advanced transportation management system is a complex array of communication technologies and may involve several political jurisdictions and agencies. Figure 12-24 gives an example of the various components of an ATMS communications network for Atlanta. In such a system, video images, voice messages, and data transmissions flow between various communications centers and data collection sources both in the field and in the transportation management and control centers. Such a system would be expected to use the latest and best communications systems including video, data transmission lines, and communications satellites. Figure 12-

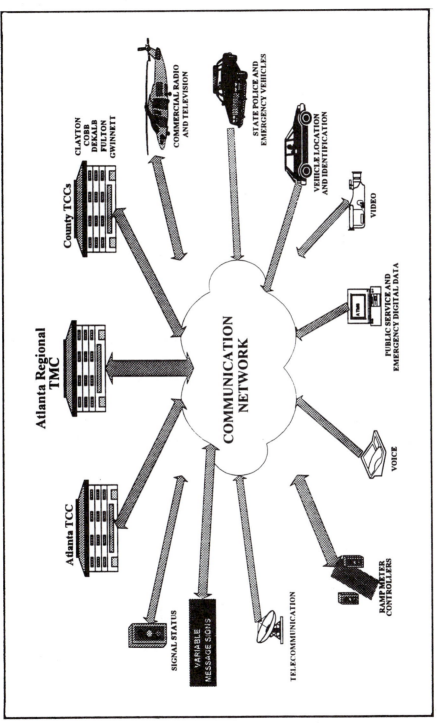

FIGURE 12-24 Principal components of the Atlanta-area ATMS. (Courtesy Georgia DOT.)

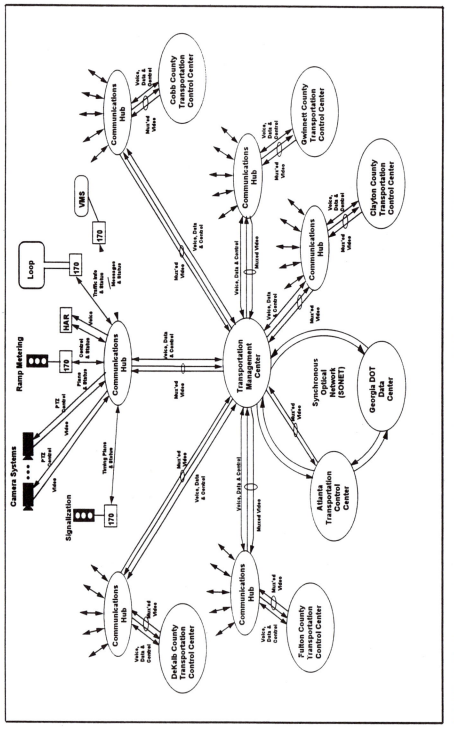

FIGURE 12-25 Flow of video, voice and data in the Atlanta-area ATMS. (Courtesy Georgia DOT.)

25 illustrates the flow of information in Atlanta's advanced transportation management system.

The management of these technologies should be integrated with that of other functions and applications, such as traveler information and electronic toll collection. Examples of components of Atlanta's traveler information system are shown in Figure 12-26.

Benefits of ATMS

Some of the expected benefits of an advanced transportation management system include

- reduced congestion.
- improved air quality.
- better-informed travelers.
- faster response to incidents such as crashes, stalls, and spills.
- preferential treatment to selected vehicles such as buses, vehicles transporting VIPs, and emergency vehicles.

12-8 Conclusion

In this chapter we have been able to outline only some of the more important aspects of traffic engineering. Certain areas, such as communications technology, are changing rapidly. The reader is referred to the references listed at the end of this chapter for a more complete treatment of these important areas of study and is advised to supplement this text, by consulting the latest materials on ATMS from the Federal Highway Administration, ITS America, and other prime sources.

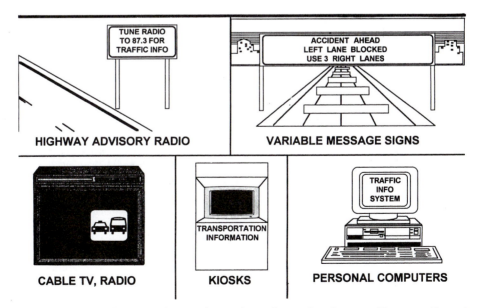

FIGURE 12-26 Atlanta's Advanced Traveler-Information System. (Courtesy Georgia DOT.)

PROBLEMS

12-1 A local street named Dysc Drive has a series of intersections controlled by four-way STOP signs (because of sight-distance problems, perhaps). The next intersection is controlled by a STOP sign on Dysc Drive but no signs on the cross street (because it has the right of way). Discuss the safety aspects of this design, putting yourself in the place of a driver proceeding along Dysc Drive in this direction.

12-2 There has been a serious crash on a section of two-lane road where there is a horizontal curve. The curve is not superelevated but has a normal crown of 1:48 (1/4 in./ft). This section of the road has been posted for a speed limit of 64 km/hr (40 mph). You have been asked to determine the recommended safe speed at which the curve may be traversed, in the direction that has the lower value. Like most other agencies, your organization will use this lower speed condition for signing both approaches.

   a)   Sketch a cross section of your road, labeling the normal crown on each lane. Assume that the road curves to the right as it goes into the paper, and indicate on the sketch whether the center of the curve is off to the right or left of your cross section. Sketch a vehicle rounding the curve in the direction, and in the lane, that will give the lower value of recommended speed. Show headlights or taillights clearly enough so that it will be clear whether the vehicle is moving into the paper or out from the paper toward the viewer.

   b)   You have measured this curve in the field to have a radius of 150 m (491 ft). Use Figure 12-4 to determine the recommended safe speed. Show clearly your calculation of the value and sign of *e*.

REFERENCES

1. *Manual on Uniform Traffic Control Devices for Streets and Highways.* U.S. Department of Transportation, Federal Highway Administration, Washington, DC (1988).

2. *Traffic Control Devices Handbook.* U.S. Department of Transportation, Federal Highway Administration, Washington, DC (1983).

3. Post, Theodore J., Alexander, Gerson, J., and Lunenfeld, Harold. *A Users' Guide to Positive Guidance,* 2nd Ed. U.S. Department of Transportation, Washington, DC (1981).

4. Ligon, Claude M., Carter, Everett C., and McGee, Hugh W. Multiway Stop Sign Removal Procedures. *Record 1010.* Transportation Research Board, Washington, DC (1985).

5. *A Policy on Geometric Design of Highways and Streets,* 1994. American Association of State Highway and Transportation Officials, Washington, DC (1994).

6. *Design and Operation of Work Zone Traffic Control.* Training Course Participant's Notebook, Federal Highway Administration, Washington, DC (1988).

7. *Accident and Speed Studies in Construction Zones.* Federal Highway Administration Report No. FHWA-RD-77-80, Washington, DC (1977).

8. *Proceedings: Special Public Hearing: Fog Accidents on Limited Access Highways.* Report PB92-917001, NTSB/RP-92/01, National Transportation Safety Board, Washington, DC (1992).

9. Evans, David L., LED Technology in Message Signs and Traffic Signals. *Compendium of Technical Papers,* pp. 48–52. 64th ITE Annual Meeting, Dallas, TX (1994).

10. Parsonson, Peter S., *Signal Timing Improvement Practices.* National Cooperative Highway Research Program, Synthesis of Highway Practice 172, Transportation Research Board, Washington, DC (1992).

11. Kell, James H., and Fullerton, Iris J. *Manual of Traffic Signal Design,* 2nd Ed. ITE/Prentice-Hall, Englewood Cliffs, NJ (1991).

12. *Traffic Control Systems Handbook.* U.S. Department of Transportation, Federal Highway Administration, Washington, DC (1985).

13. Shuman, Valerie. *Primer on Intelligent Vehicle Highway Systems.* Circular 212, Transportation Research Board, Washington, DC (1993).

CHAPTER 13

SURVEYS, PLANS, AND ESTIMATES

In the relocation or reconstruction of existing highways and the establishment of new ones, surveys are required for the development of project plans and the estimation of costs. The performance of good surveys requires well-trained engineers who have an understanding of the planning, design, and economic aspects of highway location and who are sensitive to the social and environmental impacts of highway development. As Figure 13-1 illustrates, the work of the highway location may include (1) a desk study, (2) a reconnaissance survey, (3) a preliminary survey, and (4) a final location survey. Each phase of the highway location study is discussed in some detail in the following sections.

13-1 Available Techniques

Highway surveys may be accomplished either by conventional ground survey methods or by remote sensing techniques. A variety of remote sampling devices have been suggested for highway location studies including side-look radar, infrared photography, and even multispectral devices for recording a scene at several wavelengths simultaneously. Aerial photography with either panchromatic or color film is the most commonly employed remote sensing procedure in highway location studies.

Aerial photographs provide the highway location team with both qualitative and quantitative information. By direct study of individual and stereoscopic pairs of photographs and photographic mosaics, trained professionals may glean a great deal of information about landforms, soil and drainage conditions, and the character of land use. With appropriate horizontal and vertical ground controls, accurate photogrammetric maps can be drawn at practically any desired scale, providing quantitative data on ground contours, drainage features, buildings and structures, and existing transportation facilities.

Whereas traditional techniques are still used on projects of medium and small size, the use of photogrammetric surveys on large works is almost universal, and it is becoming more frequent on smaller projects. The adaptability of photogrammetry to computer operations has enabled substantial savings in time and money over conventional methods. Conventional field surveys, described in Section 13-8, are generally used for final location surveys on projects of all sizes.

13-2 Desk Study of Area

Before any field or photogrammetric investigation is made, a great deal can be learned from a desk study of the area. The first step in highway location and

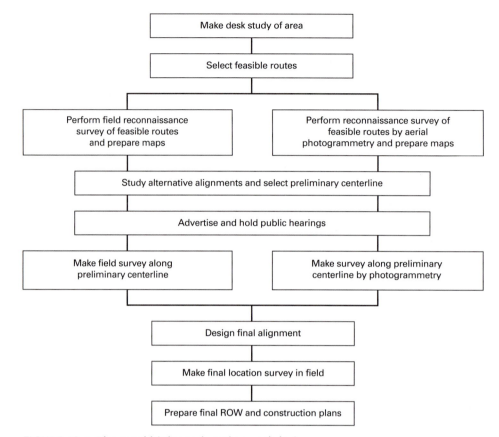

FIGURE 13-1 The rural highway location and design process.

design is to procure all available pertinent data. These data may be in the form of maps, aerial photographs, charts or graphs, and so on and may require the application of a large variety of engineering, environmental, social, and economic knowledge. The type and amount of data collected during this initial phase will, of course, vary with the functional classification of the road and the nature and size of the project. Sources of available information are listed in Table 13-1.

When all the available data have been assembled, a detailed analysis should reveal much information pertinent to the proposed project. For example, analysis of the available information may allow the engineer to determine the advisability of selecting an entirely new location or improving the existing one. After an exhaustive study of aerial photographs, topographic maps, drainage maps, soil maps, and other data is made in the office, a series of proposed locations may be selected for a field or photogrammetric investigation.

13-3 Reconnaissance Survey from Aerial Photographs

For many projects, the data on available maps and photographs will be out of date or incomplete. In some cases, there are no available base maps of suitable scale and accuracy. A reconnaissance survey from aerial photographs will amplify and verify conditions determined from the preliminary study. The function of the

TABLE 13-1 Possible Data Sources for Desk Study of Highway Locations

Engineering data
  Topographic and geological maps
  Stream and drainage basin maps
  Climatic records
  Preliminary survey maps of previous projects
  Traffic surveys and capacity studies

Environmental data
  Agricultural soil surveys indicating soil erodibility
  Water quality studies
  Air pollution studies
  Noise and noise attenuation studies
  Fish and wildlife inventories
  Historical studies

Social data
  Demographic and land-use information
  Census data
  Zoning plans and trends
  Building permit records
  Motor-vehicle registration records

Economic data
  Overall costs of previous projects
  Unit construction cost data
  Agricultural, industrial, commercial activities and trends
  Property values

survey is to provide sufficient information on the topography and culture of the area to enable the selection of one or more preliminary route locations.

Control points between the two terminal points on the route are determined and the flight lines between these control points established. It is customary for the width of coverage to be about 0.4 to 0.6 the distance between control points. From the aerial photographs a base map is produced, usually to a scale ratio of 1:1000 with 0.5- or 1-m contour intervals. Under the inch-pound unit system, a scale of 1 in. = 100 ft or 1 in. = 200 ft was generally used with a 2- or 5-ft contour interval. A stereoscopic examination of the photographs will also permit the determination of ground cover and soil conditions (Fig. 13-2). From the base maps and photointerpretation of the surveyed area, likely routes can be designated.

A location team should examine potential routes on the basis of the following requirements:

1. Traffic service for population and industrial areas.
2. Directness of route.
3. Suitability of terrain encountered.
4. Adequacy and economy of crossings at watercourses and at other transportation routes.
5. Extent of adverse social, environmental, and ecological effects.

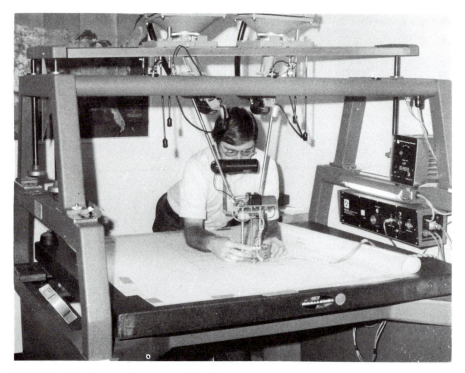

FIGURE 13-2 Stereoplotter used by the Georgia Department of Transportation.

The reconnaissance is an important portion of the work and should not be treated lightly. During this phase of the location process, the range of alternative routes is narrowed and a small number of routes are selected for further study. The use of photogrammetric base maps often enables an objective study of all possible routes with greater ease than field reconnaissance.

13-4 Selection of Preliminary Centerline

Possible alternative alignments are plotted on the base map, and from these alignments, with preliminary gradelines, the alternatives are compared for suitability. Such features as design capacity, safety, road user costs, construction costs, and maintenance costs are examined. Possible adverse impacts of the various alignments due to the dislocation of families, farms, and businesses should be carefully considered, as should the effects of the roadways on the environment and the ecology. Opportunities to provide scenic vistas and aesthetic enhancement of the area should not be overlooked. Based on the results of objective analysis, a preliminary alignment is chosen.

13-5 Mapping for Final Design

For final design and location, further photogrammetric mapping is necessary. The selected preliminary alignment is used as a guide for the strip area to be mapped. Highway design should be based on a coordinate system because of the increasing use of computers for direction and distance calculations. Therefore, at this time, prior to taking low-level aerial photographs, a baseline is established on paper and by monuments on the ground. The low-level aerial photographs are taken with markers on the ground to tie in the baseline. Distances between the baseline monuments are often determined by electronic measuring devices (EMDs). The baselines, which run between monuments of calculable coordinates, can be used in the final location survey in locating the alignment in the field.

The low-level photographs are used to produce base maps along the preliminary alignment, usually at a scale of not less than 1:1200 (1 in. = 100 ft). These base maps are more complete than those used for reconnaissance. The preliminary survey taken at a lower level can give virtually complete information on

- topography, with respect to changes in elevation and drainage characteristics, and information concerning the soil conditions in the area.
- land uses, designating type, intensity, and quality.
- transportation facilities, with respect to proximity and effect of proposed location.
- the effect of the proposed location on property within the area.

For the purposes of design, however, the preliminary survey will produce topographic base maps and information concerning property and utility location that will enable the engineer to proceed with the design of the final alignment.

13-6 Design of the Final Alignment

The process of the design of the final alignment is one requiring great skill and judgment. It is a trial-and-error process by which the most suitable alignment is obtained. The highway is fitted to the topography and land use until the engineer is satisfied that no better fitting can be achieved. This is most often accomplished with the aid of CAD software.

To facilitate computation, tangent alignments can be set at even-degree angles. This speeds computations by permitting the use of design tables, which can be generated within the computer. In urban areas, where small changes in alignment may cause property encroachment, the use of whole-degree azimuths may not be feasible.

The method of design usually followed is to adjust the alignment until a smooth, consistent alignment is achieved that is satisfactory from the viewpoint of grade, curvature, cross-sections, drainage, and stream crossings. The hand-fitted line is then converted into a defined line by fitting to it tangent lengths, transition spirals, and simple and compound curves.

13-7 Final Location Survey

Conventionally, the method of establishing the final location in the field has been by direct chaining along the tangents, first setting the points of intersection of the tangent lines (PIs) and then closing out the horizontal curves between the tangents, using deflection angles and chaining to set the curve stations. This method is still in use, although the electronic distance-measuring devices have to a great degree replaced chaining procedures. Moreover, the use of coordinates in highway location means that the computer can be used to calculate azimuths between control points on the curvilinear alignment and points of known coordinates on the baseline. In this case, the errors along the alignment are not additive. The points of curvature (PCs) and points of tangency (PTs) can be set and checked by the intersection of three azimuths instead of two. Chaining between the control points in the alignment is done to set stakes at the even

20-m stations,* and stakes on the curves can be set from the PI, PC, or PT by normal deflection angle methods.

In general, it is unnecessary for ground crews to take cross sections along the alignment. Such cross sections can be obtained directly from the photogrammetric maps with as much accuracy as ground surveys usually achieve.

13-8 Conventional Ground Techniques

Where photogrammetric surveying methods are available, or where the size or nature of project makes the use of aerial photographs unsuitable, it may be necessary to use the older, conventional ground surveys. Such instances are becoming less common, as the development of photogrammetry has led to the widespread availability of aerial photographic equipment. This can generally be used to economic advantage.

The steps involved in ground surveys are similar to those used in photogrammetry:

1. Desk study.
2. Reconnaissance survey.
3. Preliminary location survey.
4. Final location survey.

A desk study includes the preliminary steps of evaluation of all available data as covered in Section 13-2. The reconnaissance survey consists of a field investigation that usually provides a means of verification of conditions as determined from the preliminary study. For example, building symbols on maps do not indicate the true value of property under consideration, and this information can usually be secured by field investigation. A study is made of the profiles and grades of all alternative routes and cost estimates made for grading, surfacing, structures, and right-of-way. A comparison of alternative routes in this fashion will aid the final selection of the most likely location.

Next, a preliminary survey is made to gather information about all the physical features that affect the tentatively accepted route. In general, the work is carried out by a regular survey party consisting of party chief, instrument men, chainmen, and rodmen. A primary traverse or baseline is established. This is an open traverse consisting of tangent distances and deflection angles following approximately the line recommended in the reconnaissance report. When the preliminary line has been established, the topographic features are recorded. The extent to the right and left of the traverse line to which the topography should be determined will vary, and it is usually left to the judgment of the party chief in charge of the work. It will certainly never be less than the proposed width of right-of-way.

The amount of level work on a preliminary survey should be kept to a minimum. In flat areas no centerline profiles are needed; in rolling or mountainous terrain a profile of the traverse centerline is often sufficient. A few cross sections are sometimes of value in exceptionally rugged areas where centerline profiles do not give proper information.

Using the preliminary survey as a basis, a preliminary survey map is drawn. The preliminary map should show all tangents with their bearings and distances,

---

*With the traditional U.S. system of units, 100-ft stations were used.

all deflection angles, ties to property corners or state control monuments, and all witness points of the alignment. Certain topographic features are shown on the map, such as streams, water courses, lakes, hills, and ravines, and man-made features such as buildings, drainage structures, power lines, and other public facilities.

The design of the final alignment when ground survey methods have been used is essentially the same as the procedure indicated in Section 13-6. Location of the centerline requires considerable skill and judgment. Trial lines are drawn on the map and are related to control points such as road intersections, and building corners. The trial lines should avoid badly drained ground, rough or impassable areas, valuable property, or other obstructions. The use of splines and curve templates is recommended. Rough estimates of excavation and embankment may also influence the placing of the trial lines on the map. Essentially, the only difference encountered between conventional ground methods and aerial surveys is that it is more likely that with the latter the whole project will be tied to a coordinate grid system to facilitate the use of computers in determining the distances and azimuths. On the smaller projects where ground methods are still used, the older system of chaining along the centerline, tangent distances, and deflection angles may be the essential data given for the final location survey.

The final location survey serves the dual purpose of permanently establishing the centerline and collecting the information necessary for the preparation of plans for construction. The line to be established in the field should follow as closely as is practical the line drawn on the preliminary map, conforming to the major and minor control points and the alignment that was previously determined.

The first step in the final location survey requires the establishment of the centerline, which is used as a survey reference line, upon which property descriptions are based for the purpose of purchasing right-of-way. Modifications of the final location may also be made in the field as a result of incomplete or inaccurate preliminary information. Centerline stakes are placed at 20-m (100-ft) intervals, and each stake is marked with the proper station number. Curves are staked out by deflection angles, and the error of closure is determined by an independent stakeout of the PT from the PI. Particular care must be taken in the chaining, and a closure should be effected where possible on control hubs for the preliminary survey, since errors along the centerline are cumulative.

Level work is of the utmost importance, because the grade line, earthwork, and drainage are designed from the level notes. Elevations may be obtained from National Geodetic Survey (NGS), U.S. Geological Survey (USGS), or other bench marks, or elevations may be assumed for a project. Bench marks are permanent reference points, and extreme care should be used when establishing benches along the route of the proposed project. The bench marks should be placed approximately 300 m (1000 ft) apart in fairly level country and closer in hilly country. Benches should always be placed at least 60 m (200 ft) from any proposed structure, if possible. Bench-mark elevations should be established by the traverse method, by the loop system of levels, or by any other approved method where a high degree of accuracy can be obtained. The error of closure between benches should be at least third order class II, that is, not greater than 12 mm per $\sqrt{\text{km}}$.

The importance of keeping good notebooks on alignment and level surveys cannot be overemphasized. Notes taken in the field during the final location survey usually become part of the permanent record of the project.

Finally, cross-section levels are taken at even 20 m (100-ft) stations and at any intermediate points with abrupt slope changes. This information is necessary for accurate determination of existing grade line, design of drainage structures, computation of earthwork quantities, and, in general, estimation of quantities of work to be done. The manner in which the cross sections are taken and the form of notes used are fairly well established. In order to obtain a fairly representative cross section at a station, an elevation is always obtained at the centerline stake. Then elevations and distances at right angles to the centerline are recorded, first on one side and then on the other. Cross sections should also be taken at least 200 m (500 ft) beyond the limits of the point of beginning and the point of ending of a project. Centerline profile elevations should also be obtained at some distance beyond these points in order to establish proper grades for connecting projects. The final location survey is complete when all the necessary information is available and ready for the designer to use. This includes all information pertaining to (1) alignment, (2) topography, (3) bench mark levels, (4) cross sections, (5) section corner ties and other land ties, (6) plats and subdivisions, (7) soil surveys, and (8) drainage and utilities.

13-9 Automated Procedures

Much of the work involved in route location is repetitive and requires little exercise of the engineer's professional skill. Therefore, in the past few years great emphasis has been placed on increasing the individual engineer's productivity by the use of automated procedures that rely extensively on electronic computers.

The photogrammetric aerial survey methods already described are readily adaptable to automation. Digital terrain models can be stored in computer memory by linking the stereoplotter to a device that automatically records the x, y, and z coordinates of points on the stereo model. Input of a proposed alignment will result in the computation of horizontal centerline geometry, vertical profile geometry, and earthwork quantities. This allows rapid evaluation of the proposed alignment and permits the engineer to make small modifications without the tedious and expensive repetition of manual calculations.

From the stored terrain data and the input of a proposed alignment, it is possible to obtain cross-section data directly in numerical form or, by linking the computer to an automatic plotting device, to obtain cross sections in a graphical form.

By the use of more sophisticated stereoplotters, the amount of ground control work required for the aerial survey can be greatly reduced and time and cost of the survey minimized. The stereoplotter is an example of a machine that uses *instrumental aerotriangulation* in a process called "bridging," which eliminates the need for ground control points for each pair of stereo photographs forming a stereo model. The computer can also be used for analytical solutions of aerotriangulation problems. Bridging is a process whereby the horizontal and, to some degree, the vertical control on one stereo model can be extended through successive models until additional ground control is reached.

Figure 13-3 shows a Wild Aviolyt BC2 analytical compiler with an Aviotab TA10 automatic plotting table. This system provides high accuracy in photogrammetric measurements for applications in graphical compilation, digital mapping, and computer-controlled profiling.

Most highway agencies use computer programs for the adjustment of field survey traverses and for the computation of earthwork volumes from cross-

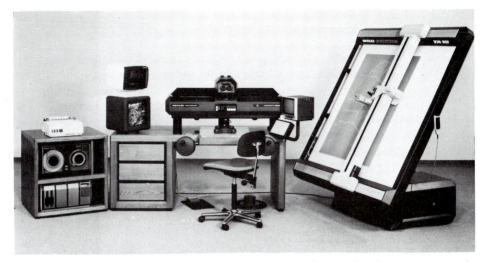

FIGURE 13-3 Wild Aviolyt BC2 analytical compiler with an Aviotab TA10 automatic plotting table. (Courtesy Wild Heerbrugg Instruments, Inc.)

section data. Modern techniques of establishing horizontal alignment rely extensively on the use of transparent templates for curve design and intersection layout. Precomputed data from these templates are compiled into design tables that can be used as input for the computation of the horizontal alignment referenced completely to a coordinate system. The use of templates and design tables relieves the designer of much computation and (possibly more important) ensures more standardization and higher design standards.

Extensive use has also been made of the computer in the layout of interchanges and bridges. The intersection or bridge is related to the coordinate system of the project. From this relation, the azimuths between any two points are readily available by machine calculation. Field layout can be done by measuring along an azimuth from a known point or by determining the point of intersection of azimuths from at least two other points. The latter method has been found successful for setting bridge piers by using three azimuths from three points, where chaining is difficult.

13-10 Problems of Highway Location in Urban Areas

In rural areas, the location of a highway is dictated chiefly by the desired endpoints of travel, topography, geology, and environmental impact. The road should serve a desired travel pattern with an alignment that provides consistent conditions within acceptable limits of curvature and grade over terrain that is capable of supporting the proposed construction without excessive costs.

In urban areas the location process is more involved, reflecting the complexities of urban conditions. Among the conditions that control urban locations are

- traffic service.
- land use.
- off-street parking.
- other transportation systems.
- topography and geology.

- sociological considerations.
- historical and environmental impacts.

The traffic planner, as a result of extensive investigations of the travel desires and existing travel patterns in an urban area, can predict the effect on travel patterns of alternative locations of proposed highways. Traffic assignment techniques can be used to analyze the probable effect of new links in the highway network. This permits the location engineer to determine how well alternative locations will fit with the existing network. Travel desire lines often can show the major elements of a desirable arterial highway system in an urban area.

Land use is a major factor in the pattern of traffic generation in an urban area. Most cities have a central commercial core surrounded by areas of residential, industrial, and commercial development. Travel patterns vary with time of day, day of week, and season of year. Trips from the home to work are heavy to the commercial downtown area and the industrial areas during the morning peak hours, with a reversed pattern during the evening rush hour. During the day, shopping trips from the residential areas to the outlying commercial areas and the center core predominate. The pattern of land use affects location in other ways. Industrial and commercial areas rely heavily on truck transportation and, therefore, need service by arterial routes. Health and safety requirements may indicate that heavily traveled routes should not be located through residential areas. Aesthetic considerations may indicate that an alignment that cuts through a public open space should be avoided if the space can be preserved by a skirting alignment. Most cities have recognized the interaction between highway location and land use and have integrated a major thoroughfare plan with the city master plan.

Parking is of major importance in urban areas. New highway locations must be considered in relation to existing or planned parking facilities for the vehicles attracted to the new facility. Ideally, the location should be close to existing and potential parking areas to minimize the amount of travel on existing streets.

Interference and interaction with other transportation systems must be considered. The aim of any new facility is to increase the overall level of service of transportation in the urban area, and a location should not necessarily disrupt existing service. Overall coordination of the various transport modes is essential in the development of the balanced transportation system so urgently needed in complex urban areas.

As in rural areas, topography and geology have a great effect on urban highway locations. Topographic barriers such as ridges, valleys, rivers, and lakes affect construction costs and determine the economy of a route. Similarly, soil and ground water conditions may render an otherwise feasible location uneconomic. The effect of topography and geology is somewhat lessened in an urban area owing to the high cost of right-of-way in certain areas. For example, the cost of a location through a heavy industrial area might greatly exceed costs involved in remedying the swampy conditions of an alternative alignment. The controls of topography remain, however, imposing grade and curvature controls on any designated alignment.

The complex of utilities that serves an urban area is usually almost totally absent from rural areas. Extensive relocation of utilities where depressed highways and underpasses are involved can add greatly to the cost of urban projects and require considerable advance planning. In certain areas within large cities, the underground maze of utility lines and subways can make the cost of a depressed highway prohibitive.

The social effects of a highway project must be of special concern in urban areas. Proposed locations should not sever residential neighborhoods or create barriers between residence and schools and other community services. Open public land should be retained where possible; park and residential land may be used for highways only if no feasible and prudent alternative to the use of such land exists. On the other hand, a highway can serve as a stabilizing force by screening industrial areas from residential communities.

PREPARATION OF PRELIMINARY PLANS

Road plans are necessary for making estimates and receiving bids for construction. The steps involved in the preparation of plans are fairly uniform in all the state highway agencies. Questions related to the type of surfacing, width of grade, preliminary right-of-way requirements, and other design elements are decided before the detailed design is started. It is also determined whether the contemplated improvement can be accomplished at one time, or whether provision should be made to complete it by stages. This stage development may, for example, limit the length to be constructed at any particular time or it may provide for additions to the highway cross section by the addition of future lanes or other improvements.

Generally speaking, plans may be prepared in two phases: preliminary plans and final plans. The first phase consists of a partial investigation and preliminary design. A field inspection is then made before final plans are prepared. Design standards as discussed in Chapter 7 are applied to all steps when designing a given project.

13-11 Metric Scales for Highway Plans

Prior to the mid-1990s in the United States, highway plans were drawn using the inch-pound unit system. With that system, plans were commonly drawn to the scale of 1 in. = 100 ft. When greater detail was desired, as in urban areas, a scale of 1 in. = 50 ft was used. Engineers will continue to need to use the conventional U.S. unit system when they refer to *as-built* plans prepared at an earlier time.

With metrication, the scales of highway plans are referred to as ratios. For example, a scale of 1:1000 in the metric system means one meter on the plan equals 1000 meters in reality.

A scale of 1 in. = 100 ft in the conventional U.S. (inch-pound) system is the same as 1/12 ft = 100 ft or a ratio scale of 1:1200. A scale of 1 in. = 50 ft in the inch-pound system is the same as 1/12 ft = 50 ft or a ratio scale of 1:600.

U.S. highway agencies have generally accepted the scale of 1:1000 for metric highway plans. Where greater detail is desired, a scale of 1:500 is suggested.

13-12 Stationing

Customarily, points along a highway route are identified by "stations," the distance from some reference point, commonly the beginning point of the project. Under the system of conventional U.S. units, one station was equivalent to 100 ft.*

---

*For example, in the conventional system of the U.S. units, station 42 + 00.00 referred to a point 4200.00 ft from the reference point.

With the metric system, a stationing concept based on 1 km is used. For example, station 6 + 375.218 indicates a point 375.218 meters from kilometer station 6 + 000.

13-13 Plan and Profile

With the information obtained from the final location survey, a plan and a profile of the proposed route are drawn on standard plan and profile sheets, as shown in Figure 13-4. The plan should show the proposed construction centerline with the limits of right-of-way and all important topographic features such as fences, buildings, streams, railroads, and other structures on the right-of-way. All survey information necessary for the establishment of the survey centerline should also be placed on the plan. This should include compass or computed bearings of all tangents, together with all curve data, including the point of beginning, point of ending, degree of curvature, and so on, and all witness information to the survey points. Bench marks, with their elevations, are also shown on the plan sheet. Where computers have been used to determine azimuths between control points of known coordinates for the purpose of final location, the coordinates of these control points should be shown on the plan.

A notation is made on the plan of the sizes and types of all existing structures and the manner in which they are to be used or removed. Other details are added, such as type and depth of ditches, slopes, right-of-way and borrow requirements, and so forth.

The centerline profile is plotted on the lower portion of the plan and profile sheet. The horizontal scale is usually the same as that used for the plan. The vertical scale, however, is distorted. Under the conventional U.S. system of units, a vertical scale of 1 in. = 5 ft or 1 in. = 10 ft was most commonly used. With the metric system, a vertical profile scale of 1:50 or 1:100 is recommended. The profile view usually shows the profile of the existing ground line and the profile of the proposed construction centerline, the tangents and vertical curves forming one continuous profile. All grade lines show the percentage of grade and elevation points where changes of grade occur, with the lengths of the vertical curves used at these points of intersection.

All existing drainage structures are generally shown on the plan and profile sections. These include culverts, catch basins, inlets, manholes, and so on. Any special information pertaining to the profile and affecting the design may also be added, such as curb grades, gutter grades, and sodded slopes.

13-14 Grades and Grade Control

All controls that affect the position of the grade line should be given thorough study. Ideal grades are those that have long distances between points of intersection, with long vertical curves between tangents to provide good visibility and smooth riding qualities.

One consideration in grade control is to try to fit the grade line to the natural ground surface as closely as possible. Good drainage is usually accomplished in this manner. Intersections with railroads and existing bridges are points that are generally fixed and the grade must be set to meet them. Intersections with trunk lines, other surfaced roads, unimproved roads, drives, trails, and field entrances may also be grade controls. The effect of soils on grade is also very important. In swamp or peat areas it is necessary to have fills high enough so that waste material can be disposed of at the sides of the roadway. The grade must also be above the water table, where possible.

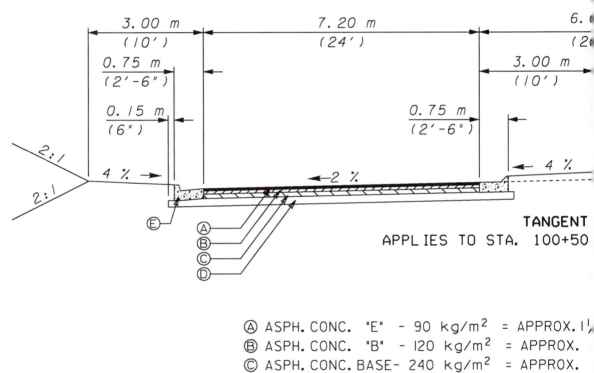

TYP
FOUR-LA

3.00 m
(10')

7.20 m
(24')

6.
(2

0.75 m
(2'-6")

3.00 m
(10')

0.15 m
(6")

0.75 m
(2'-6")

2:1

4 %

2:1

← 2 %

4 %

Ⓔ

Ⓐ
Ⓑ
Ⓒ
Ⓓ

TANGENT
APPLIES TO STA. 100+50

Ⓐ ASPH. CONC. "E" - 90 kg/m² = APPROX. 1½
Ⓑ ASPH. CONC. "B" - 120 kg/m² = APPROX.
Ⓒ ASPH. CONC. BASE- 240 kg/m² = APPROX.
Ⓓ GRADED AGG. BASE- 250mm = APPROX. 10"
Ⓔ AASHTO - 750mm CURB

NOTE: METRIC VALUES
ARE HARD CONVERTED
RATIONALIZED NUMBERS.

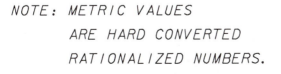

FIGURE 13-5 Typical grading and paving section.

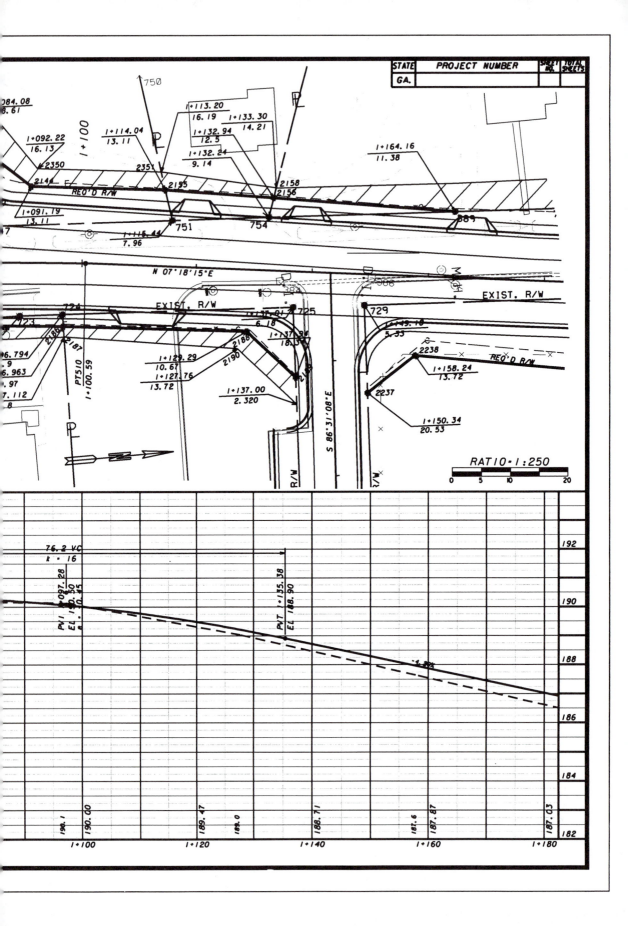

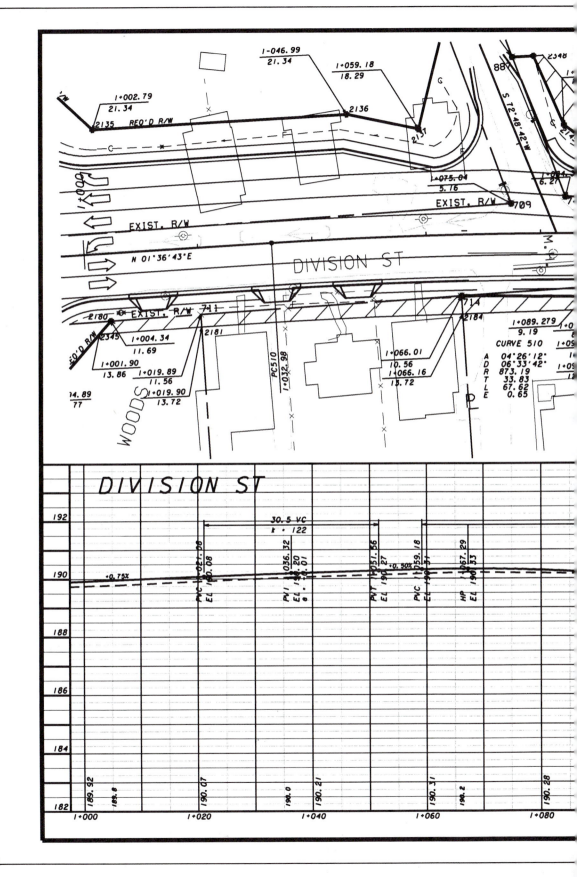

FIGURE 13-4 Typical plan and profile sheet.

ICAL
NE URBAN

)0 m
)')

7.20 m
(24')

3.00 m
(10')

3.00 m
(10')

0.75 m
(2'-6")

0.75 m
(2'-6")

0.15 m
(6")

Profile Grade

4 %%→

2 %→

← 4 %

2:1

2:1

Ⓔ

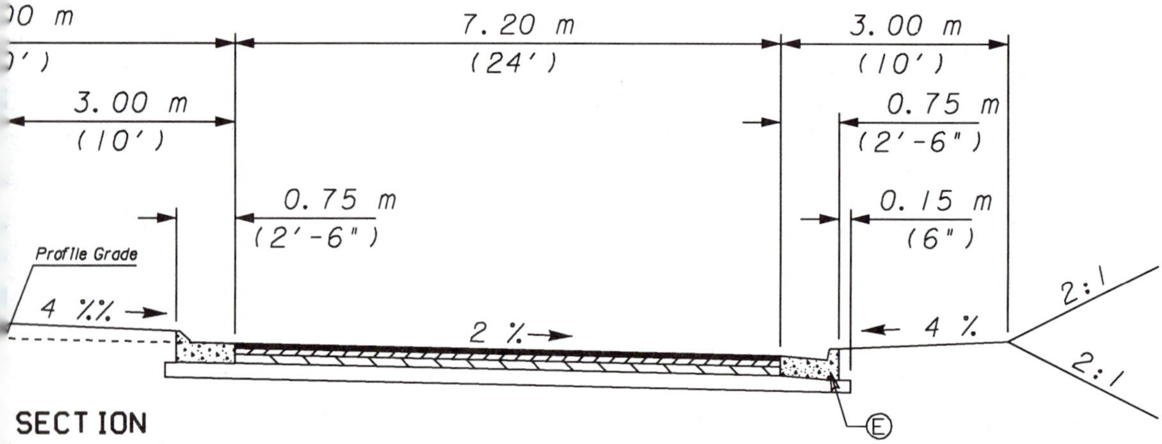

SECTION
.40 TO STA. 120+610.20

2" (165 lb/yd^2)
2" (220 lb/yd^2)
4" (440 lb/yd^2)

The placing of the grade line requires much more detailed study in urban areas than in rural areas. Changes in grade affect store entrances, driveways, and sidewalks. It is often necessary to warp or tilt the pavement to meet exisitng curbs and to provide proper drainage.

Slight distance is of fundamental importance in both rural and urban areas.

Ruling grades for the economical operation of motor vehicles have been fairly well established. These have been indicated in Chapter 7. When existing conditions indicate a new grade, care and judgment should be exercised to obtain the most economical one possible.

When establishing a grade line, an effort should be made to provide an economical balance between the quantities needed for fill sections and those required in cut sections. Changes in grade from a plus grade to a minus grade should be accomplished in cuts, and changes in grade from a minus grade to a plus grade should be placed in fills. This usually give a good design and often avoids the appearance of building hills and producing depressions contrary to the existing slope of the land. All hauls should be downhill, if possible, and not too long.

Horizontal control is usually well established when the location survey is made, although certain refinements may make it necessary to change the alignment when adjusting grade controls for vertical alignment. A combination of grades and horizontal alignment may give insufficient sight distance, poor appearance, or both. Flattening a curve may be necessary to eliminate or reduce superelevation or to lengthen the sight distance in a cut. Special efforts should be made to provide long grades and proper sight distances on curves.

When designing the grade line, consideration must be given to the grades of the side ditches. The depth and width of the ditch and its side slopes affect the width of the highway cross section and right-of-way requirements. Typical grading and paving sections, exemplified by Figure 13-5, show the lateral slopes, depths of ditches, and other details of the highway cross section. These standard sections become a part of the final design plans.

13-15 Grade Inspection of Preliminary Plans

When the grade line has been determined, final completion of the plans is delayed until a field inspection is made. In the process of preparing road plans, arrangements are usually made for two inspections in the field. The first inspection is for checking the partially prepared plans with field conditions and for supplementing the survey information with such additional data as are necessary for completing final plans and preparing a correct estimate of quantities. The field investigation can serve as a check for the general design and should reveal whether or not the width of right-of-way is sufficient for the proposed cross section; how drainage structures, ditches, and slopes are affected by field conditions; and other data that may be necessary for completing the plans. Reproductions are made of the partially completed plans for field inspection purposes.

The field investigator makes an inspection by walking the entire length of the project with a set of plans in hand and checking the various items of the plans, giving approval or making notations about recommendations for a change in design. When the inspection is completed, the plans that were used for the field investigation are returned to the designer. Where the surveys have been conducted by photogrammetric means, the amount of field inspection at this stage can be reduced by a restudy of the mosaics constructed from the aerial photographs.

A final inspection of the proposed project is made when the final plans have been completed. This inspection, called the "plans, specifications, and estimates" (PS & E) inspection, is made in order to verify the final design before the project is approved to be advertised for bids.

PREPARATION OF FINAL PLANS

After the field inspection is made with the partially completed "preliminary plans," the final design steps are carried out. These include (1) plotting the cross sections of the original ground and placing "templates" on original ground sections; (2) computing earthwork quantities; (3) preparing construction details for bridges, culverts, guardrails, and other items; (4) preparing summaries and estimates of quantities; and (5) preparing specifications for the materials and methods of construction.

13-16 Cross Sections and Templates

One important phase of design is determining the amount of earthwork necessary on a project. Earthwork includes the excavation of material and any hauling required for completing the embankment. Payment for earthwork is based on excavated quantities only and generally includes the cost of hauling the materials to the embankment. An additional item of payment called "over-haul" is often used to provide for hauling of the excavated material beyond a certain free-haul distance. The excavated material may be obtained from within the area of the highway cross section or from some distance outside the proposed highway. When excavation is obtained outside the proposed highway limits it is called "borrow excavation."

In order to determine earth excavation and embankment requirements by manual means, a section outline of the proposed highway, commonly referred to as a "template section," is placed on the original ground cross section; the areas in cut and the areas in fill are determined; and the volumes between the sections are computed. Figure 13-6 shows various conditions that may be encountered when plotting these template sections. "Cut" and "fill" are the terms that are usually used for the areas of the sections, and the terms "excavation" and "embankment" generally refer to volumes.

Cross sections are plotted on standard cross-section paper to a convenient scale. With the conventional U.S. system of units, a scale of 1 in. equals 5 ft vertically and horizontally was commonly used. A metric scale of 1:100 is recommended.

Each cross section should show the location or station of the original ground section and template section, the elevation of the proposed grade at that station, and the areas of cut and fill for each section. The computed volumes of excavation and embankment may also be placed on the cross-section sheet between two successive cross sections.

The areas of cut and fill may be measured by use of a "planimeter," a computation method using coordinates, or some other suitable method. The areas of cut and fill are usually recorded in square meters or square feet. These are then used to compute the volumes of excavation and embankment between the sections, from which a summation can be made of the total volumes of excavation and embankment for the entire project.

The average-end-area method, which has been used by a majority of state

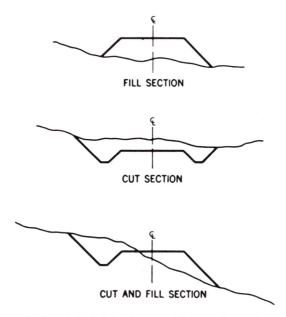

FIGURE 13-6 Original ground line and template sections.

highway agencies, is based on the formula for the volume of a right prism whose volume is equal to the average end area multiplied by the length, or

$$V = \frac{1/2\ L(A_1 + A_2)}{27} \tag{13-1}$$

where V = volume (cu yd)
A_1 and A_2 = area of end sections (ft²)
L = distance between end sections (ft)

In metric units, this equation becomes

$$V = 1/2\ L(A_1 + A_2) \tag{13-2}$$

where V = volume (m³)
A_1 and A_2 = area of end sections (m²)
L = distance between end sections (m)

Volumes computed by the average-end-area method are slightly in excess of those computed by a prismoidal formula when the sections are not right prisms. This error is small when the sections do not change rapidly, which is usually the case when a small degree of curvature is used. On sharp curves, however, a prismoidal correction should be applied.

When a section changes from a cut section to one of fill, a point is reached when a zero cut and zero fill occurs. This usually happens along an irregular line across the roadway. In order to properly compute the volumes at these locations, it is necessary to compute volumes at intermediate points. The usual practice is to determine end areas for three sections. One section would be at the centerline, where the cut and the fill equal zero. The other two sections would be at the right and left of the edge of the grade, where the change from cut to fill or fill to cut occurs.

Manual methods for computing earthwork have largely been supplanted by methods based on the use of computers. In typical cases and for earthwork quantities at the design stage, the computer input includes the cross-section data, centerline grades, and design template data with provision for superelevation, curve widening, and varying side slopes. The computer produces a tabulation of the cut and fill volumes at each station and the difference between them adjusted for shrinkage and swell, the cumulative volumes of cut and of fill, and mass diagram ordinates. Figure 13-7 shows such a tabulation.

13-17 Shrinkage

When earth is excavated and hauled to form an embankment, the freshly excavated material generally increases in volume. However, during the process of building the embankment it is compacted, so that the final volume is less than when in its original condition. This difference in volume is usually defined as "shrinkage." In estimating earthwork quantities, it is necessary to make allowance for this factor. The amount of shrinkage varies with the soil type and the depth of the fill. An allowance of 10 to 15 percent is frequently made for high fills and 20 to 25 percent for shallow fills. The shrinkage may be as high as 40 or 50 percent for some soils. This generally also allows for shrinkage due to loss of material in the hauling process and loss of material at the toe of the slope. In peat or muck areas, shrinkage should not include settlement of fills due to consolidation. A shrinkage factor of 0.8 is shown in Figure 13-7.

EARTHWORK QUANTITIES CALCULATION PROCESS
IDENTIFICATION   EARTHWORK QUANTITIES LIST FOR ROADWAY A

BASELINE STATION NUMBER	SHRINK/ SWELL FACTOR	STATION CUT (SQ-M)	STATION CUT (Cu-M)	ADJUSTED STATION CUT (Cu-M)	STATION FILL (SQ-M)	STATION FILL (Cu-M)	ADJUSTED STATION FILL (Cu-M)	MASS ORDINATE (Cu-M)
6 + 100	0.8	0.00	0	0	153.52	2932	2932	−76574
6 + 120	0.8	8.87	89	71	164.02	3175	3175	−79678
6 + 140	0.8	44.89	538	430	142.98	3069	3069	−82317
6 + 160	0.8	11.90	567	454	161.85	3048	3048	−84911
6 + 180	0.8	0.43	123	98	149.37	3111	3111	−87924
6 + 200	0.8	0.66	11	9	160.33	3096	3096	−91011
6 + 220	0.8	0.02	7	6	183.21	3434	3434	−94439
6 + 240	0.8	0.00	0	0	176.39	3595	3595	−98034
6 + 260	0.8	0.00	0	0	204.76	3810	3810	−101844
6 + 280	0.8	0.00	0	0	240.65	2627	2627	−104471
6 + 300	0.8	0.00	0	0	241.43	1976	1976	−106447
6 + 320	0.8	0.00	0	0	237.80	4791	4791	−111238
6 + 340	0.8	0.00	0	0	224.80	4625	4625	−115863
6 + 360	0.8	0.00	0	0	245.76	4704	4704	−120567
6 + 380	0.8	0.00	0	0	258.05	5037	5037	−125604
6 + 400	0.8	0.00	0	0	261.08	4670	4670	−130274
6 + 420	0.8	0.00	0	0	266.94	528	528	−130802
6 + 440	0.8	0.00	0	0	245.71	5125	5125	−135927
6 + 460	0.8	0.00	0	0	196.15	4417	4417	−140344
6 + 480	0.8	0.00	0	0	176.92	3730	3730	−144074
6 + 500	0.8	0.00	0	0	71.15	2480	2480	−146554
6 + 520	0.8	0.00	0	0	0.00	711	711	−147265

FIGURE 13-7 A tabulation of earthwork quantities.

When rock is excavated and placed in the embankment, the material will occupy a larger volume. This increase is called "swell" and may amount to 30 percent or more. The amount of swell is not important when small amounts of loose rock or boulders are placed in the embankment.

13-18 The Mass Diagram

A mass diagram is a graphical representation of the amount earthwork and embankment involved in a project and the manner in which the earth is to be moved. Its horizontal or x axis represents distance and is usually expressed in meters, feet, or stations. The vertical or y axis represents the cumulative quantity of earthwork in cubic meters or cubic yards. The quantity of excavation on the mass diagram is considered positive, and embankment is negative.

Figure 13-8 shows a hypothetical section of a mass diagram and the corresponding vertical profile plotted with the same x axis. From the figure, note that excavation sections are shown as positive slopes on the mass diagram and embankments as negative slopes. Points of zero cut/fill on the profile diagram are reflected as points of zero slope on the mass diagram. The depth of excavation or embankment on the vertical profile corresponds generally to the magnitude of the slope on the mass diagram.

The mass diagram allows a highway engineer to determine direction of haul and the quantity of earth taken from or hauled to any location. It shows "balance points," the stations between which the volume of excavation (after adjustment for "shrinkage" or "swell") and embankment are equal. Any line drawn parallel to the baseline and intersecting two points within the same curve will indicate a balance of cut and fill quantities between these two balance points. The final

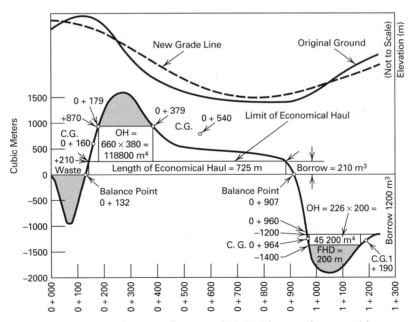

FIGURE 13-8 Example mass diagram. (NOTE: The numbers used here are for illustrative purposes only and may not be representative of values used in engineering practice.)

point on a mass diagram for a given project gives the overall net amount of earthwork for the entire project. This amount, if positive, would indicate a surplus of excavation material and a need to *waste* that quantity of material. If the final point on the mass diagram is a negative amount, it would indicate a net shortage of earthwork for the project and a need to *borrow* that quantity of earthwork material.

Some highway agencies take into account the need to haul earthwork materials especially long distances by paying contractors a premium for haul distances greater than a defined "free haul distance." In other words, the agency defines a certain free haul distance and agrees to pay the contractor on the basis of the quantity of excavation plus an amount for any overhaul. It is understood that the agreed-upon bid price for excavation includes the actual costs of excavation, the costs of hauling the material a distance not to exceed the free haul distance, and placing and compacting the earth in an embankment. For this earthwork, the unit of payment is commonly dollars per cubic meter (previously dollars per cubic yard). For any hauling beyond the free haul distance, the contractor is paid for the costs of excavation plus an amount for hauling the material any distance greater than the free haul distance. This length of haul beyond the free haul distance is called the "overhaul distance."

The free haul distance may be as low as 150 m (500 ft) and as high as 900 m (3,000 ft) or more. In fact, many state agencies no longer request earthwork bids on the basis of overhaul. Earthwork payment is based simply on the basis of the quantity of excavation, and it is understood that unit bid prices for earthwork in dollars per cubic meter or dollars per cubic yard include charges for the construction of embankments and all costs of hauling, whatever the haul length.

The overhaul distance is found from the mass diagram by determining the distance from the center of mass of the excavated material to the center of mass of the embankment. Under the conventional U.S. system of units, this distance is usually measured in 100-ft stations. Thus, a "cubic-yard-station" is the hauling of 1 cu yd of excavation one station beyond the free haul distance. In metric terms, the overhaul is the product of the amount of excavation hauled (in m^3) and the overhaul distance (in m). Thus, the product may be expressed in m^4 or some compound unit such as m^3-hectometer or m^3-kilometer.

Where it is necessary to haul material long distances, it is sometimes more economical to waste material excavated from the roadway section and borrow material from a borrow pit within the free-haul distance. The length of economical haul can be determined by equating the cost of the excavation plus the cost of the overhaul to the cost of excavation in the roadways plus the cost of excavation from the borrow pit.

If h is the length of haul in stations beyond the free-haul distance, e the cost of earth excavation, and o the cost of the overhaul, then to move 1 yd of material from cut to fill the cost will be $e + ho$, and the cost to excavate from cut, waste the material, borrow, and place 1 cu yd in the fill will be $2e$, assuming that the cost of excavation in the roadway is equal to the cost of the borrow excavation. Then

$$e + ho = 2e$$

and

$$h = \frac{e}{o} \text{ stations} \tag{13-3}$$

In metric units, if e is expressed in dollars per cubic meter and o is expressed in dollars per cubic meter-kilometer, h would be expressed in kilometers.

Columns 5 and 8 in Figure 13-7 show the cut volumes and the fill volumes, respectively, that have been adjusted for shrinkage. The cut is balanced with the fill, and the excess is added algebraically to the cumulative shown for the previous station. After the mass ordinates are computed, they are plotted to form a mass diagram.

EXAMPLE 13-1 **Mass Diagram** Consider the mass diagram shown as Figure 13-8.

Beginning at station 0 + 000 and ending at station 0 + 080, the curve of Figure 13-8 indicates that the embankment requirements exceed the available excavation; from station 0 + 080 to 0 + 132 the available excavation is greater than the embankment requirements. This can be restated to say that the embankment from 0 + 000 to 0 + 080 is constructed from the excavation between 0 + 080 and 0 + 132. The direction of haul is from the positive slope to the negative slope, in this case to the left. The earth quantities balance at station 0 + 132.

In this example, a free haul distance is assumed to be 200 m. Because the distance between the balance points 0 + 000 and 0 + 132 is less than the free haul distance, the earthwork between these stations is treated as free haul. A 200-m line is constructed at two additional sections on the diagram, illustrating free haul between the pairs of balance points: between stations 0 + 179 and 0 + 379, and between 0 + 970 and 1 + 170. Again, the direction of haul is from the positive to the negative slope, that is, between stations 0 + 179 and 0 + 379, it is to the right and between stations 0 + 970 and 1 + 170, it is to the left.

In this example, we have (arbitrarily) chosen a length of economical haul of 725 m. This indicates that it would not be economical to haul the excavation between stations 0 + 132 and 0 + 142 to the fill area between 0 + 867 and 0 + 907. It is more economical to waste the 210 m³ of excavation at one section of the project and borrow the same quantity at the other.

Continuing the analysis of the mass diagram, we are assuming that the contractor will be paid for overhaul. Perpendicular lines are drawn from the limits of economical haul line at stations 0 + 179 and 0 + 379; these vertical lines represent the volume ordinate +210 to +870, or 660 m³ of excavation between stations 0 + 142 and 0 + 179 that must be placed in the embankment between stations 0 + 379 and 0 + 867. The overhaul quantity (m⁴) is determined by multiplying the excavation, which in this case is 660 m³ by the distance between the center of mass of excavation and the center of mass of embankment, less 200 m free haul. A similar procedure is used to compute the overhaul quantity between the excavation from station 1 + 170 and 1 + 240 to the embankment from 0 + 960 and 0 + 964.

Because the last point on the mass diagram lies below the origin line, that quantity of borrow would be required to fill the embankment between stations 0 + 907 and 0 + 960.

We now briefly illustrate two methods that are commonly used to determine the center of mass of excavation and embankment when mass diagram calculations are done manually: (1) the graphical method and (2) the method of moments. Various computer programs have been written to perform these calculations using a similar approach.

The Graphical Method The volume ordinate line from $+210$ to $+870$ represents 660 m³. If a line is drawn bisecting this line and extended horizontally to intersect the curve between stations $0 + 142$ and $0 + 179$ and $0 + 379$ and $0 + 867$, the length of this line would represent approximately the average length of haul. This method is reasonably accurate when the volume curve has a fairly uniform slope but may be subject to intolerable error when the slope is not uniform.

The Method of Moments To illustrate the method of moments, we will compute the center of mass between stations $1 + 170$ and $1 + 240$. An approximate profile of this excavation is shown as Figure 13-9. Moments will be taken about station $1 + 170$. A moment is computed for each volume that makes up the total volume curve between stations $1 + 170$ and $1 + 240$. From Figure 13-9, it can be determined that the volumes and distance are as indicated below.

Station	Volume (m³)	Distance (m)	Product (m⁴)
1 + 170 to 1 + 180	58	5	290
1 + 180 to 1 + 200	87	20	1740
1 + 200 to 1 + 220	41	40	1640
1 + 220 to 1 + 240	3	60	180
Sums	189		3850

The distance from station $1 + 170$ to the centroid of the mass is

$$x = 3850/189 = 20.4 \text{ m, say 20 m}$$

The station of the center of mass is $1 + 170$ plus 20 m or $1 + 190$.

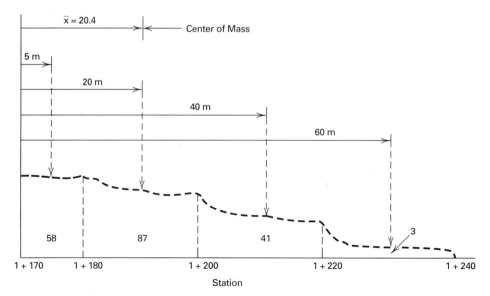

FIGURE 13-9 Illustration of method of moments.

13-19 Preparation of Construction Details

Earth excavation quantities, overhaul, and borrow requirements are generally the first items that are determined when drawing the plans for a project, and then the other details are prepared.

Many of the structures and miscellaneous items included in the project may be standard items for which designs have previously been approved and adopted for us. When such a design can be used, it is made a part of the plans by reference. If it cannot be used, the structure has to be designed to meet existing conditions.

13-20 Preparation of Quantities and Estimates

When the plans are completed, a summary is made of all the items of work, estimated quantities, and units of measurement. These are placed on one of the plan sheets. Some states provide regular quantity sheets for this purpose that show all items as per plan, with provision for inserting quantities as constructed. These sheets are useful for computing final quantities on the completed project. The summary of quantities may be arranged in groups such as grading items, drainage structures, surfacing items, and incidentals.

A summary of quantities is tabulated and an estimate made of the probable cost of the project, with an allowance for contingencies. An engineer's estimate is shown in Figure 13-10.

13-21 Preparation of Specifications

When the plans have been completed, a proposal and specifications are then prepared for the project. When this is done, the project is ready to be advertised for bids and contracts prepared for the construction of the project according to to the prepared plans. These matters are discussed in the next chapter.

13-22 Completed Final Plans

A complete set of plans will contain any number of sheets, depending on the length and complexity of the project. Most state highway departments specify a definite order in which the plan sheets are to be arranged, and this varies considerably among the states. A complete set of plans should contain the following:

1. *Title sheet.* The first sheet of a set of plans, which is known as the title sheet, shows a map of the area in which the project is located. The map should show the point of beginning and point of ending; the main topographic features; township, county, and state lines; and possible detour routes. The name and number of the project are also indicated. The index, legend, and tabulation of standard plans used are generally included in the title sheet. Provision is also made on this sheet for the affixing of signatures approving the plans that are contained in the completed set.

2. *Typical cross-section sheet.* The typical cross-section sheet shows all earthwork and finished road sections. These should include ditch sections and any treatment needed for the subgrade section.

3. *Note and summary sheet.* Notes or special provisions consist of special requirements, regulations, or directions prepared to cover the work that is not covered by the standard specifications. The summary of quantities

Summary of Quantities and Bid Unit Prices

STATE AID PROJ. NO. SAP-1135-A (2)
COUNTY HAMBLEN
LOCATED BETWEEN Ray Rd. and SR 61, 10.16 km grading and surfacing

Item	Approx. Quan.	Unit	Unit Price	Amount, $
Clearing and Grubbing	30.9	ha	LUMP	58,750
Other Clearing & Grubbing	3	ha	1,650	4,950
Unclass. Excav. & Bor.	106,000	m³	2.30	244,410
Spec. subgrade compaction	10.16	km	715.	7,262
Excav., culverts & struct.	382	m³	650.	2,483
Select. matl. for backfill	275	m³	12.	3,300
Class A concrete culverts	100	m³	420.	42,000
Bar reinforcement steel	4,800	kg	0.95	4,560
450 mm Class B conc. pipe	300	m	56.	16,800
600 mm Class B conc. pipe	15	m	92.	1,380
450 mm reinf. conc. pipe	80	m	70.	5,600
600 mm reinf. conc. pipe	410	m	120.	49,200
750 mm reinf. conc. pipe	30	m	140.	4,200
900 mm reinf. conc. pipe	58	m	200.	11,600
Culvert pipe removal	100	m	25.	2,500
Culvert pipe relaid	15	m	25.	375
Sand-cement bag rip rap	380	m²	45.	17,100
Concrete ROW markers	48	ea	35.	1,680
Sprigging	61,000	m²	0.20	12,200
Loose sod rip rap	375	m²	4.	1,500
Water for grassing	280	M. liters	5.	1,400
First app. fertilizer	9	m. tons	380.	3,450
Second app. fertilizer	980	kg	1.65	1,670
Removal exist. bridge	LUMP	LUMP	LUMP	1,000
RR warning signs, type F	2	ea	360.	720
RR stop signs, type F	2	ea	360.	720
150 mm aggreg. base course	65,800	m²	5.50	361,900
Cut-back asphalt, RC-3	218,800	liters	0.22	48,136
Aggregate for RR crossing	23	m. ton	20.	460
Bit. Mat. for RR crossing	1,500	liters	0.22	330
AC surface treatment stone	75,000	m²	0.65	48,750
AC for bituminous seal	75,000	m²	0.84	63,000
Finishing and dressing	55,000	m²	0.10	5,500
			Total	$1,028,886
			Contingency	102,889
			Grand total	$1,131,775

FIGURE 13-10 Hypothetical engineer's estimate.

is a tabulation of the items of work, quantities, and units of measurement for that particular project.

4. ***Plan and profile.*** The plan and profile sheets contain most of the details in regard to the work. These have been previously discussed. Their number depends on the length of the project.

5. ***Mass diagram.*** This sheet gives the contractor and the engineer a comprehensive graphic representation of the source and disposition of all earth moved on the project.

6. ***Drainage structures.*** All drainage structures are usually shown in detail. These will include detail drawings of culverts, inlets, manholes, and so on and will show the flow lines, sizes, and locations of the various structures. Bridge plans are generally not included in these drawings but in a separate set of plans.

7. ***Special details.*** Special details may include slab or box culverts, special intersection treatment, railroad crossings, or any special detail not covered by the standard plans or the specifications.

8. ***Utility sheet.*** On some projects, the number of utilities involved may be considerable, and a special sheet is devoted to their treatment.

9. ***Quantity sheets.*** These sheets contain all the items of work involved on a project. They have been previously discussed. They differ from the note and summary sheet by having more detail, such as stations for the location of culverts and so forth.

10. ***Cross sections.*** Prints are usually made of the cross sections used to determine earth excavation quantities and are made part of the plans. On projects of considerable length many of these sheets may be required. Many states do not attach these sheets to each set of plans but keep them at some central or division office where they are available to the contractor or others desiring this information.

When the plans are completed, a final inspection is made of PS & E (plans, surveys, and estimates), and when the necessary approvals are obtained, the project is ready for construction.

PROBLEMS

13-1 The 1200 m³ of borrow needed in Figure 13-8 can be secured from 190 m right of station 1 + 190. What is the overhaul?

13-2 Assuming that the free-haul distance is 1500 ft and the cost of excavation is $1.60/cu yd, what is the limit of economical haul when the unit price of overhaul is $0.17/cu-yd station?

13-3 Write a report on the role of global positioning system (GPS) on highway design. Describe in simple language how GPS works and its benefits and limitations.

13-4 Contact the state highway agency and request a summary of average bid unit prices for highway construction projects expressed in metric units. Report on the changes in the highway contracting process that resulted from metrication.

13-5 Given the following end areas in cut and fill, complete the earthwork calculations, using a shrinkage factor of 0.85. Prepare the mass diagram and indicate the direction of haul and whether you will have to borrow or waste any material in order to complete the project.

| | End Areas | | | End Areas | |
Station	Cut	Fill	Station	Cut	Fill
286	60	15	296	23	148
287	54	22	297	77	160
288	48	50	298	109	69
289	24	70	299	156	40
290	12	53	300	112	27
291	0	66	301	83	20
292	6	51	302	68	9
293	7	63	303	56	5
294	13	97	304	54	0
295	28	110	305	42	0

13-6 Given below are the ordinates (expressed in metric units) for the mass diagram for a highway project. No haulage is possible past station 0 + 780, the end of the project. The free haul distance is 240 m. Draw the mass diagram and (1) determine the volume of excavation and embankment; (2) calculate the quantities of overhaul (m⁴), waste, and borrow; and (3) show the directions of haul.

Station	Mass Ordinate (m^3)	Station	Mass Ordinate (m^3)
0 + 000	0	0 + 400	+2 793
0 + 020	−84	0 + 420	+3 521
0 + 040	−110	0 + 440	+4 375
0 + 060	−877	0 + 460	+4 979
0 + 080	−2 610	0 + 480	+5 743
0 + 100	−3 715	0 + 500	+6 555
0 + 120	−4 515	0 + 520	+7 322
0 + 140	−6 221	0 + 540	+10 102
0 + 160	−7 843	0 + 560	+12 882
0 + 180	−7 104	0 + 580	+14 763
0 + 200	−5 529	0 + 600	+17 842
0 + 220	−3 908	0 + 620	+20 922
0 + 240	−2 007	0 + 640	+22 401
0 + 260	−1 207	0 + 660	+24 000
0 + 280	−365	0 + 680	+20 667
0 + 300	−28	0 + 700	+17 337
0 + 320	0	0 + 720	+14 003
0 + 340	+422	0 + 740	+10 720
0 + 360	+1 127	0 + 760	+5 648
0 + 380	+1 977	0 + 780	+3 800

CHAPTER 14

CONTRACTS AND SUPERVISION

Practically all highway construction projects in the United States are public works, which are constructed with public funds. The agency authorizing this construction may be a federal, state, municipal, or county governmental unit, but the greatest number of highway construction projects today are authorized through the various state highway agencies. More than 95 percent of the construction done under state highway supervision is done by contract. The remaining 5 percent is done by the state's own forces organized and equipped to do this work. The contract system has proved beneficial, and it is very unlikely that it will be changed.

Under the contract system, plans and specifications are prepared and are submitted for competitive bids. As is usually the case when public funds are used, the methods and procedures for carrying out public works projects are prescribed by law in each state and must be strictly adhered to. Because of differences in climate, geology, geography, customs, statutes and regulations, highway specifications vary from state to state. Specification writers in the various states refer to a guide (1) published by the AASHTO, however, which provides uniformity of nomenclature, format, methods of measurement, basis of payment, and methods of tests. It is the intent to include here the procedure generally followed by most state highway agencies in preparing contractual documents and in supervising construction.

14-1 Unit Price System

Nearly all highway contracts are prepared on a unit price basis. The unit price contract includes an estimate of the number of units of each type of work and the price for each unit. The unit price contract has many advantages and is well adapted to highway construction, because usually it is not possible to determine exact quantities of some items of work when plans are prepared. Sometimes, after construction has started, unusual conditions are encountered. In either event, the various quantities can be adjusted, that is, they may be increased or decreased to the quantities actually constructed. Under the unit price system, each unit includes the materials that are to be furnished, the work to be done, the method of measurement of the work completed, and the basis on which payment is to be made. These units are described in detail in the specifications.

14-2 Lump-Sum Bids and Force-Account Work

It is not always feasible to have a breakdown of all the work on a unit price basis. It may be necessary to have minor items done on a lump-sum basis, especially when some piece of work is too intangible for unit prices to be computed. Some highway organizations now use lump-sum bidding to cover items that traditionally have been bid on a unit price basis. These items include clearing and grubbing, structural excavation, structural concrete, and the like, where estimated and final quantities can be expected to be essentially the same. In some cases, it is preferred to have this type of work done on a force-account basis, that is, the work may be done by the contractor, who is reimbursed on a "cost-plus" basis, with the work paid for at actual cost plus a percentage of the actual cost as the contractor's profit. By "actual cost" is meant the cost of labor and materials actually required on that particular operation. The percentage of profit allowed varies. The contractor is compensated for the use of equipment on a rental basis, generally at an agreed-on daily or hourly rate. Small tools are generally excluded in the cost of force-account work.

When large sums are involved, force-account work of this nature requires good supervision. Good practice dictates a close auditing of all work and materials. The audit should include certified copies of payrolls, original receipted bills of materials, and evidence of all other charges, such as freight bills. All force-account work should be checked and approved by both parties daily.

It is possible to have a contract in which the work to be done will include unit price items, a lump-sum bid item, and even provisions for a force-account item.

CONTRACTUAL DOCUMENTS

In order to put a highway construction project under contract, it is necessary to take various steps and to prepare certain documents. The contractual documents generally include (1) the advertisement, (2) instructions to bidders, (3) the proposal, (4) the agreement or contract, (5) plans, (6) the specifications, (7) bonds, and (8) all written or printed agreements and instructions pertaining to the method and manner of furnishing materials and performing the work under the contract.

14-3 The Advertisment

The advertisement or notice to contractors is a written notice to inform prospective bidders that bids are requested for certain types of work and that a contract may be entered into with the successful bidder. The advertisement is usually inserted in a newspaper or magazine that serves as a medium for supplying such information to persons interested in this particular type of construction. Advertisements or invitations for bids are also mailed to pre-qualified contractors. Highway departments are required by law to provide ample time between publication of the notice to contractors and the taking of bids. This period may vary from two to several weeks, and it should allow contractors sufficient time to prepare their estimates, obtain prices for materials, and make essential arrangements necessary for filing a proper bid. On large or complex projects, arrangements often are made for a "prebid conference," a meeting at which details of the work are discussed with prospective bidders.

The advertisement should be brief and still include the information needed by the prospective bidder. The information usually includes the following:

1. A brief description of the type of work and the location of the project.
2. The date, time, and place where bids will be received.

3. Where plans, proposals, and specifications may be seen or obtained and the amount of fee or deposit required if any.
4. Approximate quantities of the work involved.
5. Amount of bid bond, certified check, surety, or security required and provisions for its return to the unsuccessful bidder.
6. Any restrictions related to the qualifications of bidders.
7. Reservations for the rejection of any or all bids or waiver in bidding formalities.
8. A statement regarding minimum or prevailing wage rates and other conditions and regulations pertaining to labor.

14-4 Instructions to Bidders

Some state highway departments issue additional "instructions to bidders," a document in which more information is given about the procedure to be followed in the preparation and submission of bids, details of any formalities required in the acceptance or rejection of bids, and miscellaneous instructions related to the work.

This particular instrument is used to expand the information given in the advertisement, and it is sometimes used instead of an advertisement. When this is done, it is necessary to have all the information given in the advertisement contained in the instructions to bidders. The instructions to bidders may also include requirements for the submission of the bidder's experience record or prequalification data, the time for the starting and completion of the work, and a reference to any legal considerations in regard to the work.

It may be pointed out here that many highway departments omit the instructions to bidders as a separate document but include this information in other contractual documents such as the advertisement, the proposal, and their standard specifications that have been previously adopted and that satisfy any legal requirements included in the instructions to bidders.

14-5 The Proposal

The form on which the bidder submits a bid is called the "proposal" and is prepared by the highway or transportation department. The proposal is arranged so that all the items of work and the quantities are tabulated, with spaces for the bidder to insert prices and extensions for each item and a space for the total amount for all items of work. The proposal also contains the statement that the bidder has carefully examined the plans and specifications and is fully informed as to the nature of the work and the conditions related to its performance. The proposal also states that the bidder agrees to do the work in the time proposed and agrees to forfeit his "proposal guarantee" in case his bid is accepted and he fails to enter into a contract. The bidder also agrees to furnish the necessary bonds or security for carrying out the work.

The amount of the proposal guarantee (certified check, bid bond, or other security) required for submission with the proposal and the manner in which it is to be returned are also specified in the proposal. Space may also be provided for the listing of any subcontractors the bidder may require, for the proper signatures of the bidders, and for the address of their place of business.

A proposal generally includes the specifications and plans. As previously mentioned, all state highway agencies have prepared and adopted standard specifications for road and bridge construction. The standard specifications are generally in book form and include general requirements and covenants, items of work, materials, methods of measurement, and payment of quantities. The standard specifications are always referred to in the other contract documents and in this case become a part of the proposal by reference. Due to certain conditions in the design or construction of a particular project, it is often necessary to supplement existing specifications as well as to add certain special provisions necessary for the project. It is quite convenient to add the supplemental specifications and any special provisions to the proposal. This gives the bidder the information needed in order to prepare the bid.

14-6 The Agreement or Contract

The agreement is the formal part of the contract documents, and from a legal point of view it is the strongest of them all. The advertisement, instructions to bidders, and the proposal are generally classed as contract documents; but they are not, strictly speaking, parts of a contract but preliminary to it. If any statements of contractual importance are included in any of these documents, they should be made a part of the agreement or contract by reference or they must be repeated in the formal contract.

The contract must be in accord with the form prescribed by law and must contain the following:

1. The declaration of the agreement, names of the parties, their legal residence, and the date of execution of the agreement.
2. The consideration for work to be done, the items of work, quantities, and unit prices with reference to the plans, specifications, and proposal.
3. The time for starting and the date for completion of the work and provisions for damages, if any.
4. Signatures and witnesses.

14-7 Plans

Plans are approved drawings or reproductions of drawings pertaining to the work covered by the contract. They are usually complete in all details before construction is started. The plans, together with the specifications, are an essential part of the contract documents.

14-8 Specifications

The specifications are the written instructions that accompany or supplement the plans and form a guide for standards required in the prosecution of the work. These standard specifications, which are the result of experience and knowledge acquired over a period of years, cover the quality of materials, workmanship, and other technical data. Standard specifications are often revised from time to time in order to keep up with improved processes and the use of different materials. It is sometimes necessary to deviate from the standard specifications, and this is generally done by adding supplemental specifications either to the plans or to the proposal.

14-9 Bonds

All contracts require some form of security as a guaranty of faithful performance of the work. This security may be in the form of collateral deposited with the contracting agency by the contractor or it may be in the form of surety bonds.

Private individuals may serve as sureties, but as a general rule surety bonds are issued by bonding companies. Bonds issued in connection with highway construction projects are of three common types: (1) bid bonds, (2) performance bonds, and (3) lien bonds.

The bid bond, previously mentioned in connection with the advertisement and proposal, is one form of the proposal guaranty required when the contractor submits a bid. This is the contractor's guaranty that he will enter into a contract if his bid is accepted. In the event that the contractor fails to enter into a contract if his bid is accepted, the surety or bonding company is required to pay the owner damages in the amount specified in the bond. Bid bonds generally range from 5 to 10 percent of the bid amount or 5 to 10 percent of the estimated cost of the project, based on the engineer's estimate prepared for the owner.

Performance bonds, as the name implies, guarantee that the work will be completed as required by the state.

Lien bonds, or "payment bonds," are used by some organizations and guarantee that the contractor will pay all obligations for labor and materials incurred in the performance of the contract.

The amount of bonds required is generally 100 percent of the bid or contract price. Some states require performance bonds only in the amount of 100 percent of the bid price, with provisions in the bond for the payment of all bills and protection against all liens. Some states require bonds in the amount of 100 percent of the bid or contract price, but make a distribution of 50 percent for performance bonds and 50 percent for lien or payment bonds. Some states require both, each in the amount of 100 percent of the contract price.

Surety bonds are generally furnished before or at the time the contract is executed. When bonds are executed, they require the signatures of the three parties: (1) the owner, (2) the contractor, and (3) the surety.

Surety bonds are contracts between the owner and the surety. Since most highway construction contracts provide for changes and modifications in the contract, it is essential that the surety be informed of any modification of the contract between the owner and the contractor. In case the contractor fails to fulfill the terms of the contract, it becomes the responsibility of the bonding company to do so. This is usually done by letting the contractor complete the work with his own forces under the direction of an engineer supplied by the surety. This method has been found to be of maximum benefit to the owner and to the contractor in that the work is completed with a minimum of delay and expense to the owner and at the same time the loss entailed by the contractor is lessened. If this procedure is not followed, the bonding company may let a new contract for the remainder of the work or may elect to pay the penalty and let the owner complete the work, provided that the owner wishes to do so.

14-10 Supplemental Agreements

After the contract has been executed and construction work has been started, it is often necessary to make changes in the contract. Conditions commonly arising that require supplemental agreements are the time necessary to complete the work, changes in the quantities of work involved, and the addition of extra work not provided for in the original contract. The standard specifications usually provide for these conditions by granting an extension of time and authorizing any changes or extra work necessary to complete the project. All supplemental agreements must conform to the same basic requirements as the original contract with respect to competent parties, monetary consideration, legality of subject matter, and mutual agreement. All supplemental agreements should be in writing. In addition, unforeseen conditions may force an adjustment in the contract to accommodate these "changed conditions" (see Section 14-14).

14-11 Extension of Time

The contract usually designates the time when work is to begin and when it shall be completed. The time to begin work may be designated as a certain number of days after the contract has been executed. The completion date may be a specific calendar date or it may be a specific number of calendar days in which the contractor is to complete the work. In most states it is understood and agreed that time is of the essence of the contract, and failure to complete the contract on time may make the contractor liable for damages. These damages may be a definite amount per day for each day beyond the time required for completion of the work. These are known as "liquidated" damages. In lieu of liquidated damages, the owner may charge actual damages incurred, such as for engineering costs and supervision, maintenance of necessary detours, or any other direct charges caused by the delay. Commonly, standard specifications of state transportation agencies set forth a schedule of liquidated damages for delays in completion of contract work. For example, Kentucky specifies liquidated damages ranging from $60 per calendar day for contracts up to $25,000 to $1,600 per calender day for contracts of $10 million or more (2).

From a legal point of view, no work can be done beyond the date of completion, as listed in the contract. It is necessary to make a new agreement or extend the original agreement by requesting an extension of time to complete the contract. This should be done within a reasonable time before the completion date of the contract arrives. If the contractor requests an extension of time due to extra work or other extenuating circumstances beyond his control, such time should be extended without penalty. The agreement extending the time is also a formal document and requires all the necessary elements of a legal contract. When properly executed, it is as binding as the original contract.

14-12 Authorizations for Changes in the Work

It is impossible to determine exact quantities for some items of work (e.g., cubic meters of excavation) when designing a project. When a contractor bids on the various unit price items, as set up in the proposal, he is reasonably certain that the final quantities may be increased or decreased from the bid amount. To make the necessary adjustment from the quantities as stated in the contract to the quantities as actually constructed, written authorizations are made in accordance with methods of procedure set forth in the specifications.

On major items where large quantities are involved, some specifications provide for an increase of the unit price if the quantities are drastically reduced, and, conversely, if the quantities are materially increased, the unit price may be decreased. Some states require an adjustment of the unit price when the quantities vary more than 20 or 25 percent from estimated quantities and when the value of the work exceeds a certain sum.

14-13 Authorization for Extra Work

Some states make a distinction between a change authorization and an extra authorization. A "change authorization" may be defined as one in which there is a change in the quantities at the unit price as listed in the contract, while an "extra authorization" is for extra work for which no unit price is listed. The unit price for extra work is usually agreed on at the time the extra work is discussed, and it should always be approved before any extra work is started.

14-14 Changed Conditions

Highway contracts may include a "changed conditions clause" to allow the alteration of plans or character of work provided such alteration falls within the general scope of the contract. For example, specifications with such a clause would allow the agency to make changes in alignment or grade of the road or structure. Unless the alterations materially change the character or cost of the work to be performed, the altered work is paid for at the same unit prices as other parts of the work. If the character or unit costs of the work are materially changed, the specifications may permit the engineer to make an allowance in pay for the changed conditions and to extend the contract time

A typically changed conditions clause, taken from the standard specifications of the Alabama Highway Department, follows:

During the progress of the work, if subsurface or latent physical conditions are encountered at the site differing materially from those indicated in the contract or if unknown physical conditions of an unusual nature, differing materially from those ordinarily encountered and generally recognized as inherent in the work provided for in the contract, are encountered at the site, the party discovering such conditions shall promptly notify the other party in writing of the specific differing conditions before they are disturbed and before the affected work is performed.

Upon written notification, the Engineer will investigate the conditions, and if he determines that the conditions materially differ and cause an increase or decrease in the cost or time required for the performance of any work under the contract, an adjustment, excluding loss of anticipated profits, will be made and the contract modified in writing accordingly. The Engineer will notify the contractor of his determination whether or not an adjustment of the contract is warranted.

No contract adjustment which results in a benefit to the contractor will be allowed unless the contractor has provided the required written notice.

The Contractor shall carry on promptly and diligently the work involving a claim, pending a decision.

Any adjustment in compensation because of such change or changes will be made in accordance with the provisions of Article 109.05. Any adjustment in contract time because of such change or changes will be made in accordance with the provisions of Article 108.09. (3)

14-15 Subcontract

The use of a subcontract in highway work is often necessary. A subcontract may be defined as a contract for the performance of a part or all the work previously contracted for. When this is done, the subcontract should have all the elements of a formal contract. The scope of the work to be performed, the compensation to be received, and the terms and conditions of the original contract with which the subcontractor must comply should be clearly stated. The owner holds the original contractor directly responsible for the work of the subcontractor. The subcontractor, however, has the legal right to hold the owner responsible for any default of payment due him for labor or supplies by the contractor. The owner usually protects himself by having the contractor furnish the necessary bonds that guarantee payment of all labor and materials.

On state highway contracts the amount of work that a contractor may sublet is usually limited; in some cases, it is up to 50 percent of the value of the work contracted for. State highway and transportation departments usually reserve the right to approve or disapprove a subcontract, and they may require information in regard to the qualifications of the subcontractor and his ability to do the work. Copies of all subcontracts, including prices, are usually required before a subcontract may be approved.

14-16 Prequalification of Contractors

On most public works, contracts are awarded to the lowest responsible bidder and it is the problem of the state agency to determine the responsibility of the lowest bidder after the bids are taken. This is necessary because contractors with limited financial resources and limited experience may be competing with reliable contractors for an award of a contract. Such an award to an unqualified contractor often results in unnecessary delay, unsatisfactory work, and sometimes in failure to complete the work, with the corresponding extra expense to the state agency requesting bids. In order to avoid or reduce these dangers, many states have adopted prequalification laws and regulations. The fundamental purpose of the prequalification requirements is to determine beforehand whether or not the contractor is responsible and competent to bid on a project and to permit only those properly qualified to do so.

A typical procedure followed in prequalifying contractors requires the prospective bidder to submit a comprehensive experience and financial statement. From the information thus supplied, the prospective bidder is given a numerical rating that indicates the maximum amount of work in dollar volume he is permitted to have at any one time and further classifies the various types of work on which he is permitted to bid. The major rating factors for determining a prospective bidder's qualifications are his experience record, the amount of equipment he owns, and his financial resources at the time of filing his statement.

The numerical rating and classification required for each project may be given in the advertisement. Prequalification of contractors is highly desirable in that there is more time to investigate the bidders' qualifications beforehand, saving the time and expense of contractors who are not qualified to do the work. The legality of prequalification has been established by the rulings of the various courts. Opponents of prequalification object to the lengthy questionnaires and the time and expense of preparing statements, and they state that it restricts competition to those with large resources. Additional objection to prequalification based on equipment owned is raised when suitable equipment may often be rented.

14-17 State and Federal Agreements

On state projects in which the federal government is to participate in the cost, certain conditions and agreements are necessary to qualify for these funds. In the apportionment of these funds for highway purposes, each state is required to submit for approval a program of projects to be constructed on a predetermined system of federal-aid highways, the total cost of which must not exceed that state's apportionment. Plans and specifications are then prepared and a detailed estimate based on current prices is submitted to the Federal Highway Administration for approval. Approval of this agency is required before the project is advertised and before the award is made to the low bidder. After the award is made by the state to the contractor, a detailed estimate is made using contractor's unit bid prices, and it is upon this estimate that the "project agreement" is to be based. The contractor's bid, plus a contingency not to exceed 10 percent of the bid, is usually the amount of the agreement. This project agreement is the formal document or contract between the state and the federal government. There is one other formal agreement that may be mentioned here and that is the "maintenance agreement." In this document the state agency agrees to the maintenance and upkeep of the project after it is constructed. The maintenance agreement is usually made after plans are completed for the particular project and before federal approval is given to advertise the project.

14-18 Mechanics of Bidding Procedure

When plans, specifications, and estimates are completed and the necessary approvals are obtained, the project is ready for advertising. It is always preferable to have several projects ready for advertising at the same time. The advertisements are written with the required information previously discussed, and the date is set in accordance with statute requirements. Plans and proposals are then requested by those desiring to submit bids on the projects advertised. Bids may be mailed or submitted in person at the designated time and place. At large lettings, when there are many projects and bidders, much detailed work is necessary in order to check the many details that arise when bids are received. The sealed bids are opened, the bid security is checked, and the proposed details are scrutinized for irregularities. The total amount of the bid is then read aloud. Any irregularities found before the reading of the bid are usually a cause for rejection. For example, when bid bonds are requested to accompany the proposal, the bidder may neglect to submit his bid bond or it may not be in the amount requested. The omission of a unit price may also be a cause for rejection. Rejection at this point is desirable in that the bid security can be returned without undue delay.

After all bids are read aloud and the apparent low bidders are determined, all bids are checked for errors. The bids are then tabulated in the order of first, second, third, and so forth. After the low bidder is determined on each project, the detailed unit prices of the successful bidder are then read aloud. Computers are also being used successfully in this phase of the work of the state highway departments, with great savings in time and manpower. The contract is then prepared, bonds are furnished by the successful bidder, and an award is made within a reasonable time. On federal-aid projects concurrence in the award of a contract to a contractor must also be received from the federal agency. When the contractor furnishes the necessary bond, arrangements are made for the return of the other bidders' security. A majority of states return the bid security of all bidders except the lowest two or three immediately after a letting. Up to 30 days is sometimes required after the low bidder has been determined in order to make an award of a contract.

14-19 The Unbalanced Bid

On unit price contracts it is expected that each bid item will carry its proportionate share of the cost as well as the profit. It is possible for a contractor to raise certain unit prices on some items and lower unit prices on other items without changing the total of his bid. The result is usually an unbalanced bid. Generally speaking, unbalanced bids are undesirable and, when detected, are usually a cause of rejection. An example of an unbalanced bid may be shown in the bidding of the following items:

	Balanced		Unbalanced	
	Unit Price	*Amount*	*Unit Price*	*Amount*
9,000 m³ earth excavation	$ 6.00	$54,000	$ 5.00	$45,000
400 m³ rock excavation	$10.00	$ 4,000	$32.50	$13,000
Total cost		$58,000		$58,000

Assuming that when final quantities were determined the amount of earth excavation was 7,400 m³ and the rock excavation was 2,000 m³ the final cost would be as follows:

	Balanced		Unbalanced	
	Unit Price	*Amount*	*Unit Price*	*Amount*
7,400 m³ earth excavation	$ 6.00	$44,400	$ 5.00	$ 37,000
2,000 m³ rock excavation	$10.00	$20,000	$32.50	$ 65,000
Total cost		$64,400		$102,000

14-20 Right-of-Way Agreements

Contracts between a state highway agency and a property owner are often necessary in order to obtain the right-of-way for highway construction purposes. These contracts may be for the outright purchase of the land needed or as an easement for necessary grading operations. Most of these agreements are negotiated agreements between the property owner and a representative of the state highway agency.

SUPERVISION OF CONSTRUCTION

After a contract is awarded, arrangements are made by the contractor to move equipment and personnel to the job site in order to start construction operations. The time it takes to do this will depend on the distance involved. The contractor must comply with contract requirements, however, which usually state that construction work must start within so many days of the award of the contract. The manner in which supervision of construction is carried out is described in the following sections and is considered representative of the general practice of many highway agencies.

14-21 Project Supervision

The work of the project conducted under the supervision of a state highway or transportation department is usually under the administration of the project engineer. It is the engineer's duty to see that all phases of the work are carried out in accordance with plans and specifications. The engineer receives instructions from, and is usually responsible to, a district or division engineer. The project engineer is assisted by a number of trained persons, including a survey crew, inspectors, office engineer, and others. The number of people obviously depends on the length, type, and complexity of the project.

On certain federal-aid projects, Federal Highway Administration engineers make periodic inspections of the work as it progresses, as well as a final inspection. On projects performed under "certification acceptance" procedures (see Section 2-3), routine inspections and audits are performed by state personnel, but Federal Highway Administration engineers make a final inspection of the work. Their approval is required for acceptance of all federal-aid highway projects.

14-22 Duties of the Project Engineer

It is necessary that the project engineer secure a high standard of work, and this can be done only by observing whether the contractor complies with the contract and specifications. This function is usually delegated to the various inspectors on the project who are responsible to the project engineer.

In recent years, highway engineers have come to believe that absolute compliance with typical plans and specifications is neither desirable nor necessary in this type of work. Many state highway agencies are including in their specifications a "substantial compliance" clause such as the following from the State of Georgia's specifications:

> **CONFORMITY WITH PLANS AND SPECIFICATIONS:** All work performed and all materials furnished shall be in reasonably close conformity with the lines, grades, cross sections, dimensions and material requirements, including tolerances, shown on the Plans or indicated in the Specifications.
>
> Plan dimensions and contract Specification values are to be considered as the target values to be strived for and complied with as the design values from which any deviations are allowed. It is the intent of the Specifications that the materials and workmanship shall be uniform in character and shall conform as nearly as realistically possible to the prescribed target value or to the middle portion of the tolerance range. The purpose of the tolerance range is to accommodate occasional minor variations from the median zone that are unavoidable for practical reasons. When either a maximum and minimum value or both are specified, the production and processing of the material and the performance of the work shall be so controlled that material or work will not be preponderantly of borderline quality or dimension.
>
> In the event the Engineer finds the materials or the finished product in which the materials are used not within reasonably close conformity with the Plans and Specifications, but that reasonably acceptable work has been produced, he shall then make a determination if the work shall be accepted and remain in place. In this event, the Engineer will document the basis of acceptance by Contract modification which will

provide for an appropriate adjustment in the Contract Price for such work or materials as he deems necessary to conform to his determination based on engineering judgment.

In the event the Engineer finds the materials or the finished product in which the materials are used or the work performed are not in reasonably close conformity with the Plans and Specifications, and have resulted in an inferior or unsatisfactory product, the work or materials shall be removed and replaced or otherwise corrected by and at the expense of the Contractor. (*4*)

It is the project engineer's duty to see that all engineering work, such as the setting of stakes for alignment and grade, is performed well in advance of the contractor's work so that there will be no delay. It may be necessary to effect a change in design during the progress of the work, and each state defines the procedure to be followed in such matters.

The necessary reports required by the state highway or transportation department and the Federal Highway Administration are the responsibility of the project engineer. These reports include the preparation of estimates for partial payments and also for final payment.

As a representative of the highway department on a project, the project engineer must keep others properly informed as to the progress of the work; for instance, municipalities and utility companies must frequently be contacted on most construction projects. The proper authority should be informed of any additional borrow requirements or on matters involving disputes in regard to the right-of-way. Good administration of a project usually results in a well-constructed project.

14-23 Project Engineer–Contractor Relationship

The highest standards of work are developed when there is complete cooperation and accord between the contractor's employees and those of the state highway agency. It must be remembered that the contractor is entitled to a fair profit and is due every consideration consistent with the specification requirements and quality of work demanded of him.

In case of a claim by the contractor, a careful record should be made of all work in question and the information made available to those responsible for the adjustment of claims.

The contractor should make available at all times any information that may be required, such as transcripts of payrolls and receipted invoices. The project engineer should see that the contractor complies with all regulations regarding labor, equipment, and so on.

14-24 Construction and Inspection Details

All materials used on project must be inspected for compliance with applicable specifications. These materials may be inspected at the site of the project, or if quantities are sufficient to warrant it, they may be tested and approved before shipment is made to the project. When this is done, the testing is frequently done by an approved commercial testing laboratory. This does not mean that materials may not be rejected if they become damaged during shipment and cannot comply with specifications or if a defect was not detected in the inspection at the source.

Most state highway agencies maintain well-equipped testing laboratories of their own. The contractor is required to submit samples of the materials to be used for testing purposes and to name the sources of the materials. For concrete and bituminous paving mixtures, proportioning of the various ingredients is usually under laboratory control and the various details are carried out by inspectors assigned to the project.

The number and kind of inspectors needed for each project will depend on the kind and type of work to be done. The project may require the services of a grade inspector, a culvert inspector, a gravel plant inspector, a plant inspector for the preparation of various types of plant mixtures, and others. The inspector not only inspects the materials but also exercises control over the manner in which the work is done. The importance of the work of the inspector should never be underrated.

14-25 Quality Control of Materials and Construction

Quality control of materials and workmanship by tests on small samples has been the traditional practice in highway construction. Interpreting the results of such tests has been complicated by the wide variability of the test results, however. Certain variations in the quality of a highway component obey the laws of probability and occur strictly by chance. Other variations are systematic and can be attributed to assignable causes such as differences in equipment, materials, construction methods, testing conditions, and so forth.

State highway agencies are increasingly turning to the science of statistics to understand the variability of quality control tests. Statistical concepts can be used to establish specifications in terms of maximum tolerances from an established "target" value. Research studies have shown that quality control test data tend to be grouped symmetrically about the mean value, and distribution of the data can generally be satisfactorily described by the familiar bell-shaped normal curve (Fig. 14-1).

A graphical procedure employing a "control chart" is recommended for statistical quality control of highway construction. To construct a control chart, samples of some highway characteristic are randomly taken and tested. Examples of test results include subgrade densities, the percentages of asphalt in pavement mix samples, the fractions of aggregate passing a certain sieve, and so on. As Figure 14-2 illustrates, the test values are plotted as ordinates along a horizontal scale showing time or unit of production.

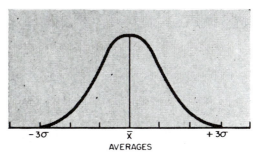

FIGURE 14-1 Distribution of chance variations in a test sample measure of quality. (SOURCE: *Public Roads,* Vol. 35, No. 11, December 1969.)

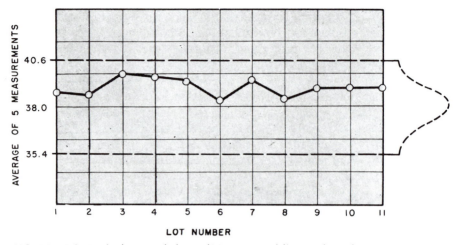

FIGURE 14-2 Typical control chart. (SOURCE: *Public Roads,* Vol. 35, No. 11, December 1969.)

When charts of individual test values are prepared, upper and lower control limits are usually established 3 standard deviations* away from a central or target value $\overline{X}$.

$$\text{Upper control limit, UCL} = \overline{X} + 3\sigma \qquad (14\text{-}1)$$

$$\text{Lower control limit, LCL} = \overline{X} - 3\sigma \qquad (14\text{-}2)$$

where σ is the standard deviation. When averages of n samples per lot are used as the control statistic, the control limits are

$$\text{UCL} = \overline{X} + \frac{3\sigma}{\sqrt{n}} \qquad (14\text{-}3)$$

$$\text{LCL} = \overline{X} - \frac{3\sigma}{\sqrt{n}} \qquad (14\text{-}4)$$

Consider the example illustrated by Figure 14-2. The test characteristic is the percentage of aggregate in a bituminous mix passing the No. 30 sieve, and the mean or target value is 38.0 percent. If the standard deviation $\sigma = 1.93$ and $n = 5$,

$$\text{UCL} = 38.0 + \frac{3 \times 1.93}{\sqrt{5}} = 40.6$$

$$\text{LCL} = 38.0 - \frac{3 \times 1.93}{\sqrt{5}} = 35.4$$

As long as the plotted test values fall within the control limits and are randomly distributed about the average, the process is said to be in control. If the

---

*Some engineers prefer a control limit of 2.33 standard deviations, which corresponds to a 98 percent probability interval.

points fall outside the control limits or are consistently above or below the average line, the reasons for the irregularities should be investigated. Granley (5) has published the following warning signals to indicate processes that are out of control:

Individual charts
   1   test value greater than 2.33 standard deviations from centerline
   3   consecutive test values greater than 1 standard deviation above or below the centerline
  11   consecutive test values on same side of centerline job mix target

Average and range charts
   1   average or range value outside control limits lines on R and $\overline{X}$ chart.
   2   consecutive averages outside 2σ limits above or below the centerline
   7   consecutive values on either side of centerline for $\overline{X}$ charts or above centerline for R charts

---

Control charts can be constructed from test data in the form of individual measurements, averages, standard deviations, and ranges (R charts), or some combination of these. Some states employ the moving average of the five most recent individual test results as the control statistic.

> Diagnosis of the reason for the process being out of statistical control is an engineering consideration. The chart can only point out that something is wrong. . . . The use of control charts will not alter the fact that the engineer has the final decision in accepting construction. The charts, however, are effective tools to visually forewarn that undesirable trends may be developing and to help the engineer and contractor decide when to take action and when not to take action. (5)

Additional information on the important topic of statistical quality control of highway construction is contained in Refs. 6 and 7.

14-26 Construction Surveys

The project engineer and his party are usually assigned to a construction project at about the same time that the plans and specifications are completed and when it is reasonably certain that bids will be received within a short time. This permits the construction party to do necessary engineering work before the contractor arrives.

The first problem in construction surveys is to check the alignment to see if it is in accord with the final plans. In many cases this means reestablishing the centerline of the entire project. It is also necessary to check all witness points of alignment. If there is a possibility that these witness points may be in the way of construction, they should be reestablished beyond construction limits. Consideration should also be given to the establishment of additional reference points on the road centerline to facilitate the instrument work when setting stakes for alignment and grade.

All bench marks should also be checked for location and elevation. Any benches that have been damaged or destroyed should be reestablished. All benches should be outside the limits of construction.

14-27 Staking Requirements

After the road centerline and the bench marks have been checked or reestablished, the major portion of the survey work consists of setting stakes for the guidance and control of construction operations. All stakes that are set by the engineer should be marked in such a manner that they can be easily interpreted.

The first stakes that are set are usually clearing and grubbing stakes. The distances out from the road centerline are generally determined by using the plans and scaling the distances to the limits of construction. The stakes may consist of small saplings cut to 1.2- or 1.5-m (4- to 5-ft) lengths, with a blaze mark facing the road centerline. These stakes are usually placed at even 30-m (100-ft) intervals.

14-28 Cross Sections and Slope Stakes

The construction survey party is responsible for all earthwork calculations. It is therefore important that they take cross sections over the entire project in many cases. Such a procedure may not be necessary in cases where cross sections (and slope stake positions) are determined from aerial photographs and by the use of computers, as discussed in Chapter 13. The original cross sections provide a record of original ground surface elevations that can never be obtained again after construction has started. The cross sections taken on the project that served as a basis for the final design may not be suitable. The period of time between the taking of the cross sections for design purposes and letting a contract for construction may vary from several weeks to two years or more. Taking cross sections just before construction operations begin eliminates one of the many possible disputes over earthwork quantities.

A slope stake is set at the point where the proposed fill slope or cut slope intersects the original ground. The stakes may be set when cross sections of the original ground are taken. Slope stakes for grading are usually set on both sides of the road at 15-m (50-ft) intervals at the station and half-station points. All information necessary for grading operations should be shown on the stake. It is sometimes necessary to offset slope stakes. When this is done, the amount of offset, which is the distance from the offset location of the stake to its intended location, is noted on the stake. The cut or fill reading is taken from the ground location of the stake, as set. Figure 14-3 shows typical cross sections and positions of slope stakes.

When setting slope stakes, the cut or fill is usually marked on the centerline stakes to serve as an aid to grading operations. Many states set slope stakes in heavy cuts only and control grading operations in other areas by the use of earth-grade stakes.

14-29 Earth-Grade Stakes

When the rough grading operations have been completed, it is generally necessary to set finishing grade stakes to control the finishing of the earth grade. These stakes are generally set on the shoulder line at every station and half station. Some operators require that the finish grade stakes be set on the road centerline instead of the shoulders. The grade stakes may be set in such a manner that the top of the stake is at grade. The stake may be left high and the amount of cut or fill noted from a keel mark on the stake.

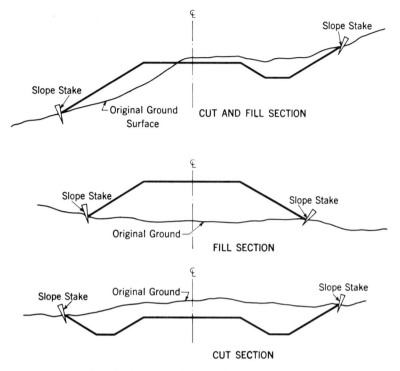

FIGURE 14-3 Sketch showing slope stakes.

14-30 Final Cross Sections

When grading operations are complete, it is necessary to take final cross sections of the completed work in order to determine the final quantities of excavation and embankment completed on the project. The final cross sections should be taken at the same points as the original sections. Final cross sections may be taken as soon as grading operations are completed. This permits the calculation of earth quantities as the work progresses and generally shortens the time necessary for the preparation of the final estimate when the project is completed.

14-31 Borrow-Pit Measurements

When additional earth for embankment sections must be obtained from borrow pits, it is necessary to take cross sections before and after the removal of earth in order to determine the quantities used. The area from which the earth is to be excavated should be staked with one or more baselines, and cross sections should be taken at 10-m (25-ft) intervals. The base lines should be well referenced so that they may be relocated for the final cross sections. Sketches of borrow-pit layout and reference lines should always be made. After the borrow earth has been removed, cross sections are again taken on the original lines. Both original and final cross sections are plotted, and the volume of each removed is determined by the average-end area method or some other acceptable method. Figure 14-4 shows a typical method of staking a borrow pit for cross-sectioning. Quantities of borrow material may also be determined by weighing loaded hauling units.

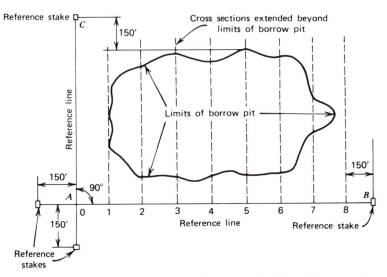

FIGURE 14-4 Method of staking and cross-sectioning a borrow pit.

14-32 Side Borrow

Most borrow pits are located at some distance from the roadway. Sometimes it may be feasible to obtain additional earth from the side of the roadway in large cuts by extending the side slopes. In order to obtain an accurate measurement of the earth, sections should be taken at 10-m (25-ft) intervals, just as in borrow pits, and the method of computing volumes may be the same.

14-33 Grade and Line Stakes for Pavements

The first stakes required for a surface paving project are generally for trenching or for the establishment of a uniform grade in accordance with the plan centerline profile. These stakes may be placed on the side of the road and may be used later for form stakes when placing concrete pavement. The form stakes are usually placed approximately 0.6 m (2 ft) from the edge of the slab unless otherwise requested by the contractor. The stakes should be set 15 m (50 ft) apart on tangents and 7.5 m (25 ft) apart on horizontal and vertical curves. Form stakes are placed on both sides of the road. On many projects, final finishing of the subgrade, base, or subbase is done by a machine with automatic controls operating on a taut wire stretched between stakes or pins; the field survey party must establish and check the position of the wire.

14-34 Culvert Stakes

The plans give the size, location, and lengths of all culverts on the project. They also give the elevations of the flow line and the angle of crossing of the road centerline. Field conditions should always be checked to see whether or not the proposed installation will be satisfactory. Figure 14-5 shows two typical culvert installations, one at right angles to the centerline and the other at an angle of less than 90°.

The stakes required for a culvert installation consist of the centerline stakes of the culvert and reference stakes that usually indicate the length of the culvert.

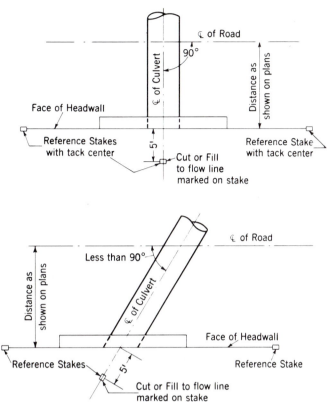

FIGURE 14-5 Method of staking culverts.

The flow-line elevations are generally given on the centerline reference stakes, as indicated in Figure 14-5. All stakes should be offset so they will not be disturbed during construction. On large box culverts the same method of staking is used.

14-35 Miscellaneous Engineering Work

The engineer may be required to furnish stakes for the location of manholes, catch basins, and inlets, with the necessary elevations for flow lines and tops of covers. Curb-and-gutter stakes and sidewalk stakes showing grade and alignment may also be required.

14-36 The Preparation of Estimates

As the work progresses, the contractor is entitled to partial payment for the work completed. These payments are generally shown on forms called "estimates" and are prepared by the project engineer. Estimates may be made monthly or bi-monthly. On lump-sum projects, the amount of the estimate will be a percentage of the whole. On unit price contracts, the individual items of work completed are listed and the value of the work is computed.

It is sometimes possible to pay for materials delivered to the site of the project or furnished for later use, such as gravel (stockpiled), reinforcing steel, and so forth. Such a procedure should be specified in the proposal or specifications, however, or definite means for providing authorizations should be made.

On partial payments the full amount is not always paid to the contractor, but a percentage, usually 10 percent, is retained. This protects the owner in case of an underrun of estimated quantities, as a guaranty for faulty work, or for minor claims for materials. On large contracts, 10 percent sometimes results in a large amount of capital being tied up for reserve purposes. Some highway departments arrange a reduction of this reserve to 5 percent when the work has been completed and accepted and pending the preparation of the final estimate. When the final estimate is complete (and certain delays in office procedure delay its processing), this reserve may be reduced to a small lump sum.

In the preparation of the final estimate, all items of work are measured as constructed in the field. If these quantities differ from those in the contract as authorized and explained under changes in the work, it is necessary to prepare authorizations that decrease or increase the quantities, as the case may be.

14-37 Final Acceptance

When all work has been completed, inspection of the whole project is made by the authorized party for the highway agency. If the project is one in which federal funds were used, the final inspection is made by an engineer of the Federal Highway Administration. When all parties indicate their acceptance, the final estimate, which is usually in the process of completion, is also approved, and recommendation for final payment is made.

During construction, it is the responsibility of the project engineer to supervise all phases of the work, approve or disapprove all authorizations for changes or extra work, and see that the work follows the schedule of operations as set up in the contract. As no work can be done beyond the completion date, the project engineer must notify the contractor in due time and prepare to extend the contract by supplemental agreements, such as an extension of time. When all the work is completed, the recommendations of the project engineer as to the contractor's performance, quality of work, and cooperation are necessary for the contractor's rating factor, as mentioned under the prequalification of contractors. The preparation of weekly and monthly reports, a history of the project, and administrative functions are always incidental to the supervision of a construction project.

PROBLEMS

14-1 At a central mixing plant, tests are made on asphaltic mix yielding the following consecutive percentages of asphalt: 6.1, 5.5, 7.1, 6.3, 5.8, 6.6, 6.3, 5.9, 6.9, and 7.3. Each of these values represents an average of three samples per lot. The central or target value is 6.3 percent. If the standard deviation of such tests is 0.5 percent, plot a control chart and determine if the process is in control.

14-2 Suppose the data given in Problem 14-1 represent single measurements, that is, $n = 1$. What can be said about the process?

14-3 A nuclear density gauge is used to determine the density of an embankment, yielding the following consecutive individual values: 1760, 1760, 1824, 1696, 1648, 1632, 1792, 1840, 1712, and 1888 kg/m^3. The central or target value is 1728 kg/m^3, and the standard deviation is 32 kg/m^3. Plot a control chart and determine if the process is out of control.

REFERENCES

1. *Guide Specifications for Highway Construction.* American Association of State Highway and Transportation Officials, Washington, DC (1993).

2. *Standard Specifications for Road and Bridge Construction.* Kentucky Transportation Cabinet, Department of Highways, Frankfort, KY (1994).

3. *Standard Specifications for Highway Construction.* State of Alabama Highway Department, Montgomery, AL (1992).

4. *Standard Specifications Construction of Roads and Bridges.* Department of Transportation, State of Georgia, Atlanta, GA (1993).

5. Granley, Edwin C. Quality Assurance in Highway Construction: Part 6, Control Charts. *Public Roads,* Vol. 35, No. 11 (1969).

6. Willenbrock, Jack H., *Statistical Quality Control of Highway Construction,* Vols. I and II. U.S. Department of Transportation, Federal Highway Administration, Washington, DC (1976).

7. *Quality Control/Quality Assurance Specification and Implementation Guide.* American Association of State Highway and Transportation Officials, Washington, DC (1993).

CHAPTER **15**

HIGHWAY MATERIALS

Each year in the United States, enormous quantities of construction materials are used for improvements to the public roadway system. Such projects require annually over 590 million tons of aggregates, 11 million tons of bituminous materials, and 19 million tons of cement, as well as vast quantities of steel, lumber, explosives, and petroleum products (1).

In this chapter, we briefly describe some of the physical characteristics and quality control tests for soils, aggregates, bituminous materials, and portland cement. Detailed material specifications and tests for these and other highway construction materials have been published by the American Association of State Highway and Transportation Officials (2).

SOILS

Basically, the engineering definition of the word "soil" is a very broad one. Soil might be defined as all the earth material, both organic and inorganic, that blankets the rock crust of the earth. Practically all soils are products of the disintegration of the rocks of the earth's crust. This disintegration, or "weathering," has been brought about by the action of chemical and mechanical forces that have been exerted on the parent rock formations for countless ages. Included among these forces are those of wind, running water, freezing and thawing, chemical decomposition, glacial action, and many others.

Soils may be described in terms of the principal agencies responsible for their formation and position. Thus a "residual soil" is one that, in its present situation, lies directly above the parent material from which it was derived. Soils formed by the action of wind are known as "aeolian soils"; a typical example of a windblown soil deposit is seen in the very considerable deposits of loess in the Mississippi Valley. "Glacial soils" occur in many parts of the United States. An example of such a soil deposit is a glacial till, which is a deposit of tightly bonded materials containing particles ranging in size from boulders down to very finely divided mineral matter.

Soils formed through the action of water are generally termed "sedimentary soils." Typical sedimentary soils are formed by the settling of soil particles (or groups of particles) from a suspension existing in a river, lake, or ocean. Sedimentary soils may range in type from beach or river sands to highly flocculent clays of marine origin.

Soils may also be described in terms related to the amount of organic material contained in them. Soils in which the mineral portion (soil particles) predominates are properly called "inorganic soils." Those in which a large amount of organic matter is contained are called "organic soils." Organic soils are usually readily identified by their dark brown to black color and distinctive odor.

One of the most important facts regarding soils and soil deposits is their normal lack of homogeneity. Due to the more or less random process of their formation, soils vary greatly in their physical and chemical composition at different locations over the surface of the earth. Although, generally speaking, soils derived from the same parent material under similar factors of geographic location, climate, and topography will be similar wherever they are found, several soil types may, and usually do, exist within a comparatively small area. Soil deposits are also characteristically varied with depth. Deposits of sedimentary soils in which layers of varying fineness were created during the process of their formation are said to be stratified, and some degree of stratification must be expected in any sedimentary soil deposit.

GENERAL SOIL TYPES

Certain general soil types are of importance to the highway engineer. No attempt will be made in this section to cover all the soil types that the engineer may encounter. Certain soil types occur generally throughout the country, however, and will be discussed here, along with the terms used in describing them.

15-1 Common Soil Types

"Sand" and "gravel" are coarse-grained soil types possessing little or no cohesion and with the particle size ranging from 80 mm for coarse gravel to 0.08 mm for fine sand (Figure 15-1). They are readily identified by visual inspection and are distinguished, generally speaking, by their relative stability under wheel loads when confined, by their high permeability, and by their low shrinkage and expansion in detrimental amounts with change in moisture content. The term "gravel" is usually applied to natural pit, river, or bank gravels consisting largely of rounded particles; "crushed gravel," or "crushed stone," is the term applied to the products of crushing larger rocks into gravel sizes.

"Silt" is the term applied to fine-grained soils of low to medium plasticity, intermediate in size between sand and clay. Silts generally possess little cohesion, undergo considerable shrinkage and expansion with change in moisture content, and possess a variable amount of stability under wheel loads. If they contain large percentages of flat, scalelike particles, such as mica flakes, they are likely to

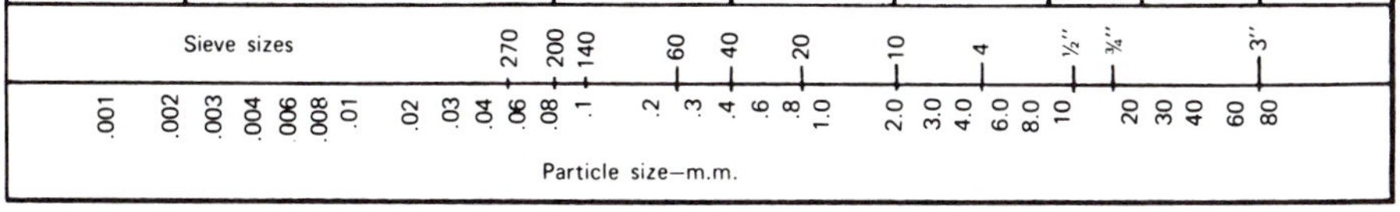

FIGURE 15-1 Grain-size classification. (AASHTO Specification M146.)

be highly compressible and somewhat elastic in nature. "Organic silts" contain appreciable amounts of decomposed organic matter and are generally highly compressible and unstable.

"Clays" are distinguished by the occurrence of very fine grains of 0.002 mm or finer. Clays generally possess medium to high plasticity, have considerable strength when dry, undergo extreme changes in volume with change in moisture content, and are practically impervious to the flow of water. "Lean clay" is the term given to silty clays or clayey silts, while fine, colloidal clays of high plasticity are called "fat clays." Clays may be further distinguished by the fact that, although they may possess considerable strength in their natural state, this strength is sharply reduced and sometimes completely destroyed when their natural structure is disturbed, that is, when they are remolded.

"Loam" is an agricultural term used to describe a soil that is generally fairly well graded from coarse to fine, that is easily worked, and that is productive of plant life. This name frequently appears in engineering literature in combination with other terms. Thus a soil may be called a "sandy loam," a "silty loam," or a "clay loam," depending on the size of the predominating soil fraction.

"Loess" is a fine-grained aeolian soil characterized by its nearly uniform grain size, predominantly silt, and by its low density. Highway cuts through loess deposits usually resemble those made in rock in that this soil will stand on a nearly vertical slope, whereas it is readily eroded by rainwater if flatter slopes are used.

"Muck" is soft silt or clay, very high in organic content, which is usually found in swampy areas and river or lake bottoms.

"Peat" is a soil composed principally of partially decomposed vegetable matter. Its extremely high water content, woody nature, and high compressibility make it an extremely undesirable foundation material.

BASIC SOIL PROPERTIES

To have an understanding of soil action, an engineer must be familiar with certain basic soil properties. We are all familiar with the basic properties of other engineering materials, such as steel, wood, and concrete. A soil engineer must have similar knowledge relative to soils. The job of the soil engineer is complicated by the fact that many soils are quite complex in nature, both physically and chemically and that soil deposits are likely to be extremely heterogeneous in character. It must be remembered that the properties of any given soil depend not only on its general type but also on its condition at the time when it is being examined.

15-2 Moisture Content

Water is an extremely important constituent of soils. The moisture content is defined as the weight of water contained in a given soil mass compared with the oven-dry weight of the soil and is usually expressed as a percentage. In the laboratory, moisture content is usually determined by selecting a small, representative sample of soil and determining the weights of the "wet" soil sample and the "oven-dry" soil. All weights are recorded in grams, and the following expression is used to determine the moisture content:

$$w(\%) = \frac{W_1 - W_2}{W_2} \times 100 \qquad (15\text{-}1)$$

where $w(\%)$ = moisture content (percent)
   W_1 = weight of "wet" soil and container (g)
   W_2 = weight of oven-dry soil and container (g)

If the void spaces in a soil are completely filled with water, the soil is said to be saturated. The moisture content of a soil may then be 100 percent or more, as might be the case in a saturated clay, muck, or peat soil.

Water in soils may be present in its normal liquid form, as when filling or partly filling the voids of a sand mass, or it may be present in the form of adsorbed water existing as films surrounding the separate soil particles or groups of particles, as in the case of the water remaining in a partially dried clay mass. The water films existing in the latter case may have properties sharply differing from those exhibited by water in its normal form. Properties of fine-grained soils are greatly dependent on the properties and behavior of the adsorbed water films.

15-3 Specific Gravity

"Specific gravity" (G), as applied to soils, is the specific gravity of the dry soil particles or "solids." The specific gravity is frequently determined by the pycnometer method, the determination being relatively easy for a coarse-grained soil and more difficult for the finer soils. Values for the specific gravity refer to the ratio of the unit weight of soil particles to the unit weight of water at some known temperature (usually 4°C) and range numerically from 2.60 to 2.80. Values of the specific gravity outside the range of values given may occasionally be encountered in soils derived from parent materials that contained either unusually light or unusually heavy minerals.

15-4 Unit Weight

The unit weight of a soil is the weight of the soil mass per unit of volume and is expressed in kilograms per cubic meter (pounds per cubic foot). As commonly used in highway engineering, the term "wet unit weight" refers to the unit weight of a soil mass having a moisture content that is anything different from zero, whereas "dry unit weight" refers to the unit weight of the soil mass in an oven-dry condition. The wet unit weight, dry unit weight, and moisture content are related by the following expression:

$$\text{dry unit weight} = \frac{\text{wet unit weight}}{\dfrac{(100 + w\%)}{100}} \tag{15-2}$$

The wet unit weight of a soil may vary from 1440 kg/m³ (90 lb/ft³) or less for saturated, organic soils to 2240 kg/m³ (140 lb/ft³) or more for well-compacted granular materials.

15-5 Shearing Resistance

Failures that occur in soil masses as a result of the action of highway loads are principally shear failures. Therefore, the factors that go to make up the shearing resistance of a soil are of importance. Shearing resistance within soil masses is commonly attributed to the existence of "internal friction" and "cohesion."

A simplified explanation of these properties is most easily accomplished by consideration of two extremely different types of soils: first, a cohesionless sand,

and second, a highly cohesive clay in which the internal friction is assumed to be negligible. In a cohesionless sand, the force required to overcome shearing resistance on any plane is assumed to be given by the expression $S_r = \sigma \tan \phi$, where σ is the normal force (stress) on the plane being considered and ϕ is the angle of internal friction. The value of the angle of internal friction is assumed to include the factors of resistance to sliding (or rolling) of the soil particles over one another and any interlocking that may have to be overcome before a slip can occur. For a dry sand, ϕ is primarily dependent on density (void ratio); the lower the void ratio the higher the value of ϕ. Grain shape is also important, as is surface texture; ϕ is higher for a rough, angular sand than for a smooth, rounded sand having the same void ratio. Gradation of a sand is also important, with ϕ being generally higher for sands that are well graded from coarse to fine. The angle of internal friction is relatively independent of the moisture content for sands; ϕ for a wet sand will be only slightly, if any, lower than ϕ for a dry sand, other conditions being the same.

In a saturated clay mass it may be assumed, for the purposes of explanation, that the angle of internal friction is equal to zero and that the resistance to sliding on any plane is equal to the cohesion, C (usually expressed in pounds per square foot). In a simple explanation, C is supposed to include both "true" cohesion, that due to intermolecular attraction, and "apparent" cohesion, that due to surface tension effects in the water contained in the clay mass. The shearing strength of most fine-grained soils decreases when their moisture content is increased and is frequently sharply reduced when their natural structure is destroyed. The interpretation of the factors influencing the shearing strength of cohesive soils is probably the most complex problem in soil mechanics, and no comprehensive explanation will be attempted here. Factors of importance include density, water content, loss of strength with remolding, drainage conditions of the clay mass subjected to stress, variation of cohesion with pressure, and variation in the angle of internal friction.

For the large majority of soils as normally encountered in the field, shearing resistance is made up of both cohesion and internal friction. For these soils the shearing resistance on any plane is frequently, although somewhat empirically, given by Coulomb's law:

$$S_r = \sigma \tan \phi + C \qquad (15\text{-}3)$$

where the symbols have the meanings previously indicated.

Shearing resistance may be evaluated in the laboratory by use of the unconfined compression test, the direct shear test, or the triaxial compression test. Samples may be tested in an undisturbed condition or under conditions similar to those expected in the field. Direct measurements of the shearing resistance of subgrade soils may be made in the field through the use of loaded circular plates (plate bearing test). The shearing resistance of subgrade soils may be estimated by *in-situ* tests such as the vane shear test, the borehole pressure test, and the cone penetration test. Various semi-empirical tests (such as the California bearing ratio test) have been developed to measure shearing resistance more or less directly in connection with the design of flexible and rigid pavements. Design methods in which these tests are used are fully described in later chapters.

15-6 Other Soil Properties

Several other soil properties, including those listed and briefly defined here, may influence the behavior of soil masses and affect the performance of highway subgrades and structure foundations. For a more complete description of these properties, the reader is referred to one of several textbooks in geotechnical or materials engineering (*3, 4*).

Permeability is the property of a soil mass that permits water to flow through it under the action of gravity or some other applied force.

Capillarity is the property that permits water to be drawn from a free water surface through the action of surface tension and independent of the force of gravity.

Shrinkage of a soil mass is a reduction in volume that occurs when the moisture content is reduced from that existing when it is partially saturated or saturated.

Swelling is a term used to describe the expansion in volume of a soil mass that accompanies an increase in the moisture content.

Compressibility is the property of a soil that permits it to consolidate under the action of an applied compressive load.

Elasticity is the property of a soil that permits it to return to its original dimensions (or nearly so) after the removal of an applied load. The resilient modulus also represents the elasticity property of a soil and is more commonly used in pavement design.

SOIL CLASSIFICATION FOR HIGHWAY PURPOSES

The objective behind the use of any soil classification system for highway purposes is to be able to predict the subgrade performance of a given soil on the basis of a few simple tests performed on the soil in a disturbed condition. On the basis of these test results and their correlation with field experience, the soil may be correctly identified and placed into a group of soils, all of which have similar characteristics and properties.

The student should realize, however, that placing a soil into the correct group in some classification system, although desirable in many respects, is not the final objective of the soil engineer. That is, classifying a soil should not be regarded as an end in itself but as a tool to further our knowledge of soil action and behavior.

The principal soil classification system in use in this country by highway engineers is the AASHTO classification system, first proposed in 1931 and later revised. Principal tests used by this and other classification systems are mechanical analysis and various routine laboratory tests. The mechanical analysis is performed on the entire sample and has as its objective the determination of the proportion of particles of various sizes in the given soil.

These routine tests are performed on the "soil binder" or fraction passing a No. 40 sieve. Most important of the routine tests or "soil constants" are the Atterberg limits, that is, liquid limit, plastic limit, and shrinkage limit. The plasticity index is also of significance and is calculated from the results of the liquid limit and plastic limit determinations. Other routine tests that may be used for classification purposes include other shrinkage factors (shrinkage ratio, volumetric shrinkage, and lineal shrinkage); the field moisture equivalent; and the centrifuge moisture equivalent. The routine tests are intended to describe definite physical properties of the soil and are performed with standardized laboratory procedures. Test procedures are briefly described in the following paragraphs. For a full description, the student is referred to the appropriate publications of the American Society for Testing and Materials (ASTM) and the AASHTO, such as Ref. 2.

15-7 Mechanical Analysis

Separation of the soil into its fractions may be done by a sieve analysis performed directly on soils that contain little or no fines, such as clean sand or a soil that may readily be separated from the coarser particles.

If the character of the fines is such that the fine material adheres to the coarser particles and is not removed by drying sieving action, the sample is prewashed and the fine material removed. Material that is retained on the No. 200 sieve during the washing process is then dried and subjected to sieving as before. This is called a "wet sieve analysis."

The practical lower limit for the use of sieves is the No. 200 sieve, which has openings 0.074 mm square. Consequently, the mechanical analysis used in many soil laboratories is a combined sieve and hydrometer analysis. The hydrometer analysis procedure is described in ASTM D-422. The results of the sieve analysis and hydrometer analysis are then combined to give the complete grain-size distribution curve for the soil being tested. One method of presenting this information is shown in Figure 15-2. Curves of four different soils are shown. Curve A is that of a uniform sand; B is that of a poorly graded gravelly soil; C is a well-graded soil, with a good distribution of particles from coarse to fine; and D is coarse aggregate such as used in concrete.

In many cases the entire grain-size distribution curve is not required. In such circumstances, a "wet sieve analysis" may be employed, and the material passing the No. 200 sieve is designated as "combined silt and clay."

15-8 Atterberg Limits

The "liquid limit" may be defined as the minimum moisture content at which the soil will flow under the application of a very small shear force. At this moisture content the soil is assumed to behave practically as a liquid. The "plasticity limit" may be defined in general terms as the minimum moisture content at which the soil remains in a plastic condition. This lower limit of plasticity is rather arbitrarily defined, and the plastic limit may be further described as the lowest moisture content at which the soil can be rolled into a thread of 3.2 mm (1/8 in.) diameter without crumbling. For the methods of determining the liquid limit and the plastic limit of a soil, the student is referred to Ref. 2.

The "plasticity index" of a soil is defined as the numerical difference between the liquid and plastic limits. It thus indicates the range of moisture content over which the soil is in a plastic condition. Sandy soils and silts, particularly those of the rock-flour type, have characteristically low PIs, while clay soils show high values of the plasticity index. Generally speaking, soils that are highly plastic, as indicated by a high value of the plasticity index, are also highly compressible. It is also evident that the plasticity index is a measure of cohesiveness, with a high value of the PI indicating a high degree of cohesion. Soils that do not have a plastic limit, such as cohesionless sands, are reported as being nonplastic (NP).

15-9 AASHTO Classification System

The AASHTO classification system was first presented in 1931 by the Bureau of Public Roads. The original system has been modified and refined and adopted

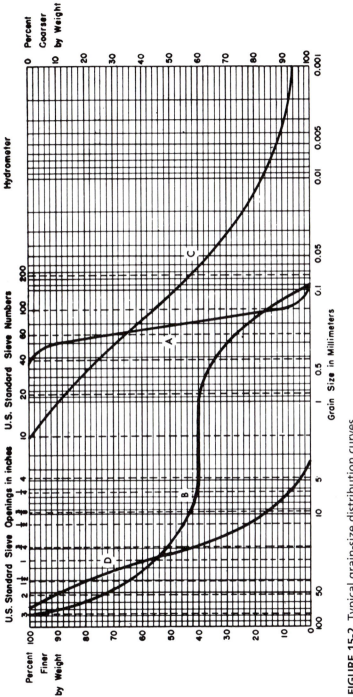

FIGURE 15-2 Typical grain-size distribution curves.

417

by the American Association of State Highway and Transportation Officials. The following description of the classification system has been quoted from AASHTO Recommended Practice M145 (2) with but minor editorial changes.

This recommended practice describes a procedure for classifying soils into seven groups based on laboratory determination of particle-size distribution, liquid limit, and plasticity index. Evaluation of soils within each group is made by means of a "group index," which is a value calculated from an empirical formula. (See Section 15-10.) The group classification including group index should be useful in determining the relative quality of the soil material for use in earthwork structures, particularly embankments, subgrades, subbases, and bases. For the detailed design of important structures, however, additional data concerning strength or performance characteristics of the soil under field conditions will usually be required.

Test Procedures

The classification is based on the results of tests made in accordance with the following standard methods of AASHTO:

Amount of material finer than 0.075 mm (No. 200) sieve in aggregate	T11
Sieve analysis of fine and coarse aggregates	T27
Mechanical analysis of soils	T88
Determining the liquid limit of soils	T89
Determining the plastic limit and plasticity index of soils	T90

Classification Procedures

The classification is made by using the test limits shown in Table 15-1. The liquid limit and plasticity index ranges for the A-4, A-5, A-6, and A-7 soil groups are shown graphically in Figure 15-3.

With required test data available, proceed from left to right in Table 15-1, and the correct group will be found by the process of elimination. The first group from the left into which the test data will fit is the correct classification. All limiting test values are shown as whole numbers. If fractional numbers appear on test reports, convert to nearest whole number for the purpose of classification. Group index values should always be shown in parentheses after group symbols as A-2-6(3), A-4(5), A-6(12), A-7-5(17), and so forth. A description of the AASHTO classification groups is given in Table 15-2.

15-10 The Group Index

The group index is calculated from the following formula:

$$\text{group index} = (F - 35)[0.2 + 0.005(\text{LL} - 40)] + 0.01(F - 15)(\text{PI} - 10)$$
(15-4)

where F = percentage passing 0.074-mm (No. 200) sieve, expressed as a whole number; this percentage is based only on the material passing the 75-mm (3-in.) sieve
LL = liquid limit
PI = plasticity index

TABLE 15-1 Classification of Highway Subgrade Materials (With Suggested Subgroups)[a]

General Classification	Granular Materials (35% or less passing No. 200)							Silt–Clay Materials (more than 35% passing No. 200)			
	A-1		A-3	A-2				A-4	A-5	A-6	A-7
Group Classification	A-1-a	A-1-b		A-2-4	A-2-5	A-2-6	A-2-7				A-7-5, A-7-6
Sieve analysis, percent passing											
No. 10 (2.0 mm)	50 max.										
No. 40 (0.425 mm)	30 max.	50 max.	51 min.								
No. 200 (0.075 mm)	15 max.	25 max.	10 max.	35 max.	35 max.	35 max.	35 max.	36 min.	36 min.	36 min.	36 min.
Characteristics of fraction passing No. 40											
Liquid limit				40 max.	41 min.	40 max.	41 min.	40 max.	41 min.	40 max.	41 min.
Plasticity index	6 max.		NP	10 max.	10 max.	11 min.	11 min.	10 max.	10 max.	11 min.	11 min.[b]
Usual types of significant constituent materials	Stone fragments, fine gravel, and sand			Silty or clayey gravel and sand				Silty soils		Clayey soils	
General rating as subgrade	Excellent to good							Fair to poor			

[a]Classification procedure: With required test data available, proceed from left to right on the chart, and correct group will be found by process of elimination. The first group from the left into which the test data will fit is the correct classification.

[b]Plasticity index of A-7-5 subgroup is equal to or less than LL minus 30. PI of A-7-6 subgroup is greater than LL minus 30 (see Fig. 15-3).

Note: See group index formula and Figure 15-3 for method of calculation. Group index should be shown in parentheses after group symbol, such as A-2-6(3), A-4(5), A-6(12), A-7-5(17), and so forth.

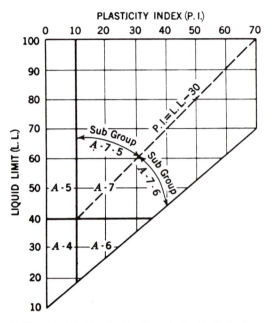

FIGURE 15-3 Liquid limit and plasticity index ranges for the A-4, A-5, A-6, and A-7 subgrade groups.

The group index should be reported to the nearest whole number. When the calculated group index is negative, it should be reported as zero. When calculating the group index of A-2-6 and A-2-7 subgroups, only the PI portion of the formula should be used. The following are examples of calculations of the group index.

Assume that an A-6 material has 55 percent passing the 0.075-mm sieve, liquid limit of 40, and plasticity index of 25. Then

$$\text{group index} = (55 - 35)[0.2 + 0.005(40 - 40)] + 0.01(55 - 15)(25 - 10)$$
$$= 4.0 + 6.0 = 10$$

Assume that an A-2-7 material has 30 percent passing the 0.075-mm sieve, liquid limit of 50, and plasticity index of 30. Then

$$\text{group index} = 0.01(30 - 15)(30 - 10)$$
$$= 3.0 \text{ or } 3$$

(Note that only the PI portion of the formula was used.)

Under average conditions of good drainage and thorough compaction, the supporting value of a material as subgrade may be assumed as an inverse ratio to its group index; that is, a group index of 0 indicates a "good" subgrade material and group index of 20 or greater indicates a "very poor" subgrade material.

15-11 Unified Soil Classification System

The Unified Soil Classification System is based on the Airfield Classification System developed by Professor A. Casagrande of Harvard University during World War II. The system was modified slightly and became the Department of the Army Uniform Soil Classification System. In turn, this system was slightly

TABLE 15-2 Description of Classification Groups

Granular Materials—Containing 35% or less passing the 0.075-mm sieve.

Group A-1. The typical material of this group is a well-graded mixture of stone fragments or gravel, coarse sand, fine sand, and a nonplastic or feebly plastic soil binder; however, this group also includes stone fragments, gravel, coarse sand, volcanic cinders, and so forth, without soil binder.

  Subgroup A-1-a includes those materials consisting predominantly of stone fragments or gravel, either with or without a well-graded binder of fine material.

  Subgroup A-1-b includes those materials consisting predominantly of coarse sand either with or without a well-graded soil binder.

Group A-3. The typical material of this group is fine beach sand or fine desert blown sand without silty or clay fines or with a very small amount of nonplastic silt.

Group A-2. This group includes a wide variety of "granular" materials that are borderline between the materials falling in groups A-1 and A-3 and the silt-clay materials of groups A-4, A-5, A-6, and A-7. It includes all materials containing 35% or less passing the 0.075-mm sieve that cannot be classified as A-1 or A-3 due to fines content or plasticity or both in excess of the limitations for those groups.

  Subgroups A-2-4 and A-2-5 include various granular materials containing 35% or less passing the 0.075-mm sieve and with a minus 0.425-mm portion having the characteristics of the A-4 and A-5 groups.

  Subgroups A-2-6 and A-2-7 include materials similar to those described under subgroups A-2-4 and A-2-5, except that the fine portion contains plastic clay having the characteristics of groups A-6 and A-7.

Silt–Clay Materials—Containing more than 35% passing 0.075-mm sieve.

Group A-4. The typical material of this group is a nonplastic or moderately plastic silty soil usually having 75% or more passing the 0.075-mm sieve.

Group A-5. The typical material of this group is similar to that described under group A-4, except that it is usually of diatomaceous or micaceous character and may be highly elastic, as indicated by the high liquid limit.

Group A-6. The typical material of this group is a plastic clay soil usually having 75% or more passing the 0.075-mm sieve. The group also includes mixtures of fine clayey soil and up to 64% of sand and gravel retained on the 0.075-mm sieve. Materials of this group usually have a high volume change between wet and dry states.

Group A-7. The typical material of this group is similar to that described under group A-6, except that it has the high liquid limits characteristic of the A-5 group and may be elastic as well as subject to a high volume change.

  Subgroup A-7-5 includes those materials with moderate plasticity indexes in relation to liquid limit and which may be highly elastic as well as subject to a considerable volume change.

  Subgroup A-7-6 includes those materials with high plasticity indexes in relation to liquid limit and which are subject to an extremely high volume change.

---

modified and adopted by the Corps of Engineers and the Bureau of Reclamation in January 1952.

Table 15-3 is the master chart for the Unified System; it contains procedures that are to be followed in identifying and classifying soils under this system.

Soils may be classified and placed in one of the 15 major soil groups in the Unified System by using either laboratory or field identification procedures. Soils are classified on the basis of (1) percentage of gravel, sand, and fines (fraction passing the No. 200 sieve); (2) shape of the grain-size distribution curve; and (3) plasticity and compressibility characteristics.

The coefficients of uniformity (C_u) and gradation (C_g) are used to judge the

TABLE 15-3 Unified Classification System

Major Divisions			Group Symbols	Typical Name	Field Identification Procedures (Excluding Particles Larger Than 3 Inches and Basing Fractions on Estimated Weights)		
1	*2*		*3*	*4*	*5*		
Coarse-grained Soils More than half of material is larger than No. 200 sieve size.	Gravels More than half of coarse fraction is larger than No. 4 sieve size.	Clean gravels (little or no fines)	GW	Well-graded gravels, gravel-sand mixtures, little or no fines	Wide range in grain sizes and substantial amounts of all intermediate particle sizes		
			GP	Poorly-graded gravels, gravel-sand mixtures, little or no fines	Predominantly one size or a range of sizes with some intermediate sizes missing		
		Gravels with fines (appreciable amount of fines)	GM	Silty gravels, gravel-sand-silt mixtures	Nonplastic fines or fines with low plasticity (for identification procedures see *ML* below)		
			GC	Clayey gravels, gravel-sand-clay mixtures	Plastic fines (for identification procedures see *CL* below)		
	Sands More than half of coarse fraction is smaller than No. 4 sieve size.	Clean sands (little or no fines)	SW	Well-graded sands, gravelly sands, little or no fines	Wide range in grain size and substantial amounts of all intermediate particle sizes		
			SP	Poorly-graded sands, gravelly sands, little or no fines	Predominantly one size or a range of sizes with some intermediate sizes missing		
		Sands with fines (appreciable amount of fines)	SM	Silty sands, sand-silt mixtures	Nonplastic fines or fines with low plasticity (for identification procedures see *ML* below)		
			SC	Clayey-sands, sand-clay mixtures	Plastic fines (for identification procedures see *CL* below)		
Fine-grained Soils More than half of material is smaller than No. 200 sieve size.	Silts and Clays Liquid limit less than 50				Identification procedures on fraction smaller than No. 40 sieve size		
					Dry strength (crushing characteristics)	Dilatancy (reaction to shaking)	Toughness (consistency near PL)
			ML	Inorganic silts and very fine sands, rock flour, silty or clayey fine sands with slight plasticity	None to slight	Quick to slow	None
			CL	Inorganic clays of low to medium plasticity, gravelly clays, sandy clays, silty clays, lean clays	Medium to high	None to very slow	Medium
			OL	Organic silts and organic silty clays of low plasticity	Slight to medium	Slow	Slight
	Silts and Clays Liquid limit greater than 50		MH	Inorganic silts, micaceous or diatomaceous fine sandy or silty soils, elastic silts	Slight to medium	Slow to none	Slight to medium
			CH	Inorganic clays of high plasticity, fat clays	High to very high	None	High
			OH	Organic clays of medium to high plasticity, organic silts	Medium to high	None to very slow	Slight to medium
Highly organic soils			Pt	Peat and other highly organic soils	Readily identified by color, odor, spongy feel and frequently by fibrous texture		

Note in column 2 left margin: *The No. 200 sieve is about the smallest particle visible to the naked eye.* And *(For visual classification, the ¼-inch size may be used as equivalent to the No. 4 sieve size.)*

(1) *Boundary classifications:* Soils possessing characteristics of two groups are designated by combinations of group symbols. For example
Adopted by Corps of Engineers and Bureau of Reclamation, January, 1952.

shape of the grain-size distribution curve of a coarse-grained soil. The term "D_{10}" which appears in the table means the grain size (diameter) that corresponds to 10 percent on a grain-size distribution curve of the type shown in Figure 15-2. D_{30} and D_{60} have similar meanings.

The influence and relationship of the LL and PI are reflected in the plasticity chart of Table 15.3.

In general terms, clays (C) plot above the "A" line of the plasticity chart and silts (M) plot below the "A" line. The silt (M) and clay (C) groups are further divided on the basis of low (L) or high (H) LL; a high LL is associated with high compressibility.

TABLE 15-3 Unified Classification System *(continued)*

Information Required for Describing Soils	Laboratory Classification Criteria
6	7

For undisturbed soils add information on stratification, degree of compactness, cementation, moisture conditions, and drainage characteristics.

Give typical name; indicate approximate percentages of sand and gravel, max. size; angularity, surface conditions, and hardness of the coarse grains; local or geologic name and other pertinent descriptive information; and symbol in parentheses

Example:

Silty sand, gravelly; about 20 percent hard, angular gravel particles $^1/_2$-in. maximum size; rounded and subangular sand grains coarse to fine; about 15 percent nonplastic fines with low dry strength; well compacted and moist in place; alluvial sand; *(SM)*

Use grain-size curve in identifying the fractions as given under field identification.

Determine percentages of gravel and sand from grain-size curve. Depending on percentage of fines (fraction smaller than No. 200 sieve size) coarse-grained soils are classified as follows:

Less than 5 percent — GW, GP, SW, SP.

More than 12 percent — GM, GC, SM, SC.

5 percent to 12 percent — Borderline cases requiring use of dual symbols

$C_u = \dfrac{D_{60}}{D_{10}}$ greater than 4

$C_g = \dfrac{(D_{30})^2}{D_{10} \times D_{60}}$ between 1 and 3

Not meeting all gradation requirements for *GW*

Atterberg limits below *A* line or *PI* less than 4

Atterberg limits above *A* line with *PI* greater than 7

Above *A* line with *PI* between 4 and 7 are *borderline* cases requiring use of dual symbols.

$C_u = \dfrac{D_{60}}{D_{10}}$ greater than 6

$C_g = \dfrac{(D_{30})^2}{D_{10} \times D_{60}}$ between 1 and 3

Not meeting all gradation requirements for *SW*

Atterberg limits below *A* line or *PI* less than 4

Atterberg limits above *A* line with *PI* greater than 7

Limits plotting in hatched zone with *PI* between 4 and 7 are *borderline* cases requiring use of dual symbols.

Give typical name, indicate degree and character of plasticity, amount and maximum size of coarse grains, color in wet condition, odor if any, local or geologic name, and other pertinent descriptive information; and symbol in parenthesis

For undisturbed soils add information on structure, stratification, consistency in undisturbed and remolded states, moisture and drainage conditions

Example:

Clayey silt, brown, slightly plastic, small percentage of fine sand, numerous vertical root holes, firm and dry in place, loess, *(ML)*

PLASTICITY INDEX chart: Comparing soils at equal liquid limit — Toughness and dry strength increase with increasing plasticity index. CH, A Line, OH & MH, CL, CL-ML, ML.

LIQUID LIMIT PLASTICITY CHART
Per laboratory classification of fine-grained soils

GW-GC, well-graded gravel-sand mixture with clay binder. (2) All sieve sizes on this chart are U.S. standard.

Further details of the classification procedure will be revealed by close study of Table 15-3. Further information is contained in Ref. 5.

15-12 Classification of Typical Soils

Table 15-4 gives the results of laboratory tests on three inorganic soils. Each soil may be classified under the AASHTO and Unified Soil Classification systems, as follows:

Soil No. 1   AASHTO system. By Eq. 15-4, group index = 1. Entering Table 15-1, and using a left-to-right elimination process, soil cannot be in one of

TABLE 15-4 Results of Laboratory Tests on Three Inorganic Soils

	Soil Number		
Sieve Size	*1*	*2*	*3*
Mechanical Analysis, percent passing, by weight			
No. 10	100.0	100.0	100.0
No. 40	85.2	97.6	85.0
No. 60	—	—	20.0
No. 200	52.1	69.8	1.2
No. 270	48.2	65.0	—
C_u	—	—	2.0
C_g	—	—	—
Plasticity Characteristics, percent, by weight			
Liquid limit	29.2	66.7	21.3
Plasticity index	5.0	39.0	NP

the *granular materials* groups because more than 35 percent passes a No. 200 sieve. It meets the requirements of the A-4 group. It is therefore A-4(1).

Unified. Soil is fine-grained because more than 50 percent passes No. 200 sieve. LL is less than 50, hence must be *ML* or *CL*, because it is inorganic. On plasticity chart it falls below the *A* line, therefore *ML*.

Soil No. 2   AASHTO system. By Eq. 15-4, group index = 27. This soil falls in the A-7 group. Referring to Figure 15-3, it falls in the A-7-6 subgroup, hence, A-7-6(27).

Unified. Soil is fine-grained because more than 50 percent passes No. 200 sieve. LL is more than 50, hence must be *MH* or *CH*. On plasticity chart, it falls above the *A* line, therefore *CH*.

Soil No. 3   AASHTO. This is one of the soils described as *granular material*. It will not meet the requirements of the A-1 group because it contains practically no fines. It does meet the requirements of the A-3 group. Equation 15-4 yields a negative group index for this soil. The classification is A-3(0).

Unified. Soil is coarse-grained, because very little passes No. 200 sieve. All passes a No. 10 sieve, hence sand. The soil contains less than 5 percent passing No. 200; therefore it must be either an *SW* or *SP*. Value of $C_u = 2$ will not meet requirements for *SW*, hence *SP*.

SOIL SURVEYS FOR HIGHWAY PURPOSES

Soil surveys are made in connection with highway location, design, and construction. Many sources of information concerning soils generally are available for the area in which a highway project is to be carried out. These sources include geological and topographic maps and reports, agricultural soil maps and reports, aerial photographs, and the results of previous soil surveys in the area. Information from such sources is of importance in two general ways. First, a study of this information will aid in securing a broad understanding of soil conditions and associated engineering problems that may be encountered. Second, such information is of great value in planning, conducting, and interpreting the results of detailed soil surveys that are necessary for design and construction. Modern techniques emphasize the use of all the available information about a given area in order to minimize the amount of detailed field and laboratory work necessary for a given project.

MOISTURE-DENSITY RELATIONSHIPS

Practically all soils exhibit a similar relationship between moisture content and density (dry unit weight) when subjected to dynamic compaction. That is, practically every soil has an optimum moisture content at which the soil attains maximum density under a given compactive effort. This fact, which was first stated by R. R. Proctor in a series of articles published in *Engineering News-Record* in 1933, forms the basis for the modern construction process commonly used in the formation of highway subgrades, bases, and embankments, earthen dams, levees, and similar structures. In the laboratory, dynamic compaction is achieved by use of a freely falling weight impinging on a confined soil mass; in the field, similar compaction is secured through the use of rollers or vibratory compactors applied to relatively thin layers of soil during the construction process. In Figure 15-4 is shown the relationship between moisture and density for a

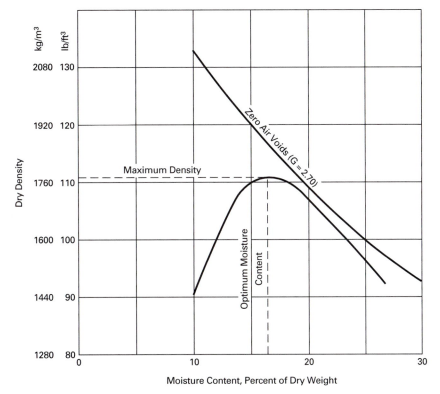

FIGURE 15-4 Moisture-density relationship for a typical soil under dynamic compaction.

typical soil subjected to dynamic compaction; the meaning of the terms "maximum density" and "optimum moisture" is clearly shown. This curve was obtained in the laboratory. In the field an attempt is usually made to maintain the soil at optimum moisture and bring the soil to maximum density or some specified percentage thereof.

In the laboratory, compaction is usually performed under what has come to be called the "standard Proctor" or "standard AASHTO" methods (T99). In the basic procedure used in this test (method A), the portion of the soil that passes a No. 4 sieve is placed in a mold 102 mm (4 in.) in diameter which has a volume of 944 cm³ ($\frac{1}{30}$ ft³). The soil is placed in three layers of about equal thickness, and each layer is subjected to 25 blows from a hammer weighing 2.5 kg (5.5 lb) and having a striking face 51 mm (2 in.) in diameter, falling freely through a distance of 305 mm (12 in.). In recent years, heavier compaction equipment has come into widespread use, with the result that under certain conditions a greater compactive effort may be required in the laboratory. An increased compactive effort that is frequently used is known as the "modified Proctor" or "modified AASHTO" compaction (T180); the soil is compacted in the Proctor mold, using 25 blows from a 4.5 kg (10-lb) hammer dropping through a distance of 457 mm (18 in.) on each of five equal layers. These are "impact" methods of compaction; some work has been done with "kneading-type" compactors, which simulate the action of a rubber-tired roller, and vibratory compaction of granular materials.

Regardless of the compactive effort used, the optimum moisture and maximum density are usually found in the laboratory by a series of determinations of wet unit weight and the corresponding moisture content. The series is started with the soil in a damp condition, at a moisture content a few percent below optimum. The soil is compacted, the wet unit weight determined, and a sample selected from the interior of the compacted soil mass for the determination of moisture content. The soil is then broken up (or a new sample of the same soil selected); the moisture content is increased by 1 or 2 percent, and the compaction process is repeated. This process is continued until a decrease is noted in the wet unit weight or until the soil definitely becomes wet and shows an excess of moisture. After the moisture content determinations are completed the dry unit weights may be calculated and plotted as shown in Figure 15-4. The relationship between the dry unit weight, wet unit weight, and moisture content is as given in Equation 15-2.

The behavior of the soil mass under dynamic (impact) compaction is of interest and may be explained in the following fashion. In the early states of the compaction run, when the moisture content is considerably less than optimum, the soil does not contain sufficient moisture to flow readily under the blows of the hammer. As the moisture content is increased, the soil flows more readily under the "lubricating" effect of the additional water, and the soil particles move closer together, with a resulting increase in density. This effect is continued until the optimum moisture content is reached, at which point the soil has achieved its greatest density. The condition at optimum may be explained in terms of the degree of saturation (the ratio between the volume of water and the volume of voids); at this moisture content as large a degree of saturation has been attained as can be reached by compaction. This limiting degree of saturation is due to the existence of air entrapped in the void spaces and around the soil particles. Further increase in moisture tends to overfill the voids but does not materially decrease the air content. As a consequence, the soil particles are forced apart and the unit weight decreases. At moisture contents well beyond optimum, the soil mass tends to become somewhat elastic, the hammer "bounces," and excess moisture is generally forced out of the bottom of the mold.

Also shown on the plot of Figure 15-4 is the "zero air voids curve" for the soil in question. Points on this curve represent the theoretical density that this soil would attain at each moisture content if all the void spaces were filled with water, that is, if the soil were completely saturated or $S = 100$ percent. The position of points on the curve may be calculated from the following equation if the specific gravity of the soil is known or may be assumed:

$$\text{dry unit weight} = \frac{wG}{1 + \dfrac{mG}{100}} \ \text{kg/m}^3 \ (\text{lb/ft}^3) \tag{15-5}$$

where w = unit weight of water, 1000 kg/m³ (62.4 lb/ft³)
G = specific gravity of solids
m = moisture content (percent)

Although the densities shown on the zero air voids curve represent theoretical values that are practically unattainable, because all the air can never be exhausted from any soil system by dynamic compaction alone, the relative positive of the moisture-density curve with respect to the zero air voids curve is of consequence, as the distance separating the two curves is indicative of the air voids remaining in the soil at various moisture contents. For example, for the soil shown, the sample at 17.8 percent moisture has the same dry weight as it has at 15.1 percent moisture, although the percentage of air voids at the lower moisture content (on the dry side of optimum) is much greater. Thus, if the soil is compacted at the lower moisture content and then subjected to inundation or "flooding," it might be expected that a larger percentage of air would be displaced by entering moisture and greater expansion would occur than if the soil were compacted at the higher moisture content.

Another item of importance about the compaction process is that maximum density and optimum moisture are a function of the compactive effort. In general terms, an increase in compactive effort results in an increase in maximum density and a decrease in optimum moisture. The effect of variation in compactive effort is shown graphically in Figure 15-5. Examination of this figure shows that if soil is to be compacted to a specified density efficiently, careful control of the moisture content is required. Suppose, for example, that the specifications required a density of 116.5 lb/ft³. If the soil were too dry (say a moisture content of 7 percent), about 70 percent (6.63/3.98 × 100 percent = 170 percent) more compactive effort would be required to achieve the specified density than would be required at a moisture content of 10.4 percent. On the other hand, as greater and greater compactive effort is used, there is a wider and wider range of moisture contents over which the soil may be compacted to a specified minimum density.

Obviously, the nature of the soil itself has a great effect on the density obtained under a given compactive effort. Different soils react somewhat differently to compaction at moisture contents that are somewhat less than optimum. Moisture content is less critical for the heavy clay than for the slightly plastic sandy and silty soils. Heavy clays may be compacted through a relatively wide range of moisture contents below optimum with comparatively small changes in dry density. On the other hand, the granular soils that have better grading and higher

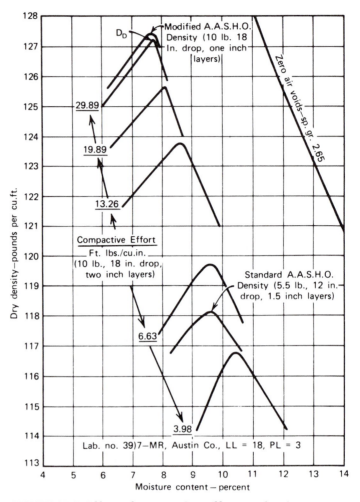

FIGURE 15-5 Effect of compacting effort on density. (SOURCE: Chester McDowell, Density Control: Its Benefits and Complexities. *Highway Research Record 177*, Highway Research Board, Washington, DC, 1967.)

densities under the same compactive effort react sharply to small changes in moisture content, with significant changes in dry density. The relatively clean, poorly graded, nonplastic sands also are relativley insensitive to changes in moisture.

Many other factors influence the results of compaction tests done in the laboratory, including the amount of coarse aggregate contained in the soil. These factors are discussed in detail in Ref. 6.

FROST ACTION IN SOILS

One of the special soil-related problems of interest in many areas of the United States is that of damage caused to roads and streets by the freezing and thawing of subgrades and bases during the winter and spring seasons. This phenomenon is discussed here for convenience and as a background for the discussion of certain design, construction, and maintenance considerations in later chapters.

Under certain conditions, very severe damage to subgrades, bases, and pavements, both rigid and flexible, may result from frost action. As a simplification, damaging action may be visualized in terms of freezing phenomena, of which an example is "frost heave," and the effects of a sudden spring thaw, which are sometimes categorically called "the spring breakup."

By frost heave is meant the distortion or expansion of the subgrade soil or base material during the period when freezing temperatures prevail and when the ground is frozen (or partially frozen) to a considerable depth. The phenomena of frost action are generally associated with the existence of a shallow water table and a frost-susceptible soil. It would be expected that excess moisture existing in a base or subgrade would freeze during a severe winter season. It might also be expected that the soil mass would expand, since water expands as it freezes. The magnitude of the severe frost heaves that occur regularly in some states cannot be explained on the basis of freezing of existing pore water alone, however. Investigations have shown that, as the water in the upper soil layers freezes, ice crystals are formed, and water may be drawn from a free water surface into the zone where freezing temperatures prevail. This water then freezes, additional water may be drawn to this level, and this process continues until layers of ice or "ice lenses" of considerable thickness may be formed. The volume increase brought about by the formation of these layers of ice is the cause of frost heaving.

It will be noted that a certain combination of circumstances (in addition to freezing temperatures) must usually exist before severe frost heave can occur. As previously mentioned, these include the existence of a shallow water table and a frost-susceptible soil. If the water table is not close to the surface, water is generally not available to the shallow soil layers, and the heave that can occur is limited to that caused by the expansion of freezing water in the voids of the soil. Since the change in volume from water to ice is only about 9 percent, this expansion would generally not be sufficient to cause severe frost damage. It might be possible, however, in somewhat unusual circumstances, that sufficient free water would be available from seepage flows or some other cause for frost heave to occur in a location where the water table is located well beneath the ground surface.

The requirement related to capillarity seems to limit detrimental frost action to soils containing a considerable percentage of fine particles such as silt or clay. Experience has shown that frost action is generally most severe in silt soils, as the upward movement of water in clay soils is usually so slow that the amount of water made available to the freezing layer is quite small. Soils that contain less than 3 percent by weight of material finer than 0.02 mm in diameter are not frost-susceptible. Soils that contain more than 3 percent of material of this size generally are susceptible to frost action, except that uniform fine sands that contain up to 10 percent of material finer than 0.02 mm generally are not susceptible to frost action.

Experience has also shown that frost heave is most severe when temperatures only slightly below freezing prevail for a considerable period of time so that frost penetration comes about somewhat gradually.

During the period of freezing temperatures, under the conditions described above, a considerable amount of water in the form of ice may accumulate in the shallow soil layers beneath a road surface. If a sudden rise in temperature or "thaw" occurs, as it frequently does in the spring in many localities, this accumulated ice may melt rapidly. Generally speaking, it would be expected that the ice in the upper soil layers would melt first, while deeper layers would remain frozen for a longer period of time. As a consequence, excess moisture may be trapped beneath the pavement and a sharp reduction in shearing strength may result. If this happens, the pavement and base may then fail under wheel loads; this type of failure is associated with the term "spring breakup." Damage to pavements and bases incurred during this period results in severe economic losses in many states.

Certain preventive measures are commonly employed to eliminate or minimize frost damage. Probably the most common solution to the problem is to remove soils subject to frost action and replace them with suitable granular backfill to the depth of the frost line. It is not always necessary to treat unsuitable soils to the full depth of frost penetration. For example, experience in Minnesota has shown that treatments to a depth of about 1 m (3 ft) are sufficient to prevent any breakup in the spring, although some slight heave may still occur. Another solution may be effected by the installation of drainage facilities in areas where frost action is likely to be detrimental, with the objective of lowering the water table or intercepting seepage flows that may cause excessive amounts of water to be accumulated in certain areas. Insulation or blanket courses may be employed to minimize the frost penetration into the subgrade; some state highway departments regulate truck traffic in the spring months to prevent damage caused by heavy wheel loads on roads known to have been weakened by frost action.

AGGREGATES

The term "aggregate" refers to granular mineral particles that are widely used for highway bases, subbases, and backfill. Aggregates are also used in combination with a cementing material to form concretes for bases, subbases, wearing surfaces, and drainage structures. Sources of aggregates include natural deposits of sand and gravel, pulverized concrete and asphalt pavements, crushed stone, and blast-furnace slag.

15-13 Properties of Aggregates

The most important properties of aggregates used for highway construction are

- particle size and gradation.
- hardness or resistance to wear.
- durability or resistance to weathering.
- specific gravity and absorption.
- chemical stability.
- particle shape and surface texture.
- freedom from deleterious particles or substances.

These properties are discussed in more detail in the following paragraphs.

15-14 Particle Size and Gradation of Aggregates

A key property of aggregates used for highway bases and surfaces is the distribution of particle sizes in the aggregate mix. The gradation of agregates, that is, the blend of particle sizes in the mix, affects the density, strength, and economy of the pavement structure.

A grain-size analysis is used to determine the relative proportions of various particle sizes in a mineral aggregate mix. To perform this analysis, a weighed sample of dried aggregate is shaken over a nest of sieves having selected sizes of square openings. The sieves are grouped together with the one with the largest openings on top and those with successively smaller openings placed underneath. The aggregate sample is shaken with a mechanical sieve shaker, and the weight of material retained on each sieve is determined and expressed as a percentage of the original sample (*4*). Detailed procedures for performing a grain-size analysis of coarse and fine aggregates are given in AASHTO Method T27 (*2*).

The grain-size analysis data are usually plotted on an aggregate grading chart, exemplified by Figure 15-6. With the aid of such a chart, engineers determine a preferred aggregate gradation and require that the gradation of aggregates used for highway projects conform to the limits of a specification band, as Figure 15-6 illustrates.

Testing sieves commonly used for highway projects are those with 2-½, 2, 1-½, 1, ¾, ½, and ⅜ in. square openings for the large fractions and those with 4, 8, 16, 30, 50, 100, and 200 meshes per inch for the smaller fractions. The latter sieves are designated No. 4, No. 10, and so on.

That portion of aggregate material that is retained on a No. 4 sieve (i.e., with particles larger than 2 mm) is known as "coarse aggregate." Material that passes a No. 4 sieve but is retained on a No. 200 sieve (particles larger than 0.075 mm) is known as "fine aggregate." Material that passes a No. 200 sieve is referred to as "fines."

15-15 Resistance to Wear

Materials used in highway pavements should be hard and resist wear due to the loading from compaction equipment, the polishing effects of traffic, and the internal abrasive effects of repeated loadings. The most commonly accepted measure of the hardness of aggregates is the Los Angeles abrasion test. The machine used in the Los Angeles abrasion test consists of a hollow steel cylinder, closed at both ends and mounted on shafts in a horizontal position (Fig. 15-7).

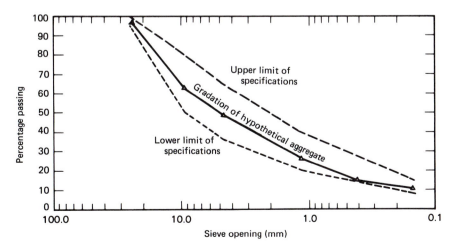

FIGURE 15-6 Aggregate gradation specification chart.

FIGURE 15-7 Los Angeles abrasion testing machine. (Courtesy The Asphalt Institute.)

A removable steel shelf extending the length of the cylinder is mounted on the interior surface of the cylinder.

To perform the Los Angeles abrasion test, a clean sample of the aggregate to be tested is placed in the cylinder along with a standard weight of steel spheres as an abrasive charge. The drum is then rotated at a speed of 30 to 33 rpm for 500 revolutions, after which the aggregate sample is removed and sieved on a No. 12 (1.70 mm) sieve. The material retained on the sieve is washed, dried to a constant mass, and weighed. The difference between the original mass and the final mass of the sample, expressed as a percentage of the original mass, is reported as the percentage of wear. A detailed procedure for this test is given by AASHTO Method T96 (2).

15-16 Durability or Resistance to Weathering

The durability of aggregates is commonly measured by a soundness test, as described in AASHTO Method T104 (2). This test measures the resistance of aggregates to disintegration in a saturated solution of sodium or magnesium sulfate. It simulates the weathering of aggregates that occur in nature.

The test is made by immersing sized fractions of the aggregate to be tested in a saturated solution of sodium or magnesium sulfate. The aggregate is then removed and dried in an oven to a constant mass. This process is repeated for a specified number of cycles, typically five. After the repeated cycles of alternate wetting and drying, the aggregate is divided into fractions by sieving, and the percentage weight loss is determined for each fraction. The percentage loss is expressed as a weighted average. For a given sieve size, the percentage weighted average loss is the product of the percentage passing that sieve and the percentage passing that sieve in the original material. The total of such values is the percent loss test value.

15-17 Specific Gravity and Absorption

The specific gravity and absorption of aggregates are important properties that are required for the design of concrete and bituminous mixes. The specific gravity of a solid is the ratio of its mass to that of an equal volume of distilled water at a specified temperature. Because aggregates may contain water-permeable voids, two measures of specific gravity of aggregates are used: *apparent* specific gravity and *bulk* specific gravity.

Apparent specific gravity, G_A, is computed on the basis of the net volume of the aggregates, that is, the volume excluding the water-permeable voids. Thus,

$$G_A = \frac{M_D/V_N}{w} \qquad (15\text{-}6)$$

where M_D = dry mass of the aggregate
   V_N = net volume of the aggregates, excluding the volume of absorbed water
   w = density of water

The bulk specific gravity, G_B, is computed on the basis of the total volume of the aggregates including the water-permeable voids:

$$G_B = \frac{M_D/V_B}{w} \qquad (15\text{-}7)$$

where V_B = total volume of the aggregates, including the volume of absorbed water

The difference between the apparent and bulk specific gravities accounts for the water-permeable voids of the aggregates. One can measure the volume of such voids by weighing the aggregates dry and in a saturated, surface dry conditions, that is, with all permeable voids filled with water. The difference between the two masses is the mass of the absorbed water, M_w. The absorption of water is usually expressed as a percentage of the mass of the dry aggregate,

$$\text{percentage absoprtion} = \frac{M_w}{M_D} \times 100 \qquad (15\text{-}8)$$

EXAMPLE 15-1   The dry mass of a sample of aggregates is 1982.0 g. The mass in a saturated, surface dry condition is 2006.7 g. The net volume of the aggregates is 734.4 cm³. Find the apparent specific gravity, the bulk specific gravity, and the percentage absorption.

The detailed procedures for determining the specific gravity and absorption of coarse aggregate and fine aggregate are given in ASTM Methods C127 and C128.

Mass of absorbed water = 2006.7 − 1982.0 = 24.7g

$$\text{Volume of absorbed water} = \frac{24.7 \text{ g}}{1 \text{ g/cm}^3} = 24.7 \text{ cm}^3$$

Bulk volume = 734.3 + 24.7 = 759.1 cm³

By Eq. 15-6, the apparent specific gravity

$$G_A = \frac{1982.0/734.4}{1 \text{ g/cm}^3} = 2.699$$

By Eq. 15-7, the bulk specific gravity

$$G_B = \frac{1982.0/759.1}{1 \text{ g/cm}^3} = 2.611$$

$$\text{percentage absorption} = \frac{24.7}{1982.0} \times 100 = 1.25\%$$

15-18 Chemical Stability of Aggregates

Certain aggregates may be unsuitable for a particular highway construction application because of the chemical composition of the aggregate particles. In asphalt mixes, certain aggregates that have an excessive affinity for water may contribute to what is known as film stripping, leading to disintegration of asphalt concrete.

An aggregate that is "hydrophobic" in nature may be said to be one which exhibits a high degree of resistance to film stripping in the presence of water. The bituminous substance in a bituminous mixture may generally be assumed to be present in the form of thin films surrounding the aggregate particles and filling, or partially filling, the void spaces between adjacent particles. These thin films of bituminous material adhere to the surface of normal aggregates and contribute to the shearing resistance of the mixture, this effect being generally considered as a part of the "cohesion" of the mix. On continued exposure to water, either in the laboratory or in the field, bituminous mixtures containing certain aggregates show a definite tendency to lose shearing resistance or "strength" because of a decrease in cohesion due primarily to replacement of the bituminous films surrounding the aggregate particles with similar films of water. Aggregates that exhibit this tendency to a marked, detrimental degree are termed "hydrophilic" aggregates, hydrophilic meaning "water-loving." Conversely, aggregates that show little or no decrease in strength due to film stripping are called "hydrophobic" or "water-hating."

In judging the relative resistance to film stripping of aggregates, several different laboratory procedures have been used, including various immersion-stripping tests, such as the ASTM D1664 Static Immersion Test, and various immersion-mechanical tests, such as the ASTM D1075 Immersion-Compression Test.

Aggregates used in portland cement concrete mixes may also cause problems related to chemical stability. In certain areas, considerable difficulty has been experienced with aggregates that contain deleterious substances which react harmfully with the alkalies present in the cement. Detrimental alkali-aggregate reactions generally result in abnormal expansion of the concrete. Methods have been devised (ASTM Methods C227 and C289) for detecting aggregates with these harmful characteristics, and suitable stipulations are included in typical specifications (for example, ASTM C33).

15-19 Other Properties of Aggregates

Specifications for aggregates used in highway construction commonly have requirements related to the particle shape, surface texture, and cleanliness of the aggregate. Specifications for aggregates used in bituminous mixes usually require that the aggregates be clean, tough, durable in nature, and free of excess amounts of flat or elongated pieces, dust, clay balls, and other objectionable material.

Similarly, aggregates used in portland cement concrete mixes must be clean and free of deleterious substances such as clay lumps, chert, silt, and other organic impurities.

Cleanliness of the aggregate is generally guaranteed by the inclusion in the specifications of requirements relative to the maximum percentages of various deleterious substances that are permitted to be present. The Sand Equivalent Test (ASTM D2419) is used to determine the relative proportions of plastic fines and dust in fine aggregate. Specific requirements in this respect vary somewhat with different agencies.

BITUMINOUS MATERIALS

Bituminous materials are used extensively for roadway construction, primarily because of their excellent binding or cementing power and their waterproofing properties, as well as their relatively low cost. Bituminous materials consist primarily of bitumen, which, according to ASTM D8, is a class of black or dark-colored solid or viscous cementitious substances composed chiefly of high-molecular-weight hydrocarbons; by definition, it is soluble in carbon disulfide.

Bituminous materials are divided into two broad categories: *asphalts* and *tars*. Asphalts are the residues of the petroleum oils. A great majority of asphalts used nowadays are the residues from the refinery of crude oils, although there are natural deposits called "native asphalt." Tars are residues from the destructive distillation of organic substances such as coal, wood, or petroleum. Tars obtained from the destructive distillation are crude tars, which must undergo further refinement to become road tars.

Asphalts have no odor, are more resistant to weathering, and less susceptible to temperature than tars, which have a pungent (creosote-like) odor and react to weathering and temperature. Asphalt will be dissolved in petroleum oils whereas tars will not. Therefore, tars have been used to seat asphalt concrete surfaces, such as fog seals, to improve the oil resistance of asphalt surfaces. Asphalts are black in color, whereas tars are usually brown-black in color.

Today, tars are not used extensively as binders for highway pavements, and therefore, they are treated only lightly here.

PRODUCTION OF ASPHALT

As mentioned earlier, asphalts are the residues, by-products, of the refinery of petroleum oils. A wide variety of refinery processes, such as the straight distillation process, solvent deasphalting process, and solvent extraction process, may be used to produce asphalt of different consistency and other desirable properties. Depending on the sources and characteristics of the crude oils and on the properties of asphalt required, more than one processing method may be employed.

15-20 Vacuum-Steam Distillation of Petroleum Oils

Figure 15-8 shows the vacuum-steam distillation process. The crude oil is heated and is introduced into a large cylinder still. Steam is introduced into the still to aid in the vaporization of the more volatile constituents (lighter fraction of the hydrocarbons) of the petroleum and to mimimize decomposition of the distillates

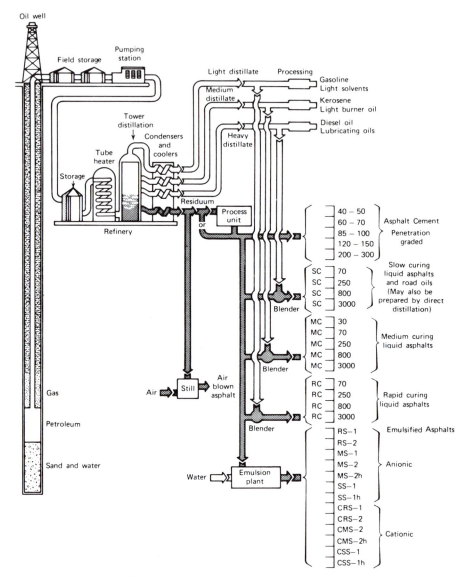

FIGURE 15-8 Simplified flow chart of recovery and refining of petroleum asphalts. (Courtesy The Asphalt Institute.)

and residues. The volatile constituents are collected, condensed, and the various fractions stored for further refining, if needed. The residues from this distillation are then fed into a vacuum distillation unit, where reduced pressure and steam will further separate out heavier gas oils. The bottom fraction from this unit is the vacuum-steam–refined asphalt cement. The consistency of asphalt cement from this process can be controlled by the amount of heavy gas oil removed. Normally, asphalt cement produced by this process is softer. As the asphalt cement cools down to room temperature, it becomes a semisolid viscous material. Consistency of the asphalt cement produced at this point can be further modified by the air-blowing, the solvent deasphalting, or the solvent extraction processes.

15-21 Air Blowing

Air blowing is normally used when the viscosity of asphalt residue from the vacuum process must be increased, as in the production of roofing asphalt. A vacuum residue from the vacuum process is pumped into a chamber where the temperature is raised to a temperature of about 260 to 300°C (500°F to 575°F). Air is introduced into the bottom of the chamber and blown through the heated asphalt residue for several hours until the residue has attained the consistency desired. Roofing asphalt is manufactured with the batch air-blowing process. Continuous air blowing, which involves a lesser degree of oxidation, is usually used for producing paving asphalt cement.

15-22 Solvent Deasphalting and Solvent Extraction

In the solvent deasphalting process, solvent (such as propane or butane) is used to strip lighter fractions of oils still from a vacuum residue. A high softening point, harder asphalt is produced in this process. This harder asphalt cement can be used to blend with vacuum asphalt to produce paving grade asphalt of different consistency or different viscosity.

A new process called the residuum oil supercritical extraction (ROSE) uses solvent (such as pentane) to mix vacuum residue under supercritical conditions, and the mixture is fed into a separator at predetermined controlled temperatures and pressures to separate the asphalt residue into different components (asphaltenes, resins, and oils). Asphaltenes or resins fraction from the ROSE process is used to blend with vacuum asphalt cement to meet the asphalt cement specification requirements.

15-23 Cutback Asphalt

Asphalt cement produced from the vacuum-steam distillation exists as a semisolid at room temperature, and usually proper workability can be attained by heating the asphalt cement to a temperature of 120 to 165°C (250°F to 330°F) to liquify it. In order for asphalt products to attain workability at room temperature, they must be rendered liquid at room temperature. There are two ways to liquefy asphalt without resorting to heat: dissolve (cut) the asphalt in solvent or emulsify it in water. Emulsified asphalt will be presented in the next section, but the use of solvent to liquify asphalt will be presented here.

When volatile solvents are mixed with asphalt cement to make a liquid product, the mixture is called "cutback asphalt." After a cutback asphalt is exposed to air, the volatile solvent evaporates, and the asphalt in the mixture regains its original characteristics (cured). Depending on the volatility of the solvent used, the rate of curing of cutback asphalt can vary from a few minutes to several days. Following are three types of cutback asphalt and the solvent used.

1. Rapid-curing (RC): gasoline or naphtha
2. Medium-curing (MC): kerosene
3. Slow-curing (SC): road oils

Cutback asphalt is commercially available in different grades, as shown in Table 15-5. The suffix numbers, for example MC-70, represent the minimum kinematic viscosity in centistokes at 60°C (140°F) for the particular grade. Specifications for RC, MC, and SC are given in ASTM D2026, D2027, and D2028,

TABLE 15-5 Typical Uses of Asphalt

Column groups: **Asphalt Cements** — Viscosity graded–original (AC-40, AC-20, AC-10, AC-5, AC-2.5); Viscosity graded–residue (AR-16000, AR-8000, AR-4000, AR-2000, AR-1000); Penetration graded (40-50, 60-70, 85-100, 120-150, 200-300). **Liquid Asphalts** — Rapid curing (RC) (70, 250, 800, 3000); Medium curing (MC) (30, 70, 250, 800, 3000); Slow curing (SC) (70, 250, 800, 3000); Emulsified Anionic (RS-1, RS-2, MS-1, MS-2, MS-2h, SS-1, SS-1h); Emulsified Cationic (CRS-1, CRS-2, CMS-2, CMS-2h, CSS-1, CSS-1h).

Type of Construction	AC-40	AC-20	AC-10	AC-5	AC-2.5	AR-16000	AR-8000	AR-4000	AR-2000	AR-1000	40-50	60-70	85-100	120-150	200-300	RC-70	RC-250	RC-800	RC-3000	MC-30	MC-70	MC-250	MC-800	MC-3000	SC-70	SC-250	SC-800	SC-3000	RS-1	RS-2	MS-1	MS-2	MS-2h	SS-1	SS-1h	CRS-1	CRS-2	CMS-2	CMS-2h	CSS-1	CSS-1h
Asphalt-aggregate mixtures																																									
Asphalt concrete and Hot laid plant mix																																									
Pavement base and surfaces																																									
Highways		X	X	X	X[a]		X	X	X	X[a]		X	X	X	X[a]																										
Airports		X	X	X			X	X	X			X	X	X																											
Parking areas		X	X	X			X	X	X			X	X	X																											
Driveways		X	X				X	X				X	X																												
Curbs	X	X				X	X				X	X																													
Asphalt-aggregate applications																																									
Surface treatments																																									
Single surface treatment				X	X									X	X		X	X	X				X	X					X	X						X	X				
Multiple surface treatment				X	X									X	X		X	X	X				X	X					X	X						X	X				
Aggregate seal				X	X				X					X	X		X	X	X			X	X						X	X						X	X				
Sand seal																	X				X	X							X	X						X	X				
Slurry seal																																		X	X					X	X
Penetration macadam																																									
Pavement bases																																									
Large voids				X				X					X					X	X					X					X	X						X	X				
Small voids														X					X										X	X						X	X				
Asphalt applications																																									
Surface treatment																																									
Fog seal																																		X[c]	X[c]					X[c]	X[c]
Prime coat, open surfaces																X	X			X	X	X												X							
Prime coat, tight surfaces																				X	X													X[c]	X[c]					X[c]	X[c]
Tack coat																X													X					X[c]	X[c]	X				X[c]	X[c]
Dust laying																					X																				
Crack filling																																									
Asphalt pavements	X[e]										X[e]																							X[d]	X[d]					X[d]	X[d]
Portland cement concrete pavements																																									

[a] For use in cold climates
[b] For use in bases only in cold climates
[c] Diluted with water
[d] Slurry mix
[e] Rubber asphalt compounds

Source: Asphalt Institute.

respectively. Cutback asphalt is increasingly being replaced by emulsified asphalt in commercial use.

15-24 Emulsified Asphalt

Emulsified asphalt is a mixture of asphalt cement, water, and an emulsifying agent. These three constituents are fed simultaneously into a colloid mill to produce extremely small globules (5–10 μ) of asphalt cement, which are suspended in the water. The emulsified agent imparts the electric charges (cationic or anionic) to the surface of the asphalt particles, which causes them to repel one another; thus the asphalt particles do not coalesce. The emulsified asphalt thus produced is quite stable and could have a shelf life of several months.

The two most common types of emulsified asphalts are anionic and cationic, a classification dependent on the emulsified agent. The anionic type contains electronegatively charged asphalt globules, and the cationic type contains electropositively charged asphalt globules.

The choice of anionic or cationic emulsions is important when considering various types of aggregates. The anionic emulsions carry a negative charge and are effective in coating electropositive aggregates such as limestone. The positive charge in a cationic emulsion reacts favorably with an electronegative aggregate, such as the highly siliceous aggregates.

When an emulsified asphalt is exposed to the air, alone or mixed with an aggregate, it "sets" or "breaks," because the asphalt globules react with the surface they are in contact with and coalesce, squeezing out the water between them. The evaporation of water is the primary mechanism that finally causes the anionic emulsified asphalt to "break." Electrochemical processes are the primary mechanisms that cause the cationic emulsified asphalt to break.

Both anionic and cationic emulsified asphalt are further graded according to their rate of setting. The following grades are available commercially:

Anionic emulsified asphalt:
    Rapid setting (RS)   RS-1, RS-2, HFRS-2
    Medium setting (MS) MS-1, MS-2, HFMS-2, MS-2h
    Slow setting (SS)   SS-1, SS-1h
Cationic emulsified asphalt:
    Rapid setting (CRS) CRS-1, CRS-2
    Medium setting (MS) CMS-2, CMS-2h
    Slow setting (SS)   CSS-1, CSS-1h

The "h" designation means a harder base asphalt cement is used in the emulsion. The "HF" designation refers to a high-float residue of the emulsion. Specifications for anionic-emulsified asphalts and cationic-emulsified asphalts are given in ASTM D977 and ASTM D2397, respectively.

Emulsified asphalt offers certain advantages in construction, particularly when used with moist aggregates or in wet weather. An emulsified asphalt does not require a solvent to make it liquid and thus is relatively pollution-free. Because emulsified asphalt has low viscosity at the ambient temperature, it generally can be used without additional heat. These factors tend to make emulsified asphalt more energy-efficient and less costly than cutback asphalt (7).

LABORATORY TESTS

A large number of different laboratory tests are performed on bituminous materials for the purpose of checking compliance with the specifications that are being used. Most of the tests that are performed in the laboratories of the various highway agencies have come to be more or less routine in nature and are conducted in accordance with methods of test established by the American Association of State Highway and Transportation Officials and the American Society for Testing and Materials. Some of the tests are intended to measure specific properties of the material, whereas others are used primarily as identification tests or in checking uniformity of the material.

In order to permit the student to obtain a better concept of the number of tests in common use, Table 15-6 has been prepared. This lists the names of the most commonly used tests and the corresponding AASHTO and ASTM designations for each method of test, as well as indicating whether the test is normally performed on asphaltic materials only, on tar products only, or on both. Some of these test methods are briefly outlined in the following pages. The reader is referred to the designated methods of test for more complete descriptions. In the following discussion an attempt to made to give some idea of the significance of the various tests. Tests performed on emulsified asphalts are mentioned separately, and no attempt has been made to discuss the testing of various miscella-

TABLE 15-6 Laboratory Tests of Bituminous Materials Used in Road Construction

Name of Test	AASHTO Designation	ASTM Designation	Applicability[a]
General			
Specific gravity	T228	D70-D76	3
Flash point (open cup)	T48	D92	1
Flash point ("Tag" open cup)	T79	—	1
Water in petroleum products, etc.	T55	D95	3
Solubility			
Solubility of bituminous materials	T44	D2042	3
Spot test	T102	—	1
Consistency			
Specific viscosity (Engler)	T54	D1665	2
Kinematic viscosity	T201	D2170	3
Absolute viscosity	T202	D2171	1
Float test	T50	D139	3
Penetration	T49	D5	1
Softening point (ring-and-ball method)	T53		3
Ductility			
Ductility	T51	D113	1
Volatility			
Distillation			
Cut-back asphaltic products	T78	D402	1
Tar products	T52	D20	2
Loss on heating	T47	D6	1
Thin-film oven test	T179	D1754	1
Tests for emulsions	T59	D244	1

[a]1, Asphaltic materials only; 2, tar products only; 3, both asphalts and tars.

neous substances derived from bituminous materials and of use in a minor way in highway work.

CONSISTENCY TESTS

Several consistency tests are commonly used in connection with the laboratory examination of bituminous materials. The preferred method for measuring the viscosity of liquids is by kinematic viscometers, as described in the following section. Consistency tests for bituminous materials in the solid or semisolid state include the float test, the penetration test, and the ring-and-ball test.

15-25 Kinematic Viscosity and Absolute Viscosity

Absolute viscosity of asphalt cements, in poises, is measured by standard test procedure AASHTO Designation T202. This test is usually performed at a temperature of 60°C (140°F). The test involves measuring the time required for a fixed volume of the liquid to be drawn through one of several specially designed capillary tubes by means of a vacuum. The absolute viscosity in poises is determined by multiplying the flow time in seconds by the viscometer calibration factor.

Kinematic viscosity is the quotient of the absolute or dynamic viscosity divided by the density, both at the same temperature (2). It is measured in dimensions of centimeters squared per second, or "stokes." For liquid asphalts, the kinematic viscosity may be measured with a gravity-flow viscometer (AASHTO Designation T201). According to the AASHTO procedure, the time for the fluid to flow between two points in a capillary tube under carefully controlled conditions of temperature and head is measured. Using the measured time in seconds and the viscometer calibration constant, it is possible to compute the viscosity of the material in fundamental units, stokes, or centistokes. A kinematic capillary test apparatus is shown as Figure 15-9.

15-26 Float Test

As previously indicated, the float test is used for determining the consistency of semisolid to solid road tars. Specified grades of heavier road tars are based on this test, performed at a temperature of either 32° or 50°C (90° or 122°F). The test is also useful in evaluating the consistency of the residues obtained in the distillation of slow-curing liquid asphalts.

This test, which is quite arbitrary in nature, involves use of the apparatus shown in Figure 15-10: The brass collar *B*, shown in Figure 15-10, is filled with the material to be tested and brought to a temperature of 5°C (41°F) by immersing it in ice water. A water bath is prepared, and the temperature of the water in the bath is brought to the desired level (32° or 50°C). The brass collar containing the chilled bituminous material is then screwed into the bottom of the float and the assembly is floated in the water bath maintained at the desired temperature. As time passes, the material in the plug softens, and eventually the water breaks through the plug of bituminous material into the upper part of the float. The time, in seconds, that elapses between the instant when the assembly is floated in the water and that when the water breaks through is the reported value. This time is held to be a measure of consistency.

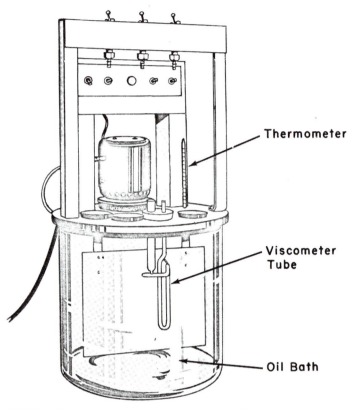

FIGURE 15-9 Apparatus for kinematic capillary viscosity test. (Courtesy The Asphalt Institute.)

15-27 Penetration

The penetration of a bituminous substance may be defined as the distance (in hundredths of a centimeter) to which a standard needle penetrates the material under known conditions of time, loading, and temperature. This test is used for evaluating the consistency of asphaltic materials; it is not regarded as suitable for use in connection with the testing of road tars because of the high surface tension exhibited by these materials and the fact that they contain relatively large amounts of free carbon.

The standard penetration test procedure involves use of the standard needle under a load of 100 g for 5 sec at a temperature of 25°C (77°F). In Figure 15-11 is shown a so-called precision-type penetrometer that may be used for laboratory measurements of penetration. Simpler and more rugged types of penetrometers may also be used. The grade of semisolid and solid asphaltic materials is usually designated by the penetration. For example, penetration ranges such as 30–40, 40–50, 50–60, 60–70, 70–85, and 85–100 may be used in specifying the desired grades of asphalt cements prepared from petroleum.

At the beginning of this chapter it was mentioned that bituminous materials could be classed as liquid, semisolid, or solid on the basis of their consistency. This classification may be arbitrarily based on the penetration test, as shown in Table 15-7.

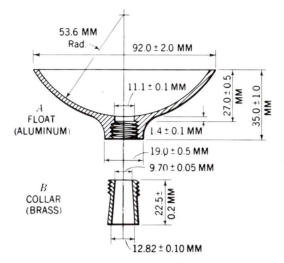

53.6 MM
Rad.

92.0 ± 2.0 MM

11.1 ± 0.1 MM

27.0 ± 0.5 MM

35.0 ± 1.0 MM

A
FLOAT
(ALUMINUM)

1.4 ± 0.1 MM

19.0 ± 0.5 MM

9.70 ± 0.05 MM

B
COLLAR
(BRASS)

22.5 ± 0.2 MM

12.82 ± 0.10 MM

FIGURE 15-10 Apparatus for float test.

15-28 Softening Point (Ring-and-Ball Method)

Because the softening of a bituminous material does not take place at any definite temperature, but rather involves a gradual change in consistency with·increasing temperature, any procedure that is adopted for determining the softening point must be of a somewhat arbitrary nature. The procedure in common use in highway materials laboratories is known as the "ring-and-ball method" and may be applied to semisolid and solid materials.

In this test procedure, the sample is melted and poured into a brass ring, $\frac{5}{8}$ in. in inside diameter and $\frac{1}{4}$ in. high. After cooling, the ring is suspended in a water bath maintained at a temperature of 5°C (41°F), in such a fashion that the bottom of the filled ring is exactly 1 in. above the bottom of the bath. A steel ball $\frac{3}{8}$ in. in diameter is placed on the surface of the bituminous material contained in the ring, and the temperature of the water is elevated at a standard rate. As the temperature is increased, the bituminous material softens and the ball sinks through the ring, carrying a portion of the material with it. The behavior of the ball is observed, and the softening point is taken to be the temperature at which the bituminous material touches the bottom of the container. The ring-and-ball method is used to determine the penetration index and, in conjunction with penetration and loading time, can be used to estimate viscoelastic properties of both the asphalt and a paving mixture. The method can also be used to advantage to predict high-temperature shear resistance of the paving mixture (3).

DUCTILITY

By ductility is generally meant that property of a material that permits it to undergo great deformation (elongation) without breaking. In regard to bituminous materials, ductility may be further defined as the distance, in centimeters, to which a standard sample or "briquette" of the material may be elongated without breaking. The test is applicable only to semisolid (or solid) asphaltic material that is melted by gentle application of heat and poured into a standard

FIGURE 15-11 Penetrometer.

mold. Dimensions of the mold are such that the minimum cross section of the briquette thus formed is exactly 1 cm². The mold is then immersed in a water bath maintained at the desired temperature of the test, which is usually 25°C (77°F).

After the material has attained the desired temperature, the sample is placed in a ductility machine. The machine is so arranged that one end of the mold is held in a fixed position while the other end is pulled horizontally at a standard rate. The behavior of the "thread" of material is noted, and elongation is continued until the thread breaks. The distance (in centimeters) that the machine has traveled is the "ductility" of the material. Many asphalt cements have a ductility of 100 or more.

TABLE 15-7 Penetration Limits of Consistency

Classification	Penetration	Load (g)	Time (sec)	Temperature (°C)
Liquid	351 or more	50	1	25
Semisolid	{ 350 or less	50	1	25
	{ 11 or more	100	5	25
Solid	10 or less	100	5	25

In interpreting the results of this test, it is generally assumed that ductility is a measure of the cementing power of the asphaltic material. Since high cementing qualities are desirable in most applications, it is held that an asphalt that is to be used as a binder should be ductile; however, it should be noted that the exact value of the ductility is not as important as the mere presence or lack of ductility.

VOLATILITY TESTS AND AGING TESTS

15-29 Distillation

The objective of a distillation test performed on a bituminous material is simply to separate the volatile from the nonvolatile substances. By fractional distillation the relative proportions of volatile materials driven off at various specified temperatures may be determined, and the residue may be tested to ascertain the properties of the residual substance contained in the original fluid material.

In testing cut-back asphaltic products, including slow-curing liquid asphalts, the apparatus shown in Figure 15-12 is used. As the distillation proceeds, the volatiles are condensed and collected in the graduated cylinder of Figure 15-12. The original volume of the material contained in the flask is 200 mL, and the temperature of the material is increased at a gradual rate. It may be desirable to know the proportion of distillate driven off at various selected temperatures. The percentage of volatiles collected to any given temperature may be determined on a volumetric basis. The residue is cooled and set aside for additional testing.

The distillation test supplies a very considerable amount of information, because it permits the testing engineer to ascertain the kind and amount of volatile material that has been used in fluxing or "cutting back" a semisolid to produce a liquid substance. Furthermore, by removal of the volatile substances, the engineer creates in the laboratory a material resembing that which will exist in the field, that is, the volatiles are removed, leaving the residual that will go to make

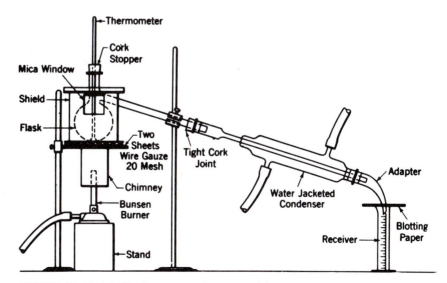

FIGURE 15-12 Distillation apparatus assembly.

up the binder in a mixture of the bituminous material and mineral aggregate. The residue may then be tested to determine whether or not it has the qualities the engineer deems desirable and necessary. Asphalt emulsions are also tested by volatility tests to determine the water content and the amount of the residue from a sample.

15-30 Loss on Heating

In determining the percentage of volatile material that will be removed from the asphaltic substance being examined under the imposed conditions of the loss on heating or "volatilization" test, 50 g of the selected material is carefully weighed into a standard flat cylindrical container. The sample is then placed in a specially constructed oven maintained at 163°C and allowed to remain there for a period of 5 hr. At the end of this period the sample is removed from the oven, cooled, and weighed. The loss in weight may then be determined, and the percentage loss on heating, based on the weight of the original sample, may then be calculated. Detailed procedures for this test are given by AASHTO Designation T47.

15-31 Thin-Film Oven Test

Relatively high temperatures are used in the plant mixing of asphalt cements and aggregates. Excessively high temperatures, however, are detrimental, hardening the mixture and reducing pavement life.

In measuring the amount of hardening of an asphalt cement that may be expected to occur during plant mixing, the standard penetration test is made before and after the thin-film oven test (AASHTO Designation T179). Hardening is recorded as the percentage of penetration of the sample before the test. A representative sample, 50 cm³ in volume, is poured into a pan of specified dimensions such that a film approximately $\frac{1}{8}$ in. thick is tested. This sample is then placed on an aluminum rotating shelf in a ventilated oven at 163°C (325°F) and rotated at a specified rate for 5 hr. It is then poured into a standard container and its penetration determined. The percentage of the penetration before the thin-film oven test is calculated and reported. Specifications usually prescribe minimum values for the percentage of retained penetration for the various grades of asphalt cement.

15-32 Flash Point

Two methods are in common use for the determination of the flash point of native and petroleum asphalts used in highway work. The flash point of cutback asphalts (RC and MC) is generally determined by use of a Tagliabue Open Cup apparatus, whereas the Cleveland Open Cup is used for flash-point determinations on other asphaltic materials. The two methods are essentially the same, although in the Cleveland Open Cup method the asphalt is heated in a metal container suspended in an air bath, whereas in the "Tag" Open Cup method heating takes place in a glass cup held in a water bath.

In both methods, the temperature of the asphaltic material is gradually increased at a uniform rate. As the temperature rises, a small, open flame is passed at specified intervals across the surface of the heated material. Volatile constituents of the asphalt are driven off as the temperature is increased. As the test proceeds, a point will be reached where sufficient volatiles will collect at the surface of the heated asphalt to briefly ignite and cause a distinct flicker or "flash" across the surface of the material. The minimum temperature at which this flash occurs is called the flash point.

15-33 Specific Gravity Test

The "specific gravity" of a bituminous material is defined as the ratio of the weight of a given volume of the material at 25°C (77°F) to that of an equal volume of water at the same temperature. The specific gravity may be determined by use of a hydrometer, by displacement, or by use of a pycnometer. The hydrometer method has been used in the evaluation of thin fluid materials, such as emulsified asphalts. In this method, the fluid is brought to the desired temperature and the specific gravity is measured directly by immersing a suitable calibrated and marked hydrometer in the fluid mass. The displacement method may be used to determine the specific gravity of bituminous materials that are sufficiently hard and solid to be handled in fragments. The method involves determination of the weight in air and weight in water of an irregular piece of the material. A simple calculation may then be employed to determine the specific gravity. The hydrometer and displacement methods are seldom used. The majority of laboratories use the pycnometer method. This method is given in ASTM D70.

Because the specific gravity of asphalt cement varies with the temperature, test results are expressed in terms of specific gravity at a given temperature for both the asphalt cement and the water used in the test.

TESTING OF EMULSIFIED ASPHALTS

Emulsified asphalts are specified on the basis of the results of a complete series of tests, some of which are of a specialized nature. Standard methods of testing emulsified asphalts are designated as AASHTO T59 and ASTM D244. Tests of the following categories are included: composition, consistency, and stability. Composition tests include water content, in which a 50- or 100-g sample is mixed with an equal volume of water-immiscible solvent, distilled, and the quantity of water determined by volume; residue by distillation, in which the percentage of weight of residue is obtained by distillation of a 200-g sample in an iron still to 260°C (500°F); and residue by evaporation, which is less frequently used. Consistency is measured by use of the standard Saybolt viscometer with the sample at 25° or 50°C (77° or 122°F) using the Furol tip (AASHTO Designation T72).

Stability tests include demulsibility, in which a 100-g sample is mixed with a dilute solution of calcium chloride and the percentage by weight of asphalt that fails to pass a No. 14 wire cloth determined; settlement, in which the difference in asphalt content of top and bottom samples is determined on a 500-mL sample after allowing it to stand for 5 days; cement mixing, in which the sample is diluted to 55 percent residue with water, mixed with cement, and the percentage by weight of coagulated material that fails to pass a No. 14 sieve determined; a sieve test, in which the percentage by weight that fails to pass the No. 20 sieve is determined; a coating test, in which the ability of a sample to coat a specified stone is judged visually after mixing; and miscibility with water, in which approximately 50 mL of sample is mixed with 150 mL of water and the visible coagulation after 2 hr observed. Tests are also performed on the residue, including specific gravity, ash content, solubility in CS_2, penetration and ductility.

A particle charge test is made to identify cationic emulsions and a pH test is performed to determine their acidity.

CLASSIFICATION OF BITUMINOUS MATERIALS

Bituminous materials are commonly grouped in various classes or grades based on consistency. The classification of these materials facilitates the development of specifications and provides the framework for the establishment of product testing and quality controls. Grading of asphalt cements will be given in the following section. Grading of cutback asphalts and emulsified asphalts are given in Section 15-23 and 15-24, respectively.

15-34 Grades of Asphalt Cements

Until about 1970, the consistencies of asphalt cements were based on the penetration test (see Section 15-27), and specifications were commonly written in those terms. The penetration grades of asphalt cement are as follows:

Asphalt Cement Grade	Penetration Range (100 g, 5 sec)
AC 40–50	40–50
AC 60–70	60–70
AC 85–100	85–100
AC 120–150	120–150
AC 200–300	200–300

The grades of asphalt cements are now more commonly bassed on a standard capillary viscometer test, AASHTO Designation T202 (2). The grades and corresponding viscosity ranges are as follows:

Asphalt Cement Grade	Viscosity, 60°C (140°F) (poise)
AC 2.5	250 ± 50
AC 5	500 ± 100
AC 10	1000 ± 200
AC 20	2000 ± 400
AC 30	3000 ± 600
AC 40	4000 ± 800

Another grading procedure for asphalt cements has been adopted by several Western states. It uses the results of viscosity tests made on the residue from a rolling thin-film oven test (AASHTO Designation T240). The grading system is as follows:

Asphalt Cement Grade	Viscosity, 60°C (140°F) (poise)
AR 10	1,000 ± 250
AR 20	2,000 ± 500
AR 40	4,000 ± 1,000
AR 80	8,000 ± 2,000
AR 160	16,000 ± 4,000

USES OF BITUMINOUS MATERIALS

The bituminous materials that have been described in the preceding portions of this chapter find many and varied uses in highway construction and maintenance. (See Table 15-5.) Additional information on the nature, uses, and testing of asphalt is given in Ref. 8 and other publications of the Asphalt Institute.

PORTLAND CEMENT

Portland cement is a material that reacts chemically with water by a process called hydration to form a stonelike mass. It was patented by an English mason named Joseph Aspdin in 1824. He named his product portland cement because it produced a concrete that resembled the natural limestone on the Isle of Portland.

Portland cement is composed of four principal compounds: tricalcium silicate, dicalcium silicate, tricalcium aluminate, and tetracalcium aluminoferrite. Raw materials used in the manufacture of portland cement include lime, iron, silica, alumina, gypsum, and magnesia. A wide variety of natural and industrial materials may be used as sources of raw materials in the manufacture of portland cement.

When mixed with water, sand, and gravel, portland cement produces concrete for pavements, bridges, drainage structures, buildings, and other structures. The process by which portland cement concretes are made is described in Chapter 20.

In the manufacture of portland cement, raw materials are crushed and blended to produce the desired proportions of minerals and then burned in a large rotary kiln. The high temperatures 1400° to 1650°C (2550 to 3000°F) in the kiln produce rocklike particles called clinker. The clinker is then cooled, blended with gypsum, and ground to a fine powder. The resulting pulverized material is ready for use as portland cement.

The blend of the constituent compounds in portland cement determines how rapidly it sets or hardens in a concrete mix, how much heat is generated by the hardening process, and how it reacts with other chemicals in soils and ground waters.

ASTM provides specifications for eight types of portland cement in ASTM Designation C150. Three of these cements are frequently used in highway construction: types I, II, and III.

Type I is what might be termed standard or "normal" portland cement and is intended for use in general concrete construction where the cement is not required to have special properties. Type I cement is normally supplied by the manufacturer, unless one of the other types is specified.

Type II cement is also regarded as a standard type of portland cement and is used for general concrete construction. It is specifically recommended for use in situations in which the concrete will be exposed to moderate sulfate action or where a moderate heat of hydration is required. Types I and II are both widely used by highway agencies at the present time in paving and structural work.

Type III cement is "high early strength" cement. It differs from the standard types described in that concrete made from it attains, in a much shorter period of time, compressive and flexural strengths that are comparable to those attained by concrete in which the same amount of one of the standard types is used. It is useful, for example, in paving operations in which the surface must be opened to traffic very soon after construction is complete.

Specifications for air-entraining cements cover three types that are comparable to types I, II, and III, except that they contain air-entraining agents that are interground with the cement at the mill. The properties of air-entrained concretes made by using cements of these types will be discussed more fully in Chapter 20, but the use of such a cement may be said to result in the trapping of a small percentage of air which is present in the form of small, disconnected bubbles, distributed uniformly throughout the mix. Air-entrained concretes are, among other things, more durable than comparable concrete mixtures that do not contain air. A number of state highway agencies permit (or require) the use of air-entraining cement, while others secure air-entrainment by the use of an admixture that is incorporated into the concrete during the mixing operation.

Common units of measurement of cement quantity are the sack, which weighs 94 lb (43 kg) and has an approximate loose volume of 1 ft³ (0.028 m³), and the ton. The barrel, which is equivalent to four sacks of 376 lb (170 kg) of cement, is seldom used today as a unit of measurement. Shipment of cement is sometimes made in sacks or bags but more commonly in bulk quantities.

PROBLEMS

15-1 Given the following information, calculate the group index and classify each of these soils into the proper subdivision of the AASHTO classification system.

Sample No.	Sieve Analysis			Characteristics of Soil Binder	
	Percent Passing				
	10 (2.00 mm)	40 (0.425 mm)	200 (0.075 mm)	Liquid Limit	Plasticity Index
1	100.0	97.5	56.1	62.3	12.7
2	100.0	73.1	5.4	20.1	NP
3	100.0	46.2	15.9	16.9	NP
4	77.2	37.1	28.2	33.1	6.8
5	100.0	100.0	87.6	60.2	21.7
6	100.0	100.0	83.7	54.2	33.6
7	100.0	56.3	19.1	24.8	10.6
8	100.0	96.3	75.6	33.7	8.9
9	100.0	100.0	83.1	50.6	8.6
10	100.0	100.0	95.0	30.3	11.3
11	100.0	73.7	29.0	42.9	12.7
12	37.1	21.1	8.6	12.3	NP

15-2 Classify each of the soils of Problem 15-1 under the Unified Soil Classification System. The following additional data are available for certain of the samples.

Sample No. 2	C_u = 3.5	
Sample No. 4	Passing No. 4 sieve	86.2%
	Passing ⅜-in. sieve	100.0%
Sample No. 12	Passing No. 4 sieve	49.9%
	Passing ½-in. sieve	80.6%
	Passing 1-in. sieve	100.0%
	C_u = 5	
	C_g = 2	

15-3 Given the following information from a compaction test performed in the laboratory by the standard Proctor compaction procedure, draw the moisture-density curve and determine optimum moisture and maximum density for this soil:

weight of mold = 2456 g

volume of mold = $\frac{1}{30}$ ft^3

Trial No.	Weight of Compacted Soil Plus Mold (g)	Moisture Content (percent)
1	4136	9.4
2	4205	11.2
3	4308	13.1
4	4408	13.9
5	4398	15.8
6	4354	17.9

15-4 Assuming that the soil of Problem 15-3 has a specific gravity of 2.74, make the necessary computations and plot the "zero air voids curve" on the drawing prepared in Problem 15-3.

15-5 The dry mass of a sample of aggregates is 1206.0 g. The mass in a saturated, surface dry condition is 1226.8 g. The volume of the aggregates, excluding the volume of absorbed water, is 440.6 cm^3. Calculate the bulk specific gravity, the apparent specific gravity, and the percentage absorption.

15-6 The following weights are recorded during the determination of the specific gravity of a bituminous material by the pycnometer method. Calculate the specific gravity of this substance.

Weight of pycnometer, empty	34.316 g
Weight of pycnometer, filled with water	60.000 g
Weight of pycnometer, filled with bituminous material	58.202 g
Temperature (all determinations)	25°C (77°F)

15-7 The following are the results of laboratory tests performed on a sample of medium-curing liquid asphalt (MC800) for the purpose of checking compliance with the specifications. Consult Ref. 2 and determine if this material complies with AASHTO specifications.

Flash point	49°C (120°F)
Kinematic viscosity at 140°F	805 centistokes
Distillation	
To 437°F	15.0%
To 500°F	34.5%
To 600°	45.0%
Residue from distillation to 680°F, by volume	75%
Tests on residue from distillation	
Penetration (standard conditions)	110
Ductility, 77°F	100 cm
Solubility in CCL_4	99.1%

15-8 Prepare a report on the significance of the following tests for bituminous materials:

- Penetration test.
- Kinematic viscosity test.

- Ductility test.
- Solubility in carbon disulfide test.
- Thin-film oven test.

REFERENCES

1. *Federal Aid Highway Construction Usage Factors.* Federal Highway Administration, Washington, DC (1994).

2. *AASHTO Materials,* Part I, Specifications, Part II, Tests, 16th ed. American Association of State Highway and Transportation Officials, Washington, DC (1993).

3. Sowers, G. F. *Introductory Soil Mechanics and Foundations,* 4th ed. Macmillan, New York (1979).

4. Atkins, Harold N. *Highway Materials, Soils, and Concretes,* 2nd ed. Reston Publishing Co., Reston, VA (1983).

5. *Materials Testing.* Joint publication of the U.S. Army, Navy, and Air Force, TM5-530/NAVFAC MO-330/AFM 89-3 Washington, DC (August 1987).

6. *Factors Influencing Compaction Test Results.* Highway Research Bulletin No. 319, Highway Research Board, Washington, DC (1962).

7. *A Basic Asphalt Emulsion Manual.* Manual Series No. 19, 2nd ed., The Asphalt Institute, Lexington, KY.

8. *A Brief Introduction to Asphalt and Some of Its Uses.* Manual Series No. 5, The Asphalt Institute, College Park, MD (1984).

CHAPTER 16

FLEXIBLE PAVEMENT DESIGN

*Generally speaking, pavements (and bases) may be divided into two broad classi-
fications or types: rigid and flexible. As commonly used in the United States, the
term "rigid pavement" is applied to wearing surfaces constructed of portland-
cement concrete. A pavement constructed of concrete is assumed to possess
considerable flexural strength that will permit it to act as a beam and allow it to
bridge over minor irregularities, which may occur in the base or subgrade on which
it rests; hence the term "rigid." Similarly, a concrete base that supports a brick or
block layer might be described as "rigid." The structural design of this type of
pavement deserves special consideration and is discussed in detail in Chapter 20.*

16-1 General Discussion of Flexible Pavements

All other types of pavements have traditionally been classed as "flexible." A
commonly used definition is that "a flexible pavement is a structure that maintains intimate contact with and distributes loads to the subgrade and depends on
aggregate interlock, particle friction, and cohesion for stability" (*1*). Thus, the
classical flexible pavement includes primarily those pavements that are composed
of a series of granular layers topped by a relatively thin high-quality bituminous
wearing surface. Typically, the highest quality materials are at or near the surface.
It should be pointed out that certain pavements that have an asphalt surface may
behave more like the classical "rigid" pavements, for example, pavements that
have very thick asphalt surfaces or that have base courses composed of aggregate
treated with asphalt, cement, or lime-fly ash. For convenience of presentation,
however, these pavements will be considered to be in the flexible class and their
design will be considered in this chapter. In Figure 16-1 is shown a typical flexible
pavement cross section. The principal elements of this structure are shown to be
the pavement that is composed of a "wearing surface," base, subbase (not always
used), and subgrade. The wearing surface and the base often comprise two or
more layers that are somewhat different in composition and that are put down
in separate construction operations. On many heavy-duty pavements, a subbase
of select material is often placed between the base and subgrade. The wearing
surface may range in thickness from less than 25 mm (1 in.) in the case of a
bituminous surface treatment used for low-cost, light-traffic roads to 150 mm (6
in.) or more of asphalt concrete used for heavily traveled routes. The wearing
surface must be capable of withstanding the wear and abrasive effects of moving
vehicles and must possess sufficient stability to prevent it from shoving and
rutting under traffic loads. In addition, it serves a useful purpose in preventing
the entrance of excessive quantities of surface water into the base and subgrade
from directly above. For some heavily traveled highways, a 13 mm to 18 mm

453

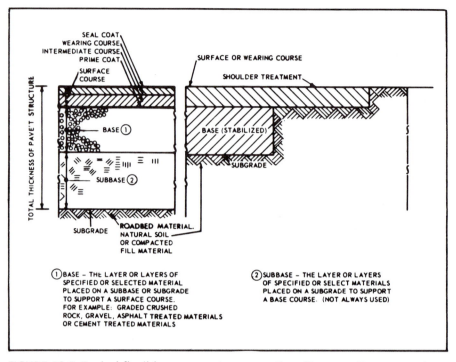

FIGURE 16-1 Typical flexible pavement cross section. (Courtesy Transportation Research Board.)

thickness of highly drainable open-graded friction course is placed on top of the wearing course for the purpose of improving skid resistance, minimizing hydroplaning effects at high speeds, and improving wet night visibility.

The base is a layer (or layers) of very high stability and density. Its principal purpose is to distribute or "spread" the stresses created by wheel loads acting on the wearing surface so that the stresses transmitted to the subgrade will not be sufficiently great to result in excessive deformation or displacement of that foundation layer. The base must also be of such character that it is not damaged by capillary water and/or frost action. Locally available materials are extensively used for base construction, and materials preferred for this type of construction vary widely in different sections of the country. For example, the base may be composed of gravel or crushed rock or it may be a granular material treated with asphalt, cement, or lime-fly ash stabilizing agents.

A subbase of granular material or stabilized material may be used in areas where frost action is severe, in locations where the subgrade soil is extremely weak or where a construction working table is needed. It may also be used, in the interests of economy, in locations where suitable subbase materials are cheaper than base materials of higher quality.

The subgrade is the foundation layer, the structure that must eventually support all the loads that come onto the pavement. In some cases this layer will simply be the natural earth surface. In other and more usual instances it will be compacted soil existing in a cut section or the upper layer of an embankment section. In the fundamental concept of the action of flexible pavements, the combined thickness of subbase (if used), base, and wearing surface must be great enough to reduce the stresses occurring in the subgrade to values that are not sufficiently great to cause excessive distortion or displacement of the subgrade soil layer.

16-2 Status of Thickness Design

Before World War II, the determination of the combined thickness of flexible pavement and base required in any set of circumstances was largely a matter of judgment and experience. This is attributed (2) to the fact that (1) this method appeared to be giving satisfactory results, (2) much of the basic knowledge required for a more scientific approach to the problem was not available, and (3) the methods of stage construction then in wide use did not seem to require the evolution of more scientific design methods. In the period immediately prior to the war and during its early stages, however, the necessity for economically designing and constructing a truly enormous mileage of airport runways and access roads to military installations focused attention on this design problem. Efforts directed toward the solution of the problem of economical design of flexible pavements and bases were further emphasized by the increased cost of road building during and after the war years.

As a consequence, many government agencies entered on comprehensive investigations that led to development of a large number of different design methods. There was (and is) no universal agreement among practicing engineers as to which of these is the best.

The start of construction of the Interstate System in 1956 greatly accelerated efforts to improve pavement design methods. The AASHO road test (Section 16-10) was the most extensive field test in road-building history. One major phase of the road test, which was completed in 1961, was concerned with flexible pavements. Since that time, the results of the road test have been distributed widely.

The AASHTO Interim Guide for Design of Flexible Pavement Structures was originally published in October 1961. Based on a review of this guide and research and experience since that time, an interim guide was published in 1972 (3). In 1977, it was reported (4) that 32 state highway agencies used the interim guide directly for pavement design. This interim guide was subsequently revised, and a new *AASHTO Guide For Design of Pavement Structures* was published in 1986 (5).

For many years, some engineers have wanted to design flexible pavements on an analytical basis, that is, by applying the theory of elasticity or similar concepts to the behavior of the layered system that is a flexible pavement. They have been hampered by the complexity of the problem and by lack of certain parameters to fit into the design equations. At present, aided by the accumulation of vast amounts of experimental data and extensive research activity, implementable flexible pavement designs based on analytical methods are available. Concepts of some mechanistic-based design methods are described in the AASHTO Guide (5). The Asphalt Institute method, discussed in Section 16-9, incorporates some of the analytic design concepts.

16-3 Elements of Thickness Design

From a somewhat simplified viewpoint, the principal factors entering into the problem of the thickness design of flexible pavements are

- traffic loading.
- climate or environment.
- material characteristics.

A number of other elements must also be considered in order to arrive at a final thickness design. These include cost, construction, maintenance, and design period. Thus, the student should realize that the design process is complex, and it is highly unlikely that any extremely simple method of approach will prove entirely successful under all conditions.

16-4 Traffic Loading

Protection of the subgrade from the loading imposed by traffic is one of the primary functions of a pavement structure. The designer must provide a pavement that can withstand a large number of repeated applications of variable-magnitude loadings. The primary loading factors that are important in flexible pavement design are

- magnitude of axle (and wheel) loads.
- volume and composition of axle loads.
- tire pressure and contact area.

The magnitude of maximum loading is commonly controlled by legal load limits. Traffic surveys and loadometer studies are often used to establish the relative magnitude and occurrence of the various loadings to which a pavement is subjected. Prediction or estimation of the total traffic that will use a pavement during its design life is a very difficult but obviously important task. Most design procedures provide for an increase in traffic volume on the basis of experience by using some estimated growth rate.

In many design methods (5, 6), the total estimated or projected magitude and occurrence of the various traffic loadings are converted to the total number of passes of the equivalent standard axle loading, usually the equivalent 80-kN (18-kip) single axle load (ESAL). The total number of ESALs are used as the traffic loading input for design of the pavement structure.

16-5 Climate or Environment

The climate or environment in which a flexible pavement is to be established has an important influence on the behavior and performance of the various materials in the pavement and subgrade. Probably the two climatic factors of major significance are temperature and moisture.

The magnitude of temperature and its fluctuations affect the properties of certain materials. For example, high temperatures cause asphaltic concrete to lose stability, whereas at low temperatures asphaltic concrete becomes very hard and stiff. Low temperature and temperature fluctuations are also associated with frost heave and freeze–thaw damage. Granular materials, if not properly graded, can experience frost heave. Likewise, the subgrade can exhibit extensive loss in strength if it becomes frozen and subsequently thaws. Certain stabilized materials (lime, cement, and lime-fly ash treated) can suffer substantial damage if a large number of freeze–thaw cycles occur in the material.

Moisture also has an important influence on the behavior and performance of many materials. Moisture is an important ingredient in frost-related damage.

Subgrade soils and other paving materials weaken appreciably when saturated, and certain clayey soils exhibit substantial moisture-induced volume change.

16-6 Material Characteristics

Proper design of flexible pavement systems requires a thorough understanding of important characteristics of the materials of which the pavement is to be composed and on which it is to be founded. Depending on the nature of the design procedure, the required material characteristics may vary, but in general the following are desirable:

1. **Asphalt surface.** Strength or stability (possibly repeated load properties).
2. **Granular base and subbase.** Gradation, strength, or stability (shear strength and/or possibly repeated load properties).
3. **Treated or stabilized layers.** Strength (flexural, compressive) and repeated load properties such as fatigue.
4. **Subgrade.** Strength or stability, soil classification, and possibly repeated load properties.

Various standard test methods are available for determining the desired properties. Many of the testing procedures are covered in ASTM and AASHTO Testing Standards (*7, 8*) whereas methods for conducting repeated load-type tests on paving and subgrade materials are discussed in ref. 9. Some of the traditional testing methods are covered in this book. For example, the Marshall stability test for asphalt mixtures is discussed in Section 19-10. The California bearing ratio (CBR) and the resilient modulus tests are presented in Sections 16-7 and 16-8. Shear strength is discussed briefly in Section 15-5.

16-7 California Bearing Ratio Test

The basic testing procedure employed in the determination of the CBR was developed by the California Division of Highways before World War II and was used by that agency in the design of flexible pavements. The basic procedures of this test were adopted by the Corps of Engineers of the U.S. Army during the early stages of the war and served as a basis for the development of design curves that were used for determining the required thickness of flexible pavements for airport runways and taxiways. Certain modifications were made in the test procedure that had been used in California. The modified method adopted by the Corps of Engineers has come to be regarded as the standard method of determining the CBR. The brief description given here will be limited to the testing of disturbed samples in the laboratory.

The selected sample of subgrade soil is compacted in a mold that is 152 mm (6 in.) in diameter and 152 to 178 mm (6 to 7 in.) high. The moisture content, density, and compactive effort used in molding the sample are selected to correspond to expected field conditions. After the sample has been compacted, a surcharge weight equivalent to the estimated weight of pavement and base is placed on the sample, and the entire assembly is immersed in water for 4 days. At the completion of this soaking period the sample is removed from the water and allowed to drain for a period of 15 min. The sample, with the same surcharge imposed on it, is immediately subjected to penetration by a piston 1.95 in. in diameter moving at a speed of 0.05 in./min. The total loads corresponding to penetrations of 0.1, 0.2, 0.3, 0.4, and 0.5 in. are recorded.

A load-penetration curve is then drawn, any necessary corrections made, and the corrected value of the unit load corresponding to 0.1-in. penetration determined. This value is then compared with a value of 1000 lb/in² required to effect the same penetration in standard crushed rock. The CBR may then be calculated by the expression

$$\text{CBR (\%)} = \frac{\text{unit load at 0.1-in. penetration}}{1000}(100) \qquad (16\text{-}1)$$

Figure 16-2 shows the apparatus required for conducting the CBR. It should be noted that this penetration test may also be performed in the field or on "undisturbed" samples. In certain cases the bearing ratio is computed at 0.2-in. penetration rather than at the standard 0.1 in. The value for standard crushed rock at 0.2-in. penetration is 1500 lb/in². Various modifications of the basic test procedure have been adopted by highway agencies that use a CBR-based method of design.

FIGURE 16-2 Apparatus for California bearing ratio test. (Courtesy Soiltest, Inc.)

16-8 Resilient Modulus

The resilient modulus is the equivalent "elastic modulus" of the materials in the pavement structure. It is well known that most materials that comprise flexible pavements are not elastic and exhibit inelastic behaviors such as permanent deformation and time-dependency (viscoelastic behavior). If the stress exerted on the material is small compared to its strength, however, and the exertion is repeated many times, the strain under each load application is nearly the same and is proportional to the stress; thus, it can be considered elastic. The latest versions of the AASHTO design method and the Asphalt institute design method have used the resilient modulus as the material property input for the subgrade soil.

The procedure for determining the resilient modulus of subgrade soils is specified by AASHTO in "T274 Resilient Modulus of Subgrade Soils" (8). Figure 16-3 shows the test setup. The soil sample, 71 mm in diameter by 152 mm high, is placed in the triaxial chamber and subjected to a prescribed confining pressure and an axial repeated loading, typically 0.1 sec of loading and 1 to 3 sec of unloading. The axial deformation (and the strain) of the soil sample due to the

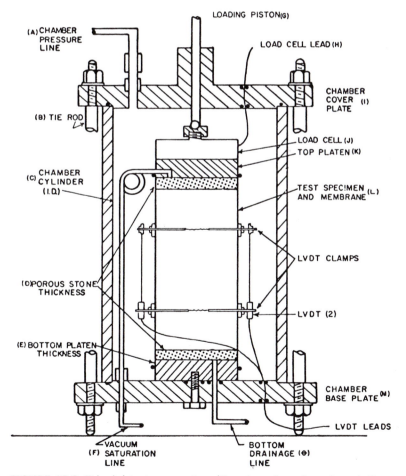

FIGURE 16-3 Triaxial test apparatus. (Courtesy American Association of State Highway and Transportation Officials.)

cyclic loading is measured by a pair of linear variable displacement transducers (LVDT). The recoverable axial strain (the maximum strain minus the permanent strain) in a complete load–unload cycle is used to calculate the resilient modulus by dividing the applied deviator stress (axial stress minus the confining pressure) by the measured recoverable axial strain. More detailed information regarding the resilient modulus testing procedure and the interpretation are given in the AASHTO T274 method and in Ref. 5.

It is quite time-consuming to determine the resilient modulus of a subgrade soil from the test procedure described above. Various empirical correlations between the resilient modulus and the other soil properties, such as the one

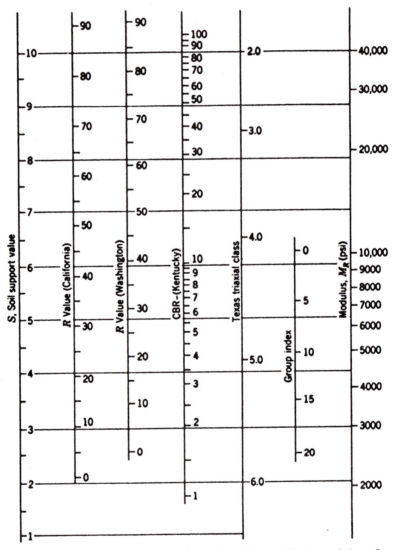

FIGURE 16-4 Correlation chart for estimating resilient modulus of subgrade soils (1 psi = 6.9 kPa). (SOURCE: Van Til, C. J., et al. Evaluation of AASHTO Interim Guides for Design of Pavement Structures. *NCHRP Report 128,* Transportation Research Board, Washington, DC, 1972.)

shown in Figure 16-4, are available and can be used as a guide to estimate the resilient modulus of a subgrade soil.

16-9 The Asphalt Institute Method

The Asphalt Institute has published a pavement thickness design manual (6) which characterizes asphalt pavement as a multilayered elastic system. Using established theory, experience, and test data, the Institute's engineers developed a structural thickness design method suitable for a variety of asphaltic pavements.

The method is based on two assumed stress–strain conditions:

1. The wheel load, W, is transmitted to the pavement surface through the tire at a uniform vertical pressure P_0. The stresses are then spread through the pavement structure to produce a reduced maximum vertical stress, P_1 at the subgrade surface (Fig. 16-5).
2. The wheel load, W, causes the pavement structure to deflect, creating both compressive and tensile stresses in the pavement structure (Fig. 16-6). In developing the design procedure, Asphalt Institute engineers calcualted induced horizontal tensile strains, ϵ_t, at the bottom of the asphalt layer and vertical compressive strains, ϵ_c, at the top of the subgrade (Fig. 16-7).

Thus, the method addresses two of the most commonly occurring traffic-related modes of flexible pavement distress: fracture or cracking of the asphalt-treated layer, especially the surface layer, and distortion or rutting of the subgrade and the other layers of the pavement system.

A computer program called DAMA was used to determine the thicknesses for the two strain criteria. The materials in the pavement layers were characterized by a modulus of elasticity and a Poisson ratio. Specific values of these characteristics were selected on the basis of experience and extensive laboratory tests. Two thicknesses, one for each critical strain value, were calculated for various combinations of subgrade and loading conditions. The larger of the two values was used to prepare pavement design charts such as that shown as Figure 16-8.

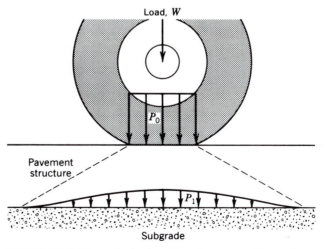

FIGURE 16-5 Spread of wheel load pressure through pavement structure. (Courtesy The Asphalt Institute.)

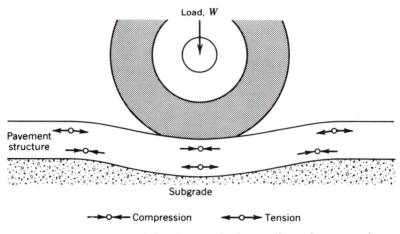

FIGURE 16-6 Pavement deflection results in tensile and compressive stresses in pavement structure. (Courtesy The Asphalt Institute.)

The Asphalt Institute's thickness design manual (*6*) includes design charts for the following six types of pavement structures at three asphalt surface temperatures (7°C, 15.5°C, and 24°C):

1. Full-depth asphalt concrete with an asphaltic concrete for both the surface and the base.
2. Emulsified asphalt mix (type 1) constructed with processed, dense-graded aggregate and an asphaltic emulsion.

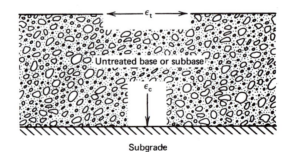
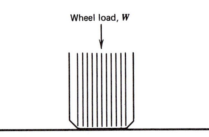

FIGURE 16-7 Strains in a flexible pavement.

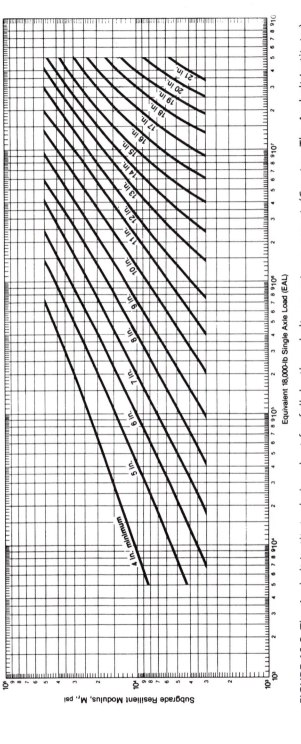

Subgrade Resilient Modulus, M_r, psi

Equivalent 18,000-lb Single Axle Load (EAL)

FIGURE 16-8 The Asphalt Institute design chart for full depth asphalt concrete pavement. (Courtesy The Asphalt Institute.)

3. Emulsified asphalt mix (type 2) made with semiprocessed, crusher-run, pit-run, or bank-run aggregates and an asphaltic emulsion.

4. Emulsified asphalt mix (type 3) made with sands or silty sands and an asphaltic emulsion.

5. Asphalt concrete over an untreated aggregate base with thicknesses of 150 mm and 300 mm (6 in. and 12 in.).

Traffic Analysis

The Asphalt Institute recommends that the effects of traffic on pavement structural design be expressed in terms of the number of equivalent 18,000-lb single-axle loads (ESAL). This approach, which is similar to that recommended by AASHTO (5) involves several steps:

First, estimate the number of vehicles of different types (such as passenger cars, single-unit trucks, and multiple-unit trucks of various sizes and configurations) expected to use the proposed pavement over the design period. During the design period, the pavement is assumed to support the cumulative effects of traffic and provide a satisfactory riding quality without major rehabilitation. The designer should make allowance for traffic growth, basing the growth rate on historic records for the given road or comparable facility. Assuming a compounded rate of growth, the total volume of traffic expected during the design period is

$$T = \left[\frac{(1 + r)^n - 1}{r} \right] T_1 \qquad (16\text{-}2)$$

where T_1 = traffic volume during the first year
   r = rate of growth expressed as a fraction
   n = design period (years)

Second, estimate the percentage of total truck traffic expected to use the design lane. (The design lane is the lane expected to receive the most severe service.) This estimate can be made from traffic observations for the specific facility or from Table 16-1.

Third, for each weight class, determine the "truck factor," the number of ESALs contributed by the passage of a vehicle. Truck factors can be computed or estimated from Table 16-2. The truck factors are computed by multiplying the number of axles in each weight class by an appropriate "load equivalency factor," summing the products for the various weight classes, and dividing the sum by the total number of vehicles involved. The load equivalency factor is the number of ESALs contributed by the passage of an axle. Typical load equivalency factors are given in Table 16-3.

TABLE 16-1 Percentage of Total Truck Traffic in Design Lane

Number of Traffic Lanes (Two Directions)	Percentage of Trucks in Design Lane
2	50
4	45 (35–48)[a]
6 or more	40 (25–48)[a]

[a]Probable range.

TABLE 16-2 Distribution of Truck Factors for Different Classes of Highways and Vehicles in the United States

Vehicle Type	*Highway System Type*		
	Interstate Rural	*Other Rural*	*Urban*
Single-unit trucks			
Two-axle, four-tire	0.02	0.02	0.03
Two-axle, six-tire	0.19	0.21	0.26
Three-axle or more	0.56	0.73	1.03
Tractor-semitrailers			
Three-axle	0.51	0.47	0.47
Four-axle	0.62	0.83	0.89
Five-axle or more	0.94	0.98	1.02

Source: *Thickness Design—Asphalt Pavements for Highways and Streets.* Manual Series No. 1, The Asphalt Institute, Lexington, KY (1991).

EXAMPLE 16-1

Computation of Equivalent 18,000-lb Load Applications During the first year of service, a pavement is expected to accommodate the following numbers of vehicles in the classes shown. Estimate the ESALs.

Vehicle Type	*No. of Vehicles*		*Truck Factors*		*Product*
Single-unit trucks					
Two-axle, four-tire	87,600	×	0.02	=	1,750
Two-axle, six-tire	23,600	×	0.19	=	4,480
Three-axle or more	4,400	×	0.56	=	2,460
Tractor semis and combinations					
Three-axle	2,100	×	0.51	=	1,180
Four-axle	7,300	×	0.62	=	4,530
Five-axle or more	50,200	×	0.94	=	47,190
			ESAL = Sum =		61,590

EXAMPLE 16-2

Design ESAL for 20-Year Design Period If the traffic using the pavement grows at an annual rate of 4 percent, determine the design ESAL for a 20-year design period.

By Eq. 16-2

$$\text{design ESAL} = \left[\frac{(1 + 0.04)^{20} - 1}{0.04} \right] 61{,}590 = 1{,}834{,}000$$

Note that if the traffic is expected to grow nonuniformly among weight classes, Eq. 16-2 should be applied to each weight class using appropriate rates of growth.

Materials Evaluation

In the Asphalt Institute pavement design procedure, the subgrade strength is characterized by its modulus of elasticty or "resilient modulus." The procedure for determining the resilient modulus of subgrade soils was presented in Section 16-8.

TABLE 16-3 Typical Load Equivalency Factors

Gross Axle Load (lb)[a]	Single Axles	Tandem Axles
5,000	0.00500	
10,000	0.0877	0.00688
15,000	0.478	0.0360
20,000	1.51	0.1206
25,000	3.53	0.308
30,000	6.97	0.658
35,000	12.50	1.23
40,000	21.08	2.08
45,000	34.00	3.27
50,000	52.88	4.86
55,000		6.93
60,000		9.59
65,000		12.96
70,000		17.19
75,000		22.47
80,000		28.99

[a]1 lbf = 4.4482 N.

Source: Thickness Design—Asphalt Pavements for Highways and Streets. Manual Series No. 1, The Asphalt Institute, Lexington, KY (1991).

The Asphalt Institute has also published equations by which it is possible to estimate the resilient modulus from the standard CBR or R test results (6). In metric units, the equations are

$$M_r \text{ (kPa)} = 10,340 \text{ CBR}$$
$$M_r \text{ (kPa)} = 793 + 3826 \text{ (R value)}$$

In conventional U.S. units the equations are

$$M_r \text{ (psi)} = 1500 \text{ CBR}$$
$$M_r \text{ (psi)} = 115 + 555 \text{ (R value)}$$

The Asphalt Institute method recommends that the design subgrade resilient modulus M_r be based on the expected level of traffic, expressed in ESALs. To ensure a more conservative design, lower values of M_r are used for higher volumes of anticipated traffic. Specifically, it is recommended that the resilient modulus be obtained for six to eight samples of the subgrade. The results are then arranged in ascending order and plotted as a cumulative distribution (Fig. 16-9). The design subgrade resilient modulus is then chosen from the curve as the value that 60, 75, or 87.5 percent of the values are equal to or greater than. The applicable percentile depends on the traffic level (ESAL) as shown in Table 16-4.

16-10 The AASHO Road Test

The AASHO road test* was a $27 million project undertaken cooperatively by 49 of the states, the District of Columbia, Puerto Rico, the Bureau of Public

---

*At the time of the test, AASHTO was called the American Association of State Highway Officials (AASHO).

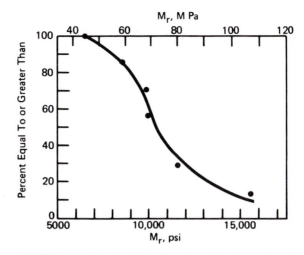

FIGURE 16-9 Typical distribution of subgrade resilient modulus test results. (Courtesy The Asphalt Institute.)

Roads, and various industry groups; the test was administered by the Highway Research Board.

Construction began in April 1956, and test traffic began in October 1958. Except for some special tests, traffic ceased on the test road in November 1960.

The test road was located near Ottawa, Illinois, about 80 miles southwest of Chicago. Major portions of the test dealt with flexible pavements, rigid pavements, and short-span bridges. Only one subgrade soil, an A-6, was used in the pavement tests.

The test sections comprised four major loops and two smaller ones. Each loop was a segment of a four-lane divided highway with the parallel roadways connected by a turnaround at each end of the loop; tangent sections of the large loops were 6800 ft long.

Each tangent was built as a succession of pavement sections arranged in such a way that pavement design could be varied from one section to another. The minimum length of any section in the main loops was 100 ft.

In the principal flexible pavement test sections, the surface course was bituminous concrete; the base course, a well-graded crushed limestone; and the subbase, a uniformly graded sand-gravel mixture.

Major design factors in the principal experiments were surface, base, and subbase thicknesses. In the main factorial experiments, three levels of surface

TABLE 16-4 Subgrade Design Limits

Traffic Level ESAL	Design Subgrade Percentile Value
10^4 or less	60
Between 10^4 and 10^6	75
10^6 or more	87.5

Source: Thickness Design—Asphalt Pavements for Highways and Streets. Manual Series No. 1, The Asphalt Institute, Lexington, KY (1991).

thickness existed in combination with three levels of base thickness; each of these nine combinations existed in combination with three levels of subbase thickness. Over the test road as a whole, surface thickness ranged from 1 to 6 in., base thickness from 0 to 9 in., and subbase thickness from 0 to 16 in.

Special experiments on the flexible pavements included three other design variables—base type, bituminous surface treatment, and shoulder paving. Besides the crushed limestone, bases included a well-graded uncrushed gravel, a bituminous plant mixture, and a cement-treated aggregate.

Test traffic included both single- and tandem-axle vehicles with 10 different axle arrangement–axle load combinations. Single-axle loads ranged from 2,000 to 30,000 lb, tandem-axle loads from 24,000 to 48,000 lb. Each pavement section was tested with one of the 10 combinations; each section was subjected to thousands of load repetitions before it was taken out of the test.

Hundreds of thousands of "bits" of data were gathered by investigators during the road test. These included observations and measurements of pavement condition (cracking and patching done to keep the section in service), longitudinal profile, and transverse profile (to determine rutting and other transverse distortions). Other measurements included surface deflection under loaded vehicles moving at creep speed, deflections at different levels in the pavement structure under vehicles operated at various speeds, curvature of the pavement under vehicles operated at various speeds, pressures transmitted to the surface of the subgrade soil, and temperature distribution in the pavement layers.

Data gathered in the test were subjected to exhaustive analysis by the road test staff and by other interested organizations.

16-11 The Pavement Serviceability Concept

One of the products of the AASHO road test was the "pavement serviceability concept." In essence, this involves the measurement, in numerical terms, of the behavior of the pavement under traffic, its ability to serve traffic at some instance during its life (*10*).

Such an evaluation can be made on the basis of a systematic but subjective rating of the riding surface by individuals who travel over it. Or pavement serviceability can be evaluated by means of certain measurements made on the surface, as was done at the test road. The serviceability rating was on a scale of 0 to 5 with 0 corresponding to a Very Poor rating and 5 representing a Very Good rating.

For flexible pavements, researchers at the test road established that the serviceability index (p) at any time is a function of roughness or slope variance in the two wheel paths, the extent and type of cracking (and patching) of the pavement, and the pavement rutting displayed at the surface. The slope variance is an expression of variations in the longitudinal profile, or longitudinal roughness.

At the start of the test, researchers at the test road determined the average initial serviceability index to be 4.2 for the flexible pavements. In general, the value of the index declined gradually under traffic. When the serviceability index (as measured every 2 weeks) dropped to 1.5 for any section, that section was taken out of test. A value of p of, say, 2.5 is an intermediate one between initial construction and failure to render adequate traffic service.

Analysis of relationships among the flexible pavement variables considered in the test resulted in a series of equations, none of which is given here. Figure 16-10 is one way of showing the relationships among the thickness index, axle load,

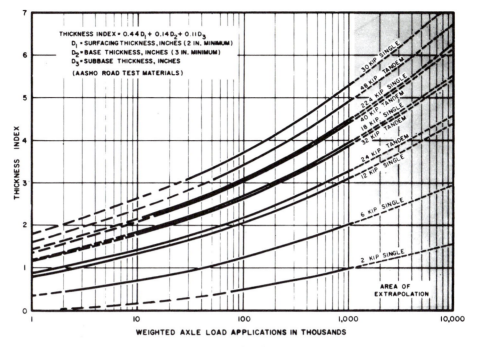

FIGURE 16-10 AASHO road test relationship between thickness index and axle loads at $p = 2.5$. (Courtesy Transportation Research Board.)

and number of load applications. The thickness index expresses the total effectiveness of the layered pavement structure (Fig. 16-10).

As an example of the use of this chart, assume that it is desired to determine a pavement structure that would survive 1 million applications of a 22,400-lb single-axle load before its serviceability index dropped to 2.5. From Figure 16-4, the thickness index is about 4.5. Many combinations of asphaltic concrete surface, base, and subbase will meet the conditions of the thickness index equation. One such combination is 4 in. of asphaltic concrete, 10 in. of crushed stone base, and 12 in. of sand-gravel subbase. Figure 16-10 is not intended for use in design, but it can serve as a basis for the development of design procedures such as the one explained in Section 16-12.

16-12 AASHTO Flexible Pavement Design Method

One of the major objectives of the AASHO road test was to provide information that could be used to develop pavement design criteria and design procedures. In 1961, coincident with the release of the road test results, the AASHTO Design Committee released the "AASHTO Interim Guide for the Design of Rigid and Flexible Pavements." After the Guide has been used for several years, the AASHTO Design Committee evaluated and revised the Interim Guide in 1972 (*3*). Further evaluation of the Guide was undertaken in 1983, and it was determined that some improvments could be made. Thus, the current version, "AASHTO Guide for the Design of Pavement Structures—1986" was issued (*5*).

This current version incorporates the various design inputs, including traffic, reliability, subgrade soil property, environmental effects, and performance criteria into the design equation and the design chart as shown in Figure 16-11 to determine the structural number, a combined structural capacity of the pavement, required for the pavement. The combinations of the layer thickness and the property of the materials (layer coefficients) constituting the pavement layers are then determined from the structural number. Important facts relative to the design inputs and the layer coefficients are described in the following sections.

Traffic

The total load applications due to all the mixed traffic within the design period are converted to the 18-kip ESAL, $^{\sim}W_{18}$, using the axle load equivalency factors for each axle weight group provided in the Guide. If the estimated traffic to be used in design can be broken down into axle load groupings, the number of load applications in each group can be multiplied by the equivalence factor to determine the number of 18-kip axle loads that would have an equivalent effect on the pavement structure (see Table 16-3). The total 18-kip ESAL applications expected on the highway must be further broken down into the design 18-kip ESAL applications in the design traffic lane by taking into consideration the directional distribution and the lane distribution factors. The following equation may be used to determine the traffic (W_{18}) in the design lane.

$$W_{18} = D_D \times D_L \times W'_{18}$$

where D_D = a directional distribution factor
D_L = a lane distribution factor

No. of Lanes in Each Direction	Percent 18-kip ESAL in Design Lane
1	100
2	80–100
3	60–80
4	50–75

W'_{18} = the cumulative two-direction 18-kip ESAL

Reliability

The concept of incorporating the reliability factors into the design procedure was developed to ensure that the various design alternatives would allow for inherent design and construction variabilities and perform as they were intended in the design period. The reliability design factor accounts for chance variations in both traffic prediction and the performance prediction; therefore it provides a predetermined level of assurance (R) that pavement sections will survive the period for which they were designed. For a given reliability level, the reliability factor is a function of the overall standard deviation (S_o), which accounts for standard variation in materials and construction, the chance variation in the traffic prediction, and the normal variation in pavement performance. The standard deviations of 0.45 and 0.35 respectively are suggested by the Guide for flexible and rigid pavements.

Environmental Effects

The AASHTO design equations were developed from the results of the road tests in a two-year period. The long-term effects of temperature, moisture, and material aging on pavement performance could not be directly accounted for

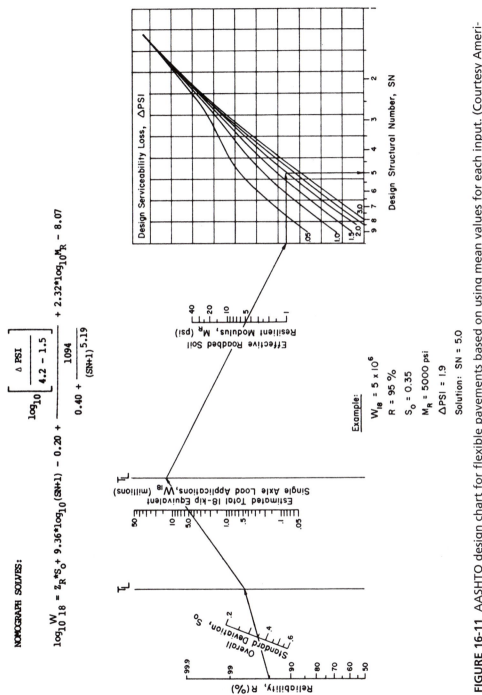

NOMOGRAPH SOLVES:

$$\log_{10} W_{18} = z_R * S_O + 9.36 * \log_{10}(SN+1) - 0.20 + \frac{\log_{10}\left[\dfrac{\Delta PSI}{4.2 - 1.5}\right]}{0.40 + \dfrac{1094}{(SN+1)^{5.19}}} + 2.32 * \log_{10} M_R - 8.07$$

Design Serviceability Loss, ΔPSI

Design Structural Number, SN

Effective Roadbed Soil
Resilient Modulus, M_R (psi)

Estimated Total 18-kip Equivalent
Single Axle Load Applications, W_{18} (millions)

Overall Standard Deviation, S_O

Reliability, R (%)

Example:

$W_{18} = 5 \times 10^6$

R = 95 %

S_O = 0.35

M_R = 5000 psi

ΔPSI = 1.9

Solution: SN = 5.0

FIGURE 16-11 AASHTO design chart for flexible pavements based on using mean values for each input. (Courtesy American Association of Highway and Transportation Officials.)

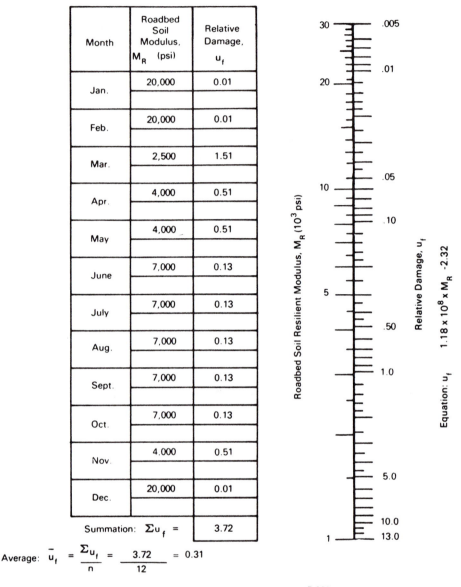

Month	Roadbed Soil Modulus, M_R (psi)	Relative Damage, u_f
Jan.	20,000	0.01
Feb.	20,000	0.01
Mar.	2,500	1.51
Apr.	4,000	0.51
May	4,000	0.51
June	7,000	0.13
July	7,000	0.13
Aug.	7,000	0.13
Sept.	7,000	0.13
Oct.	7,000	0.13
Nov.	4,000	0.51
Dec.	20,000	0.01
Summation: $\Sigma u_f =$		3.72

Average: $\bar{u}_f = \dfrac{\Sigma u_f}{n} = \dfrac{3.72}{12} = 0.31$

Effective Roadbed Soil Resilient Modulus, M_R (psi) = $\underline{5,000}$ (corresponds to $\bar{u}_f$)

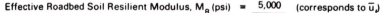

FIGURE 16-12 Chart for estimating effective roadbed soil resilient modulus for flexible pavements designed using the serviceability criteria. (Courtesy American Association of State Highway and Transportation Officials.)

from the road test data. Also, if the effects of swell clay and frost heave of a subgrade soil on the performance of the pavement in a specific region are significant, the loss of serviceability over the design period should be estimated and added to that due to traffic loads.

Serviceability

Initial serviceability and terminal serviceability indexes must be established to compute the change in serviceability (ΔPSI) in the design equation given in

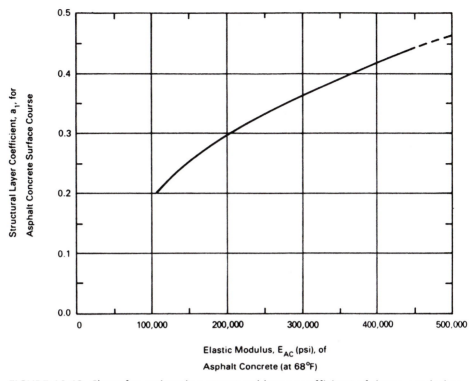

FIGURE 16-13 Chart for estimating structural layer coefficient of dense-graded asphalt concrete based on the elastic (resilient) modulus. (Courtesy American Association of State Highway and Transportation Officials.)

Figure 16-11. The initial serviceability index is a function of pavement construction quality. Typical values from the AASHO road test were 4.2 and 4.5 respectively for flexible and rigid pavements. The terminal serviceability index is the lowest index that is tolerable for a pavement before it requires rehabilitation. A terminal index of 2.5 is suggested by the AASHTO Guide for major highway pavements, whereas a value of 2.0 suffices for other pavements. In addition, the change in serviceability (ΔPSI) should also include the loss of serviceability during the design period due to the potential subgrade swelling and frost heave.

Effective Roadbed Soil Resilient Modulus

The basis for subgrade soil property in the current version of the Guide is resilient modulus. The procedure for determining the resilient modulus has been described in Section 16-8 in this chapter. An effective roadbed soil resilient modulus is then established that is equivalent to the combined effect of the subgrade resilient modulus of all the seasonal resilient moduli. Figure 16-12 is a worksheet given in the Guide for estimating effective roadbed resilient modulus. A year is divided into 24 periods, and the resilient modulus of the roadbed soil in each period is determined and entered in the second column in the figure. The corresponding relative damage value from the resilient modulus value is determined from the vertical scale, and the averaged relative damage value is calculated (0.31 in this example). The averaged relative damage value is then used to determine the

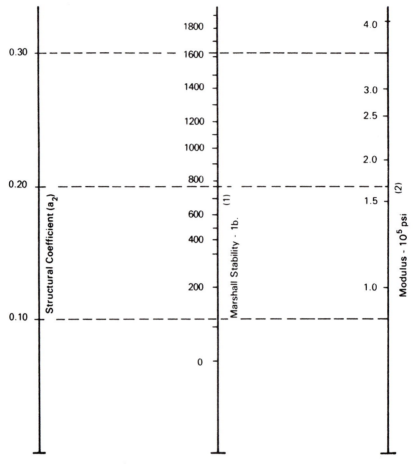

(1)   Scale derived by correlation obtained from Illinois.
(2)   Scale derived on NCHRP project (3).

FIGURE 16-14 Variation in a_2 for bituminous-treated bases with base strength parameter. (Courtesy American Association of State Highway and Transportation Officials.)

effective resilient modulus of the roadbed soil (M_r = 5,000 psi), again by use of the vertical scale.

Determination of the Required Structural Number

Figure 16-11 represents the nomograph for determining the design structural number (SN) required for the specific input conditions. The nomograph is constructed from the equation shown on top of the figure. The required inputs are

estimated future traffic (W_{18}).

reliability (R).

overall standard deviation (S_o).

effective roadbed soil resilient modulus (M_r).

design serviceability loss (ΔPSI).

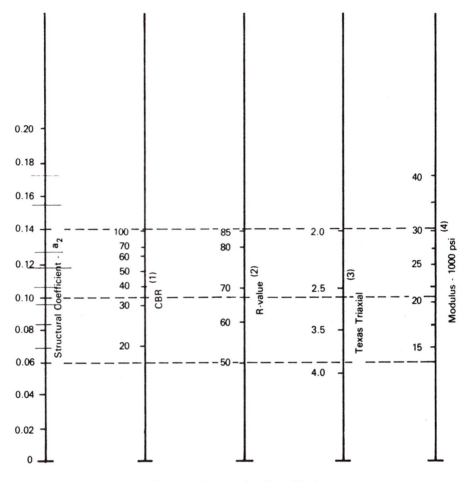

(1) Scale derived by averaging correlations obtained from Illinois.
(2) Scale derived by averaging correlations obtained from California, New Mexico and Wyoming.
(3) Scale derived by averaging correlations obtained from Texas.
(4) Scale derived on NCHRP project *(3)*.

FIGURE 16-15 Variation in granular base layer coefficient (a_2) with various base strength parameters. (Courtesy American Association of State Highway and Transportation Officials.)

With the input data in the example given in Figure 16-11, an SN = 5.0 can be determined from the figure or from the equation.

Selection of Pavement Thickness Designs

Once the structural number is determined, it is necessary to determine the thickness of the various layers in a flexible pavement that will provide the required load-carrying capacity that corresponds to this design structural number, according to the following equation:

$$SN = a_1 D_1 + a_2 D_2 m_2 + a_3 D_3 m_3 \qquad (16\text{-}3)$$

where a_i = layer coefficient of layer i
D_i = thickness of layer i (in.)
m_i = drainage modifying factor for layer i

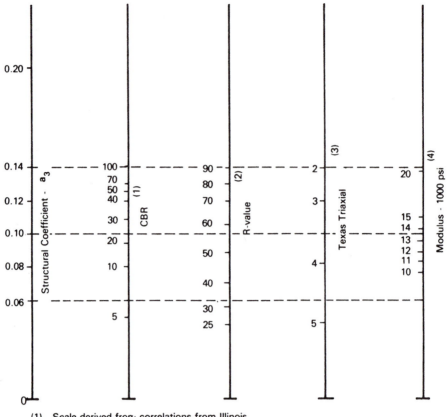

(1) Scale derived from correlations from Illinois.

(2) Scale derived from correlations obtained from The Asphalt Institute, California, New Mexico and Wyoming.

(3) Scale derived from correlations obtained from Texas.

(4) Scale derived on NCHRP project *(3)*.

FIGURE 16-16 Variation in granular subbase layer coefficient (a_3) with various subbase strength parameter. (Courtesy American Association of State Highway and Transportation Officials.)

Layer Coefficients (a_i) The AASHTO flexible pavement layer coefficient is a measure of the relative effectiveness of a given material to function as a structural component of the pavement. The layer coefficients of different materials are presented below (5).

Asphalt concrete surface course. Figure 16-13 presents a chart that can be used to determine the layer coefficient of a dense-graded asphalt concrete surface course based on its elastic modulus at 68°F.

Bituminous treated bases. Figure 16-14 presents a chart that can be used to estimate the layer coefficient of a bituminous-treated base material based on its elastic modulus or its Marshall stability value.

Granular base and subbase layers. Figure 16-15 and Figure 16-16 can be used to estimate the layer coefficient of a granular base material and a granular subbase material based on different laboratory test results of the material.

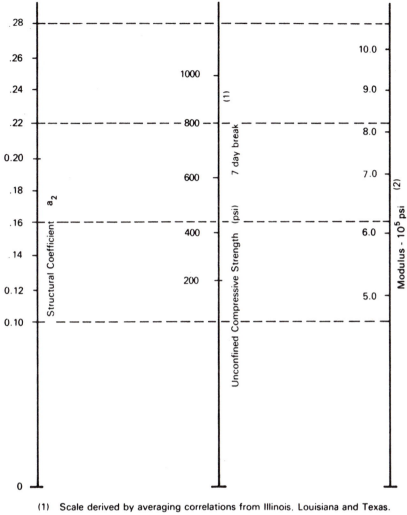

(1) Scale derived by averaging correlations from Illinois, Louisiana and Texas.
(2) Scale derived on NCHRP project (3).

FIGURE 16-17 Variation in *a* for cement-treated bases with base strength parameter. (Courtesy American Association of State Highway and Transportation Officials.)

TABLE 16-5 Recommended m_i Values for Modifying Structural Layer Coefficients of Untreated Base and Subbase Materials in Flexible Pavements

| Quality of Drainage | Percent of Time Pavement Structure is Exposed to Moisture Levels Approaching Saturation | | | |
	Less Than 1%	*1%–5%*	*5%–25%*	*Greater Than 25%*
Excellent	1.40–1.35	1.35–1.30	1.30–1.20	1.20
Good	1.35–1.25	1.25–1.15	1.15–1.00	1.00
Fair	1.25–1.15	1.15–1.05	1.00–0.80	0.80
Poor	1.15–1.05	1.05–0.80	0.80–0.60	0.60
Very poor	1.05–0.95	0.95–0.75	0.75–0.40	0.40

Cement-treated bases. Figure 16-17 presents a chart that can be used to estimate the layer coefficient of a cement-treated base from its unconfined compressive strength or the elastic modulus.

Drainage Modifying Factor (m_i)   Table 16-5 presents the drainage coefficients recommended in the AASHTO Guide for untreated base and subbase materials in flexible pavements. The coefficients depend on the quality of drainage and percentage of time the pavement structure is saturated. The quality of drainage is measured by the length of time it takes for water to be removed from base or subbase.

General Procedure for Selection of Layer Thickness   The procedure for selecting the pavement layer thickness usually starts from the top, as shown in Figure 16-18.

1. Using E_2 as M_r and Figure 16-11, determine the structural number SN_1 required to protect the base and compute the thickness of layer 1 (D_1):

$$D_1 \geq SN_1/a_1$$

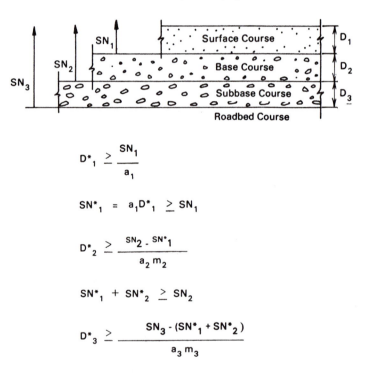

1) a, D, m and SN are as defined in the text and are minimum required values.

2) An asterisk with D or SN indicates that it represents the value actually used, which must be equal to or greater than the required value.

FIGURE 16-18 Procedure for determining thicknesses of layers using a layered analysis approach. (Courtesy American Association of State Highway and Transportation Officials.)

2. Using E_3 as M_r and Figure 16-11, determine the structural number SN_2 required to protect the subbase and compute the thickness of layer 2 (D_2):

$$D_2 \geq (SN_2 - a_1 D_1)/a_2 m_2$$

3. Using the roadbed soil resilient modulus and Figure 16-11, determine the structural number SN_3 required to protect the roadbed soil and compute the thickness of layer 3 (D_3):

$$D_3 \geq (SN_3 - a_1 D_1 - a_2 m_2 D_2)/a_3 m_3$$

16-13 Conclusion

In recent years, a great deal of progress has been made in developing more reliable pavement structural design methods. Formerly, design procedures relied heavily on subjective engineering judgments based on evaluations of pavement performance under service conditions. As more is learned about the mechanical properties of paving and subgrade materials and the mechanics of pavement failure, engineers are turning to more rational approaches to pavement design. The efforts of researchers to refine and improve pavement design methodology are certain to result in pavements that are more economical and long lasting.

REFERENCES

1. *Standard Nomenclature and Definitions for Pavement Components and Deficiencies. Special Report No. 113,* Highway Research Board, Washington, DC (1970).

2. *Highway Practice in the United States of America.* Public Roads Administration, Washington, DC (1949).

3. *AASHTO Interim Guide for Design of Pavement Structures—1972.* American Association of State Highway and Transportation Officials. Washington, DC (1972).

4. Van Til, C. J., et al. *Evaluation of AASHO Interim Guides for Design of Pavement Structures.* National Cooperative Highway Research Program Report No. 128, Washington, DC (1972).

5. *AASHTO Guide for Design of Pavement Structures,* American Association of State Highway and Transportation Officials, Washington, DC (1986).

6. *Thickness Design—Asphalt Pavements for Highways and Streets.* Manual Series No. 1, The Asphalt Institute, Lexington, KY (1991).

7. *Annual Book of Standards: Bituminous Materials, Soils, Skid Resistance,* Part 11. American Society for Testing Materials, Philadelphia, PA.

8. *Standard Specification for Transportation Materials and Methods of Sampling and Testing.* American Association of State Highway Officials, Washington, DC (1993).

9. *Test Procedures for Characterizing Dynamic Stress—Strain Properties of Pavement Materials. Special Report No. 162.* Transportation Research Board, Washington, DC (1975).

10. *The AASHO Road Test: Pavement Research. Special Report No. 61E.* Highway Research Board, Washington, DC (1962).

11. *International Conference on the Structural Design of Asphalt Pavements.* Preprint volume (supplement), University of Michigan, Ann Arbor (1962).

CHAPTER 17

EARTHWORK OPERATIONS AND EQUIPMENT

Nearly all highway construction jobs, especially those in new locations, involve a consderable amount of earthwork. In general terms, earthwork operations are those construction processes that involve the soil or earth in its natural form and that precede the building of the pavement structure itself.

Basic earthwork operations may be classified as clearing and grubbing; roadway, borrow, structural, and several special forms of excavation; the formation of embankments; and finishing operations. Any or all of these construction processes may be performed on a given highway project, and they may overlap to a certain extent. This chapter includes a description of the various procedures used in this phase of highway construction. It should be remembered that the discussion which follows will deal largely with average practice and that the specific procedures followed by any one organization may differ considerably from those given here. It should also be noted that the construction specifications which are discussed are intended to be typical of those used by highway agencies in the direction and control of contract work.

Before describing typical earthwork procedures, let us briefly consider the principal tools that are used in earthwork operations.

17-1 Equipment Used in Earthwork Operations

In modern highway practice in the United States, most earthwork operations are accomplished by the use of a large number of highly efficient and versatile machines. The development and application of such machines in the past five decades has been an outstanding phase of highway progress. Machines have been developed that are capable of efficient and economically performing every form of earthwork. Some machines are tailored to perform a specific operation or series of operations, while others have a wide variety of uses. The proper selection and efficient use of these machines are an important part of the highway scene.

To the beginner in this subject, the large number and almost infinite variety of machines used in earthwork operations are undoubtedly bewildering. Not only are there many different basic types of machines and modifications of them, but there are many manufacturers engaged in the production of machines of similar characteristics and uses. Space does not permit an extensive description of equipment used in highway earthwork operations. The most common types of equipment are listed in Table 17-1, along with an indication of typical uses. Various

TABLE 17-1 Typical Applications of Highway Construction Equipment

Equipment Type	Typical Uses
Crawler tractor	Various pushing/pulling operations such as push-loading scrapers
Bulldozer units	Ripping and loosening of rocky earth in preparation for excavation; earthmoving operations for short distances, especially in rough, rocky ground and on steep slopes; tree removal
Rubber-tired tractor scraper units	Earthmoving operations—digging, loading, transporting, dumping, and spreading earth material
Farm tractor with front-end loader	Miscellaneous operations associated with earthwork, such as filling trenches, excavation in limited areas, and cleaning up along aggregate stockpiles
Shovel	Excavation at or above existing ground level in firm or hard material, e.g., excavating material from cut slope
Backhoe	Excavation below existing ground in firm or hard material, as in digging trenches
Crane with clamshell	Placing and handling aggregates at a stockpile; dredging operations
Carne with dragline	Excavation in soft materials at or below existing ground level
Crane with hook block	Lifting various heavy objects
Trucks and wagons	Various hauling operations
Motor grader	Shaping of subgrades, shoulders, ditches, and backslopes; maintenance of construction roads
Steel-wheeled roller, vibratory compactor	Compaction of gravels and gravel-sand mixtures
Pneumatic-tired roller	Compaction of wide range of soils of low to moderate plasticity including silty gravels, sands, silty sands, and clayey sands
Sheep's-foot roller	Compaction of soils of moderate to high plasticity including silts, clays, clayey sands, silt-clay mixtures, clayey gravels, and silty gravels.
Supercompactors (pneumatic-tired rollers weighing 20–50 tons)	All types of soil

classes of equipment commonly used in earthwork operations are illustrated in Figures 17-1 and 17-2.

For additional information on equipment used in earthmoving operations, the reader should consult with equipment manufacturers or dealers and refer to the various construction magazines.

CLEARING AND GRUBBING

Clearing and grubbing is generally the first operation to be undertaken on any project involving earthwork, and thus it precedes any excavation.

Clearing and grubbing may be defined as the removal of trees, stumps, roots, snags, down timber, rotten wood, rubbish, and other objectionable material from

FIGURE 17-1 Common types of earthwork equipment. (*a*) Bulldozer. (*b*) Rubber-tired front-end loader. (*c*) Rubber-tired tractor scraper unit. (*d*) Truck being loaded by shovel. (Courtesy Terex Division, General Motors Corporation.)

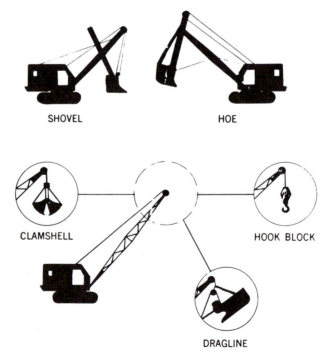

FIGURE 17-2 Shovel-crane units.

an area marked on the plans or otherwise designated by the engineer. The area designated generally includes the limits of proposed excavation and embankment as well as all other areas where the obstructions that are to be removed would interfere with the conduct of the proposed work.

"Clearing" refers to the removal of materials above existing ground surface, and "grubbing" means the removal of roots, stumps, and similar objects to a nominal depth below the surface. Frequently, clearing and grubbing constitute a single contract item that includes the removal of topsoil to a shallow depth.

In cut areas, in addition to the removal of vegetation of other obstructions from the surface of the ground, specifications generally require that all stumps and roots be removed to a depth of not less than 0.3 m (1 ft) below the proposed grade. This requirement also usually holds in embankment areas where the height of the fill is less than about 1.5 m (4 or 5 ft). In embankment sections, when the embankment height is to be more than 1.5 m (5 ft), trees and stumps may be left in place and cut off at ground level or at a height of 75 to 150 mm (3–6 in.) above the existing ground surface. Some agencies require that all large trees 450 mm (18 in.) or more in diameter be removed, regardless of the height of the fill. It is usually provided that excavations made for the purpose of removing stumps and similar obstructions be filled to a level consistent with the surrounding ground before construction proceeds.

Large trees usually receive special consideration and may be left in place at the discretion of the engineer. In such cases trees that overhang the roadway are usually required to be properly trimmed to furnish a clearance of 4.3 to 4.9 m (14–16 ft.). Frequently, areas within the limits of the right-of-way but beyond the limits of excavation and embankment are subjected to "partial" or "selective" clearing in which certain trees and shrubs are left standing. Provisions are usually made for the storing and salvage of marketable timber cleared from the rights-of-way. Debris removed from the designated area is generally disposed of by burning or, if burning is prohibited, by burying it in waste disposal pits or other approved areas. It should also be noted that certain organizations include the removal and disposal of buildings, fences, and so on in clearing and grubbing, whereas others include a special contract item when obstacles of this nature are to be removed.

Measurements of this item for the purpose of payment were formerly made in acres. In SI units, the bid unit should be 1 hectare. Where only clearing operations are involved, special payment may be made for large trees, that is, so much per tree. Clearing and grubbing operations are generally performed largely by machines, with the bulldozer and its various attachment being the basic devices, or by companion tools such as rock rakes, rippers, and tree-'dozers. A considerable amount of hand labor may also be necessary, and liberal use is made of power saws and similar tools.

EXCAVATION

In the interests of simplicity the following discussion will deal with three types of excavation: roadway and drainage excavation, excavation for structures, and borrow excavation. Before beginning the discussion of each of these types, it is desirable to discuss briefly the subject of "classification of excavation."

17-2 Classification of Excavation

It seems obvious that the costs of excavation of materials that require blasting for removal, such as solid rock, will generally be greater than those encountered in the removal of earth materials. Similarly, the removal and disposal of muck and other unsuitable materials may require special treatment and cost more than ordinary earth excavation. Highway agencies usually find it advantageous to divide excavation into four separate classes. These classes are generally termed "rock excavation," "common or earth excavation," "unsuitable excavation," and "borrow excavation"; they are defined by AASHTO (*1*) as follows:

Rock excavation. Material that cannot be excavated without blasting or the use of rippers and all boulders or other detached stones, each having a volume of 2 cu yd* or more.

Common excavation. Excavation and disposal of all materials of whatever character encountered in the work not otherwise classified.

Unsuitable excavation. The removal and disposal of deposits of saturated or unsaturated mixtures of soil and organic matter not suitable for embankment material.

Borrow excavation. Excavation of approved material required for construction of embankments (discussed further in Section 17-5).

This method of classification has certain disadvantages, as borderline cases frequently arise in which there may be some dispute as to the proper classification of a portion of the work. Many organizations obviate this difficulty by using the term "unclassified excavation" to describe the excavation of all materials, regardless of their nature. The latter procedure is now preferred by a majority of highway agencies.

17-3 Roadway and Drainage Excavation

Roadway and drainage excavation, as defined here, means the excavating and grading of the roadway and ditches, including the removal and disposal of all excavated material and all work needed for the construction and completion of the cuts, embankments, slopes, ditches, approaches, intersections, and similar portions of the work.

All excavated materials that are suitable are used in constructing the elements of the roadway structure, including embankments, shoulders, subgrade, slopes, backfill when required for structures, and so on. Provision is made for the disposal of unsuitable or surplus excavated material. It must again be emphasized that, under the definitions that we have given here and in accordance with normal practice in this country, payment made for excavation items is not just payment for excavating the earth. On the contrary, it frequently includes nearly all the work needed in many jobs for forming the earthwork portions of the highway structure. This, again, would normally include such items as the formation of embankments and the finishing operations that are to be discussed later. Practice is not uniform in this respect, however, and some agencies pay separately for compaction, addition of water during compaction, fine grading, and other items.

---

*1.5 m³

Many miscellaneous provisions are generally included in specifications for roadway and drainage excavation. Provison is usually made for the removal of vegetation if the area has not previously been cleared and grubbed and also for the salvage of portions of the excavated materials that are to be used in miscellaneous sections of the roadway. The latter item, for example, might provide for the use of suitable subgrade materials, saving of topsoil for use on side slopes, and work of a similar nature. When unsuitable materials are encountered in cut or embankment sections, provision is generally made for their removal and replacement with satisfactory materials.

The specifications may also include instructions for grading the roadbed, intersections, and so forth in accordance with the required alignment, grade, and cross section. In connection with this portion of the work, the roadway is usually required to be well drained at all times during the course of construction operations. Provision is also made for the grading of shoulders, slopes, ditches, and so forth in accordance with the plans, sometimes including requirements relative to the flattening, rounding, and warping of side and ditch slopes where applicable. When rock is encountered in cut sections, it is generally required that it be excavated to a depth of at least 150 mm (6 in.) below the grade line and that suitable backfill material then be used in bringing the cut section up to the proposed grade. Another provision frequently included is that of maintenance of the graded roadbed until final acceptance of the work or until construction of the base or surface course begins. Provisions under this item might also include the scarification and compaction of cut sections, although they would not normally include subgrade soil stabilization.

Measurement of excavated material is usually based on a determination of the volume, in original position, of material actually excavated and disposed of as directed. Volumes are usually determined by the average-end area method, and payment is made for the number of cubic meters (or cubic yards) of roadway and drainage excavation and for overhaul.

Construction methods used in performing the operations required in this type of excavation may involve nearly every one of the types of equipment previously mentioned. For example, tractor-scraper units are normally used in the basic operation of excavating, hauling, and spreading ordinary soils; if hard, compacted materials are encountered, rippers may be required for breaking up the material so that the scrapers may load. Scrapers with crawler tractors are used for relatively short hauls or in rough terrain; rubber-tired units are used for longer hauls where suitable haul roads can be established. In the case of rock excavation, drilling and blasting will be necessary, and a shovel may be needed to load the blasted rock into trucks or rock wagons for transportation to its disposal point. A dragline is frequently used for excavation in swampy land, and a backhoe for trench excavation. In excavation of this type 'dozer units may be extremely useful in performing excavations involving very short hauls, for spreading dumped materials and for similar operations. A motor grader may be used for shaping ditches, trimming slopes, and maintaining haul roads, as well as for a host of other functions.

17-4 Excavation for Structures

Structural excavation refers to the excavating of material in order to permit the construction of pipe culverts, concrete box culverts, foundations for bridges, retaining walls, and practically all other structures that may be required in a particular job. Payment is usually made separately for this item, except in the case of structures like manholes and catch basins, where the payment made for completion of the structure includes payment for excavation. Some organizations also include in this item the satisfactory placing of backfill around the completed structure.

Specifications of the various state highway agencies covering this item provide for the satisfactory removal of all material, regardless of its nature, which is encountered in structural excavation. In certain cases, particularly where large excavations are required, as in the case of the construction of piers and abutments for large bridges, it may be necessary to employ caissons or cofferdams to facilitate the excavation process. The construction and subsequent removal of these items are also generally provided for in the specifications. Structural excavation is measured and not paid for by the cubic meter (formerly cubic yard); the volume on which payment is based is usually the actual volume of material excavated in order to place the given foundation structure. The pay quantity is generally limited, however, to a volume computed by the imagined placing of planes parallel to the lines of the foundation but located 0.3 to 0.6 m (1–2 ft) outside these lines, as shown on the plans. Provision is also frequently made for an increase in depth of the foundation, with consequent increased payment to the contractor.

Suitable materials taken from excavations for structures are used either in backfilling around the completed structure or in other portions of the roadway structure. Unsatisfactory or surplus excavated materials are, of course, wasted. Requirements related to the placing of suitable backfill are usually quite rigid, particularly with regard to the production of a satisfactory density in the backfill. This density may be obtained by rolling or vibrating the material.

Both machines and hand methods are used in the performance of structural excavation, with more hand labor being required in this operation than in the other types of excavation described in this chapter. When working in average soils, the clamshell is particularly adapted to this type of work because of its ability to work to vertical lines and in close proximity to forms and protective devices such as the bracing and sheathing required in deep excavations. Smaller units of excavating machinery, such as the tractor shovel, also find application in this work. The bulldozer, which is present on nearly every earthwork operation, can also be used to good advantage in excavating and backfilling for structures. The vibratory and tamping tools previously discussed are very useful for compacting backfills, especially in restricted spaces.

17-5 Borrow Excavation

When sufficient material for the formation of embankments and other elements of the roadway structure is not available from excavation performed within the limits of the right-of-way, additional suitable material is generally taken from borrow pits. The contracting agency may indicate and furnish the contractor with suitable borrow pits, or the contractor may obtain specified and suitable materials from locations of his own choosing. Specifications related to borrow excavation are generally about the same as those related to roadway excavation; that is, they provide for satisfactory excavation of material from an indicated area and for incorporation of that material into the roadway structure. Additional requirements usually pertain to the condition of the borrow pit during and after construction. For example, it may be stipulated that when construction operations are complete, borrow pits are to be left in such a condition that water will not collect or stand in them. Payment for borrow excavation is generally by volume in cubic meters (or cubic yards) of borrowed material, measured in place (original volume) and actually incorporated into the designated sections of the work.

The opening and working of borrow pits is frequently a large-scale operation that requires the best in mass-production methods of earth moving. If the bulk of the material is to be incoporated into embankment, and if it is soil as distinguished from rock or rocklike materials, there is little doubt but that borrow-pit excavation in most cases may be most efficiently effected by the use of scraper units. This is, of course, because of the inherent nature of the scraper unit, which, with one operator and one power unit, will dig, load, haul, and spread material. Either crawler or rubber-tired tractor units may be used; the latter units are generally preferred where long hauls above about 350 m (1200 ft) for large units are to be made. Where rock or cemented soils are to be excavated, the shovel is usually the best tool unless the material can be ripped economically; the shovel loads into trucks or wagons that haul the material to the fill, where the load is dumped and spread by a 'dozer or a motor grader. Shovel-truck groups are capable of very large production in suitable circumstances, but the interdependence of the units involved makes their comparison with scraper units generally unfavorable for this operation, except when the excavation is being made in materials or under conditions unsuited to scraper operations.

EMBANKMENTS

Embankments are used in highway construction when it is required that the grade line of the roadway be raised some distance above the level of the existing ground surface in order to maintain design standards or prevent damage to the highway through the action of surface or ground water. Both rolled-earth and hydraulic-fill embankments are built, with the vast majority of highway embankments being of the former type. From the standpoint of average practice it is also likely that the large majority of fills used in highway construction are 4.5 m (15 ft) or less in height. Many fills are less than 1 m (3.3 ft) in height, and fills as high as 117 m (383 ft) have been reported (2).

Before discussing construction procedures related to embankments, it seems desirable to discuss briefly some of the basic factors involved in the design of such structures. These design elements may be classed as height, stability of slopes, stability of foundations, and the selection of embankment materials.

17-6 Height

As has been indicated, the height of an embankment is generally fixed by considerations related to the general location of the highway in the area. Thus an embankment may be needed to maintain the grade to some fixed point, as in the approaches to a bridge or a large culvert structure, or to maintain desirable grades in rolling or mountainous country. Similar considerations related to alignment may also force the use of embankments rather than the possible use of side-hill sections. In rolling country it is frequently possible to reduce embankment construction to a minimum on secondary roads by the simple expedient of following closely the shape of the existing ground. Such a procedure is usually not desirable on primary routes because of sight distance requirements.

In low-lying areas where the water table is at or close to the surface of the ground, the minimum height of embankment is frequently established by the desirability of preventing the intrusion of ground water into the subgrade and base. In such cases the elevation of the top of the subgrade is generally required to be at least 0.6 m (2 ft) above the water table, and it may be considerably more when soils subject to capillarity are used in the construction of the embankment or in areas subject to frost. If free water is expected in the area crossed by the embankment, the minimum distance above the expected water level may be further increased in the interests of protecting the embankment and pavement structure.

17-7 Embankment Slopes

Essentially, the cross section of a highway embankment consists of a flat, horizontal top section with generally symmetrical slopes on either side that begin at the top and continue until they intersect the natural ground surface. The width of the top section depends on the required pavement and shoulder dimensions and is typically a minimum of 12 m (40 ft) for a two-lane roadway.

Generally, flat side slopes are preferred to satisfy standards and to facilitate maintenance. Typical standard slopes established on the basis of such criteria are shown in Table 17-2. At certain locations, it may be necessary to depart from the standard cross section because of restrictions in rights-of-way or unstable soil conditions.

Generally speaking, embankments constructed of high-quality materials (e.g., A-1 and A-3 classifications) can be constructed with slopes as steep as 1.5 to 1 without concern for soil stability. For other soil classifications, a maximum slope of 2 to 1 has been recommended, and where the embankment will be subjected to flooding, a 3 to 1 slope is preferred (*3*).

All major fills should be subject to stability analyses and designed in accordance with established principles of soil mechanics.

17-8 Embankment Foundations

As has been indicated, a portion of the analysis required for the design of a highway embankment consists of checking the stability of the side slopes or the body of the embankment itself. The design should also include an examination of the soil beneath the embankment proper, or the embankment "foundation." An embankment may fail because the stresses imposed on the underlying soil

TABLE 17-2 Typical Standard Design Side Slopes for Major Class Two-Lane Highway with 3-m Stabilized Shoulder[a]

Height of fill (m)	Distance from edge of shoulder	
	0–6 m	*Beyond 6 m*
≤3	6:1	6:1
>3 ≤9	6:1	4:1
>9	2:1	2:1

[a]Assumed design hourly volume is less than 850.

Source: Adapted from *Construction of Embankments.* NCHRP Synthesis of Highway Practice No. 8, Washington, DC (1971).

layer due to the weight of the fill are greater than the shearing resistance of the foundation soils. In such a case, the underlying soil layer generally would flow laterally with resulting subsidence of the fill. The embankment on which a pavement surface has been placed may also fail in its supporting function by continued settlement due to consolidation of the underlying soil layers. Either of these situations is likely to occur when a high fill is founded on cohesive soil or where the foundation material is a soft, compressible, fine-grained soil, such as an organic silt or clay, peat, or muck. Where any doubt exists as to the stability of the embankment foundation, the shearing stresses that will be created in the foundation should be compared with the available shearing resistance of the soils involved, the consolidation characteristics of the layer determined, and a settlement analysis made. The measurement of the shearing resistance and the consolidation characteristics must both be based on the laboratory testing of undisturbed samples taken from the foundation soil.

Construction procedures required for improvement of the structural properties of the foundation are generally involved and costly. Relocation of the route should be considered in every possible case where extremely weak soils are encountered to considerable depth. In the event that relocation is not feasible, several methods have been evolved for the treatment or displacement of the underlying soil layer. These methods are discussed in detail in later portions of this chapter.

17-9 Selection of Embankment Materials

Many different soils may be satisfactorily used in the construction of rolled-earth highway embankments. In terms of the AASHTO classification system outlined in Chapter 15 and in general statement, materials of the A-1, A-2-4, A-2-5, and A-3 groups should be used when available. Soils of the A-2-6, A-2-7, A-4, A-5, A-6, and A-7 groups are generally regarded as being less desirable for embankments, and when such materials must be used, special attention should be given to the design and construction of the embankment (*4*).

The A-1 soils are highly desirable for inclusion in embankments since they may be compacted to a high degree of stability and density by the use of normal compaction equipment. A-2 soils are also generally suitable for this purpose, although they require a little more careful control of the compaction process during construction. The A-3 group consists of cohesionless sands that can generally be satisfactorily used in embankment construction, although they cannot be satisfactorily compacted by the use of sheep's-foot rollers. Adequate compaction may be obtained by the use of pneumatic-tired rollers, construction equipment, or a vibratory compactor.

With regard to groups A-4 and A-6, the sandy and silty soils included in these groups, as well as some of the inorganic clays, may be satisfactory for inclusion in embankments under certain conditions. Such conditions include low height of fill, careful control of the compaction process, and utilization in areas where the moisture content of the soil might be expected to remain the same or less during service than that which was utilized in the construction process. These soils are more difficult to compact than those of the first three groups because of their high moisture-retaining characteristics. The moisture content of these soils during construction must generally be maintained within relatively narrow limits in order to secure adequate density and stability. The elastic soils included in the A-5 and A-7 groups are generally regarded as unsatisfactory for embankment construction. In addition, the clay soils of the A-7 group are subject to very high volume change with change in moisture content. Despite some of the faults exhibited by several of the soils included in the A-4, A-5, A-6, and A-7 groups, conditions are sometimes such that these are the only soils available within a reasonable haul distance and their use is therefore dictated by considerations of economy. These soils must all be carefully handled during the construction process.

CONSTRUCTION OF ROLLED-EARTH EMBANKMENTS

17-10 Formation

Rolled-earth embankments are constructed in relatively thin layers of loose soil. Each layer is rolled to a satisfactory degree of density before the next layer is placed, and the fill is thus built up to the desired height by the formation of successive layers or "lifts." The majority of state highway agencies at the present time require layers to be 150 to 300 mm (6 to 12 in.) thick before compaction begins, when normal soils are involved. Specifications may permit an increase in layer thickness where large rocks are being used in the lower portion of a fill, up to a maximum thickness of 600 mm (24 in.).

The layers are required to be formed by spreading the soil to approximately uniform thickness over the entire width and length of the embankment section at the level concerned. "End-dumping" from trucks without spreading is usually specifically prohibited. An exception to this latter requirement may be permitted when the embankment foundation is of such a character that it cannot support the weight of spreading and compacting equipment. In such case the lower portion of the fill may be placed by end-dumping until sufficient thickness is developed to permit the passage of equipment. Where unstable soils compromise the foundation, the special construction measures described later in the chapter may be required.

Two general construction methods are used in embankment formation. The first of these is the direct dumping and spreading of the soil by scraper units in one operation. This, of course, is a function for which these units are ideally suited. The second method involves the dumping of the material in the proper location by trucks or wagons. The dumped material must then be spread to the required uniform thickness by 'dozer units or occasionally by motor graders. In addition, on very short fills, 'dozer units might conceivably form the embankment, working alone by moving material short distances from cut into embankment sections.

17-11 Compaction

Requirements of the various highway agencies related to the compaction of soils in embankments (and subgrades) have undergone considerable change since World War II. So also have various elements of the understanding of the behavior of soils undergoing compaction and the end results of such compaction. Considerable variation in requirements for compaction exists among the various agencies.

The general nature and purpose of compaction are well understood, however. The general notion is that soils should be compacted by rolling, at or near optimum moisture content, to some percentage of the maximum density established through the use of a known laboratory compactive effort. It has been established that certain advantages may be expected to accrue from the use of this compaction process. Included among the more important of these advantages are the increased shearing resistance or "stability" of the soil, its decreased permeability, and the minimizing of future settlement of the embankment itself (but not necessarily settlement due to consolidation or shearing failure of the embankment foundation).

Present specifications of the various state highway departments related to compaction are not uniform. A large majority of the states require compaction to a certain minimum percentage of the maximum density, as determined by the standard AASHTO laboratory compaction procedure (Method T99). (See Chapter 15.) The most common requirement, employed by about 50 percent of state highway agencies, is 95 percent of the maximum dry density obtained in the AASHTO T99 test (5). In some states, maximum compaction requirements are stated in terms of the maximum laboratory density; for example, for a soil with a maximum density of 1599 kg/m^3 (99.9 lb/ft^3) or less, the required density in the field must be 100 percent of the laboratory value; for 1600 to 1918 kg/m^3 (100–119.9 lb/ft^3), 95 percent; and so on. A few agencies still specify the number of passes to be made by specified equipment. Many specifications contain requirements related to the size and nature of compaction equipment. In recent years, the tendency has been toward the incorporation of "end-result" specifications for compaction, with desired end results specified but leaving the choice of equipment up to the contractor. Moisture content during compaction is controlled, sometimes by general specification statements, sometimes by numerical values (e.g., ± 2 percent of optimum moisture).

The establishment of compaction requirements by the design engineer is a matter that requires careful study of all the factors involved. A detailed discussion of this and other matters related to the construction of embankments is given in Ref. 5.

State highway department specifications represent a compromise among the many variables and reflect requirements that are reasonable for the conditions which exist in each state. On large or unusual projects, special provisions may be written and used to control the work. General recommendations for density requirements for highway embankments are contained in Table 17-3.

17-12 Control of Compaction

Generally speaking, the field compaction process is controlled by making relatively frequent checks of the density and moisture content of the soil that is undergoing compaction. The measured density is the wet unit weight; the dry unit weight is calculated on the basis of this figure and the measured moisture content. The dry unit weight may then be compared with the compaction curve for the soil and compactive effort involved to see if the density being obtained in the field meets the requirement established in the laboratory. If different soils are being placed in different sections of the fill, it is understood that the curve (or figure) that is being used to check the rolling process is the proper one for the soil concerned.

Three methods have been used widely by highway agencies for determining the density of the soil. These may be designated as the sand, balloon, and heavy oil methods. In each of these the density determination is begun by carefully

TABLE 17-3 Recommended Minimum Requirements for Compaction of Embankments and Subgrades

Class of soil (AASHTO M145)	Density, Percent of AASHTO Standard T99	
	Embankments	Subgrade
A-1, A-3, A-2-4, A-2-5	100	95
A-2-6, A-2-7, A-4, A-5, A-6, A-7	95[a]	95[a]

[a]AASHTO recommends compaction within 2 percent of the optimum moisture content. Use of these materials requires special attention to design and construction of embankments.

Source: Standard Specifications for Transportation Materials. Designation M57-80, American Association of State Highway and Transportaiton Officials, Washington, DC (1982).

excavating a cylindrical hole in the soil layer; the hole is usually about 100 mm (4 in.) in diameter and the full depth of the layer. All the material taken from the hole is carefully saved, placed in a sealed container, and weighed as quickly as possible. The weight determination may be made immediately on a field balance if desired. The volume of the hole may then be determined by the use of sand, a balloon apparatus, or heavy oil. Some use has also been made of small undisturbed samples taken with cylindrical drive sampler; in such cases, wet unit weights may be used for the comparison.

The apparatus required for the sand density determination is shown in Figure 17-3. In using this apparatus, the area around the hole is carefully leveled, the jar—with funnel attached and filled with "standard" sand—inverted over the hole, and the valve opened. The sand flows into and fills the hole. When the hole is filled the valve is closed, and the jar and the remaining sand weighed. The weight of sand required to fill the hole may then be determined. Since the same procedure is carefully followed in each determination and the same dry sand is used, a previously established relationship between a given weight of sand and

FIGURE 17-3 Sand-density equipment. (Courtesy Soiltest, Inc.)

the volume occupied by this amount of sand may be used to ascertain the volume of the hole. Because the weight of soil taken from the hole and its volume are then known, the wet unit weight may be calcualted. A balloon filled with water may also be used to determine the volume of the hole, and the volume of the hole measured directly. This volume may also be obtained by filling the hole with a heavy oil of known specific gravity. The latter procedure is more widely used for checking the densities of bases than embankments.

If the soil is known to be at or near optimum moisture content, the density check may be made by comparing the measured wet unit weight with the wet unit weight established by laboratory procedure. Generally, however, the procedure used is to make an additional moisture determination so that this factor may be checked and the dry unit weight determined. If a moisture content determination is deemed necessary, then speed is of the essence. The determination may be made by rapid drying over a field stove. Drying is also frequently accomplished by adding alcohol, or some other volatile solvent, to the soil and then removing the water by igniting the solvent. Time is seldom available for the use of more accurate methods because the decision as to whether or not a satisfactory density has been obtained must be made quickly so that construction operations will not be delayed.

Many highway agencies use nuclear devices to measure in-place density of soils. Reference 6 gives the following description of the principle of nuclear density gauges:

> When gamma rays are emitted from a radioisotope source in proximity to a surface they interact with the material and are scattered or absorbed. The count of the gamma rays emerging from the surface at some point is influenced by the density and the composition of the material. A typical gauge consists of a gamma-ray source, a detector with associated counting electronics, and shielding between the two to prevent direct transmission of the gamma rays from the source of the detector. A wide variety of gauge configurations is possible, involving source energy and intensity, type and efficiency of detector, and source-detector separation. The most universally employed method of determining density with a gamma-ray gauge is by use of a calibration curve prepared from the empirically determined relationships between density and response for each individual instrument. The calibration curve for a particular instrument is originally obtained by plotting the response measured by the gauge for a set of calibration standards of known density.

As Figure 17-4 illustrates, there are two basic types of nuclear gauges: (1) the surface or backscatter type and (2) the transmission type. The latter type has a probe, the transmission source, that is lowered into the ground through an access tube driven into the soil. One type of nuclear gauge is shown as Figure 17-5.

Nuclear equipment may also be used to measure moisture content. A neutron moisture gauge consists of a fast-neutron source and a low-neutron detector. Fast neutrons emitted from the device are slowed principally by collisions with hydrogen atoms. Assuming that most of the hydrogen present is in the form of water, a calibrated gauge can be used to measure the moisture content.

The principal advantage of nuclear gauges is speed. Such devices require as little as one-fifth the time required for conventional tests. They also do not disturb the soil, can be used on a wide range of materials, and require little operator training. There is little danger from the radioactivity, but the devices must be calibrated carefully.

If the density determined as outlined above is equal to or greater than that

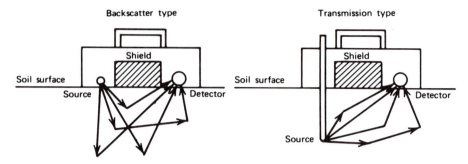

FIGURE 17-4 Representative gamma-ray paths for density gauges. (SOURCE: Gardner, R. P., and Roberts, K. F. Density and Moisture Content Measurements by Nuclear Methods. NCHRP Report No. 43, Highway Research Board, Washington, DC, 1967.)

required, the rolling may generally be judged to be satisfactory and construction of another lift may proceed. If the density is lower than that specified, then additional rolling may be required or the moisture content may have to be adjusted. If these measures fail, then it is reasonable to assume that the rolling equipment being used is not capable of producing the required density or that the moisture-density relationship that is being used is not the one for the soil that is actually being compacted.

The procedure that has been outlined for the control of compaction may seem unduly cumbersome and complicated. Actually, after compaction of a given section has begun and the initial adjustments have been made, the situation is usually considerably simplified. The inspector will, for example, quickly learn to judge the moisture content of the soil by appearance and feel. The inspector may also quickly decide that, with the proper moisture content being maintained, a certain number of passes of the roller—say six or eight—will accomplish the required densification. The job may then be simplified into a count of the roller trips, with occasional density checks being made as judgment dictates.

The number and frequency of density and moisture checks that are required for adequate control are, of course, functions of job conditions. On an average basis, perhaps one check will be required every hour on a fill of moderate length. Some other standard may be applied, such as a minimum of one test per 750 m³ (1000 cu yd) of compacted material. Some soils are more sensitive to moisture change than are others. For example, sands usually have a moisture-density curve that shows a sharp peak, with the optimum moisture content being clearly defined, whereas clay soils usually have a flatter curve so that a small variation in moisture content does not greatly affect the density. It is impossible, generally speaking, to maintain the moisture content exactly at optimum for any substantial length of time, so that some tolerance must be permitted in the moisture content.

17-13 Compaction Equipment

Principal factors involved in the selection of compaction equipment for use on a given project include the specifications under which the work is to be carried out and the choice of equipment available to the contractor. More than one type of equipment and compaction procedure can often be used successfully for a given

FIGURE 17-5 A nuclear moisture density meter. (Courtesy Soiltest, Inc.)

set of circumstances; however, there usually is one combination that will give the most economical result.

A detailed discussion of the most efficient and economical use of a given type of equipment for a given project is beyond the scope of this text. A more complete treatment of this topic is given by Ref. 5. Pertinent factors that must be considered include thickness of lift, total weights and unit pressures exerted on the soil, required number of passes, operating speeds, soil type and moisture content, and others. On large projects, relationships among the variables involved may often be resolved by the construction of test fills prior to the start of actual construction or by trial during early phases of construction. For economical operation, there must be a proper balance between rolling equipment and hauling equipment on the job.

Additional water required for compaction is generally added by ordinary pressure distributors. Rotary speed mixers may be used to mix the water and soil in some instances. Aeration of soils that are too wet for compaction may be accomplished by the use of agricultural tools, such as harrows and disks, by turning the soil with blade graders, or by rotary speed mixers with the hood up. The problem also may be handled by "sandwiching" alternative layers of wet and dry materials. In many cases aeration is accomplished by simply allowing the soil to dry of its own accord.

REINFORCED EARTH EMBANKMENTS

Reinforced earth has been defined as a construction material composed of soil fill strengthened by the inclusion of rods, bars, fibers, or nets which interact with the soil by means of frictional resistance (7). The development of the concept of reinforced earth is credited to Henri Vidal, a French engineer (8, 9).

Figure 17-6 illustrates a common type of reinforced structure. Its components are (1) the soil backfill, typically a cohesionless free-draining material such as sand; (2) reinforcing members consisting of long thin strips called ties, which are arranged in a regular pattern; and (3) a covering called "the skin," which maintains the integrity of the sand at the outface wall.

The earliest reinforced earth system used galvanized steel strips for the ties and galvanized steel semicylindrical elements approximately 3 mm in wall thickness for the skin (9). Such systems depend on the frictional resistance between the grains of the fill mass and the reinforcing ties.

Since the introduction of the patented Vidal reinforced earth system in the mid-1960s, other soil reinforcement systems have been developed using bearing rather than friction for the stress transfer or a combination of bearing and frictional forces. These systems consist of metallic bar mats or polymeric tensile-resisting elements arranged in rectangular grids placed in horizontal planes in the back fill. Wire mesh has been used in a similar manner, as well as continuous sheets of geotextile material laid between layers of backfill (10). A variety of

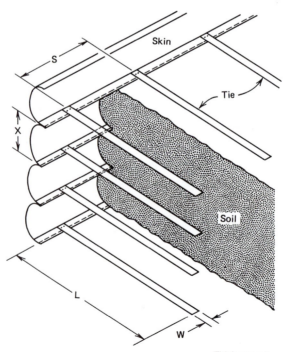

FIGURE 17-6 Components and key dimensions of a typical reinforced earth wall. (Courtesy American Society of Civil Engineers.)

coverings or "facing elements" have also been used, including shotcrete, geotextiles, seeded soil, precast concrete panels, and welded wire mesh.

Reinforced earth systems facilitate the support of sections of highway along steep slopes (Fig. 17-7). Such systems are especially advantageous in urban areas where rights-of-way are limited and land is expensive. Reinforced earth structures may cost as little as one-half as much as fills constructed behind reinforced concrete walls.

Reinforced earth fills can be constructed on relatively weak foundations because the loads are spread over a large area and can be built to great heights. Reinforced earth structures as high as 21 m (70 ft) have been constructed.

For more information on the design of reinforced earth embankments the reader is advised to consult Refs. 7 through 12.

SPECIAL TREATMENT OF EMBANKMENT FOUNDATIONS

The foundations of embankments that are constructed in swampy areas—particularly where peat or other highly organic soils are encountered—frequently require special treatment if failures of the embankment or pavement structure due to consolidation or displacement of the underlying soil layer are to be avoided. The methods in general use may be classified as follows:

1. Gravity subsidence.
2. Partial or total excavation.
3. Blasting.
4. Jetting.
5. Vertical sand drains.
6. Reinforcement with engineering fabrics.

Various combinations of these methods may also be used in specific locations.

17-14 Gravity Subsidence

In some instances, a fill may simply be placed on the surface of an unsatisfactory foundation soil and allowed to settle as it will with no special treatment of the

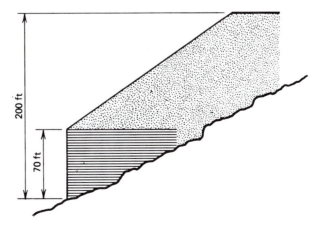

FIGURE 17-7 Typical application of a reinforced earth wall.

underlying soil. Where soils of high compressibility and low shearing resistance are present in the embankment foundation, this method is obviously unsuitable for any except secondary roads that carry extremely low volumes of traffic. The fill may settle rapidly during or shortly after the construction period, owing to shearing failure and displacement of the foundation soil. Settlement due to consolidation may continue for many years. In either case the satisfactory placing of any wearing surface is virtually impossible, and a long period of continuous maintenance must be anticipated, as more and more soil must be added to the fill as time passes until the embankment eventually comes to a stable condition.

In some cases a surcharge of extra height of fill has been used to speed up settlement; the surcharge is later removed. A variation of this method was used by the Louisiana Department of Highways to build a section of interstate highway near New Orleans. They used a "rolling" surcharge over a deposit of peat and muck 3 to 6 mm (10–20 ft) thick. In this process 1.2 m (4 ft) of sand surcharge of a total of 1.7 m (5½ ft) was placed at one end of the fill and allowed to remain in place for a specified time; the sand then was removed and placed at the other end of the fill. Thus, the fill "rolled" slowly ahead in a series of steps.

17-15 Partial or Total Excavation

Where the undesirable soil is of shallow depth up to 2.4 to 3.7 m (8 to 12 ft) and is underlaid by a soil of satisfactory character or by rock, many highway agencies excavate the entire depth of unsatisfactory soil. In New York State, total excavation has been carried out to a depth of 10.7 m (35 ft). The excavated area is then backfilled with suitable granular material. This method is expensive, of course, but has the advantage of permitting the immediate placing of a pavement on the completed fill. When such treatment is used, the backfill is composed of sand and gravel of suitable gradation.

A procedure sometimes used on secondary roads where deep swamp deposits are encountered consists of partial excavation of the foundation soil. That is, the soil is excavated to a depth of several feet and backfill is placed as before. In this operation it is essential that the backfill be placed as excavation proceeds. This method is open to the same criticisms that are applicable to the method which we have called "gravity subsidence." Displacement of the deep underlying soil may still occur under the weight of the backfill and embankment, and settlements due to consolidation are also to be expected, although they should be somewhat smaller than if the compressible soil had not been partially excavated. It is again usually impossible to place any permanent type of wearing surface without the expectation of very high maintenance costs.

17-16 Blasting

Explosives have been widely used to aid in the placing of embankments in swampy areas. When this method is used, the embankment is generally constructed to a level considerably above the final grade and the displacement of the underlying soil is accelerated by the detonation of deep charges of explosives. The blasts serve a dual purpose in that some of the material may be displaced by the force of the explosions, and the remainder of the unstable soil is liquefied to a greater or lesser extent and can thus be displaced more readily by the weight of the overlying fill. As the underlying soil is displaced, the fill subsides; and it may then be finished to the required elevation and cross section. This method requires careful control but should result in the formation of an embankment that will be stable in a relatively short period of time.

17-17 Jetting

The process of "jetting" involves the pumping of water into the underlying soil layer in order to liquefy it and thus aid in the displacement of the layer by the weight of the embankment. Jetting is carried out in conjunction with the placing of the embankment material. In typical installations, a single line of jets is placed along one side of the embankment. The jets are spaced 3 to 7.6 m (10 to 25 ft) apart, and they extend about two-thirds of the way into the underlying layer. The fill settles or "subsides," as before, and must be brought to the desired cross section after a stable condition has been attained. Trenches may be excavated along one or both sides of the embankment to assist in the displacement of the weak soil layer. In some cases the embankment material, rather than the underlying soil, may be jetted; this results in an increase in the weight of the fill.

17-18 Vertical Sand Drains

Another method used to treat troublesome embankment foundations uses vertical sand drains. Vertical sand drains generally consist of circular holes or "shafts" from 450 to 600 mm (18 to 24 in.) in diameter, which are spaced from 2 to 6 m (6 to 20 ft) apart on centers beneath the embankment section and are carried completely through the layer of compressible soil. The holes are fashioned by rotary drilling or by the use of a hollow or plugged-end mandrel. The holes are backfilled with suitable granular material. A "sand blanket" 1 to 1.5 m (3 to 5 ft) thick is generally placed at the top of the drains (and the existing soil) extending across the entire width of the embankment section. The embankment is then constructed by normal methods on top of the sand blanket.

The basic purpose served by the installation of vertical sand drains is the acceleration of consolidation of the compressible soil layer. Consolidation is, of course, due to the forcing of water from the voids of the soil, and the rate of consolidation is dependent, among other things, on the distance which the water has to travel to escape. Without sand drains the water may escape only at the top of the soil layer if there is an underlying layer of impervious material, or at the top and bottom of the layer. If the layer is of considerable thickness, settlements due to consolidation may occur over a period of many years. Vertical sand drains speed up the consolidation process by providing many shortened paths through which the water may escape. Vertical drainage at the top (or top and bottom) continues, and in addition radial horizontal drainage takes place. Water that is thus forced into the sand drains moves up the drains and out through the sand blanket. In addition to the fact that radial drainage is provided, this method takes advantage of the fact that many fine-grained sedimentary soils have a permeability that is several times greater in a horizontal direction than in a vertical direction. This fact also contributes to the accelerated rate of consolidation. Still another advantage lies in the fact that a rapid increase in shearing strength usually accompanies the accelerated consolidation process so that the probability of lateral flow is diminished and the settlement of the fill confined to that caused by consolidation.

Phenomenal results have been reported from the use of this method on many jobs in different sections of the country. In many cases, even where high fills and deep deposits of compressible soil were involved, 90 percent or more of the ultimate settlement has occurred during the construction period or very shortly thereafter. This behavior makes possible the placing of a permanent-type surface within a few months after the fill is completed. Costs of this method have been generally reported as somewhat less than those of most of the methods previously described. In Figure 17-8 is shown a profile of an installation of vertical sand drains used in connection with the construction of an embankment across a swampy area encountered in the construction by the New York Department of Public Works of a highway near Syracuse, N.Y. (*13*). Fills up to 13.7 m (45 ft) high were built on surface organic soils underlaid by soft clays to a depth of 10.7 m (35 ft).

17-19 Fill Foundation Reinforcement with Engineering Fabrics

In recent years, engineers have found that engineering fabrics can be used to advantage in the construction of low fills over swampy or marshy areas. Other names for engineering fabrics include geotextiles or geotechnical fabric. These products are permeable textile cloths or mats made from a variety of artificial fibers. Engineering fabrics are generally lightweight, possess moderate to high tensile strength, and withstand large amounts of elongation without rupture (*14*).

Typically, the engineering fabric is placed on the weak foundation and overlaid with the embankment fill. This increases the bearing capacity of the foundation and allows a higher fill to be constructed on it. The use of a fabric in such circumstances has several advantages: (1) it permits construction of embankments on otherwise unsuitable foundations; (2) it eliminates lost construction time due to failures and delays inherent in soft ground construction; and (3) it results in considerable savings of fill material compared to displacement construction methods (*14*).

FINISHING OPERATIONS

"Finishing operations" refer to the final series of operations that are required to complete the earthwork operations involved in a typical highway project. These operations are taken here to include such things as trimming of shoulders, side, and ditch slopes; the fine-grading operations required to bring the earthwork sections to their final grade and cross section; the compaction of cut sections; and similar items. Actually, most of these are not separate operations that are performed after all other earthwork operations are completed; rather they are generally carried along and performed as the job approaches completion.

The tool most widely used in the performance of the majority of finishing operations is the motor grader. Its long wheelbase and ability to work to close tolerances make it ideal for this application. 'Dozers, scraper units, and gradalls are also sometimes used in finishing operations. In cut sections, excavations and shaping of the cross section to the dimensions shown on the plans may be all that is required; however, many specifications call for scarifying and compaction of the soil. The specifications for compaction, generally speaking, are the same as those for the compaction of embankments. There is some tendency to require greater densities in cut sections and in the top layer of embankment sections than are required in normal embankment construction. The required scarification

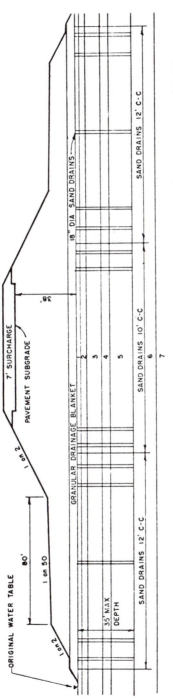

FIGURE 17-8 Typical section of sand drain installation. (Courtesy Lyndon H. Moore, New York Department of Public Works.)

of cut sections may be accomplished by rippers, disks, and harrows of various types or by scarifier attachments on motor graders. The required depth of scarification and compaction in these sections is generally from 200 to 300 mm (8 to 12 in.). The compaction operation, as before, may be accomplished through the use of various types of rollers, with the sheep's-foot roller the predominating tool. The section would generally be shaped to final dimensions through the use of a motor grader.

It should be noted that the finishing operations and other earthwork items that have been described are in many cases practically all the work involved in construction prior to the placing of the base and surface. In other cases, additional finishing operations may be required before this phase of construction can begin. For example, the subgrade soil may require stabilization. The placing of high-type pavements and bases, such as portland-cement concrete, usually involves additional subgrade preparation beyond that which has been described in this section. These items and others of a similar nature are discussed in later chapters.

PLANNING AND EXECUTION OF CONSTRUCTION PROJECTS

Because the majority of highway construction is performed by contract, the principles stated in the following sections are primarily intended to apply to the planning and execution of contract work. Most of the items that are discussed are also applicable to work which is performed by force-account. Many contractors prepare tentative job plans as a part of their bid computations, and on large projects it may be required that the job plan be submitted with the bid. In any event, a comprehensive job plan should be prepared by the successful bidder immediately after the contract is awarded and before construction operations begin. Space does not permit lengthy exposition of the many details that enter into the successful planning and conduct of the earthwork operations involved in a typical highway project, but some of the steps that are necessary are presented here.

The first step in the preparation of a schedule of construction operations is a detailed analysis of the amount of work of different types involved in the job. By this is meant such items as the number of acres to be cleared and grubbed, the number of cubic meters (or cubic yards) of roadway excavation, the haul distances, and so on. The quantities used are generally those of the engineer's estimate. The plans are carefully examined and the location, extent, and conditions affecting the excecution of each step of the work determined. A detailed step-by-step analysis is then generally made on the equipment, labor, and material requirements for each phase of the project.

In establishing the equipment estimate sheet, for example, it would be necessary to tabulate each operation and the quantity of work involved in that operation, to select the units of equipment to be assigned to this portion of the job, to estimate the hourly production rate for each of the machines to be used, and to calculate the total number of equipment hours required for completion of the operation. If the time to be consumed in this operation was limited and if sufficient equipment was available, a sufficient number of machines would simply be assigned to the operation to complete it in the given time. The labor required for each operation would probably be estimated at the same time, whereas material estimates would probably be made separately.

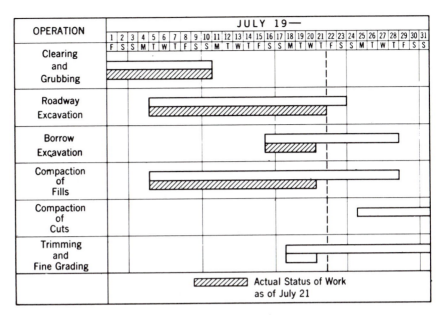

FIGURE 17-9 Schedule of construction operations.

17-20 Job Scheduling

On the basis of the information tabulated in the work estimates previously referred to, a detailed construction schedule may be prepared. This schedule may be presented in several different forms, and one method of presentation is shown in Figure 17-9. As will be noted, each of the general operations involved in the project is tabulated, and the time assigned for completion of each operation is shown by the plain bar. The sequence of these operations is also shown. As indicated by the chart, the progress that has been made on the project at any particular time is also shown by the crosshatched bars. Using this sheet, the contractor may quickly determine the status of the work.

It will be noted that certain operations overlap, as would be expected. As a matter of fact, the separate operations must be carefully coordinated and balanced so that they will be completed in the proper sequence and so that equipment on the job will be used to the fullest possible extent. Although the schedule that is illustrated here is for earthwork operations only, a similar schedule would be prepared for every phase of the complete project.

In addition to the construction schedule, contractors on many jobs prepare detailed sheets that show equipment, labor, and material requirements for every day of the project. Preparation of these sheets enables the contractor to further plan operations so as to make use of the available equipment and labor in the best possible way, and they help to maintain a smooth flow of materials, fuel, lubricants, and so on to the job.

17-21 Critical Path Method

Highway contractors also generally use the critical path method, developed in the mid-1950s by engineers from DuPont and the Univac Division of Sperry Rand, for construction project scheduling and control. State highway agencies also use the method for planning and programming, scheduling of design activities, and other activities.

CPM is an analysis and management tool that is useful in planning and scheduling any project made up of various subunits of work (as are most highway projects), some of which can be done sequentially from start to finish of the project (*15*). Many engineers feel that the CPM diagram has advantages over the more conventional bar chart described in the preceding section.

The basic concept of CPM is that, among the many operations that form a complete project, there must be one series of sequential operations that takes longer than any other possible series. The shortest possible time in which the project can be completed is the time needed for the longest series. The critical series is the "critical path." Determination of the critical path is in itself an exercise in logic that promotes good management of the project.

Determination of the critical path is aided by the construction of an "arrow diagram," such as the one shown in Figure 17-10. Detailed instructions for the arrow diagram and its applications to highway problems are contained in Ref. 16.

In the planning phase of CPM, the first step in drawing the arrow diagram is to break the project into its subunits (activities). On paper, an activity is represented by an arrow drawn from left to right. At each end of the arrow is placed a circle or node to represent an "event." The head end (right end) of the arrow merely indicates the flow of work; the arrow is not to scale.

In CPM terminology, the event at the tail of the arrow is the "i" event; it represents the instant of time at which this event may start and the instant at which the preceding event must end. The event at the head of the arrow is "j" event; it represents the instant at which the activity indicated by the arrow ends and subsequent activities may begin.

A logical analysis of a project may produce the general form of the diagram of Figure 17-10. In this diagram, activities A, B, and F may proceed simultaneosly with C and E. Note, however, that E cannot begin until C is complete, and H and I, although they can be done concurrently, cannot begin until A, B, E, and F are complete.

Note also that activities D, G, and J must follow one another in that order. The dashed line between nodes 4 and 7 is a "dummy" activity put there only to show a logic restraint. In this case, it means that K cannot proceed until J is complete (as well as H). It is made necessary on the diagram because of the fact that the arrow for H and I could not be drawn to nodes 4 and 7 at the same time. Activities are commonly designated by connected nodes, for example, 1 to 3; in that case, additional dummy activities must be used to keep the tabulations straight.

In Figure 17-10, the remaining steps, K and L, must follow one another, whereas M can be done concurrently.

In the usual approach, the drawing of a generalized diagram and a check of its logic, forward and backward, complete the "planning" phase of the analysis. The "scheduling" phase may then begin, working with the basic diagram.

The first step in the scheduling phase is to estimate the time required for each activity. The usual procedure is for persons experienced in the work to estimate the normal time required for each operation. At this stage, no attention is paid to other than normal resources of humans, time, machinery, and money. Later, adjustments can be made if desired to obtain the most efficient utilization of one

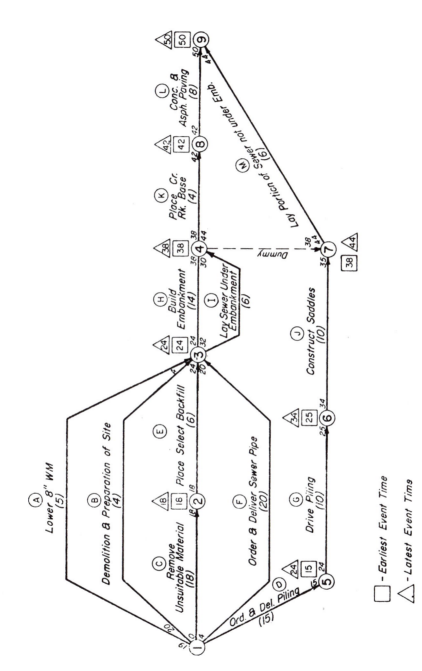

FIGURE 17-10 Typical critical path diagram for a simple highway project. (Courtesy Douglas L. Jonas.)

□ - Earliest Event Time

△ - Latest Event Time

or more of these resources. The normal time, in days, for each of the operations of Figure 17-10 is shown in parentheses along each arrow; for example, 20 days are alloted for activity F.

Study of the diagram will reveal that, because activity C takes 18 days, activity E cannot begin before that time has elapsed. It follows that, because activity E takes 6 days, activity H cannot begin until 24 days after the start of activity C. In CPM terminology, the earliest start time for activity E would be the end of the 18th day; this is called the "earliest event time" and is shown in a square on the diagram. The "earliest finish time" for E would be 24 days.

Note also that each of the activities concurrent with C and E (i.e., A, B, and F) requires less than 24 days to complete; respectively, they take 5, 4, and 20 days to do. Putting it another way, free time, called "float," is available for each of the activities.

The same sort of analysis can be followed from start to finish of the project to find the earliest finish time for the entire project (in this case, 50 days). Examination also will show that one sequence of activities has no float time; this sequence is, of course, the critical path. In Figure 17-10, the critical path is the horizontal line connecting nodes 1, 2, 3, 4, 8, and 9 (activities C, E, H, K, and L).

One further step can be taken at this stage: one can start at the right end of the diagram, with the earliest finish time, and work backward through the diagram to determine the latest time that each activity could start or finish without affecting the project completion date.

Basic construction of the CPM diagram for a simple project thus is a relatively simple matter; calculations may be done manually. On a complex project with hundreds of subitems, the CPM diagram can be highly complex; a computer often is used to speed calculations.

If the time of 50 days is satisfactory, the CPM diagram of Figure 17-10 may be used to control the project. Detailed equipment, labor, and material schedules may be prepared on a day-to-day basis. If job conditions change, the CPM diagram may have to be adjusted to fit the changed conditions. If the time of completion is to be advanced ("crashed"), it is obvious that it is most fruitful to crash the activities on the critical path; crashing noncritical activities will not advance the completion date.

As mentioned previously, one of the greatest values of the CPM approach is in resources planning. Detailed resources planning is beyond the scope of this text; however, its importance cannot be overemphasized. Obvious applications are to make the most efficient use of equipment to use skilled labor in the most productive way, and to determine the optimum job schedule from the standpoint of overall cost. The contractor or owner can, if desired, determine the cost of crashing a project schedule and compare the cost with anticipated benefits. In all these approaches, a computer is an essential tool for a complex project.

REFERENCES

1. *Guide Specifications for Highway Construction.* American Association of State Highway and Transportation Officials, Washington, DC (1993).

2. Smith, Travis, and Kleiman, William F. Behavior of High Earth Embankment on U.S. 101. *Highway Research Record No. 345* (1971).

3. *Compaction of Embankments, Subgrades, and Bases.* Highway Research Bulletin No. 58 (1952).

4. *Standard Specifications for Transportation Materials,* Part I. American Association of State Highway and Transportation Officials, Washington, DC (1993).

5. *Construction of Embankments.* National Cooperative Highway Research Program Synthesis of Highway Practice 8, Highway Research Board, Washington, DC (1971).

6. Gardner, R. P., Dunn, W. L., McDougall, F. H., and Lippold, W. J. *Optimization of Density and Moisture Content by Nuclear Methods.* National Cooperative Highway Research Program Report No. 125, Washington, DC (1971).

7. Lee, Kenneth L., Adams, Bobby Dean, and Vagneron, Jean-Marie J. Reinforced Earth Retaining Walls. *Journal of the Soil Mechanics Division,* Vol. 99, No. SM10, American Society of Civil Engineers, New York (October 1973).

8. Vidal, H. The Principal of Reinforced Earth. *Highway Research Record 282,* Transportation Research Board, Washington, DC (1969).

9. Vidal, H. Reinforced Earth Retaining Wall. *Civil Engineering,* Vol. 40, No. 2, American Society of Civil Engineering, New York (February 1970).

10. Mitchell, James K., and Villet, William C. B. *Reinforcement of Earth Slopes and Embankments.* National Cooperative Highway Research Program Report 290, Transportation Research Board, Washington, DC (June 1987).

11. Ingold, T. S. *Reinforced Earth.* American Society of Civil Engineers, New York (1982).

12. Yamanouchi, Toyotoshi, Miura, Norhhiko and Ochiai Hidetoshi, *Proceedings of the International Geotechnical Symposium on Theory and Practice of Earth Reinforcement,* Fukuoka Kyushu, October 5–7, 1988. A. A. Balkema, Rotterdam, Brookfield (1988).

13. Moore, Lyndon H. Summary of Treatments for Highway Embankments on Soft Foundations. *Highway Research Record 133* (1966).

14. Haliburton, T. Allan, Lawmaster, Jack D., and McGuffey, Verne C. *Use of Engineering Fabrics in Transportation-Related Applications.* Prepared for Federal Highway Administration, Washington, DC (October 1981).

15. Jonas, Douglas L. Reducing Equipment Needs With CPM. *Western Construction.* (November 1964).

16. Reid, Douglas M. *Project Management System (PMS) Through the Use of the Critical Path Method (CPM).* Federal Highway Administration, Washington, DC (1974).

CHAPTER 18

BASES, SUBBASES, AND LOW-COST SURFACES

In this chapter and in the following two chapters, we turn our attention to the design and construction of the subbases, bases, and wearing surfaces that constitute the pavement structure. The sequence of the material on pavements generally follows a line of increasing costs. This chapter focuses on relatively low-cost pavements, ranging from soil-stabilized roads to bituminous surface treatments and low-cost bituminous plant mixtures.

SOIL STABILIZED ROADS

18-1 Introduction

In a broad sense, soil stabilization may be defined as the combination and manipulation of soils, with or without admixtures, to produce a firm mass that is capable of supporting traffic in all weather conditions.

If a stabilized soil is to be truly "stable," then it must have sufficient shearing strength to withstand the stresses imposed on it by traffic loads in all kinds of weather without excessive deformation. In addition, if the stabilized soil mixture is to be used as a wearing surface, then it must be capable of withstanding the abrasive effects of traffic.

In this branch of highway design and construction the emphasis is placed on the effective utilization of local materials, with a view toward decreased construction costs. In some areas the natural soils are of an unfavorable character and require modification through the use of suitable mineral constituents such as gravel or crushed stone or clay binder. In still other areas, admixtures such as bituminous materials, portland cement, salt, or lime are used for effective stabilization. The type and degree of stabilization required in any given instance are largely a function of the availability and cost of the required materials as well as the use that is to be made of the stabilized soil mixture.

The primary use of stabilized soil mixtures at the present time is in base and subbase construction. A stabilized soil base or subbase may provide the support for a relatively thin wearing surface that will be subjected to light or moderate amounts of traffic, or it may function as a base for a high-type pavement that will be subjected to very heavy volumes of traffic. Certain stabilized soil mixtures, including granular soil stabilized mixtures, serve as wearing surfaces on light-traffic roads.

Stabilized soil mixtures also lend themselves readily to the process of "stage construction," which involves the gradual improvement of the individual units of a highway system as the demands of traffic increase. Thus a properly designed stabilized soil mixture might function briefly as a wearing surface, receive a thin bituminous surface treatment as traffic increases, and eventually serve as a support for a high-type bituminous pavement that will serve a very heavy volume of traffic. Certain naturally occurring soils may require only compaction and drainage for stabilization. Other soils require treatment in various fashions and with various materials in order to satisfactorily perform their intended function. Four general classes of stabilization are briefly described in the following pages: (1) soil-aggregate roads and "granular stabilization," (2) bituminous soil stabilization, (3) soil-cement roads, and (4) stabilization by the addition of salt, lime, and various other chemicals.

18-2 Soil-Aggregate Roads and Granular Stabilization

Soil-aggregate roads consist of a substantial layer of a properly proportioned and blended mixture of soil and aggregate compacted to form a road that is capable of supporting traffic in all weather conditions. The AASHTO definition of soil-aggregate (dense-graded aggregate) is "natural or prepared mixtures consisting predominantly of stone, gravel, or sand and containing silt-clay" (*1*). In addition to serving as wearing courses, generally for light traffic, soil-aggregate mixtures are very widely used as bases and subbases.

In this discussion, the term "aggregate" refers to the portion of a granular mixture or natural soil that is retained on a No. 200 sieve, including, by the AASHTO definitions, "stone," "gravel," and "sand." The function of aggregate in a soil-aggregate mixture is primarily to contribute internal friction. The most suitable aggregates, generally speaking, are those that are well graded from coarse to fine. One approach to the problem of gradation is to establish limits for the various sizes of particles that will result in a combination which gives maximum density; high density, in turn, is associated with high stability. It is not wise to lay too much stress on gradation, however, as many aggregates have been used that are not well graded in the usual sense. Specifications concerning gradation for this purpose are usually quite tolerant, with the idea, again, of making the best possible use of locally available materials.

The "fine-soil fraction" is that portion of the mixture that passes a No. 200 sieve, that is, silts and clays. The function of this portion of the soil-aggregate mixture is to act as a filler for the remainder of the mixture and to supply cohesion that will aid in the retention of stability during dry weather, whereas the swelling of the clay contained in this mixture serves to retard the penetration of water in wet weather. The nature and amount of this fine material must be carefully controlled if the mixture is to function properly, because an excessive amount of clay and silt may result in excessive volume change with change in moisture content. The properties of the fine-soil fraction are frequently determined by measurements of the plasticity index, or PI. As will be recalled, the PI is the difference between the liquid limit and the plastic limit. Both of these latter tests are customarily performed on the portion of the soil that passes a No. 40 sieve; this is frequently called the "soil binder."

It must again be noted that, for satisfactory results, the soil-aggregate mixture must be properly proportioned. In some cases natural soils may have satisfactory properties and thus may require only compaction and drainage for stabilization. Far more frequently, however, the existing soil will be missing some of the important soil elements. Thus a granular soil or "aggregate" may require the addition of fine material for binder, or a fine-grained soil may require the addition of aggregate for stability and to prevent excessive volume change. In certain cases the soil existing in the roadbed may only require the addition of a small amount of material that may be brought in, distributed along the roadway in the proper amounts, and blended with the existing soil to produce a satisfactory soil-aggregate mixture. In other cases the desired soil-aggregate mixture may be compounded in the proper proportions at a central mixing plant and then deposited on top of the existing roadway soil. In any case the properties of the final mixture are generally judged and controlled by the mechanical analysis, the liquid limit, and the PI. In addition, because the mixture is generally required to be compacted to a high degree of density, the moisture-density relationships for the mixture must be determined for a given compactive effort so that the field compaction process may be properly controlled.

Before presenting detailed recommendations as to the composition of satisfactory soil-aggregate combinations, it seems desirable to discuss briefly the difference in requirements for the composition of mixtures intended for use as bases and those for use as wearing surfaces. If the mixture is to withstand the abrasive effects of traffic, it should have more binder than is required in a base course so that it will be more impervious to precipitation that falls on the surface of the road and can, to a certain extent, replace by capillary moisture what is lost by evaporation. Bases do not require as high a proportion of binder soil, because they may generally be required to have high stability and low capillarity so that they do not tend to soften because of the accumulation of capillary moisture. This consideration becomes of special importance when stage construction is being employed, during which the mixture may serve as a wearing surface for a time and then as a base, so that the mixture must be capable of performing both functions satisfactorily during its service life. Where frost action is a factor, the percentage of material passing the No. 200 sieve may be reduced to prevent damage to a base or subbase.

Specifications for Granular Soil Stabilized Mixtures

Many different specifications are used for these soil-aggregate mixtures by different highway agencies. Indicative of the requirements are those given in Table 18-1, which is contained in Ref. 1.

It is recommended that the fraction passing the No. 200 sieve not be greater

TABLE 18-1 Gradation Limits for Soil-Aggregate Subbase, Base, and Surface Courses

Sieve Designation Grading	A	B	C	D	E	F
2 in. (50 mm)	100	100	—	—	—	—
1 in. (25 mm)	—	75–95	100	100	100	100
⅜ in. (9.5 mm)	30–65	40–75	50–85	60–100	—	—
No. 4 (4.75) mm	25–55	30–60	35–65	50–85	55–100	70–100
No. 10 (2.00 mm)	15–40	20–45	25–50	40–70	40–100	55–100
No. 40 (0.475 mm)	8–20	15–30	15–30	25–45	20–50	30–70
No. 200 (0.075 mm)	2–8	5–20	5–15	5–20	6–20	8–25

Source: AASHTO Materials, Part I, Specifications, 16th ed. American Association of State Highway and Transportation Officials (1993).

than two-thirds of the fraction passing the No. 40 sieve (*1*). For bases, subbases, and temporary wearing surfaces, the plasticity index should not exceed 6 and the liquid limit should not exceed 25. Maximum size of aggregate for wearing surfaces is recommended as 1 in. In other words, gradings A and B of Table 18-1 are not recommended for wearing surfaces. The quality of the coarse aggregate is controlled by specifying that the percentage of wear by the Los Angeles abrasion test shall not exceed 50. When a mixture of this type is to be used as a wearing surface for several years, the minimum amount of material passing the No. 200 sieve should be 8 percent, the maximum liquid limit is 35, and the PI should be from 4 to 9.

Use of Calcium Chloride

Calcium chloride is sometimes used in connection with the contruction of soil-aggregate mixes in order to expedite the compaction process by slowing the rate of evaporation of moisture from the mixture during compaction. The incorporation of small amounts of calcium chloride will in many instances produce increased density for a given compactive effort. In wearing surfaces, it aids in maintaining the moisture content at that desired for maximum stability, prevents raveling of the surface, and reduces dust.

Calcium chloride is a hygroscopic material; that is, it has the ability to attract moisture from the air and other sources. Two types of calcium chloride are used for road building and maintenance: (1) regular flake calcium chloride (type 1), which contains a minimum of 77 percent of $CaCl_2$, and (2) concentrated flake, pellet, or other granular calcium chloride (type 2), which has a minimum of 94 percent $CaCl_2$ (*1*). In new construction, calcium chloride should be added at the rate of 7 to 10 lb of type 1 or 5.6 to 8 lb of type 2 per ton of aggregate (*2*).

Selection and Proportioning of Soil Elements

It is, of course, frequently desired to use the existing roadbed soil in the stabilized mixture in the interests of economy. Information regarding this material, in both cut and embankment sections, will generally be available from the detailed soil survey for the project under consideration. The results of the mechanical analysis and routine soil tests will then be available for each of the soils encountered in the section concerned before this phase of construction begins. The availability and location of granular and binder materials suitable for use in stabilization will also generally be known.

Assuming that the nature and location of suitable materials are known, the problem is to determine the proportion of "borrow" material that must be combined with the existing soil to produce a granular soil-stabilized mixture that will meet the specifications of the organization concerned. It should also be noted that in many cases the entire soil-aggregate mixture may be imported and placed on top of the existing subgrade or previously constructed subbase; the material may come from a single source or may be a blend of materials from two or more sources. Trial combinations of materials are usually made in the laboratory on the basis of the mechanical analyses of the materials concerned. In other words, calculations are made in order to determine the sieve analysis of the combined materials, and the proportions of the various components are adjusted to fall within the limits of the specified gradation.

Numerical proportioning is a simple process for two materials. For example,

TABLE 18-2 Example of Numerical Proportioning Percent Passing (By Weight)

	Subgrade Soil (A)	Borrow Material (B)	1A, 1B	1A, 3B
Mechanical Analysis				
1 in. (25 mm)	—	100	100	100
¾ in. (19.0 mm)	—	90	95	93
⅜ in. (9.5 mm)	100	80	90	85
No. 4 (4.75 mm)	98	62	80	71
No. 10 (2.00 mm)	96	45	70	58
No. 40 (0.475 mm)	63	13	38	25
No. 200 (0.075 mm)	38	1	20	10
Plasticity Characteristics				
Liquid limit (%)	35	20	31	25
Plasticity index (%)	10	2	9	7

assume that the materials of Table 18-2 are available and are to be blended to meet the requirements of Table 18-1 (grading D).

By simple trial, it can be established that any blend between the ranges of one part A and one part B to one part A and three parts B will be satisfactory, that is, will meet the grading requirements of Table 18-1. Using a greater amount of material A will abrogate the requirement related to the maximum passing the No. 10 sieve. At the other end of the scale, use of an excessive amount of borrow material will abrogate the maximum passing the No. 40 sieve. Presumably, the borrow material is more expensive than the existing soil; hence, it would be desirable to use as little of this material as possible. In other words, use the 1 to 1 mix.

The plasticity index of the proposed mixture is then determined in the laboratory and compared with the specified limits. If this value is satisfactory, the blend may then be assumed to be satisfactory. If the plasticity characteristics of the first mixture are not within the specified limits, then additional trials must be made until a satisfactory blend is obtained. The proportions thus secured may then be used in the field-construction process.

In certain cases, where the agency is familiar with the soils concerned, materials of known PIs may be combined in trial proportions established by a design "formula" in order to produce a mixture of given plasticity characteristics. An example of such an equation is as follows:

$$PI \text{ (mixture)} = \frac{\{(\%A)(\% \text{ passing No. 40, A})(PI, A) + (\% B)(\% \text{ passing No. 40, B})(PI, B)\}}{(\%A)(\% \text{ passing No. 40, A}) + (\%B)(\% \text{ passing No. 40, B})}$$

$$(18\text{-}1)$$

Such proportioning will furnish an approximation of the PI that will serve until an actual test is made. For the 1 to 1 mixture of Table 18-2, the approximate value is

$$PI = \frac{50(63)10 + 50(13)2}{50(63) + 50(13)} = 8.6$$

A similar equation may be used to determine an approximate value of the liquid limit; in this case, the result is 31. In any event, a mixture is generally sought that will be well graded from coarse to fine and that will have the desired plasticity. The laboratory examination is generally completed by the determination of the compaction characteristics of the stabilized mixture so that the compaction process may be satisfactorily controlled in the field.

Graphical methods are also available and are particularly useful in the proportioning of more than two materials.

Construction Procedures

Three basic procedures are used in the construction of soil-aggregate roads. These methods, which are equally applicable to the production of bases and wearing surfaces as well as to the less frequent purposes of subbase construction and subgrade improvement, are as follows:

1. Road mix construction.
2. Traveling plant construction.
3. Central plant construction.

In all these methods the basic processes are the same: the soil elements must be properly selected and proportioned; the soils pulverized and uniformly blended; water and chemicals, if desired, added and uniformly mixed with the soil; and the blended materials spread in a thin layer of uniform thickness, properly compacted and finished. The several steps may be performed separately, or certain of the steps may be combined in one operation.

Road Mix Construction

In the basic process of road mixing, the proper proportions of soil elements that are to form the base or wearing surface are mixed directly on the surface of the subgrade or subbase. This may be done by alternately blading the material from one side of the roadway to the other. Special high-speed rotary mixing machines may also be used and will generally provide a more uniform mix.

After the soil elements have been properly blended, the mix must be spread in a uniform layer for compaction. If the moisture content of the mixture is less than the optimum moisture content, water must be added and blended with the soil to form a uniform mixture. After the water has been added, the soil is spread in a uniform layer and compacted. Thickness of layer is quite variable. For example, in the Midwest common practice is to spread the material in thin lifts (4 in. maximum compacted thickness) by a motor grader and begin rolling immediately; each layer is compacted before the next layer is spread. Where large rollers or vibratory compactors are used, material may be spread in much thicker layers, up to about 8 in. loose thickness. Initial compaction of a single layer may be accomplished by the use of sheep's foot rollers. More commonly, pneumatic-tired rollers are used. Vibratory compactors are very effective on the more granular soil-aggregate mixtures. Following initial compaction, the surface is generally shaped to the proper cross section by a blade grader and final rolling done by pneumatic-tired or smooth-wheel rollers.

Traveling Plant Construction

When a traveling plant is used in this type of construction, the construction process is not greatly different from the road mix method previously described. However, the operations of mixing the various soil elements, calcium chloride,

FIGURE 18-1 A pair of single-pass soil stabilizers process materials for a granular base. (Courtesy The Koehring Company).

and water are accomplished by the traveling plant, usually in a single pass of the machine. A photograph of a traveling plant is shown as Figure 18-1.

Central Plant Method

Many organizations that build a large mileage of granular soil stabilized roads have found it desirable to use central mixing plants for the production of stabilized soil mixtures. Usually mixtures produced in this fashion cost slightly more than those formed by road-mixing methods, but certain advantages accrue from the use of this scheme. Advantages claimed for the central plant method include greater uniformity of the mix, greater ease of control of the proportions of the mixture, greater ease in supplying water to the mix, and fewer delays due to bad weather.

Plant setups vary with the type of aggregates and mixture and with the equipment available. In most plants, the aggregates are carried by conveyor belt into one or more storage bins, whether they are pit or crusher run or a combination of coarse and fine. From the storage bin, the material is conveyed to a pugmill or other mixing equipment. Where clay is to be added to coarser aggregates, a separate setup may be necessary for pulverizing and feeding this material to the mixer.

Usually, calcium chloride is spread on the aggregate in the conveyor through a small hopper. Water is added at the top of the conveyor through spray bars. Most authorities in this field recommend the use of a pugmill mixer. Suitable pugmills are available from various manufacturers. When mixing is complete, the material is discharged into trucks for transportation to the job site. At the job site, steps of spreading, compaction, finishing, and curing are generally the same as those described for in-place construction, although spreading often is done with a self-propelled aggregate spreader or with a bituminous paver.

18-3 Bituminous Soil Stabilization

Bituminous materials are used in conjunction with soils (and soil-aggregate mixtures) for two general purposes. In one application the bituminous material may supply cohesion to the stabilized soil mixture, as in the case of the stabilization of sands or very sandy soils. On the other hand, bituminous material may be incorporated into a natural or artificial soil mixture for the purpose of "waterproofing" it. That is, enough bituminous material is added to reduce the detrimental effects of water that may enter the soil during its service life. In some cases the bituminous material may perform the dual functions of supplying cohesion and necessary waterproofing. Bituminous soil-stabilized mixtures are most frequently used for base construction.

Material Requirements A wide variety of soils may be stabilized with bituminous materials, including well-graded soils, sands, and even clays, although generally speaking, the more fine material included in the mixture, the greater will be the amount of bitumen required for satisfactory waterproofing. Very fine, plastic soils generally cannot be stabilized economically with bituminous materials because of inherent difficulties of pulverization and mixing.

Soils that are stabilized with a bituminous material may contain a relatively large percentage of fines, with 25% or greater passing the No. 200 sieve. The fines may possess some plasticity, but for best results it is recommended that the plasticity index be not greater than 10 for subgrades and not more than 6 for bases (*3*).

In addition to the soil itself, the components of a soil-bitumen mixture usually are water and bituminous material. Water is used to facilitate compaction of the mixture and to aid in the dispersion of the bituminous material uniformly throughout the mix. Typically, the amount of bituminous material used for bituminous stabilized soils varies from 4 to 7 percent. The sum of the percentages of water and bituminous material used should not exceed that which will fill the voids of the compacted mixture. Many different types of bituminous materials have been used in this type of stabilization, including rapid-curing, medium-curing, and slow-curing liquid asphalts. Medium-setting and slow-setting grades of asphalt emulsions increasingly have been chosen in recent years.

The choice of a bituminous material for use in a given project is primarily dependent on local experience and comparative costs.

Of the cutbacks, rapid-curing grades are recommended for use in extremely sandy soils or those containing a minimum of silt and clay particles. As the percentage of silt and clay increases, a medium-curing cutback is recommended because it will mix better and give a more homogeneous mixture. Slow-curing liquid asphalts have been used successfully with soils that contain 30 to 40 percent silt and clay, particularly in arid regions. As to grade of cutback asphalt, the general rule is "use the heaviest asphalt that can be readily worked into the soil." Asphalt emulsion should be of mixing grade. General recommendations for asphalt emulsion grades are shown in Table 18-3.

Design of Mixture The proportions of soil, water, and bituminous material are generally determined in the laboratory. Soil from the project is usually trial-mixed with various asphalt types at different percentages and tested by some

TABLE 18-3 Recommended Grades of Emulsified Asphalts for Bituminous-Aggregate Mixtures

Type of Construction	ASTM Specification D977					ASTM Specification D2397 (Cationic)			
	MS-1, HFMS-1	MS-2, HFMS-2	MS-2h, HFMS-2h	SS-1	SS-1h	CMS-2	CMS-2h	CSS-1	CSS-1h
Bituminous-aggregate mixtures									
For pavement bases and surfaces									
Plant mix (hot) (D3515)			X[a]						
Plant mix (cold)									
Open-graded aggregate,		X				X	X		
Dense-graded aggregate			X	X	X			X	X
Sand				X	X			X	X
Mixed-in-place									
Open-graded aggregate	X	X				X	X		
Dense-graded aggregate				X	X			X	X
Sand				X	X			X	X
Sandy soil				X	X			X	X
Slurry seal				X	X			X	X

[a]Specification D3515 permits the use of other emulsion grades by note: "Grades of emulsion other than MS-2h may be used where experience has shown that they give satisfactory performance."

Source: A Basic Asphalt Emulsion Manual. Manual Series No. 19, The Asphalt Institute (1979).

type of strength test to find the best mixture (*4*). Reference 5 gives additional information on mix design methods for liquid asphalt mixtures.

Construction Methods Methods of mixing soil-bitumen materials are generally the same as those previously described for granular soil stabilized mixtures. An additional step, of course, involves the application of the necessary bituminous material and its uniform distribution through the mixture. In the case of road mixing, the bituminous material, which sometimes has been heated moderately to the specified temperature for application, is applied in measured amounts by a pressure distributor, along with water applied separately, and the entire mix thoroughly blended by motor graders disk harrows, rotary speed mixers, and so on. Where a traveling or central mixing plant is used, the measured amount of bituminous material is incorporated in the plant.

When aeration is complete, the material is spread in a layer of uniform thickness ready for compaction. The thickness of the layer may vary from 50 to 150 mm. (2 to 6 in.), depending on job conditions. Compaction usually is done with a sheep's-foot roller or with pneumatic-tired rollers. Compaction is continued until a specified density (expressed in terms of laboratory density) is achieved. After compaction is complete, it may be required that some "curing" time elapse before the next layer is placed so that the water content after construction may be maintained at the desired figure. It should also be noted, at this stage, that the mixture may be compacted without the addition of water, as such, especially when asphalt emulsions are being used. Rolling of the final layer may be accomplished by the use of steel-wheel or pneumatic-tired rollers, and the base must be finished to the final section shown on the plans.

It should be emphasized that this type of soil-stabilized mixture is generally only satisfactory for base construction. Placing of a wearing surface would then generally follow shortly after completion of the base and before any considerable amount of traffic would be permitted to use the road.

Oiled Earth Surfaces

Oiled earth surfaces are quite different from the other types of bituminous soil stabilization that have been discussed. No mechanical mixing is involved in this construction process, as the liquid bituminous material is simply applied to a properly prepared natural soil surface. The material penetrates a short distance into the soil layer, thus preventing dust and forming a thin "stabilized" surface that will be capable of supporting a limited number of light vehicles. Over a period of time, with proper annual retreatment, a stabilized surface may be built up to a thickness of several inches. Its principal use has been in the construction of very lightly traveled local or "service" roads. This type of construction also has been widely used in areas where cracked or uncracked residual oils are cheaply available.

Material Requirements The soil, of course, will be that naturally existing in the section concerned, and the method is principally applicable to silt and clay soils. The bituminous material that is used should generally be a product of low viscosity which contains a wide range of volatile materials or "light fractions." The material that has probably been most widely used is slow-curing liquid asphalt, generally grade SC-70 or SC-250. Medium-curing cutbacks, MC-70 and MC-250, have also been successfully used. It should be noted here that a large number of cracked residuals have also been used, as in the construction of access roads in oil fields. Slow-setting asphalt emulsions are also suitable for oiled earth surfaces. The amount of bituminous material or "road oil" used in initial treatment during the first year is generally about 4.5 L/m² (1 gal/yd.²). Applications during successive years are somewhat less.

Construction Methods The natural soil surface must be adequately prepared by crowning for drainage and by compaction. The surface to which the oil is applied should be free from dust and should be slightly moist in order to facilitate penetration of the oil. Generally speaking, treatment should follow immediately after the final shaping (grading) operation. The oil is usually applied in two or three applications, the total amount applied being as previously indicated. Some time should be allowed to elapse between applications, and traffic should not be allowed on the road until the oil has completely penetrated the surface. After each application, the surface should be compacted under traffic or by the use of pneumatic-tired rollers.

18-4 Soil-Cement Bases

A type of soil stabilization that has enjoyed increasing popularity in recent years involves the incorporation of portland cement, in amounts generally varying from 7 to 14 percent by volume of the compacted mixture, with naturally occurring or artificially created soils or soil-aggregate mixtures. This type of construction is generally employed in the formation of base courses, usually with thicknesses varying from 100 to 150 mm (4 to 6 in.). A soil-cement mixture may serve as a base for a thin-wearing surface that will be subjected to light or medium traffic or as a support for a high-type flexible or rigid pavement. Bases of this type have been successfully used in the construction of city streets, and soil-cement is also suitable for use in driveways, shoulders, parking areas, some airport runways, and so on. In some areas, particularly California, this general type of construction is termed "cement-treated base."

Modern use of soil-cement began in South Carolina in 1933 with the construction of experimental sections, and the quantity of soil-cement used in highway construction has steadily increased since that time. Many millions of square yards of soil-cement have been constructed in the United States.

Material Requirements Nearly all subgrade soils, with the exception of those that contain high percentages of organic material, may be stabilized through the use of portland cement. Soils that contain high amounts of fine material, such as silts and clays, generally require large percentages of cement for successful stabilization. Gradation requirements for soils to be used in soil-cement mixtures are nearly nonexistent, because practically any soil may be used. Sandy and gravelly soils with 10 to 35 percent silt and clay combined have the most favorable characteristics. Glacial and water-deposited sands and gravels, crusher-run limestone, *caliche,* limerock, and almost all granular materials are good if they contain 55 percent or more material passing the No. 4 sieve. Exceptionally well-graded materials may contain up to 65 percent material retained on the No. 4 sieve and still have sufficient fine material for adequate binding. The maximum size of aggregate in soil-cement mixtures should not be more than 75 mm (3 in.). Materials in old gravel or crushed stone roads or bases make excellent soil-cement. An important factor in determining the suitability of a given soil for soil-cement construction is the ease with which the soil may be pulverized. Soils that contain large amounts of clay will be more difficult to pulverize, will require more cement, and will demand more careful control during the construction process.

The cement used is generally standard portland cement, types I and IA. The third ingredient of soil-cement mixtures is water, which is necessary to aid in the compaction of the loose mixture and for the hydration of the cement in the mix. In this respect, soil-cement mixtures are somewhat similar to concrete in that water is a necessary ingredient, and the loss of excessive amounts of water during the curing period must be prevented if the mixture is to harden properly. Practically any normal source of water may be used, although it should be quite clean and free from excessive amounts of organic matter, acids, or alkalies. Fresh water is, of course, normally used, although seawater has been used in a few instances. The proper amounts of cement and water to be used on a given project are determined by a series of laboratory tests that are briefly described in the following paragraphs. Information secured in the laboratory will also be used in the control of construction, particularly in the control of the compaction of the mixture.

Laboratory Tests Detailed laboratory testing procedures for soil-cement have been developed by the Portland Cement Association (6) and adopted by AASHTO and American Society for Testing and Materials (ASTM). Three general procedures are in existence, depending on the time available for testing and the size of the project.

On major projects, when time is available for testing, and with all soils, a detailed procedure to establish the satisfactory minimum cement content is recommended. As a first step in the laboratory procedure, the subgrade soil encountered in the project would be subjected to routine classification and placed in one of the groups of the AASHTO classification system. With this information as a background, cement contents to be used in the preparation of trial mixtures would then be established on the basis of past experience. For example, it might be decided that mixes would be prepared using 8, 10, and 12 percent of cement by volume. It should be emphasized that the percentage of cement used here is expressed in terms of the volume of the compacted mixture rather than by weight, that is, the percentage of a sack of cement per cubic foot of compacted soil cement. Thus, 10 percent by volume means that there is 10 percent of 94 lb, or 9.4 lb, of cement in 1 ft³ (150 kg/m³) of finished, compacted roadway.

Moisture-density relationships for soil-cement mixtures containing the selected amounts of cement are then determined, using a procedure entitled, "Methods of Test for Moisture-Density Relations of Soil-Cement Mixtures," ASTM Designation D558. The procedure used is very similar to that employed in the standard AASHTO compaction procedure. The percentage of cement required for satisfactory stabilization is based on two series of durability tests, and the moisture-density tests supply information necessary for the formation of the durability test specimens containing the desired amounts of cement. The durability tests used involve the subjecting of the various soil-cement mixtures to cycles of freezing and thawing and of wetting and drying. The test methods used are "Methods of Wetting-and-Drying Test of Compacted Soil-Cement Mixtures," ASTM Designation D559, and "Methods of Freezing-and-Thawing Test of Compacted Soil-Cement Mixtures," ASTM Designation D560.

The detailed laboratory procedures involved in the durability tests are much too lengthy to be presented here; however, they may be described generally. For each of the durability tests and for each of the selected cement contents, two specimens are prepared by compaction at optimum moisture to maximum density. Both samples are cured for 7 days before testing is begun. After curing, both samples are subjected to 12 cycles of alternate wetting and drying. At the end of each cycle, one specimen is vigorously brushed, using a standard procedure. Moisture and density determinations are made on the other, unbrushed specimen during each test cycle. At the end of the test, the specimens are oven-dried to constant weight and the weight losses determined. A similar procedure of alternate freezing and thawing is used to test the two specimens prepared for that test. Although the cement requirements are principally established on the basis of the results of the described durability tests, in some cases the compressive strengths of the test mixtures also are determined. The following are the recommendations of the Portland Cement Association relative to the behavior of the test specimens (6):

1. Losses during 12 cycles of either the wet-dry or freeze-thaw test shall conform to the following standards:
   Soil groups A-1, A-3, A-2-4, and A-2-5, not over 14 percent.
   Soil groups A-2-6, A-2-7, A-4, and A-5, not over 10 percent.
   Soil groups A-6, and A-7, not over 7 percent.
2. Compressive strengths of soil-cement test specimens should increase with age and with increases in cement content in the ranges of cement content producing results meeting the requirements given above.

A short-cut test method has been developed (6) by the Portland Cement Association for determining cement factors for sandy soils. The procedure is designed for use on smaller projects, particularly those where testing facilities and manpower are limited; it results in safe, but not necessarily minimum, cement contents for these soils. The only tests required are grain size analysis, moisture-density tests, and compressive strength tests. Results of the tests are used in conjunction with charts based on past experience involving thousands of test specimens to establish safe cement content.

Construction Methods The basic steps in the construction of a soil-cement base, assuming that the subgrade beneath the base requires no special treatment or has previously been brought to the desired condition, may be listed as follows:

1. Pulverizing of the soil that is to be processed.
2. Spreading of the required amount of cement and mixing with the soil.
3. Addition of the required amount of water and its incorporation into the soil-cement mixture.
4. Thorough compaction, including final rolling and finishing.
5. Curing of the completed soil-cement base.

Soil-cement mixtures may be "processed" by road-mixing equipment, by traveling plants, or by central mixing plants. In the large majority of cases traveling plants are used, including those of the flat type, the windrow type, and rotary speed mixers. The operation of traveling and central mixing plants in other types of soil stabilization has been discussed earlier in this chapter. Their use in this type of construction is not greatly different from that which has been previously described.

Pulverization With soils that are difficult to process, the soil that is to be used must be thoroughly pulverized before cement is added. If the existing soil is to be used it sometimes must be scarified to the desired depth by use of a ripper or a scarifier attachment on a motor grader. If imported material is being used, it must be hauled in and spread on the existing soil to the required depth. The soil may then be pulverized by the use of offset disk harrows, gang plows, or rotary speed mixers. It is generally required that the soil, with the exception of gravel or stone, be pulverized until at the time of compaction 100 percent of the soil-cement mixture will pass a 25 mm (1-in.) sieve and at least 80 percent will pass No. 4 sieve. It is especially important that the moisture content of fine-grained soils be within rather narrow limits during pulverization. For some soils the correct moisture content for pulverization will be at or near the optimum moisture content. Aeration of a wet soil or addition of water to a dry soil may be necessary, and in some cases steps may be taken to protect the pulverized soil from moisture change before the next step in the construction process is begun.

Incorporation of Cement The proper amount of cement may be spread on the surface of the soil either by hand or by mechanical means. On small jobs, bags of cement are "spotted" by hand along the surface in rows of predetermined spacing. The bags are then opened and the cement spread in uniform transverse rows, also by hand. Spreading is completed by the use of a spike-toothed harrow. On most projects, bulk cement is used and a mechanical spreader is used. When spreading of the cement has been properly completed, dry mixing is ready to begin.

Mixing In the road-mix process, blending of the soil and cement may be done by the use of such tools as gang plows, disk harrows, spring-tooth cultivators, and rotary tillers, but usually it is done with a traveling plant. When a traveling plant is used, mixing, of course, is carried out within the plant; exact processing procedure will vary with the type of equipment used. With a central plant, mixing is carried out in the plant and the mixture hauled to the job site in trucks; such a procedure is used on projects involving borrow soils. Careful control is necessary in a road mix operation in order that the mixing be carried to the proper depth and that a uniform mix be secured. The equipment available for this purpose will generally be operated in a train, with additional units being added to the train as construction proceeds, to supply water and complete the wet mixing operation. In many cases dry mixing and wet mixing are not essentially separate operations but are carried along together in an integrated series of operations. The proper amount of water is added, and the soil, cement, and water are thoroughly blended to a uniform mixture ready for compaction. The amount of water added is generally enough to increase the moisture content to 1 or 2 percent more than the optimum desired for compaction. This is done because of the loss of water that may occur during road-mixing operations as compared with laboratory mixing. In other words, if a slight excess of water is incorporated during mixing, the soil may then generally be expected to be very close to optimum by the time compaction is ready to begin.

Compaction and Finishing Initial compaction of a soil-cement base is generally achieved by the use of sheep's-foot rollers. After the base has been compacted to within 50 or 75 mm (2 or 3 in.) of the surface, the layer is brought to shape with a motor grader. Rolling then generally continues until satisfactory density has been secured to within about 25 mm (1 in.) of the surface. In very sandy soils, of course, compaction may not be possible with sheep's-foot rollers, and densification may then be accomplished by the use of pneumatic-tired rollers. Steel-wheel rollers have been used with granular soils, and newer types of compaction equipment, including vibratory compactors, grid, and segmented rollers are finding increasing use. After the soil-cement mixture has been compacted as described, the rollers are taken off and the surface brought to final shape. The final shaping is done in a variety of ways, including steps to remove the compaction planes left in the surface by the rollers. Final compaction is secured by rolling with pneumatic-tired rollers alone or in combination with steel rollers weighing from about 2.7 to 10.9 metric tons (3 to 12 tons), depending on the soil involved. Rolling is continued until a tightly packed surface is secured. During all the rolling operations that have been described, it is essential that the moisture content be maintained at optimum. Frequent checks on the density and moisture content are, of course, necessary.

Curing The water contained in the soil-cement mixture is necessary to the hardening of the cement, and steps must be taken to prevent loss of moisture from the completed base by evaporation. Curing is done by the application of a light coat of bituminous material in most cases; commonly used bituminous materials are RC-250, MC-250, RT-5, and emulsified asphalt (RS-2). Rate of application varies from about 0.7 to 1.4 l/m^2 (0.15 to 0.30 gal/yd^2). A bituminous wearing surface should be placed on the completed soil-cement base as soon as practical.

A sketch showing typical operations for soil-cement road construction is shown as Figure 18-2. Additional information on the construction of soil-cement roads may be obtained from the Portland Cement Association.

18-5 Miscellaneous Materials Used in Soil Stabilization

Primarily because of the dictates of economy, many materials other than those that have been discussed here have been used for soil stabilization at various times and in various places. A few of the more important of these materials are discussed in the following paragraphs. The discussion is not intended to be complete but will serve to indicate the wide range of materials that may be used in construction of this sort.

Lime

The incorporation of small amounts of hydrated lime has proved to be effective in the improvement of certain plastic clay soils. Lime has been used—principally in states along the Gulf Coast—to reduce the plasticity, shrinkage, and swell of clay soils while at the same time somewhat increasing their bearing capacity. In essence, its use has permitted the upgrading of certain marginal and submarginal soils into satisfactory base and subbase materials.

The use of lime also has certain advantages from the standpoint of the construction process by improving the workability of plastic soils (making them easier to pulverize). It tends to waterproof the soil to some extent and allow it to dry out more quickly when saturated, thus speeding construction.

The quantity of lime used in subgrade treatment is generally from 3 to 6 percent. Quantities as low as 1 percent have been used, whereas amounts much above 6 percent are not economical. Depth of treatment is usually 150 mm (6

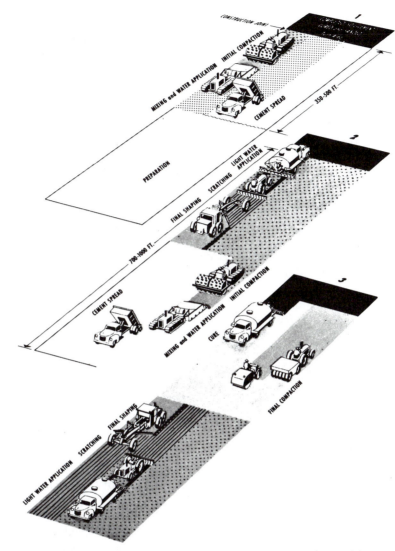

FIGURE 18-2 Sketch of soil-cement processing operations with multiple-shaft traveling mixing machine. (Courtesy Portland Cement Association.)

in.). The subgrade soil is scarified and pulverized, followed by spreading of lime, usually with a mechanical spreader or a bulk hauler. Enough water is added to bring the moisture content to 5 percent or more above optimum, and the water is distributed by a rotary speed mixer. The lime-soil mixture is allowed to cure for periods of 1 to 7 days. Mixing and pulverizing then continue until all the material will pass a 1-in. sieve and at least 60 percent will pass a No. 4 sieve. Compaction is done by pneumatic rollers or vibratory compactors, and the compacted layer is allowed to cure for 3 to 7 days before the next layer is placed. On some projects, the lime is added to the soil in the form of a slurry.

Construction procedure is quite similar to that used for soil-cement.

More detailed information on lime stabilization is contained in *Transportation Research Circular No. 180,* published by the Transportation Research Board

(TRB). Research on the use of a combination of lime and flyash for soil stabilization is described in the TRB's National Cooperative Research Program publication *Synthesis of Highway Practice 57.*

Salt

Rock salt has been successfully used in the stabilization of soils for bases and wearing surfaces subjected to very light traffic. The procedure is very similar to that described for mixtures in which calcium chloride is used. Recommended quantity of salt is generally from about 0.5 to 1.0 kg/m² (1 to 2 lb/yd²) per 25 mm (1 in.) of loose thickness.

Other Materials

The use of lignin binder, which is produced as a by-product in the manufacture of paper from wood by the sulfite process, has been reported experimentally, and a limited mileage of roads has been constructed with this material. Similarly, various silicates of soda have been used experimentally to improve certain clay soils.

A number of resinous waterproofing materials, including Stabinol and Vinsol, have been used experimentally; none is in widespread use. Considerable study has been devoted to a variety of resinous bonding materials for soil stabilization. Most successful of these have been a combination of aniline and furfural, and calcium acrylate; both offer considerable promise for the future but are currently too expensive for ordinary civilian use.

MACADAM ROADS

Crushed-stone roads first constructed in Europe during the latter part of the eighteenth and the early part of the nineteenth centuries were the forerunners of modern macadam roads. Construction of this type was introduced in France by Pierre-Marie Tresaguet about 1765 and in England by Thomas Telford shortly after 1800. Both of these early engineers used large pieces of broken stone in the lower course and placed a layer of finer crushed stone as a wearing surface. Telford advocated the use of large, flat pieces of stone, which were hand placed to form the base course. John McAdam, a famous Scottish road builder and engineer, insisted on the use of smaller stone—about 1.5 in. maximum size—for the entire thickness of the "pavement." The first roads of this type were built by him in England early in the nineteenth century. Both Tresaguet and McAdam laid emphasis on the grading and drainage of the subgrade and the crowning of the crushed-rock surface to aid in removal of rainfall from the surface of the road. The technique of macadam construction improved rapidly over the years, and by the end of the nineteenth century many miles of this type of surface had been built and were in service as rural roads and city streets in the United States. With the advent of the motor vehicle, the basic macadam or broken-stone surface became less suited to the demands of traffic, and this type of pavement assumed less importance in highway construction. Roads of the original macadam type were (and are) basically unsuited for use as wearing surfaces carrying large numbers of rapidly moving pneumatic-tired vehicles as they become quite rough and dusty under this type of traffic.

The term "macadam" originally referred to a road surface or base in which crushed or "broken" stone was mechanically "keyed" or locked by rolling and cemented together by the application of stone screenings and water. Later, when "bituminous macadam" roads were built by using a bituminous material as the binder, the original type of road came to be known as a "water-bound macadam." These two types of roads, the most prominent classes of macadam roads, are defined as follows:

> *Water-bound macadam* is a layer composed of broken-stone (or crushed gravel or slag) fragments that are bound together by stone dust and water applied during construction, in connection with consolidation of the layer by a heavy roller or a vibratory compactor. This type of macadam road closely resembles those so widely used in the early days of road building. Now water-bound macadam roads are seldom constructed. Detailed material requirements and construction procedures for such roads are therefore not given here.

> *Bituminous macadam* is crushed-stone or crushed-slag base or wearing surface in which the fragments are bound together by bituminous material; the aggregate layer is compacted, and bituminous material is applied to the surface of the layer. The bituminous material then penetrates into the voids of the compacted layer and serves to bind the fragments together. This type is frequently termed simply "penetration macadam."

18-6 Bituminous Macadam

As indicated, the term "bituminous macadam" refers to a type of macadam "pavement" in which the aggregates are bonded together by bituminous material. Both wearing surfaces and bases are constructed by this method and, at least as far as primary highway systems in the United States are concerned, this is the most important and extensively used of the macadams. Bituminous macadams have been most widely developed in areas in which suitable aggregates are readily available. A penetration macadam road, described below, is the basic type of bituminous macadam in which "hard" stone is used and the bituminous binder is asphalt cement or one of the heavier grades of road tars.

Material Requirements

Detailed requirements for materials used in bituminous macadams vary considerably from state to state, and the following sections are intended only to indicate requirements that are typical of those in current use.

Aggregates The aggregate that is used is crushed stone or crushed slag. Very hard stone, such as traprock, is used in many areas, although crushed limestone and sandstone are also used. The aggregate is required to be quite uniform in character and free from deleterious amounts of dust, flat or elongated pieces, and other objectionable elements. It may be subject to requirements for soundness. Another typical requirement is that the maximum loss in the Los Angeles abrasion test be 40 percent for wearing surfaces and 50 percent for base courses. In states where unusually hard stone is abundant, the maximum loss permitted may be as low as 25 percent. Some few states still base this requirement on the results of the Deval abrasion test. Crushed slag is generally required to weigh not less than 1120 kg/m^3 (70 lb/ft.^{3}).

The gradation of the aggregate for penetration macadam courses best suited

TABLE 18-4 Example Gradation (Penetration Macadam)

Sieve Designation	Percentage Passing (by Weight)	
Coarse aggregate		
2½ in. (63 mm)	100	
2 in. (50 mm)	90–100	
1½ in. (37.5 mm)	35–70	
1 in. (25.0 mm)	0–15	
½ in. (12.5 mm)	0–5	
	Viscous Binders	*Fluid Binders*
Choke aggregate		
1 in. (25.0 mm)	100	
¾ in. (19.0 mm)	90–100	100
½ in. (12.5 mm)	20–55	90–100
⅜ in. (9.5 mm)	0–15	40–75
No. 4 (4.75 mm)	0–5	5–25
No. 8 (2.36 mm)		0–10
No. 16 (1.18 mm)		0–5

Source: Annual Book of Standards, Designations D448 and D693, American Society for Testing and Materials (Reapproved 1993).

for the work depends on the size of surface voids in the course and fluidity of the bituminous binder. Table 18-4 gives an example of aggregate gradations that have been recommended for such courses (7).

Bituminous Materials The bituminous materials used in this type of macadam are heated to make them liquid at the time of application to the compacted aggregates. Asphalt cements of penetration grades 85 to 100 and 120 to 150 are used, and they are heated to temperatures that generally do not exceed 177°C (350°F) and applied at temperatures between 135° and 177°C (275° and 350°F). Some organizations allow use of heavier grades of cutback asphalts in this type of pavement when finer aggregates are used. Road tars have also been used, with grades RT-10, -11, and -12 being preferred. These latter materials are heated to about 121°C (250°F) and applied at temperatures from about 79° and 121°C (175° to 250°F).

Construction Methods

The construction of a penetration macadam may be generally broken down into the following basic steps:

1. Spreading and rolling of coarse aggregate.
2. Initial application of bituminous material.
3. Spreading and rolling of key aggregate.
4. Application of seal coat.

Spreading and Rolling of Coarse Aggregate Coarse aggregate generally is placed between well-compacted shoulders with vertical faces by means of mechanical spreaders. The material must be placed in a uniform loose layer of the required thickness. Precautions are taken to avoid segregation of the aggregate and to prevent the material from becoming mixed with dirt or other foreign materials during and after the spreading operation. Before rolling begins, the layer is subjected to careful visual inspection, and excessively large and/or flat or elongated pieces are removed. Areas that show an excess of fine material are also subject to correction, and the layer is generally brought to a uniform condition.

When spreading of the course aggregate has been satisfactorily completed, the layer is rolled with a three-wheel roller weighing not less than 10 tons. Rolling is continued until the surface is thoroughly compacted and the stone keyed together. Material that crushes under the roller in such a way as to interfere with the proper penetration of the bituminous material is removed, and weak spots are corrected and irregularities removed as rolling continues. The finished surface of the coarse layer must be true to the cross section shown on the plans and at the desired elevation. An excessive amount of rolling is not desirable, because enough surface voids must remain to permit ready penetration of the bituminous binder. Figure 18-3 shows a close-up view of a properly compacted layer of coarse slag.

Initial Application of Bituminous Material The initial application of bituminous material is then made by means of a pressure distributor, with the material at the correct temperature and in the amount prescribed by the engineer. The quantity recommended by the Asphalt Institute is 7.9 to 10.2 L/m^2 (1.75 to 2.25 gal/yd^2) for a compacted thickness of 89 to 100 mm ($3\frac{1}{2}$ to 4 in.); this would correspond to a first application of 190 to 217 kg/m^2 (350 to 400 lb) of aggregate (specific gravity, 2.55 to 2.75) and a second application of 19 to 27 kg/m^2 (35 to 50 lb per square yard). It is important that the bituminous material be spread uniformly over the layer in the correct amounts. Precautions are usually taken to prevent application of excessive amounts of bitumin at the points where successive applications overlap. This may be accomplished by spreading tar paper over the end of the preceding spread of bituminous material so that the sprays may be opened over the paper and uniform distribution over the untreated section assured.

The pressure distributor is the key machine on many types of bituminous

FIGURE 18-3 Air-cooled blast-furnace slag shown compacted on the left and bonded with an application of hot asphalt by the penetration method on the right. (Courtesy Macadam Pavements, Inc.)

paving work, including macadam, surface treatment, and road-mix construction. The machine must be capable of applying an accurately measured amount of bituminous material over the distance required to empty the load, regardless of changes that may occur in grade or direction. Distributors are available in several sizes, with capacities from 800 to 5500 gal.

A typical distributor of this type consists of a rubber-tired truck, truck-drawn trailer, or semitrailer on which is mounted an insulated tank or one that is equipped with heating facilities. Quite frequently an oil burner located at the bottom and rear of the tank is used to supply heat. Heat from the oil burner passes through the interior of the tank, close to the bottom, and out at the top of the tank through a number of heating flues. The distributor is also supplied with an engine-driven pump, so designed that it may handle products ranging from light "oils" that are fluid at air temperature to semisolid materials that are fluid at temperatures of 177°C (350°F) or more.

The bituminous material is applied to the surface of any desired area through a system of nozzles mounted on a spray bar at the rear of the tank. The fluid material is pumped under pressure through this spray bar, which may be of varying width, with the minimum width being about 1.8 m (6 ft) and the maximum being 7.3 m (24 ft). Most modern distributors have a circulating system that keeps the material flowing when it is not being sprayed. The distributor is further equipped with a thermometer installed in the tank and with a hose connection to which a single nozzle may be attached so that the material may be readily applied to a limited area or at a specific point.

When application is to be made through the spray bar and nozzles, rather simple calculations may be made to relate the variables involved in the problem of applying the fluid material at a uniform, specified rate. Assuming that the width of the spray bar, the quantity (gallons) handled by each revolution of the pump, and the speed of the pump are known, the speed at which the truck must be driven to secure a given rate of application may be readily obtained. The rate of application is controlled by a "bitumeter," which comprises a small wheel beneath the truck, connected by a cable to a dial in the cab; the dial usually shows the speed in feet per minute but may show gallons per square yard directly. A photograph of a bituminous distributor is shown in Figure 18-4.

The computation of the application rate of a bituminous distributor assumes that all the material pumped goes through the nozzles. Many distributors have circulating spray bars so that part of the material pumped returns to the distributor. For a given pump speed or pressure, a given bituminous material, and a given temperature, a given flow through the nozzles will result. Charts or tables showing these data assist the inspector or driver in the selection of vehicle speed. The application rate is then verified by the distance traveled.

Spreading and Rolling of Key Aggregate Immediately after the initial application of bituminous material is made, the surface is lightly covered with key aggregate in sufficient quantity to prevent the roller wheels from sticking. The layer is then rolled until thoroughly compacted. The key aggregate is then spread over the surface in the desired quantity so that the surface voids of the layer will be completely filled but there will be no excess aggregate remaining on the surface. The surface is then rolled with the same equipment as before. Simultaneously with the rolling, the key aggregate may be lightly broomed to effect an even distribution of the material, and additional small amounts may be added as needed. Vibratory compactors are used in some states in conjunction with rollers.

FIGURE 18-4 A bituminous pressure distributor. (Courtesy E. D. Etnyre Company.)

This process is continued until the key aggregate is firmly embedded in the bituminous material and the surface is firmly compacted. Figure 18-5 shows a properly bonded bituminous macadam after the application and rolling of the key aggregate. The layer is then ready for the application of the seal coat if it is to function as a wearing surface; if it is to be a base, the seal coat usually is omitted.

Application of Seal Coat Before placing of the seal coat is begun, the surface is swept, usually by means of a broom drag, for the purpose of either removing all loose material or uniformly distributing any loose key aggregate that may remain after the previous rolling. Bituminous material is then applied to the surface in the amount determined by the engineer and is immediately covered with the fine aggregate (stone chips). Rolling and brooming are then begun again and continued until the cover stone is firmly bonded and completely fills the voids in the finished surface. The required amount of rolling cannot be definitely stated, and this process is continued until the surface is completed to the satisfaction of the supervising engineer. Pneumatic rollers may be used on this final layer in some states. The completed surface must, of course, be correct to grade and cross section. The finished surface is tested by templates and straightedges, with deviations up to 6 mm in 3 m ($\frac{1}{4}$ in. in 10 ft) being permitted in the majority of states performing this work. The surface may be opened to traffic soon after the final rolling is complete.

PRIME COATS AND TACK COATS

When a bituminous wearing surface is to be constructed, the foundation layer is first sprayed with a liquid bituminous material. Depending on the nature of the foundation layer, the treatment is called either a "prime coat" or a "tack coat."

FIGURE 18-5 Close-up of bituminous penetration macadam showing the texture after the first and second penetration applications of bituminous material and choke stone have been applied. (Courtesy Macadam Pavements, Inc.)

18-7 Prime Coats

A prime coat is an application of liquid bituminous material to a previously untreated base or wearing surface. The bituminous material penetrates the surface and is generally completely absorbed. A prime coat serves several definite purposes: First, the prime coat serves to promote adhesion or "bond" between the base and wearing surface; this is probably its most important function. Second, it serves to consolidate the surface on which the new treatment is to be placed, and this may aid in maintaining the integrity of this underlying layer during the construction of a surface treatment or some other type of wearing surface. Third, it may function to a certain extent as a deterrent to the rise of capillary moisture into the wearing surface.

Materials

The material used in the prime coat should be liquid asphalt or road tar of low viscosity. It is important that the material used have high penetrating qualities, and it should be of such a nature that it will leave a high-viscosity residue in the void spaces of the upper part of the surface that is being primed. Of the asphaltic materials, grades MC-30 and MC-70 are widely used, respectively, in the priming of dense, tightly bonded surfaces and surfaces having a more open texture. Slow-setting emulsions, grades SS-1, SS-1h, CSS-1, and CSS-1h, are recommended for penetrable surfaces (8).

The amount of bituminous material that is used varies over a considerable range, depending on the material selected and job conditions. Generally the amount of bituminous material required for priming is from 0.9 to 2.3 L/m^2 (0.20 to 0.50 gal/yd^2) of surface. The lower limit applies to tight, fine-grained surfaces; the upper to loose, open-textured surfaces.

Construction Methods

After the foundation layer has been swept to remove loose material, dust, and other foreign matter, the prime coat is applied with a pressure distributor. Best results are obtained when the surface is dry or only slightly damp. The material should be uniformly applied at the recommended spraying temperature. Priming generally should be done only when the air temperature in the shade is above 13°C (55°F) and rising or above 16°C (60°F) if falling.

18-8 Tack Coats

A "tack coat" is defined as a single application of bituminous material to an existing bituminous, portland-cement concrete, brick, or block surface or base. The purpose of a tack coat is to provide adhesion between the existing surface and the new bituminous wearing surface. Liquid asphalts of grades RC-70 through RC-250 are frequently used as tack coats, although medium-curing liquid asphalt of grade MC-250 has also been used. Road tars are used to some extent for this purpose, with grade RT-2 being used by some organizations and the heavier grades, such as RTCB-5 and RTCB-6, by others. Emulsified asphalts diluted with water are specified by some agencies.

The quantity of bituminous material that is normally required for a tack coat is quite small, with the stipulated amount generally ranging from 0.19 to 0.38 L/m² (0.05 to 0.10 gal/yd²).

Construction Methods

Tack coats, as has been indicated, are very frequently applied to old surfaces of various types as a first step in the construction of a new bituminous wearing surface. In many cases the old surface will require extensive correction and remedial treatment before construction of the new surface may begin because of the rough or inadequate condition of the old surface. As in the case of a prime coat, it is absolutely essential that the surface to which the material is applied be clean and dry. In most cases the surface is swept clean of all loose material just before the application is made. If asphalt emulsion is used, the surface may be slightly damp without harmful effect. The bituminous material is then applied at the specified rate with a pressure distributor, with the material being maintained at the proper temperature for application. The treated surface is then protected from traffic and allowed to dry until it reaches the proper stage of stickiness or "tackiness" for the application of the bituminous wearing surface. Any breaks that may occur in the tack coat may be corrected by the application of additional bituminous material at these points.

SINGLE- AND MULTIPLE-LIFT SURFACE TREATMENTS

Single- and multiple-lift surface treatments have been widely used in many sections of the country for the surfacing of primary and secondary roads that carry light to moderate amounts of traffic. The thickness of a single surface treatment is usually from 12 to 20 mm (½ to ¾ in.) and may be as much as 64 mm (2.5 in.) when multiple-lift treatments are used. A surface constructed by this method is frequently quite "open" in character and has excellent nonskid and visibility characteristics. This type of surface is generally so thin that it has little load supporting value in itself and must depend on an entirely adequate base and subgrade to function satisfactorily.

Material Requirements

Materials that are used in this type of surfacing are extremely varied, with a large number of different aggregates and bituminous materials being successfully used in different areas. In the interests of somewhat simplifying the coverage of this subject, it has been decided to present the detailed material requirements of only one organization, the Federal Highway Administration (*9*), in the belief that these will be typical of those in current use in this country.

Aggregates The aggregates that are used in this type of surface are crushed stone, crushed gravel, and crushed slag. The aggregates are required to be clean, hard, tough, and durable in nature, free from an excess amount of flat or elongated pieces, dust, clay films, clay balls, and other objectionable material. They are also required to be hydrophobic in nature, as measured by a stripping test (AASHTO T182).

Six aggregate gradations are included in the specification (*9*). They are designated gradings A through F and are given in Table 18-5.

Bituminous Materials A wide variety of bituminous materials may be used for surface treatments including medium- and rapid-curing cutback asphalt, rapid-setting emulsified asphalt, penetration grade asphalt cement, and road tar.

Composition of Surface Treatments

The FHWA specification (*9*) contains quantities of materials and indicated sequences of operations for five different wearing surfaces, using six different aggregate gradings and a variety of bituminous materials (Tables 18-6 and 18-7). The surfaces range from a double surface treatment (Designation AT-35) to a four-lift treatment (Designation AT-110). The designation refers to the quantity of aggregate per square yard. In the table, the indicated quantities of aggregate are based on a bulk specific gravity of 2.65 and must be corrected if the specific gravity of the aggregate used is more than 2.75 or less than 2.55. The total quantity of aggregate used in each of the surfaces given in the table, corrected for specific gravity and except for some stockpiling if desired, is fixed, whereas the amount of bituminous material actually used may be varied somewhat to meet job conditions.

TABLE 18-5 Aggregate Gradations for Surface Treatments

Sieve Designation	Percentage Passing (by weight)					
	A	*B*	*C*	*D*	*E*	*F*
1½ in. (37.5 mm)	100	—	—	—	—	—
1 in. (25.0 mm)	90–100	100	—	—	—	—
¾ in. (19.0 mm)	20–55	90–100	100	—	—	—
½ in. (12.5 mm)	0–10	20–55	90–100	100	—	—
⅜ in. (9.5 mm)	0–5	0–15	40–70	85–100	100	100
No. 4 (4.75 mm)	—	0–5	0–15	10–30	85–100	85–100
No. 8 (2.36 mm)	—	—	0–5	0–10	10–40	60–100
No. 100 (0.150 mm)	—	—	—	—	—	0–10

TABLE 18-6 Approximate Quantities of Material and Sequences of Operations for Bituminous Surface Treatments Using Cutback Asphalt or Asphalt Cement

Aggregate Gradings and Sequence of Operations	*Treatment Designation*				
	AT-35	*AT-50*	*AT-60*	*AT-70*	*AT-110*
First Course— Apply asphalt material gal/yd² (L/m²)	0.22(0.99)	0.25(1.13)	0.15(0.68)	0.30(1.36)	0.20(0.92)
Spread aggregate lb/yd² (kg/m²)					
Grading D	25(13)				
Grading C		35(19)			
Grading B			40(21)	50(27)	
Grading A					70(38)
Second Course— Apply asphalt material gal/yd² (L/m²)	0.13(0.58)	0.25(1.13)	0.30(1.36)	0.35(1.57)	0.40(1.80)
Spread aggregate lb/yd² (kg/m²)					
Grading E	10(6)	15(8)			
Grading D			12(7)	20(11)	
Grading C					20(11)
Third Course— Apply asphalt material gal/yd² (L/m²)			0.15(0.68)		0.20(0.92)
Spread aggregate lb/yd² (kg/m²)					
Grading E			8(5)		12(7)
Fourth Course— Apply asphalt material gal/yd² (L/m²)					0.20(0.92)
Spread aggregate[a] lb/yd² (kg/m²)					
Grading F					8(5)
TOTALS— Asphalt material gal/yd² (L/m²)	0.35(1.57)	0.50(2.26)	0.60(2.72)	0.65(2.93)	1.00(4.56)
Aggregate lb/yd² (kg/m²)	35(19)	50(25)	60(33)	70(38)	110(61)

[a]After final aggregate spread, apply additional cover aggregate, grading F, to all areas with unabsorbed asphalt materials.

Source: Standard Specifications for Construction of Roads and Bridges on Federal Highway Projects. Federal Highway Administration, Washington, DC (1992).

TABLE 18-7 Approximate Quantities of Material and Sequences of Operations for Bituminous Surface Treatments Using Emulsified Asphalt

Aggregate Gradings and Sequence of Operations	Treatment Designation				
	E-35	E-50	E-60	E-70	E-110
First Course— Apply asphalt material gal/yd² (L/m²)	0.45(2.02)	0.35(1.57)	0.45(2.02)	0.50(2.26)	0.40(1.80)

Spread aggregate lb/yd² (kg/m²)
  Grading D
  Grading C
  Grading B
  Grading A
	25(13)	30(17)	36(19)	40(21)	70(38)

Second Course— Apply asphalt material gal/yd² (L/m²)	0.25(1.13)	0.25(1.13)	0.25(1.13)	0.25(1.13)	0.45(2.02)

Spread aggregate lb/yd² (kg/m²)
  Grading E
  Grading D
	10(6)	10(6)	16(9)	20(11)	20(11)

Third Course— Apply asphalt material gal/yd² (L/m²)		0.25(1.13)	0.25(1.13)	0.25(1.13)	0.25(1.13)
Spread aggregate lb/yd² (kg/m²) Grading E		10(6)	8(5)	10(6)	12(7)
Fourth Course— Apply asphalt material gal/yd² (L/m²)					0.25(1.13)
Spread aggregate[a] lb/yd² (kg/m²) Grading F					8(5)
TOTALS— Asphalt material gal/yd² (L/m²) Aggregate lb/yd² (kg/m²)	0.70(3.15) 35(19)	0.85(3.83) 50(29)	0.95(4.28) 60(33)	1.00(4.52) 70(38)	1.35(6.08) 110(61)

[a]After final aggregate spread, apply additional cover aggregate, grading F, to all areas with unabsorbed asphalt materials.

Source: Standard Specifications for Construction of Roads and Bridges on Federal Highway Projects. Federal Highway Administration, Washington, DC (1992).

Specifications for surface treatments have also been developed and published by the Asphalt Institute (*10*).

Construction Methods—Single- and Multiple-Lift Surface Treatments

It will be assumed in this discussion that the surface is to be applied to a granular-type base that has been previously primed, so the application of the prime coat is not included in the procedure presented here. The surface should be free from dust and dry, unless emulsified asphalt is used, when the surface may be slightly damp. No depressions should exist in the base course into which the fluid bituminous material may flow when applied to the surface and collect at these points in excess quantities.

The construction of surface treatments of this type is quite simple in basic procedure, with the surface being built up in lifts through the successive application of bituminous material and progressively smaller aggregate spread in the correct amounts. The layer formed by each application of bitumen and aggregate is rolled, and this basic procedure is continued until the specified number of lifts has been completed.

The initial application of bituminous material to the prepared base is made by a pressure distributor in the specified amount and at the proper temperature of application. It is essential that this application be uniform over the desired width and length of treatment. The desired amount of aggregate is then immediately spread over the bituminous material by means of approved mechanical spreading devices, including self-propelled spreaders. Normally, trucks are used to transport the aggregate to the point of application, and specifications covering this type of work usually require that the trucks be brought into position by traveling over previously spread aggregate. They are not allowed to pass over the portion of the surface on which no cover aggregate has been placed. A special spreading device is normally used to ensure that the aggregate is spread uniformly and in the prescribed quantity.

Rolling begins immediately after the aggregate has been spread. Both pneumatic-tired and steel-wheel rollers are used in this type of work, either alone or in combination with one another. The Asphalt Institute recommends pneumatic rollers because of the more uniform pressure they exert. Where pneumatic-tired rollers are used, they are operated at comparatively low tire pressures in order to avoid displacement and crushing of the aggregate.

The weight of steel-wheel rollers that is more generally used in this type of work is 4.5 metric tons (5 tons), and the weight may vary from 2.7 to 7.3 metric tons (3 to 8 tons) under different specifications. Rolling of the aggregate is continued until the aggregate particles are firmly embedded in the bituminous material and sufficiently well bonded that they will not pick up under traffic, but it is discontinued after the asphalt has taken its initial set. Excessive rolling, which results in crushing of aggregate particles or breaking of the asphalt bond, is, of course, undesirable.

If a single-surface treatment is being constructed, this is the end of the construction process and the surface may be opened to traffic. If the wearing surface is to be a multiple-surface treatment, construction continues by the application of another "shot" of bituminous material, which is immediately covered by another spread of aggregates, and the layer is shaped and rolled as before. This process is continued until the desired number of lifts is completed. Surface treatment work is most successfully done in warm weather and should not be attempted when air temperatures are at 13°C (55°F) or less.

18-9 Seal Coats

In line with the concepts that have been advanced in this chapter, a seal coat is a very thin single-surface treatment that is usually less than 13 mm ($\frac{1}{2}$ inch) thick. Seal coats are applied as a final step in the construction of many types of bituminous wearing surfaces, their primary purpose in this application being to waterproof or "seal" the surface. They are also used to rejuvenate or revitalize old bituminous wearing surfaces, to "nonskid" slippery surfaces, and to improve night visibility. A seal coat usually consists of a single application of bituminous material that is covered by a light spreading of fine aggregate or sand. The selection of the bituminous material and cover aggregate to be used in a given case is usually dependent on the type of wearing surface that is being sealed as well as on the nature of locally available materials. Specifications for seal coats are usually not given separately but are made part of the specification for each particular type of wearing surface. A large number of different bituminous materials are used in this work, and aggregate gradings vary widely in different sections of the country.

18-10 Slurry Seal Coats

The slurry seal coat is a mixture of well-graded, fine aggregate, mineral filler (if needed),* emulsified asphalt, and water (8). It is used principally for the resealing of old bituminous wearing surfaces. The slurry seal coat is quite thin, typically 3 to 6 mm (1/8–1/4 in.).

A common aggregate specification used for medium-textured surfaces calls for all of the aggregate to pass a 4.75-mm (No. 4) sieve and 5 to 15 percent to pass a 75-m (No. 200) sieve (9). At least half of the material must be a crushed product that meets the requirements of the sand equivalent test and a film stripping test.

General purpose slurry seal coats usually require an application rate of 5.5 to 8 kg/m² (10–15 lb/yd²) based on the mass of dry aggregate. The residual asphalt content, percent of dry aggregate, generally ranges from 7.5 to 13.5 percent (8).

The surface to be treated is swept and cleaned prior to application of the seal coat. The work may be done only in warm, dry weather. Before application of the slurry, the surface is sprinkled with water. In some cases, a dilute asphalt emulsion prime coat has been used.

The slurry seal coat is usually prepared in a specially designed traveling mixing plant. The slurry goes from the mixing plant into an attached rectangular spreader box that spreads the slurry across the surface and controls the width. Following the spreader box is an adjustable screed that controls the thickness of the application and squeegees the slurry into the cracks of the old pavement. Figure 18-6 is a sketch of a typical slurry seal mixer.

Traffic should be kept off the new surface for a period of 2 to 4 hr, depending on the weather, thickness of the seal coat, and other factors.

---

*The mineral filler, consisting of hydrated lime, portland cement, fly ash, or limestone dust is almost always needed to aid in stabilizing and setting the slurry (8).

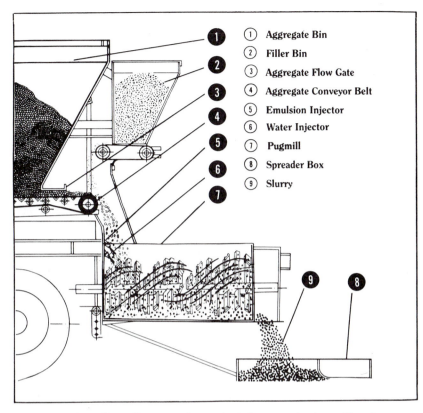

1. Aggregate Bin
2. Filler Bin
3. Aggregate Flow Gate
4. Aggregate Conveyor Belt
5. Emulsion Injector
6. Water Injector
7. Pugmill
8. Spreader Box
9. Slurry

FIGURE 18-6 A flow diagram of a typical slurry seal mixer. (Courtesy AKZD Nobel Asphalt Applications, Inc.)

ROCK ASPHALT SURFACES

"Rock asphalt," as the term is applied in highway work in the United States, refers to sandstone or limestone which is naturally impregnated with bitumen. This material should not be confused with native asphalts like Trinidad Lake Asphalt, which are primarily naturally occurring bitumens containing a large percentage of mineral matter. Rock asphalt is primarily mineral matter containing a small amount of natural bituminous binder. The amount of bitumen contained in rock asphalts that have been used in highway construction in this country has generally varied from about 3 to 12 percent.

Rock asphalt occurs in a number of states and has been produced on a commercial basis and in sizable quantity in Kentucky, Texas, Oklahoma, and Alabama. Material of this nature produced in these states generally requires some modification for use in highway work. Both hot and cold paving mixtures have been produced through the use of these rock asphalts. These paving mixtures are used as wearing surfaces on new or old bases, in the reconstruction of old pavements, and for maintenance operations such as patching. Their use at the present time is largely confined to the states in which they are found or in neighboring areas.

ROAD MIX SURFACES

The term "road mix" refers to a type of construction in which the mineral aggregates and bituminous material are intimately blended by mixing on top of the existing surface or base. The upper portion of the existing layer may or may not be included as part of the road mix surface, as circumstances warrant. The surfaces produced by this type of construction are also sometimes designated as "mixed-in place" surfaces.

Formerly, road mix bituminous wearing surfaces were widely used in this country for the surfacing of both primary and secondary roads and to a lesser extent for the surfacing of city streets. The initial development of this type of surface was probably primarily due to the need for the construction of a large mileage of surfaced roads at low cost, using local materials in areas of relatively sparse population, as in certain of the western states. This type of construction is seldom used today in the United States for wearing surfaces.

BITUMINOUS PLANT MIX SURFACES USING LIQUID BINDERS

In many "low-cost" bituminous plant mixtures, liquid bituminous materials—including liquid asphaltic products, asphalt emulsions, and road tars—are used as the binder.

The development of these intermediate surfaces was a natural outgrowth of certain difficulties encountered in the road mix process. Preparation of the paving mixture at a central plant offers such advantages as more careful proportioning of the ingredients, more uniform and thorough mixing with consequent production of more uniform mixtures, less dependence on favorable weather conditions, and the use of more viscous bituminous materials. Mixtures that are being included here under the classification of intermediate-type plant mixes range in nature from those that differ only very slightly from the road mix surfaces to those that very closely resemble the high-type mixtures discussed in the next chapter. They also cover a fairly wide range in cost, being generally slightly more expensive than comparable road mix surfaces and somewhat less expensive than high-type mixtures. The more expensive of the intermediate mixes described here differ from high-type mixtures in the following respects:

1. Softer bituminous materials that require only a moderate amount of heating (or no heating, as in the case of asphalt emulsions) are used in the intermediate types, whereas the majority of the high-type mixtures uses semisolid binders and are mixed and laid at elevated temperatures.
2. The nature and grading of the aggregate components of a high-type mixture and the composition of the final mixture are more carefully controlled than for the intermediate type.
3. Requirements related to the density and stability of the compacted mixture are generally much more rigid for the high-type pavement than for the low-cost plant mix surface.

Specifications related to the composition and control of the low-cost bituminous plant mixtures show extreme variation, with all sorts of mixtures that would fall into this classification being used by different highway organizations in line with their particular experience and to fit their particular needs.

Consider now the equipment that is used in the construction of plant mix surfaces. The same (or similar) equipment is employed in the preparation of the high-type surfaces described in the next chapter, and this fact will be emphasized at various points in the following discussion.

18-11 Central Mixing Plants

The term "central mixing plant" refers to the plant or "factory" at which the bituminous paving mixture is produced, in a process beginning with the aggregates and bituminous materials and ending with the discharge of the mixture into hauling units for transportation to the job site.

Central mixing plants for the production of bituminous paving mixtures are frequently described as being portable, semiportable, or stationary in nature. The term "portable" is applied to relatively small units that are self-contained and wheel-mounted, and it is also applied to larger mixing plants in which the separate units are themselves easily moved from one place to another. The term "semiportable" is reserved for those plants in which the separate units must be taken down, transported on trailers, trucks, or railroad cars to a new location, and then reassembled, which process may require only a few hours or several days, depending on the plant involved. "Stationary plants" are those that are permanently constructed in one location and are not designed to be moved from place to place. Portable and semiportable plants are much more numerous than stationary plants and are widely used in the construction of rural highways. The capacities of these two types of plants range up to about 400 tons of mixture per hour.

Two general types of central mixing plants are in common use: (1) batch plants and (2) drum mix plants. In the batch type of plant, the correct amounts of aggregate and bituminous material, determined by weight, are fed into the mixing unit of the plant; the "batch" is then mixed and discharged from the mixer into trucks before additional materials are introduced. In the drum mix plants, the aggregates are proportioned prior to entry into the mixing drum by means of precision belt feeders which control the amount of each class of aggregate entering the drum. The drum mixer not only dries the aggregate but also blends the aggregate and asphalt, discharging the mixture in a continuous flow into a surge silo, where it is temporarily stored and later loaded into trucks. Consider now a more complete description of these two types of mixing plants.

Batch Plants

One of the basic units contained in a batch bituminous paving plant is the drier. This unit is a necessary part of all "hot-mix" plants and is also generally used in the preparation of low-cost plant mixtures. The drier may not be needed for some mixes of the latter type, as, for example, those in which asphalt emulsion is used as the binder. The drier consists primarily of a large (from 3–10 ft in diameter and 20–40 ft long) rotating cylinder that is equipped with a heating unit at one end, usually a low-pressure air atomization system using fuel oil. The drier is mounted at an angle to the horizontal, with the heating element being located at the lower end. Hot gases from the burner pass from the lower end up the cylinder and out at the upper end. The so-called cold aggregate is fed into the upper end of the drier, picked up by steel angles or blades set on the inside face of the cylinder, and dropped in "veils" through the burner flame and hot gases, and it moves down the cylinder because of the rotating action and the force of

FIGURE 18-7 Drier used in central mixing plant. (Courtesy Iowa Manufacturing Company).

gravity. The hot aggregate then discharges from the lower end of the drier, generally onto an open conveyor or enclosed "hot elevator" that transports it to the screens and storage bins mounted in conjunction with the mixing unit. A drier of this conventional type is shown in Figure 18-7.

In the preparation of the hot mixtures described in the next chapter, the temperature of the aggregates may be raised to 325°F or more, and practically all the moisture in them removed. In the preparation of "cold" mixtures, such complete drying may be neither necessary nor particularly desirable. The moisture content may be reduced to about 1 percent in many cases, and excessively high temperatures are not permissible in the dried aggregates when they are to be combined with, for example, cutback asphalt. The dried aggregates may be cooled in various ways, including movement on an open conveyor to the next step in the process, the use of another drier (no heat) as a cooling unit, longer retention time in the storage bins, and so on.

As indicated, the dried aggregates are generally transported from the drier into a unit that contains screens, storage bins, and aggregate proportioning devices (Fig. 18-8). In this unit the dried aggregates pass over vibrating screens that separate them into the desired number of fractions, usually from two to four. Each fraction is then passed into a storage bin or hopper ready to be used in the mixture. The capacity of the storage bins varies with the size and type of plant being used. Where the aggregates are to be recombined by weight, each storage bin is equipped with a discharge gate which is located directly over the so-called weigh box. In proportioning the batch in this type of plant, the required weight of each size of aggregate is placed into the weigh box. When the desired aggregate batch has been obtained, the contents of the weigh box are then discharged into the mixer.

Facilities must also be provided for the storage, heating, and proportioning of

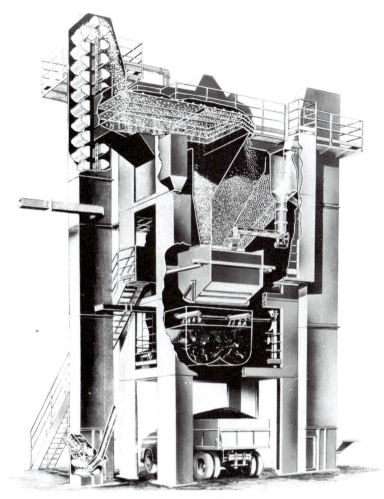

FIGURE 18-8 Flow of material through a batch-type hot-mix plant. (Courtesy Barber-Greene Company.)

the bituminous material that is to go into the mixture. When heating is necessary, the bituminous material is heated in a storage tank, generally by steam, hot oil, or electrical heaters. Devices are provided for the measurement and control of the temperature of the bituminous material and for agitating the heated mass. The heated material is fed into an "asphalt bucket," with facilities being provided for securing the correct weight of material for each batch and for emptying the bucket into the mixing unit.

The final basic unit that will be described here is the mixer itself. The mixer that is most commonly used is a twin pugmill of the type shown in Figure 18-9. In a batch-type plant this unit is mounted directly beneath the weigh box and asphalt bucket and high enough so that it may discharge the mixture into a truck or other hauling unit. The aggregates and bituminous material are dumped into the mixer through the top and the unit is then closed. The mixing is accomplished by the mixing blades mounted on the two shafts, which rotate in opposite direc-

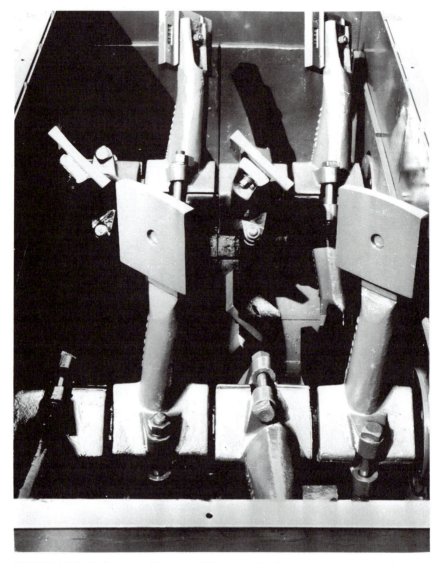

FIGURE 18-9 Twin pugmill mixer. (Courtesy Barber-Green Company.)

tions. The space between blades is generally about $\frac{3}{4}$ in. Mixing is continued for a fixed length of time, and the mixture is then discharged into the waiting truck.

Drum Mix Plants

Although batch plants are widely used, few of these plants are being sold today. They are gradually being replaced by the less expensive drum mix plants with special control systems that permit production of a variety of mixes. A sketch showing the major components of a drum mix plant is shown as Figure 18-10, while a flow diagram is indicated in Figure 18-11.

Simply stated, drier drum mixing is a process in which asphalt mixtures are produced in a plant without aggregate gradation screens, hot bins, and a pugmill mixer. The basic plant consists of

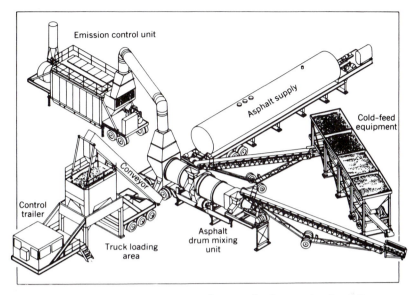

FIGURE 18-10 Major components of a recycle drum mix plant. Reclaimed asphalt pavement is fed from the two rear compartments of the cold-feed unit and mixed with virgin aggregate in the drum mixing unit. (Courtesy Cedarapids, Inc.)

1. aggregate cold-feed bins.
2. conveyor and aggregate weight system.
3. drum mixer.
4. liquid asphalt storage tank and pump.
5. hot mix conveyor.
6. mix surge silo.
7. control van.
8. dust collection system.

Gradation is controlled in drum mix plants prior to entry of the aggregate into the mixing drum. This is normally accomplished with a multiple-bin cold-feed system equipped with precision belt feeders that control the quantities of each aggregate (*11*). On the basis of a sieve analysis for the aggregate in each bin, the operator is able to calibrate the feed system to determine the proper belt speed and gate openings that will produce the desired aggregate blend.

Drum mix plants normally have a continuous weighing system on the cold-feed conveyor belts. Because the aggregate being fed to the drum is not dry, it is necessary to measure accurately the moisture content and to determine the weight of the dry aggregate to ensure that the amount of asphalt delivered to the drum is appropriate for the amount of dry aggregate.

From the cold-feed conveyor the aggregate is fed into the drum mixer, where it is dried and heated. Specially designed shelves or flights attached to the interior of the rotating drum pick up the aggregates and drop them in a uniform veil through the burner flame. In the lower portion of the drum mixer, liquid asphalt is added and thoroughly blended with the aggregates.

The drum mix plant produces a continuous flow of asphalt mix, which is temporarily stored in a surge silo, from which it is loaded into trucks.

FIGURE 18-11 Asphalt drier drum mix plant. (Courtesy The Asphalt Institute.)

The operation of drum mix plants is monitored and controlled by instruments in a control van.

18-12 Spreaders and Finishers

Some low-cost plant mixtures may be spread by the use of long-base blade graders or multiple-blade drags. This procedure may be desirable when some aeration of the mix is necessary to remove some of the volatiles contained in the bituminous binder, as might be the case when a dense-graded aggregate is used with slow-curing liquid asphalt. In the majority of cases, however, plant mixes are placed by use of a "bituminous paver" that spreads the mixture in a uniform layer of the desired thickness and shapes or "finishes" the layer to the desired elevation and cross section, ready for compaction. The use of a machine of this type is desirable in the case of open-graded plant mixtures and other mixtures that use highly viscous liquid binders. They are also widely used in the placing of the hot mixes described in the following chapter, which must be placed and finished rapidly so that they may be compacted while still hot. Bituminous pavers have not been previously described and will be briefly discussed in the following paragraphs.

The typical modern bituminous paver combines the functions of spreading and finishing in one machine. A machine of this type operates without the use of side forms, being supported on crawler treads or wheels. The machine is of sufficiently long wheelbase to eliminate the necessity for forms and to minimize irregularities occurring in the existing base or surface. These machines will process pavements from a fraction of 25 mm up to 254 mm (1 in. up to 10 in.) in thickness over a width of 1.83 to 4.27 m (6 to 14 ft), with working speeds generally ranging from 3 to 21 m/min (10 to 70 ft/min); one model has working speeds up to 31 m/min (102 ft/min), although most construction specifications do not permit this high a laydown speed.

The mixture that is to be placed by the paver is dumped into a receiving hopper from trucks or other hauling units. Material is fed from the hopper toward the finishing section of the machine and is spread and agitated by means of screws or other agitator-distributors that serve the purposes of ensuring uniformity of the spread mixture, spread it uniformly the full width of processing, and loosen or "fluff up" the material. The spread material is then struck off at the desired elevation and cross section by one or more screeds. In some cases a single cutter bar may be used to strike off and compact the mixture to some extent. In other machines, oscillating or vibratory screeds are used to strike off and initially compact the mixture. Several machines of this type employ a tamping mechanism in conjunction with the screed to provide additional compaction. Cutter bars and screeds are usually provided with heating units to prevent pickup of the material during the spreading and finishing operations. All machines of the type that have been described are fully adjustable to ensure a uniform flow of material through the machine and to produce a smooth, even layer of the desired thickness and cross section.

Most bituminous pavers are now fitted with electronic screed control systems, which were introduced in 1961. The various systems differ in detail but perform the same general functions. A sensor operates on a reference profile, senses changes in the position of the floating screed element of the paver or of the reference profile, and then applies any necessary correction to the angle of the screed so that the surface being laid off will be parallel to the reference profile.

FIGURE 18-12 Sensor operates from 6-m (20-ft)-long ski to guide bituminous paver. (Courtesy Iowa Manufacturing Company.)

Usually, the reference profile controls the longitudinal profile of the surface at one side of the machine and a slope sensor controls transverse slope, hence the pavement cross section; however, both sides of the paver can be controlled by reference profiles if desired.

Any of several reference profiles can be used, depending on the smoothness

FIGURE 18-13 A paver using a "Leveler" device that averages longitudinal grade errors. (Courtesy Barber-Greene Company.)

of the base or other surface on which the pavement is being laid, surface smoothness and thickness requirements, and so on. Reference profiles can be fixed or mobile. A typical fixed reference profile is a taut wire or string line carefully stretched between stakes or pins set at close spacings, usually 7.6 m (25 ft) or less. The taut wire or string line reference is rarely used today because of the high cost of erecting it. A moving reference profile is attached to the side of the paver; it may be a long (up to 12 m) rigid or articulated ski that incorporates a taut wire on which the sensor rides, a rolling string line (Fig. 18-12), or a short shoe that rides, for example, on the new surface of an adjacent lane. Figure 18-13 shows a paver using a "Leveler" device that averages longitudinal grade errors.

Some pavers also have automatic control of material feed to the screed; sensing elements at the ends of the augers interlock with the material conveyors to control the flow.

PROBLEMS

18-1 It is desired to combine material A, which is the soil existing in the roadbed, with material B, which may be obtained from nearby borrow sources at low cost, to form a stabilized soil-aggregate surface course conforming to the specification recommended by AASHTO, Table 18-1, grading C. The sieve analyses of materials A and B are shown in the following tabulation. Determine approximately the limiting proportions of A and B that should be used to produce a stabilized mixture of the desired gradation. That is, determine the greatest proportion of A that can be used in a mixture and still meet the specifications, then the smallest proportions of A that will meet the same objective. Plot the gradation curves of the two blends and select the mixture you think best fits the specifications. What other requirements would normally be included in the specification for this material?

		Percentage Passing (by weight)	
Sieve Designation		Material A	Material B
2 in.	(50 mm)	100	—
1 in.	(25 mm)	100	100
³⁄₈ in.	(9.5 mm)	68	45
No. 4	(4.75 mm)	42	28
No. 10	(2.0 mm)	30	17
No. 40	(0.425 mm)	28	7
No. 200	(0.075 mm)	11	3

18-2 For the materials of Problem 18-1, the following are the plasticity characteristics. Material A: liquid limit, 40, plasticity index, 12. Material B: liquid limit, 13, plasticity index, 3. Estimate the liquid limit and plasticity index of the aggregate blend selected in Problem 18-1. Assume 80% A, 20% B combination.

18-3 Assuming that the compacted mixture of Problem 18-1 has a dry unit weight of 118 lb/ft³, determine how many tons of material B will be required per mile of road, assuming that the surface is to be 24 ft wide and will have compacted thickness of 7 in. Estimate the number of cubic yards (loose measure) of this material that will be required per mile, assuming that the loose material weighs 105 lb/ft³. Assume % B = 20%.

18-4 If the compacted mixture of Problem 18-1 has a dry unit mass density of 1800 kg/m³, estimate the mass in kilograms of material B required per kilometer of road, given road width = 7.0 m and surface thickness = 0.25 m. If the loose material has a mass density of 1700 kg/m³, how many cubic meters of loose material will be needed?

18-5 Assume that a pressure distributor has a spray bar that is 12 ft long. If the pump discharge is at the rate of 260 gallons per minute, with the pump operating at the proper speed for the material that is being used, determine the rate of application of the bituminous material in gallons per square yard when the distributor is traveling at the rate of 1020 ft/min. Also determine the reading that must be maintained on the truck tachometer to maintain this rate of application if the tachometer is calibrated to read 100 times the truck speed in miles per hour. Assume that all the pumped material passes out through the spray bars and is applied to the road surface.

18-6 A pressure distributor that has a spray bar 3.5 m wide travels 20 km/hr while spraying at a rate of 15 L/sec. Determine the rate of application in liters per square meter.

18-7 A new granular-type base is to be covered with a double-surface treatment, meeting the requirements of FHWA Type AT-50, as outlined in the text. Assuming that the base is to be 24 ft wide, compute the quantities of bituminous material and crushed stone required for 1 mile of pavement if the specific gravity of the stone is 2.70. Estimate the cost of a double-surface treatment of this type in your area.

18-8 Determine the quantities of bituminous material and aggregate required for 1 km of double bituminous surface treatment meeting the requirements of FHWA Type AT-35, Table 18-6. The pavement is to be 7.0 m in width, and the specific gravity of the aggregate is 2.68.

18-9 Visit a bituminous paving plant in your area and write a report detailing the layout and equipment being used. Where do the aggregates come from? How are they fed to the plant? How is the mixture transported to the place where it is to be used?

18-10 Consult a highway construction agency in your area and determine the smoothness requirements being applied to new asphalt pavements. Are contractors using electronic screed controls to meet these requirements? How do the controls work?

REFERENCES

1. *AASHTO Materials, Part I, Specifications,* 16th Ed. American Association of State Highway and Transportation Officials, Washington, DC (1993).

2. Shearer, W. L. Calcium Chloride Stabilization. In *Materials for Stabilization.* Education and Information Guide HC-100A, American Road Builders Association, Washington, DC (1976).

3. *FHWA Implementation Package 80-2.* Federal Highway Administration, Washington, DC.

4. Johnson, James C. Asphalt Stabilization. In *Materials for Stabilization.* Education and Information Guide HC-100A, American Road Builders Association, Washington, DC (1976).

5. *Mix Design Methods for Liquid Asphalt Mixtures.* Supplement to MS-2 (Misc. 74-2), The Asphalt Institute, College Park, MD (1974).

6. *Soil–Cement Laboratory Handbook.* Portland Cement Association, Skokie, IL (1992).

7. *Book of ASTM Standards,* Designation D448 and Designation D693, American Society for Testing and Materials, Philadelphia, PA (Reapproved 1993).

8. *A Basic Asphalt Emulsion Manual.* Manual Series MS-19, 2nd Ed., The Asphalt Institute, Lexington, KY.

9. *Standard Specifications for Construction of Roads and Bridges on Federal Highway Projects.* Federal Highway Administration, Washington, DC (1992).

10. *Asphalt Surface Treatments—Specifications.* Educational Series No. 11, The Asphalt Institute, Lexington, KY (1982).

11. *Asphalt Plant Manual.* Manual Series No. 3, 5th Ed., The Asphalt Institute, Lexington, KY.

CHAPTER 19

HIGH-TYPE BITUMINOUS PAVEMENTS

The bituminous wearing surfaces discussed in this chapter represent the highest types of bituminous pavements that are in current use in this country. They are widely used both on rural highways and on city streets that are subjected to large volumes of traffic and severe service conditions. Properly designed and constructed surfaces of this general type are capable of carrying almost unlimited volumes of passenger, mixed, or truck traffic, provided only that they are supported by adequate foundation structures. The majority of these surfaces might be expected to have an economic life of 20 years or more.

All of the high-type bituminous mixtures discussed in this chapter are so-called "hot-mixes" that are prepared in central mixing plants. The thickness of these surfaces commonly vary from slightly less than 25 mm (1 in.) to 100 mm (4 in.) or more, depending on the type of surface and its purpose. These mixtures are, generally speaking, marked by the use of carefully selected and graded aggregates in conjunction with semisolid bituminous binders. The composition of the mixture is more rigidly specified and controlled than for other types of bituminous surfaces, and the preparation, placing, and finishing of the mixture are more carefully controlled.

DESIGN OF HIGH-TYPE BITUMINOUS PAVING MIXTURES

It is the purpose of this section to set forth some of the underlying principles on which the design of a high-type bituminous paving mixture is based. Statements made in this section are intended to apply specifically to the design of the hot-mixed, hot-laid bituminous concrete mixtures decribed in this chapter, although they may also generally apply to those mixtures that do not contain coarse aggregate, such as sheet asphalt and sand asphalt.

The two fundamental properties of a bituminous paving mixture that are held to be of the utmost importance are stability and durability. By "stability" is meant that property of the compacted mixture that enables it to withstand the stresses imposed on it by moving wheel loads without sustaining substantial permanent deformation. By "durability" is meant that property of the compacted mixture that permits it to withstand the detrimental effects of air, water, and temperature change. For successful results the pavement must be both durable and stable during its entire service life.

Both stability and durability are intimately related to the density of the mixture. The density of a given mixture is frequently expressed in this work in terms of the "voids" in the mixture, meaning, in general, the amount of space in the compacted mixture that is not filled with aggregate or bituminous material, that is, is filled with air. Thus a dense mixture would have a low percentage of voids, while a loose mixture would have a high percentage of voids, all other things being the same.

As will be explained, the extent of voids in a compacted mix is determined in large part by the percent of asphalt cement in the mix. A fundamental goal of the mix designer is to determine the "best" or optimum asphalt content that will provide the required stability and durability, as well as additional desirable properties such as impermeability, workability, and resistance to "bleeding." Detailed procedures to be followed in the determination of the optimum asphalt content will be given later, but certain general principles can be stated here.

Consider a blend of the mineral aggregates alone, without any bituminous material, assuming that the mineral aggregates have been compacted to a density equivalent to that which will be attained in the field process. This dense aggregate alone will possess a certain density and a relatively low stability. As small quantities of bituminous material are added to the mixture, both the density and stability will increase. At this stage the bituminous material would generally be present in the form of thin films surrounding, or partially surrounding, the aggregate particles. If more bituminous material is added, a point will be reached at which the aggregate particles are completely covered with bituminous material; additional bituminous material would serve to fill the void spaces between the aggregate particles. For many mixtures, both stability and density would increase until the voids in the mixture had been completely filled. The addition of any more bituminous material would result in a decrease in both density and stability, as the aggregates would generally be forced apart by the excess bituminous material.

It thus appears that the most desirable bitumen content for many mixtures would be that which would just fill the voids in the compacted mixture. This concept is, however, modified by certain practical considerations. In the first place, any bituminous material expands with an increase in temperature. Thus if the voids in the mixture were completely filled at the time of placing, any increase in temperature subsequent to that time would result in overfilling of the voids in the mixture with consequent "bleeding" of the pavement and loss of stability. Similarly, the density of the compacted mixture may increase under the action of traffic until it exceeds that used in the design. If this happened, an excess of bituminous material might again be present. As a result of these conditions and others, a compromise is frequently made in the selection of the optimum bitumen content. Assuming that all mixtures must possess adequate stability that would have, on the one hand, more than a certain *mimimum* void content and, on the other, less than a certain *maximum* void content. The selection of the maximum void content that could be permitted would generally be based on durability considerations; air and water permeability may also be considered. For example, the percentage of bituminous material chosen for a given mix might be such that the compacted mixture would have a percentage of voids (calculated as explained later) that was not less than two or more than five. A final fundamental requirement for the design of a mixture is that it be economical.

To summarize, the objectives of asphalt paving mix design are to select and proportion materials to produce a mix that has

- sufficient asphalt to ensure a durable pavement.
- sufficient mix stability to satisfy the demands of traffic without distortion or displacement.
- sufficient voids on the total compacted mix to allow for a slight amount of additional compaction under traffic and a slight amount of asphalt expansion due to temperature increases without flushing, bleeding, and loss of stability.
- a maximum void content to limit the permeation of harmful air and moisture into the mix.
- sufficient workability to permit efficient placement of the mix without segregation and without sacrificing stability and performance.
- for surface mixes, proper aggregate texture and hardness to provide sufficient skid resistance in unfavorable weather conditions. (*1*)

19-1 Classification of Hot-Mix Paving

According to The Asphalt Institute (*1*):

Asphalt paving mixes may be designed and produced from a wide range of aggregate blends, each suited to specific uses. The aggregate composition typically varies in size from coarse to fine particles. Many different compositions are specified throughout the world—the mixes designated in any given locality generally are those that have proven adequate through long-term usage and, in most cases, these gradings should be used.

For a general classification of mix compositions, The Asphalt Institute recommends consideration of mix designations and nominal maximum size of aggregate: 37.5 mm (1-1/2 in.), 25.0 mm (1 in.), 19.0 mm (3/4 in.), 12.5 mm (1/2 in.), 9.5 mm (3/8 in.), 4.75 mm (No. 4), and 1.18 mm (No. 16) as specified in the American Society for Testing and Materials (ASTM) Standard Specification D3515 for *Hot-Mixed, Hot-Laid Bituminous Paving Mixtures*. The grading ranges and the asphalt content limits of these uniformly-graded dense mixes generally agree with overall practice but may vary from the practice of a particular local area.

A broad but widely used class of asphalt paving mixes is referred to as asphaltic concrete. Asphaltic conccrete may be generally defined as an intimate mixture of coarse aggregate, fine aggregate, mineral filler, and asphalt cement. These mixtures may be called "open graded," "coarse grade," "dense graded," or "fine graded," depending on the gradation of the aggregates. Other classes of hot asphalt paving mixtures include sheet asphalt and sand asphalt, which contain no coarse aggregate. Except for a brief description of sheet asphalt and sand asphalt at the end of this chapter, the remaining discussion will focus on the design and construction of asphaltic concrete pavements.

Mixes may also be designated according to use in the layered pavement system. "Surface mixes" constitute the upper layer, whereas "base mixes" are used for the layer directly above the subbase or subgrade. *Leveling mixes* are those used in an intermediate layer, often to eliminate irregularities in contour of an existing surface prior to construction of a new layer.

19-2 Materials for Asphaltic Concrete Paving Mixes

The desirable qualities of a bituminous paving mixture are dependent to a considerable degree on the nature of the aggregates used. "Coarse aggregate" may be defined as that portion of the aggregate material contained in a bituminous paving mixture retained on a No. 10 sieve. "Fine aggregate" is that portion of the aggregate material which passes a No. 10 and is retained on a No. 200 sieve. It should be noted that the Asphalt Institute (*1*) uses the No. 8 sieve as the dividing line between coarse aggregate and fine aggregate. "Filler" is that mineral material that passes a No. 200 sieve. The function of the coarse aggregate in contributing to the stability of a bituminous paving mixture is largely due to interlocking and the frictional resistance of adjacent aggregate particles to displacement. Similarly, the fine aggregate or "sand" contributes to stability through interlocking and internal friction, so particles that are sharp and angular in nature are preferred. Sands are also held to contribute to stability by their function in filling the voids in coarse aggregate. The mineral filler is largely visualized as a void-filling agent, with limestone dust being most widely used for this purpose. Most commercial materials used as fillers contain only a small proportion that is retained on the 75 μm (No. 200) sieve.

Aggregates that are well graded from coarse to fine are generally sought in high-type bituminous paving mixtures. Well-graded materials tend to produce the most dense mixtures and therefore the most durable, requiring minimum bitumen content for satisfactory results. In most specifications, each of the separate aggregates—coarse, fine, and filler—is required to be well graded from coarse to fine as is the combination of these materials. The concept simply is that in a well-graded aggregate each smaller size or "fraction" of aggregate serves to fill the voids in the next larger one, with the result that a very dense aggregate combination may be secured.

The coarse aggregates most frequently used in asphaltic concrete paving mixtures are crushed stone, crushed gravel, and crushed slag. Uncrushed gravel, which may also be used in some cases, is required to consist largely of rough-textured particles. The coarse aggregates are, as was indicated in Chapter 15, required to be hard, tough, clean, and durable in nature, free from an excess of flat or elongated pieces, clay material, and other deleterious substances. Aggregates known to polish under traffic are not used in surface courses.

Crushed stone and gravel are generally permitted to have percentages of wear in the Los Angeles abrasion test of not more than 40, although this figure is increased to 50 in some instances. They are also sometimes subjected to requirements for soundness (AASHTO M76); requirements in this respect frequently call for weight loss incurred when the material is subjected to five cycles of the sodium (or magnesium) sulfate soundness test (AASHTO Method T104) not to exceed 12 or 15 percent. Crushed slag is usually subject to the same requirements related to abrasive resistance as stone and gravel; in addition, it is generally required to weigh not less than 1120 kg/m³ (70 lb/ft³). Certain additional requirements related to crushed and uncrushed gravels are found in some specifications. General requirements related to coarse aggregates to be used in asphaltic concrete base courses may be somewhat less severe than those indicated above.

Requirements for sands to be used in asphaltic concrete mixtures are usually quite general in nature. Sands are generally required to be composed of tough, clean, rough-surfaced, and angular particles, free from excess amounts of clay and other deleterious materials. Stone screenings may also be used as fine aggregate and are sometimes required to be produced from coarse aggregates that meet the requirements given above. The crushed aggregates previously listed may, of course, contain a considerable amount of material that passes a No. 10 sieve and is classed as fine aggregate. The fine aggregate for a given mixture may consist of sand alone, stone or slag screenings alone, or a blend of sand and stone or slag screenings.

A variety of materials are used for mineral filler, when required, and specifications in use by a given agency generally list the specific materials that may be used for this purpose. Commonly used materials include limestone dust, dolomite dust, portland cement, slag dust, and other similar inert mineral substances. The mineral filler is generally required to be dry and free from lumps. It is desirable that this material be definitely hydrophobic in nature.

The bituminous material used in the preparation of hot-mix, hot-laid asphaltic concrete mixtures is semisolid asphalt cement. The nature and quality of the asphalt cement are controlled by specifications like those discussed in Chapter 15. Generally speaking, the more viscous grades of asphalt cement such as AC-20 or AC-40 are recommended for highways that will carry heavy traffic in hot climates. Grade AC-10 would be suitable for pavements designed to serve medium to light traffic, except in cold climates, where AC-5 or AC-2.5 would be a better choice.

19-3 Requirements Related to Aggregate Gradation and Composition of Mixtures

Separate grading requirements are usually given for the separate aggregate components of the mixture—coarse aggregate, fine aggregate, and filler—and then an overall specification is given pertaining to the composition of the paving mixture itself, including the bituminous material. The specifications of the Kentucky Department of Highways are shown as Table 19-1 to illustrate typical differences in aggregate gradiations for a base, binder (or leveler), and surface course.

The various specifications cited are intended to represent the requirements of typical specifications related to aggregate gradings and mix compositions. The percentages of the various sizes of aggregates and the amount of bituminous material to be used in a given mixture vary to some extent within the limits of each specification. Before the preparation of the mixture can actually begin, a

TABLE 19-1 Aggregate Grading Requirements for Class I Plant Mix Bituminous Pavements—Kentucky Department of Highways (2)

| Sieve Size | *Percentage Passing, by Weight* | | |
	Base	*Binder*	*Surface Course*
37.5 mm (1½ in.)	100		
25 mm (1 in.)			
19 mm (¾ in.)	70–98	100	100
12.5 mm (½ in.)	—	—	80–100
9.5 mm (⅜ in.)	44–76	57–85	55–80
4.75 mm (No. 4)	30–58	37–68	35–60
2.36 mm (No. 8)	21–45	25–52	20–45
1.18 mm (No. 16)	14–35	15–38	10–32
0.600 mm (No. 30)	8–25	9–27	5–21
0.300 mm (No. 50)	5–20	5–20	3–14
0.150 mm (No. 100)	3–10	3–10	2–7
0.075 mm (No. 200)	2–6	2–6	

TABLE 19-2 Job-Mix Formula Tolerances (2)

Material Passing	Tolerance Percentage Points
12.5-mm (½-inch) sieve and above	±6
9.5, 4.75, 2.36, 1.18-mm (⅜-in., and Nos. 4 and 8, and 16) sieves	±5
0.600 mm (No. 30)	±4
0.300 mm (No. 50)	±4
0.150 mm (No. 100)	±2
0.075 mm (No. 200)	±1½
Asphalt content	±0.3

single figure must be definitely established for each of the aggregate fractions contained in the specification and for the bituminous material. In other words, the composition of the mix must be definitely established. The process of doing this, which is frequently called the "design" of the mixture, results in the establishment of a definite "job-mix formula," which will then be applicable to the mixture being used on a particular project. Considerations related to the determination of the job-mix formula will be given in the following sections.

Once the job-mix formula has been established, most specifications allow small variations from it during production of the mix. The purpose of these tolerances is to obtain a uniform mixture in reasonable compliance with the specifications, since it is not possible to produce the exact job mix every time. A representative job-mix tolerance specification appears in Table 19-2.

19-4 Determination of the Job-Mix Formula

The procedure that is presented here for the determination of the job-mix formula is intended to indicate the procedure that might be followed by an imaginary highway agency in determining the composition of a typical dense-graded asphaltic concrete surface mixture. Because detailed procedures in the determination of the job-mix formula vary so much, it is not deemed wise to try to present typical or "average" practice in this respect. The procedure indicated is generally similar to that followed by some agencies and has been somewhat simplified in the interests of clarity and brevity. Two general steps are involved in this process: the selection and combination of aggregates to meet the limits of the specification being used and the determination of the optimum asphalt content.

19-5 Selection and Combination of Aggregates

Relatively little will be said here about the selection of the aggregates to be used in a given paving mixture, although this is a very important phase of the overall design process. It should be noted, however, that an agency's road specifications may have very strict requirements for paving projects and take into account local factors and conditions.* In the normal procedure, both coarse and fine aggregates

---

*The state of Alabama (3), for example, states that, in the case of the wearing surface, the use of carbonate stone such as limestone, dolomite, or aggregates tending to polish under traffic will not be permitted.

available in the vicinity of the proposed work are sampled and carefully examined for compliance with the individual specifications for these materials. In some cases more than one suitable aggregate for each type will be found, and the selection of the particular ones to be used may then be largely a matter of economics. In other cases no suitable single aggregate may be locally available, in which case aggregates from several different sources may have to be blended to meet the requirements of the specifications, or imported aggregates may have to be used. Materials for use as mineral filler must be similarly examined and selected. Regardless of what the selection of aggregates may involve on a given project, all possible aggregates should be carefully examined, and those selected should be suitable for the purpose and economically available. The aggregates finally selected for use must then be combined to provide an aggregate mixture that will comply with the limits established for the composition of the mixture.

The procedure used in determining the properties of coarse aggregate, fine aggregate, and mineral filler (and the optimum asphalt content) may best be illustrated by a discussion of the procedure to be followed in the design of a definitely specified asphaltic concrete mixture. This procedure will be followed in the subsequent discussion. It is assumed that the composition of the mixture to be designed (for which a job-mix formula is to be established) must be within the limits of the specifications in Table 19-3.

It will be assumed that the aggregates available for use in this mixture have been thoroughly tested and examined for compliance with the specifications related to these materials. Sieve analyses of the aggregates that can be most economically used in this case, as determined by AASHTO methods of tests T27 and T37 (mineral filler), are shown in Table 19-4.

The problem now is to determine the proportions of the separate aggregates that must be used in order to give an aggregate combination that will meet the requirements of the specification previously given.

Not only must the proportions selected be within the specification but they must be far enough from its extremes to provide room for the job-mix tolerance,

TABLE 19-3　Mineral Aggregate and Mix Composition

Passing Sieve Designation	*Retained on Sieve Designation*	*Percent (by Weight)*
¾ in. (19.0 mm)	½ in.	0–6
½ in. (12.5 mm)	⅜ in.	9–40
⅜ in. (9.5 mm)	No. 4	9–45
No. 4 (4.75 mm)	No. 10	8–27
Total coarse aggregate	No. 10	50–65
No. 10 (2.00 mm)	No. 40	6–22
No. 40 (0.475 mm)	No. 80	8–27
No. 80 (0.177 mm)	No. 200	5–17
No. 200 (0.75 mm)	—	5–8
Total fine aggregate and filler	(Passing No. 10)	35–50
Total mineral aggregate	—	100
Total mix		
Total mineral aggregate		92–95
Asphalt cement		5–8
Total mix		100

TABLE 19-4 Sieve Analysis of Aggregates

Percent (by weight)		Aggregate Type		
Passing Sieve Designation	Retained on Sieve Designation	Coarse Aggregate	Fine Filler	Mineral Filler
¾ in. (19.0 mm)	½ in.	5	—	—
½ in. (12.5 mm)	⅜ in.	32	—	—
⅜ in. (9.5 mm)	No. 4	37	—	—
No. 4 (4.75 mm)	No. 10	22	7	—
No. 10 (2.00 mm)	No. 40	4	28	—
No. 40 (0.425 mm)	No. 80	—	39	5
No. 80 (0.177 mm)	No. 200	—	24	30
No. 200 (0.075 mm)	—	—	2	65
Total		100	100	100

so that when it is added or subtracted the mixture will not be outside the original specification master range. This is done by trial. It is obvious that the fine aggregate and coarse aggregate only, in any combination, cannot meet the requirements of the specification for total mineral aggregate, as the specification requires from 5 to 8 percent of material passing a No. 200 sieve and the coarse aggregate contains none of this material, while the fine aggregate contains only 2 percent. Obviously, some mineral filler must be used in the mixture; for the first trial, the amount of mineral filler is arbitrarily set at 8 percent. The total coarse aggregate in the mix must be from 50 to 65 percent, and this figure is again more or less arbitrarily set at 52 percent. The remaining 40 percent must be fine aggregate.

Calculations made using the indicated proportions, which are necessary in determining the sieve analysis of the combined aggregates, are shown in Table 19-5. A comparison of the figures in the last column of this table with the requirements of the specification will show that this combination of aggregates meets the stipulated requirements. This combination will therefore be judged satisfactory, and no additional trials will be made here. Additional trials would have to be made if the combination had not been satisfactory. Various graphical methods have been devised that make the blending process somewhat simpler

TABLE 19-5 Calculations for Sieve Analysis

Passing Sieve Size	Retained on Sieve Size	Percent (by weight)			
		Coarse Aggregate	Fine Aggregate	Mineral Filler	Total Aggregate
¾ in.	½ in.	0.52 × 5 = 2.6	—	—	2.6
½ in.	⅜ in.	0.52 × 32 = 16.6	—	—	16.6
⅜ in.	No. 4	0.52 × 37 = 19.2	—	—	19.2
No. 4	No. 10	0.52 × 22 = 11.4	0.40 × 7 = 2.8	—	14.2
No. 10	No. 40	0.52 × 4 = 2.2	0.40 × 28 = 11.2	—	13.4
No. 40	No. 80	—	0.40 × 39 = 15.6	0.08 × 5 = 0.4	16.0
No. 80	No. 200	—	0.40 × 24 = 9.6	0.08 × 30 = 2.4	12.0
No. 200	—	—	0.40 × 2 = 0.8	0.08 × 65 = 5.2	6.0
Total		52.0	40.0	8.0	100.0

than the arithmetic process used here. Quite frequently no possible combination of the aggregates initially selected will prove to be satisfactory, in which case others must be selected or the available ones improved by blending. This situation can be avoided with experience. Additional trials might also be made in an effort to improve the gradation of the total aggregate; these also will not be considered here. The combination selected will be used in the preparation of the trial mixes required for the determination of the optimum asphalt content.

19-6 Determination of Optimum Asphalt Content

The steps remaining in the process of determining the job-mix formula for a given asphaltic concrete surface course are directed toward the determination of the optimum asphalt content to be used in the mix. The asphalt content established as part of the job-mix formula must, of course, be within the limits of the given specification relative to composition of the mixture.

The laboratory procedure used in determining the optimum asphalt content involes the preparation of trial mixtures, using the selected aggregates and various percentages of asphalt within the limits of the mix specification. Each trial mixture is prepared in a manner that is intended to secure a very high density. Densities that are secured in this fashion generally represent the ultimate densities that are practically attainable either in the laboratory or in the field. The density, stability, and other properties of each trial mixture are then determined and the results tabulated. Applicable criteria may then be applied to determine the optimum asphalt content.

Design criteria related to the laboratory compacted specimens frequently require that the density of satisfactory mixes be not less than a certain percentage of the "theoretical maximum density." By "theoretical maximum density" is meant the density, usually expressed in terms of the bulk specific gravity, of a voidless mixture composed of the same ingredients in the same proportions used in the actual mixture. Design criteria for density, stability, and other properties that are used by a given organization have been established by extensive laboratory and field experience and investigation and are generally applicable only to mixtures that are to be used as wearing surfaces, being regarded as unnecessary to the design of base and binder course mixtures in many cases. Although we are discussing asphaltic concrete, the same or similar design criteria may be applied to the design of the other high-type mixtures described in this chapter.

Two mix design methods are in widespread use: the Marshall method (Section 19-7) and the Hveem method (Section 19-12).

19-7 Marshall Method of Design

The Marshall method of design was originally developed by Bruce Marshall, formerly of the Mississippi Highway Department, and improved by the U.S. Army Corps of Engineers. The Marshall test procedures have been standardized by the American Society for Testing and Materials (ASTM) and published as ASTM D1559. The method is applicable only to hot mixtures using penetration grades of asphalt cement and containing aggregates with a maximum size of 25 mm (1 in.) or less.*

---

*A modified method has been proposed to allow testing of paving mixtures containing aggregates with maximum sizes up to 38 mm (1.5 in.).

The Marshall method uses standard test specimens 64 mm (2.5 in.) high and 102 mm (4 in.) in diameter. They are prepared by use of a standard procedure for heating, mixing, and compacting the asphalt-aggregate mixtures. Density-voids determinations and calculations are made on the compacted specimens by methods and equations presented in Section 19-8.

19-8 Preparation of Test Specimens (Marshall Method)

In preparation of test specimens for the Marshall method, it is essential that the gradation of the aggregate and the quantity of asphalt be very carefully controlled. Considerable care must be exercised in the entire conduct of laboratory testing operations so that test results will be reliable and reproducible. Tests should be planned on the basis of 1/2 percent increments of asphalt content, with at least two asphalt contents above the expected design value and two or more below. Reference 1 gives a computational formula that provides an estimate of the expected design asphalt content based on the gradation and absorption characteristics of the aggregate. In the example given, it will be assumed that, because the possible range in asphalt content is from 4 to 6 percent of the total mix, trial mixtures will be prepared with asphalt contents over this range at intervals of 1/2 percent. In other words, work will be begun on the assumption that mixtures will be prepared that contain 5.0, 5.5, 6.0, 6.5, and 7.0 percent of asphalt.

As a first step in the procedure, the aggregates with the proper gradation are thoroughly dried and heated. Sufficient mixture is generally prepared at each asphalt content to form at least three specimens. Each specimen will require approximately 1.2 kg (2.7 lb) of mixture.

The asphalt and the aggregates are heated separately and then mixed. The mixing and compaction temperatures are established by determining the temperatures to which the asphalt must be heated to produce viscosities, respectively, of 170 ± 20 and 280 ± 30 centistokes kinematic.

The mixture is placed in the mold, mixed with a mechanical mixer or by hand with a trowel, and compacted. The specimens are compacted with a standard hammer device, which weighs 4.5 kg (10 lb) and is designed to drop from a height of 457 mm (18 in.). A compactive effort of 35, 50, or 75 blows is specified, depending on the traffic category. The compactive effort is applied to each side of the specimen. After compaction, each specimen is subjected to a density-voids analysis and then tested for stability and flow.

19-9 Density-Voids Analysis

The density-voids analysis for an asphalt mix begins with the determination of the bulk specific gravity,* G_{mb}, of the compacted mixture. The G_{mb} is determined in accordance with ASTM Designation D2726, and it involves determining the mass of the specimen in air and the mass of the same specimen in water. The same method may be used to determine the specific gravity and density of a sample cut or cored from a completed pavement. With very open or porous mixtures, the sample may be coated with paraffin before the density determina-

---

*The bulk specific gravity of extremely porous samples may be determined using paraffin-coated specimens which are tested in accordance with ASTM Designation D1188.

tion. The bulk specific gravity of the compacted mixture (uncoated specimen) may be expressed as follows:

$$G_{mb} = \frac{W_a}{W_a - W_w} \qquad (19\text{-}1)$$

where G_{mb} = bulk specific gravity of compacted mixture
W_a = mass of test specimen in air (g)
W_w = mass of test specimen suspended in water (g)

For example, assume that a test specimen weighs 3041.2 g in air and 1713.2 g in water. The bulk specific gravity of the specimen would be

$$G_{mb} = \frac{3041.2}{3041.2 - 1713.2} = 2.290$$

This value of G_{mb} means that the unit weight of the compacted mixture would be 2.29(62.4) = 142.9 lb/ft³. In metric units, the unit weight would be 2.29 (1000 kg/m³) = 2290 kg/m³.

The determination of the percentage of air voids in a compacted mix is complicated by the fact that a portion of the asphalt in the mix may be absorbed by the aggregate (Fig. 19-1).

In asphalt paving technology, three different types of specific gravity are discussed. These are apparent, bulk, and effective specific gravities. Water absorption is normally used to determine the quantity of permeable voids in the aggregate,

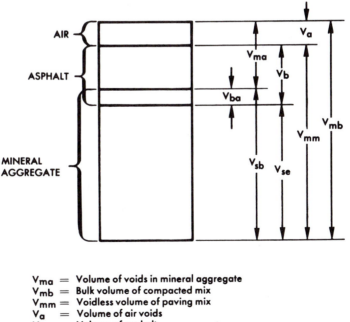

V_{ma} = Volume of voids in mineral aggregate
V_{mb} = Bulk volume of compacted mix
V_{mm} = Voidless volume of paving mix
V_a = Volume of air voids
V_b = Volume of asphalt
V_{ba} = Volume of absorbed asphalt
V_{sb} = Volume of mineral aggregate (by bulk specific gravity)
V_{se} = Volume of mineral aggregate (by effective specific gravity)

FIGURE 19-1 Representation of volumes in a compacted asphalt specimen. (Courtesy The Asphalt Institute.)

for use in specific gravity computations (ASTM Methods C127 and C128). The principal difference between bulk and apparent specific gravities is that the volume of the permeable voids is included in the volume of the aggregate for bulk specific gravity and is excluded in the volume of the aggregate for apparent specific gravity. In all cases, the bulk specific gravity is less than the apparent specific gravity. Recognizing that the absorption of asphalt by the aggregate is different from that of water, various organizations have devoted considerable effort to the development of methods of determining the effective specific gravity for use in asphalt mix calculations. The effective specific gravity is normally somewhere between the bulk and apparent values. At this time ASTM D-2041 has established a method for determining the maximum theoretical density of an asphaltic concrete mixture. This value attempts to give the maximum density that otherwise would have to be calculated from the specific gravities of the ingredients. This procedure makes it possible to more accurately calculate the percentage of voids in a paving mix.

The Asphalt Institute recommends the following procedure for the calculation of the percentage of voids in the mineral aggregate (VMA) and the percentage of air voids in the compacted mixture, P_a. A more complete description of this procedure is given in Ref. 1.

Step 1. Calculate the Bulk Specific Gravity of Aggregate When the total aggregate is made up of separate fractions of coarse aggregate, fine aggregate, and mineral filler, all having different specific gravities, the bulk specific gravity of the total aggregate, G_{sb}, is calculated by the following equation:

$$G_{sb} = \frac{P_1 + P_2 + P_3}{\dfrac{P_1}{g_1} + \dfrac{P_2}{g_2} + \dfrac{P_3}{g_3}} \qquad (19\text{-}2)$$

where P_1, P_2, P_3 = percentage by weight of coarse aggregate, fine aggregate, and mineral filler, respectively
g_1 = bulk specific gravity of coarse aggregate, measured by method ASTM C127 or AASHTO T85
g_2 = bulk specific gravity of fine aggregate, measured by method ASTM C128 or AASHTO T84
g_3 = apparent specific gravity of mineral filler, measured by method ASTM D854 or AASHTO T100

Step 2. Calculate the Effective Specific Gravity of Aggregate The effective specific gravity of the aggregate, G_{se}, includes all void spaces in the aggregate particles except those that absorb asphalt. It is calculated as follows:

$$G_{se} = \frac{P_{mm} - P_b}{\dfrac{P_{mm}}{G_{mm}} - \dfrac{P_b}{G_b}} \qquad (19\text{-}3)$$

where P_{mm} = total loose mixture (percentage by total weight of mixture = 100 percent)
P_b = asphalt (percentage by total weight of mixture)
G_{mm} = maximum specific gravity of paving mixture (no air voids), determined in accordance with ASTM D2041
G_b = specific gravity of asphalt, determined by ASTM70 or AASHTO T228

Step 3. Calculate the Asphalt Absorption of the Aggregate Absorption is expressed as a percentage by weight of aggregate. The absorption, P_{ba}, is calculated by the following equation:

$$P_{ba} = 100 \frac{G_{se} - G_{sb}}{G_{sb}G_{se}} G_b \qquad (19\text{-}4)$$

where P_{ba} = absorbed asphalt (percentage by weight of aggregate)

Step 4. Calculate the Effective Asphalt Content of a Paving Mixture The effective asphalt content, P_{be}, is the portion of the total asphalt content that remains as a coating on the outside of the aggregate particles after a portion of asphalt is lost by absorption into aggregate particles. It is calculated by Equation 19-5:

$$P_{be} = P_b - \frac{P_{ba}}{100} P_s \qquad (19\text{-}5)$$

where P_{be} = effective asphalt content (percentage by total weight of mixture)
$\quad\quad\ \ P_s$ = aggregate (percentage by total weight of mixture)

Step 5. Calculate the Percentage of Voids in the Mineral Aggregate (VMA) in the Compacted Paving Mix *VMA*, the voids in the mineral aggregate, is defined as the intergranular void space between the aggregate particles in a compacted paving mixture, including the air voids and the effective asphalt content. It is expressed as a percentage of the total volume. The *VMA* is based on the bulk specific gravity of the aggregate and is expressed as a percentage of the bulk volume of the compacted paving mixture. Therefore, it can be calculated by subtracting the volume of the aggregate determined by its bulk specific gravity from the bulk volume of the compacted paving mixture. This leads to the following equation:

$$VMA = 100 - \frac{G_{mb}P_s}{G_{sb}} \qquad (19\text{-}6)$$

where VMA = voids in mineral aggregate (percentage of bulk volume)

Step 6. Calculate the Percentage of Air Voids in the Compacted Mixture The air voids in a compacted paving mixture consist of the small air spaces between the coated aggregate particles. The percentage of air voids in a compacted mixture, P_a, can be computed as follows:

$$P_a = 100 \frac{G_{mm} - G_{mb}}{G_{mm}} \qquad (19\text{-}7)$$

where P_a = air voids in compacted mixture (percentage of total volume)

Step 7. Calculate the Percentage of Voids Filled with Asphalt The voids in the mix are often expressed in terms of the percentage of the total voids in the mineral aggregate that are filled with asphalt, *VFA*, which can be determined by the following equation:

$$VFA = \frac{100\,(VMA - V_a)}{VMA} \qquad (19\text{-}8)$$

This quantity does not account for any asphalt absorbed by the aggregate.

A typical worksheet for the analysis of a compacted paving mixture is shown as Figure 19-2.

Worksheet for Volumetric Analysis of Compacted Paving Mixture
(Analysis by weight of total mixture)

Sample: _____   Date: _____

Identification: _____

Composition of Paving Mixture

		Specific Gravity, G			Mix Composition, % by wt. of Total Mix, P					
						Mix or Trial Number				
				Bulk		1	2	3	4	5
1. Coarse Aggregate	G_1		2.716		P_1			47.4		
2. Fine Aggregate	G_2		2.689		P_2			47.3		
3. Mineral Filler	G_3		--		P_3			--		
4. Total Aggregate	G_s	--	--		P_s			94.7		
5. Asphalt Cement	G_b	1.030	--		P_b			5.3		

	Equation	1	2	3	4	5
6. Bulk Sp. Gr. (G_{sb}), total aggregate	19-2			2.703		
7. Max. Sp. Gr. (G_{mm}), paving mix ASTM D2041	—			2.535		
8. Bulk Sp. Gr. (G_{mb}), compacted mix ASTM D2726	—			2.442		
9. Effective Sp. Gr. (G_{se}), total aggregate	19-3			2.761		
10. Absorbed Asphalt (P_{ba}), % by wgt. total agg.	19-4			0.8		

CALCULATIONS

11. Effective Asphalt Content (P_{be}) =
$$\text{Line 5 } P_b - \frac{(\text{Line 10} \times \text{Line 4 } P_s)}{100}$$
19-5 → 4.5

12. VMA =
$$100 - \frac{\text{Line 8} \times \text{Line 4 } P_s}{\text{Line 6}}$$
19-6 → 14.4

13. Air Voids (V_a) =
$$100 \frac{\text{Line 7} - \text{Line 8}}{\text{Line 7}}$$
19-7 → 3.7

14. VFA =
$$100 \frac{\text{Line 12} - \text{Line 13}}{\text{Line 12}}$$
19-8 → 74.3

FIGURE 19-2 Worksheet for analysis of compacted paving mixture based on weight of total mixture. (Courtesy The Asphalt Institute.)

19-10 Stability-Flow Test (Marshall Method)

The Marshall stability-flow test measures the maximum load resistance and cor-responding deformation (or flow) of a standard test specimen at 60°C (140°F) when subjected to a load by a standardized test procedure. It is intended to simulate pavement failure under the worst conditions expected in field condi-tions.

Loading is done with the specimen lying on its side in a split loading head at a constant rate of deformation, 51 mm (2 in.)/min, until failure occurs (Fig. 19-3). The total number of Newtons (lb) required to produce failure is recorded as the Marshall stability value. The deformation (or flow) at maximum load is recorded and expressed in units of 1/100 in. For example, if the specimen is deformed 0.18 in., the flow value is 18.

19-11 Interpretation of Marshall Test Data

When testing is complete, six plots are prepared, as illustrated in Figure 19-4, from which a preliminary mix can be determined. As a starting point, the Asphalt Institute suggests that the median of the percent air voids from Table 19-6 (4

FIGURE 19-3 The Marshal stability test. (Courtesy The Asphalt Institute.)

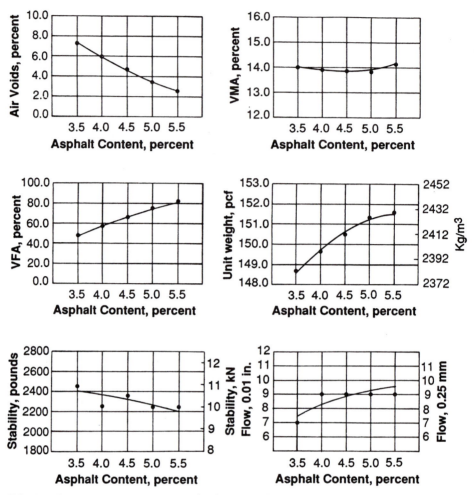

FIGURE 19-4 Test property curves for hot-mix design data by the Marshall method. (Courtesy The Asphalt Institute.)

percent) be used. The mix properties for a percent air voids of 4 percent are then compared to the design criteria (Table 19-7). If all of the design criteria are not met, some adjustment or compromise may be necessary or another trial mix evaluated. If all of the criteria are met, the asphalt content corresponding to 4 percent air voids is chosen as the preliminary design value. Further adjustment of this value may be needed to provide a mix design that will satisfy the requirements of a specific project, taking into account such considerations as traffic, paving season, climate, pavement structure, and so on.

Additional guidance on the selection of a final mix design is given in Ref. 1.

EXAMPLE 19-1   **Determination of a Preliminary Mix Design** Determine a preliminary asphalt content, stability, flow, VMA, and VFA from the curves in Figure 19-4 and determine if that design meets the design criteria for a heavy traffic area. The normal maximum particle size of aggregate is 25 mm (1 in.).

TABLE 19-6 Design Criteria for the Marshall Method

Marshall Method Mix Criteria[1]	Light Traffic[b] Surface & Base		Medium Traffic[b] Surface & Base		Heavy Traffic[b] Surface & Base	
	Min	Max	Min	Max	Min	Max
Compaction, number of blows each end of specimen	35		50		75	
Stability, N	3336		5338		8006	
(lb)	(750)	—	(1200)	—	(1800)	—
Flow, 0-.25 mm (0.01 in.)	8	18	8	16	8	14
Percent air voids	3	5	3	5	3	5
Percent voids in mineral aggregate (VMA)			See Table 19-7			
Percent voids filled with asphalt (VFA)	70	80	65	78	65	75

1. All criteria, not just stability value alone, must be considered in designing an asphalt paving mix. Hot mix asphalt bases that do not meet these criteria when tested at 60°C (140°F) are satisfactory if they meet the criteria when tested at 38°C (100°F) and are placed 100 mm (4 in.) or more below the surface. This recommendation applies only to regions having a range of climatic conditions similar to those prevailing throughout most of the United States. A different lower test temperature may be considered in regions having more extreme climatic conditions.
2. Traffic classifications
   - Light   Traffic conditions resulting in a Design EAL $<10^4$
   - Medium Traffic conditions resulting in a Design EAL between 10^4 and 10^6
   - Heavy   Traffic conditions resulting in a Design EAL $>10^6$
3. Laboratory compaction efforts should closely approach the maximum density obtained in the pavement under traffic.
4. The flow value refers to the point where the load begins to decrease.
5. The portion of asphalt cement lost by absorption into the aggregate particles must be allowed for when calculating percent air voids.
6. Percent VMA is to be calculated on the basis of the ASTM bulk specific gravity for the aggregate.

Source: Modified from *Mix Design Methods For Asphalt Concrete and Other Hot-Mix Types.* Manual Series No. 2, 6th Ed., The Asphalt Institute.

Solution From the curves, at 4 percent air voids, the mix properties are as follows. (The design criteria from Table 19-7 are also shown.)

		Design Criteria	
		Minimum	Maximum
Asphalt content (%)	4.7	—	—
Stability (N)	10.1	8	—
Flow (0.01 in.)	9	8	14
VMA (%)	14	12	—
VFA (%)	68	65	75

The mixture would be acceptable for heavy traffic areas.

19-12 Hveem Method of Design

The Hveem method of design of paving mixtures was developed under the direction of Francis N. Hveem, formerly Materials and Research Engineer in the

TABLE 19-7 Minimum Percent Voids in the Mineral Aggregate (VMA) for the Marshall Method

Nominal Maximum Particle Size[1,2]		Minimum VMA, percent		
		Design Air Voids, Percent[3]		
mm	*in.*	*3.0*	*4.0*	*5.0*
1.18	No. 16	21.5	22.5	23.5
2.36	No. 8	19.0	20.0	21.0
4.75	No. 4	16.0	17.0	18.0
9.5	3/8	14.0	15.0	16.0
12.5	1/2	13.0	14.0	15.0
19.0	3/4	12.0	13.0	14.0
25.0	1.0	11.0	12.0	13.0
37.5	1.5	10.0	11.0	12.0
50	2.0	9.5	10.5	11.5
63	2.5	9.0	10.0	11.0

1 = Standard Specification for Wire Cloth Sieves for Testing Purposes, ASTM E11 (AASHTO M92)
2 = The nominal maximum particle size is one size larger than the first sieve to retain more than 10%
3 = Interpolate minimum VMA for design air void values between those listed.

Source: Modified from *Mix Design Methods For Asphalt Concrete and Other Hot-Mix Types,* Manual Series No. 2, The Asphalt Institute.

California Division of Highways (now California Department of Transportation). It is generally applicable to paving mixtures using either penetration grades of asphalt or liquid asphalts and containing aggregates up to 25 mm (1 in.). It is applicable not only to the high-type paving mixtures described in this chapter but also to some of the intermediate mixes of Chapter 18. It has been used principally in the design of dense-graded asphaltic mixtures. The method has been standardized as ASTM D1560 and D1561.

Determining Approximate Asphalt Content

The first step in the Hveem method is to determine the "approximate" asphalt content through determination of the centrifuge kerosene equivalent (CKE). Factors determined from the CKE test are combined with calculated surface area. The surface area of a given aggregate, or blend of aggregates, is calculated by the use of the gradation and a table of surface area factors. Surface area factors, expressed in units of square meters per kilogram (square feet per pound), to be applied to total percentages passing the various sizes of sieves are as follows:

	m^2/kg	ft^2/lb
4.75 mm (No. 4)	0.41	2
2.36 mm (No. 8)	0.82	4
1.18 mm (No. 16)	1.64	8
0.600 mm (No. 30)	2.87	14
0.300 mm (No. 50	6.14	30
0.150 mm (No. 100)	12.29	60
0.075 mm (No. 200)	32.77	160

The surface area of an aggregate of known gradation can be calculated from Table 19-8. The percentage passing each sieve (expressed as a decimal) is simply multiplied by the applicable surface area factor and the products added. It is assumed that the surface area of the material coarser than the 4.75 mm (No. 4) sieve is 0.41 m²/kg (2 ft²/lb).

To determine the CKE in a laboratory, a special centrifuging device is used. A representative sample of 100 g of aggregate passing the 4.75 mm (No. 4) sieve is saturated with kerosene and subjected to a centrifugal force of 400 times gravity for 2 min. The amount of kerosene retained, expressed as a percentage of the dry aggregate is the CKE.

The "surface capacity" of the coarse aggregate portion of the mixture is determined by first placing into a metal funnel 100 g of dry aggregate that passes the 9.5-mm (3/8-in.) sieve and is retained on the 4.75-mm (No. 4) sieve. The sample and funnel are immersed in SAE No. 10 lubricating oil at room temperature for 5 min. The funnel and sample are removed from the oil and drained for 15 min. at a temperature of 60°C (140°F). The sample is then weighed and the amount of oil retained computed as a percentage of the dry aggregate weight.

To determine the estimated design asphalt content, use the following procedure:

1. Use the CKE value and the chart of Figure 19-5 to determine the surface constant for fine material (K_f).
2. Use the percentage of oil retained in the surface capacity test for coarse aggregate and Figure 19-6 to determine the value of the surface constant for coarse material (K_c).
3. Use the values of K_f and K_c and Figure 19-7 to determine the surface constant for the combined fine and coarse aggregate, K_m. $K_m = K_f +$ correction to K_f.
4. Determine the approximate asphalt content (bitumen ratio) for the mixture based on cutback asphalts of RC-250, MC-250, and SC-250 grades. With the values determined for the surface area of the aggregate, its specific gravity, and K_m, the Case 2 procedures on the chart shown as

TABLE 19-8 Calculation of the Surface Area of the Aggregate

Sieve Size	Percent Passing	×	S.A. Factor m²/kg (ft²/lb)	=	Surface Area m²/kg (ft²/lb)
19.0 mm (3/4 in.)	100 }*		.41 (2)		.41 (2)
9.5 mm (3/8 in.)	90				
4.75 mm (No. 4)	75		.41 (2)		.31 (1.5)
2.36 mm (No. 8)	60		.82 (4)		.49 (2.4)
1.18 mm (No. 16)	45		1.64 (8)		.74 (3.6)
600 μm (No. 30)	35		2.87 (14)		1.00 (4.9)
300 μm (No. 50)	25		6.14 (30)		1.54 (7.5)
150 μm (No. 100)	18		12.29 (60)		2.21 (10.8)
75 μm (No. 200)	6		32.77 (160)		1.97 (9.6)

Surface Area = 8.67 m²/kg (42.3 ft²/lb)

*Surface area factor is .41 m²/kg (2 ft²/lb) for any material retained above the 4.75mm (No. 4) sieve.

Source: Mix Design Methods For Asphalt Concrete and Other Hot-Mix Types, Manual Series No. 2, Sixth Edition, The Asphalt Institute.

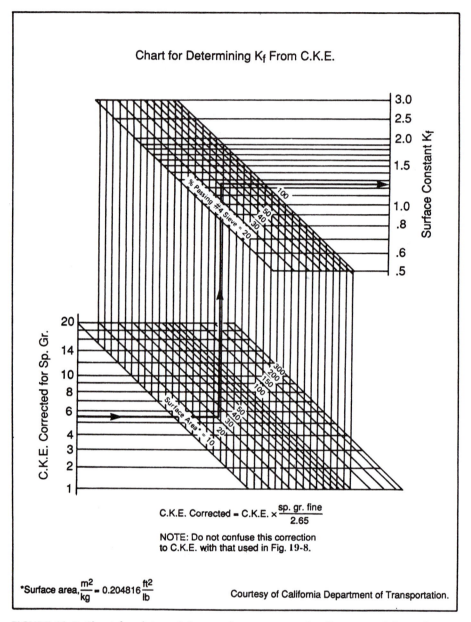

FIGURE 19-5 Chart for determining surface constant for fine material, K_f, from CKE Hveem method of design. (Courtesy The Asphalt Institute.)

Figure 19-8 are used to determine the "oil ratio." The oil ratio is the pounds of oil per 100 pounds of aggregate and applies directly to oil of SC-250, MC-250, and RC-250 grades. A correction to this value for heavier cutbacks or paving asphalts should be made using Figure 19-9.

EXAMPLE 19-2 **Use of the Hveem Design Charts** In order to illustrate this procedure, consider a mixture containing aggregates that have the following properties:

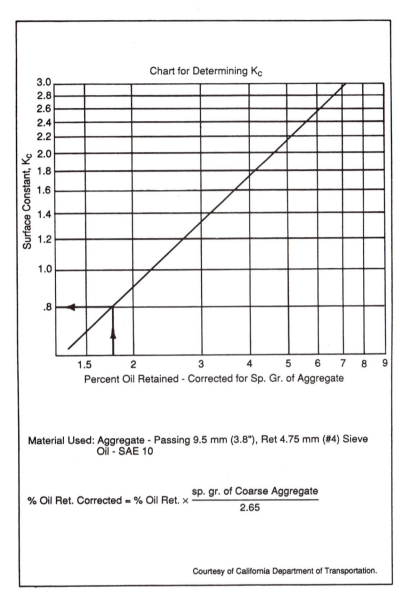

FIGURE 19-6 Chart for determining surface constant for coarse material, K_c, from C.K.E., Hveem method of design. (Courtesy The Asphalt Institute.)

Specific gravity, coarse aggregate	2.45
Specific gravity, fine aggregate	2.64
Precentage passing No. 4 sieve	45

$$\text{Average specific gravity} = \frac{100}{\dfrac{55}{2.45} + \dfrac{45}{2.64}} = 2.53$$

Surface of aggregate, m²/kg (ft²/lb)	6.6 (32.4)
CKE	5.6

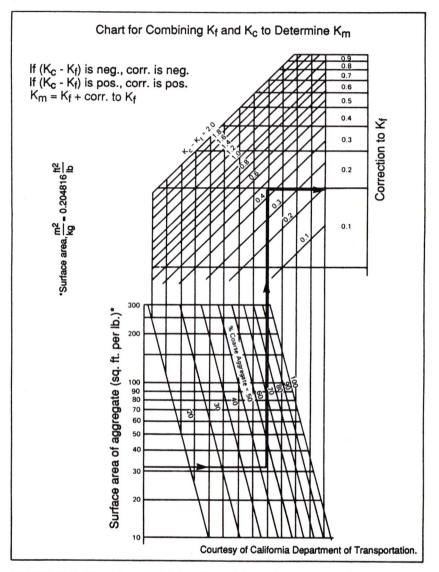

FIGURE 19-7 Chart for combining K_f and K_c to determine surface constant for combined aggregate, K_m, Hveem method of design. (Courtesy The Asphalt Institute.)

Percentage oil retained, coarse aggregate =   1.9
  Corrected for specific gravity, this value is =   1.7
  (See Fig. 19-6.)

The mixture is to contain AC-10 viscosity grade asphalt cement.
From Figure 19-5, $K_f = 1.25$.
From Figure 19-6, $K_c = 0.8$.
From Figure 19-7, $K_m = 1.15$.
From Figure 19-8, and using case 2, the oil ratio is 5.2 percent.
From Figure 19-9, the optimum asphalt content (bitumen ratio) for AC-10 asphalt is 6.1 percent by weight of dry aggregate.

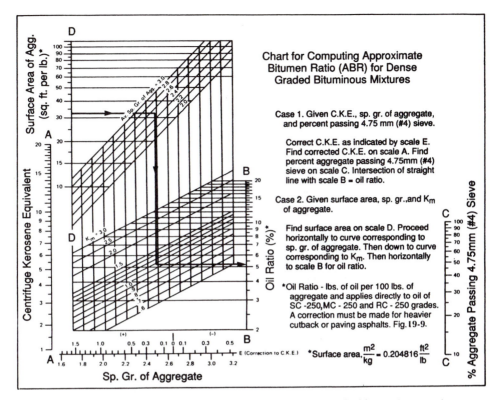

FIGURE 19-8 Chart for computing oil ratio for dense-graded bituminous mixtures, Hveem method of design. (Courtesy The Asphalt Institute.)

Stabilometer Tests

To design a paving mixture by the Hveem method, engineers prepare a series of stabilometer test specimens for a range of asphalt contents, both above and below the estimated optimum asphalt content indicated by the CKE procedure. In addition, two swell test specimens are prepared at the estimated optimum asphalt content.

Samples are prepared by mixing heated aggregates and asphalt in accordance with a standardized procedure. As part of the procedure, the mixture is cured for 2 to 3 hours at a temperature of 146°C (295°F). After curing, the mixture is heated to 110°C (230°F) preparatory to compaction. The compaction of the specimens is accomplished by use of a mechanical compactor that consolidates the material by means of a kneading action resulting from a series of individual impressions made with a ram, the face of which is shaped as a sector of a 101.6-mm (4-in.) circle. At full pressure, each application of the ram subjects the material to a pressure of 3.45 MPa (500 lb/in.²) over an area of approximately 2000 mm² (3.1 in.²). A detailed compaction procedure is followed, including final application of a static load to the specimen. Cylindrical specimens are formed 101.6 mm (4 in.) in diameter and approximately 63.5 mm (2.5 in.) high.

The compacted test specimens are used in the performance of four tests: swell test, stabilometer test, bulk density determination, and cohesiometer test.

In the swell test, the mold and compacted specimens are placed in a deep aluminum pan. A tripod arrangement is used to record the change in height of the specimen during the test. The upper portion of the mold is partially filled

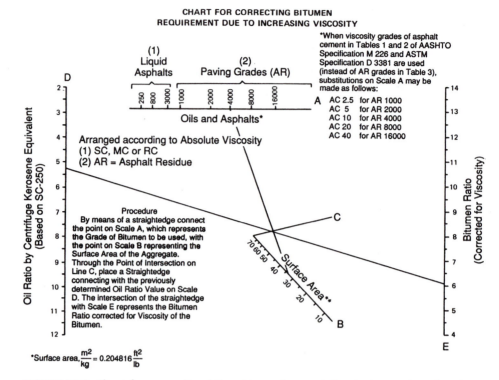

FIGURE 19-9 Chart for correcting bitumin requirement due to increasing viscosity of asphalt, Hveem method of design. (Courtesy The Asphalt Institute.)

with water, which is allowed to remain on top of the specimen for 24 hr. At the end of that time, the change in height of the specimen, or "swell," is determined. A measurement is also made of the amount of water that moves through the specimen during the 24-hr period, or the "permeability" of the sample.

Stabilometer test specimens are heated to 60°C (140°F) before testing. A schematic drawing of the Hveem stabilometer is shown in Figure 19-10; the device was described briefly in Chapter 16. In testing asphaltic mixtures, the stability value resulting from the test is dependent on the ratio between the horizontal and vertical pressures at 2.76 MPa (400 lb/in.²). Detailed procedures for conducting the stabilometer test are given in Ref. 1.

Bulk density determinations are made on the stabilometer test specimens after completion of the stabilometer tests and as soon as the specimens have cooled to room temperature using standard AASHTO procedures (*4*).

Design criteria for hot-mix asphaltic paving mixtures by the Hveem method are as follows:

	Light Traffic	*Medium Traffic*	*Heavy Traffic*
Stabilometer value	30+	35+	37+
Swell	Less than 0.762 mm (0.030 in.)		

An effort is also made to provide a minimum percentage of voids in the total mix of approximately 4 percent. In the application of these criteria, the optimum

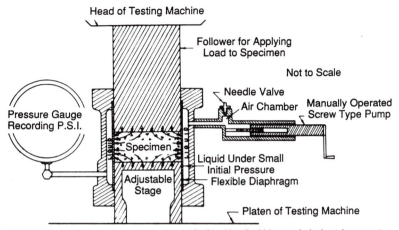

Head of Testing Machine

Follower for Applying
Load to Specimen

Not to Scale

Needle Valve

Air Chamber

Manually Operated
Screw Type Pump

Pressure Gauge
Recording P.S.I.

Specimen

Liquid Under Small
Initial Pressure

Adjustable
Stage

Flexible Diaphragm

Platen of Testing Machine

NOTE: The specimen is given lateral support by the flexible sidewall, which transmits horizontal pressure to the liquid. The magnitude of the pressure can be read on the gauge.

FIGURE 19-10 Diagrammetric sketch of Hveem stabilometer (not to scale). (Courtesy The Asphalt Institute.)

asphalt content of the mix should be the highest percentage of asphalt the mix will accommodate without reducing the stability or void content below minimum values. Generally, the voids in the total mix should approach the 4 percent minimum; however, in cases where the mixture shows a marked decrease in stability for slight increases in asphalt content, it is advisable to adjust the aggregate grading or reduce the asphalt content to provide a less critical mixture with more air voids.

The Hveem method of mix design is a lengthy and complex procedure. Here we have provided an abbreviated description of the most significant steps that are involved in the procedure. Readers interested in more detail are referred to The Asphalt Institute's Manual Series No. 2 (*1*).

19-13 Construction Methods for Hot-Mix, Hot-Laid Asphaltic Concrete

All the mixtures described in this chapter are, of course, proportioned and mixed in central mixing plants. Construction procedures discussed here are intended to be generally applicable to all the hot-mix, hot-laid pavements described in this chapter and are also largely applicable to the cold mixtures that are discussed. The necessary construction steps are described in some detail in this section; deviations from the basic procedure indicated will be mentioned in subsequent sections of this chapter. The specific procedures described here are primarily intended to be applicable to the construction of wearing surfaces, although the same procedures are also generally applicable to the construction of base and leveling courses. The fundamental steps in the construction of a high-type bituminous pavement may be listed as follows:

1. Preparation of the mixture.
2. Preparation of base or leveling course.
3. Transportation and placing of the surface-course mixture.
4. Joint construction.
5. Compaction and final finishing.

19-14 Preparation of Mixture

The essential elements of a central mixing plant have been described in Chapter 18. Plants used in the preparation of high-type hot mixtures are essentially the same as those previously described, except that certain additional units may be added to the plant and the entire setup geared to the necessity for volume production of uniform hot mixtures with very close control over the proportioning and mixing steps.

Central mixing plants for the preparation of high-type bituminous paving mixtures may be of continuous or batch type. Figure 19-11 shows the flowchart for a continuous-flow plant. The gradation unit shown is a four-bin element, providing for four sizes of aggregate. This is an older type of continuous plant which uses a pugmill type of mixer as a final step in the mixing process. These units are increasingly being replaced by drum drier plants, which were described in Section 18-14.

Refer again to Figure 18-8, which shows a cutaway view of the heart of a batch-type hot-mix plant. Operation of this plant, as with certain plans produced by other manufacturers, is almost completely automatic. Batch plants are widely used for asphalt production; however, like the older models of continuous plants, batch plants are gradually being replaced by drum drier plants. A photograph of a drum drier mixing plant is shown as Figure 19-12.

Two of the units that are frequently included in a hot-mix plant and that have not previously been mentioned are the dust-collecting system and a device for feeding mineral filler or "dust" into the mixture. The dust collector is generally operated adjacent to and in conjunction with the aggregate drier and is necessary for efficient plant operation in many cases. The collector serves to eliminate or abate the dust nuisance that exists around the plant and is especially important in plants that are located in or adjacent to municipalities. Modern dust-collection systems are highly efficient; frequently the fine material collected, if uniform in gradation, is fed back into the plant at the entrance to the hot elevator. To meet air pollution control requirements, many urban plants have "wet wash" systems in which exhaust from the dust collector enters a tower and passes through a series of water sprays that remove the remaining dust.

Mineral filler that is added to the mixture is not normally passed through the aggregate drier, this material being fed by a separate device directly into the mixing unit in some cases or into the aggregate batching unit. Separate feeder units are also used in the preparation of cold mixtures to supply liquefier, other fluxes, or hydrated lime to the mix. Central mixing plants for asphaltic mixtures are quite complex in nature, and no further detailed description is deemed desirable in this text. The description of the basic units given in the preceding chapter and the information given in this chapter should suffice to give the reader a good understanding of the basic nature of these plants.

Control of the uniformity of hot mixes of this type is important, as any appreciable change or variation in gradation or asphalt content will be reflected in a change in some other characteristic of the mix. The control of the plant to produce the job mix chosen is a cooperative effort of the plant operator and inspector and requires continual inspection and careful control.

Sampling and testing are among the most important functions in plant control. Data from these tests are the tools with which to control the quality of the product. Samples may be obtained at various points in the plant to establish if the processing is in order up to those points. A final check of the mixture is

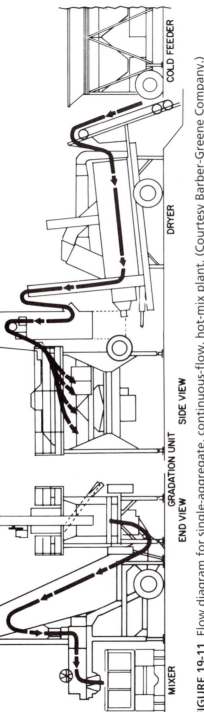

FIGURE 19-11 Flow diagram for single-aggregate, continuous-flow, hot-mix plant. (Courtesy Barber-Greene Company.)

FIGURE 19-12 Photograph of a drum-drier mixing plant. (Courtesy Cedarapids, Inc.)

needed to determine its uniformity, its gradation, and its asphalt content. An extraction test measures the asphalt content and provides aggregate from which the gradation may be determined.

The results of extractions and gradations should fall within the job-mix tolerance specified; if they do not, corrective measures must be taken to bring the mix within the uniformity tolerance. Adjustments are frequently necessary, particularly during the early stages of plant operation or if changes occur in the aggregates. For a complete discussion of this phase of construction, see Ref. 5.

The temperature-viscosity relationship for the asphalt should be used to arrive at a proper mixing temperature, since asphalts vary as to source and type and grade. Generally speaking, higher viscosities are usually more suitable for coarse aggregate mixes and the lower viscosities for fine aggregate mixes. Many agencies now specify a mixing temperature, based on viscosity, as part of the job-mix formula and allow a variation of $\pm 20°F$ from that figure.

It is likely that a majority of mixtures of this type are now prepared at temperatures between 121° and 135°C (250° and 275°F) and the overall range is perhaps 107° to 163°C (225° to 325°F). Modern plants incorporate recording temperature gages that provide a continuous record of asphalt and aggregate temperatures.

19-15 Temperatures and Mixing Times

As has been indicated, the aggregate and bituminous material are handled separately in the plant, prior to their combination in the mixing unit. The bituminous material is uniformly heated to a specified temperature in a tank or "kettle," whereas the aggregates are heated and dried in the drier unit. The temperature of the asphalt is carefully controlled in order to prevent overheating or "burning" of the material, with consequent destruction of certain desirable qualities. The temperature of the asphalt prior to its entrance into the mixer is generally from

107° to 177°C (225° to 350°F), and many modern specifications place the upper limit of this temperature at 149° or 163°C (300° or 325°F).

In batch-type plants, the dry aggregates are proportioned into the mixer unit and thoroughly manipulated for a brief time prior to the introduction of the bituminous material. Time for this dry-mixing operation generally would be from 15 to 20 sec. The bituminous material is then introduced into the mixer and the material mixed for a period that generally is a minimum of 45 sec. Total time in the mixer in a batching plant would then generally be from 55 to 70 sec and may be longer if necessary to produce a uniform and homogeneous mix. Similar times are required in continuous mixing units. Time in the mixer should be held to a minimum consistent with proper mixing.

19-16 Automatic Control Systems

Several states now require automatic controls on hot-mix plants used to produce mixtures on state highway contracts, and contractors in other states have installed automatic controls even though they are not required. Batch plants may be manual, semiautomatic, or automatic depending on the degree of automation. In manual plants, air cylinders or hydraulic cylinders, actuated by the operator by electric switches, control bin gates, fines feeders, asphalt supply and spray valves, the weigh box discharge gate, and the pugmill discharge gates (5).

In semiautomatic batch plants, the various operations constituting each mixing cycle are under automatic control. Thus, the quantities of asphalt and aggregate introduced into the mixes, the mixing times, the sequencing of the mixing functions, and the operation of the pugmill discharge gate are controlled automatically.

The fully automatic batch plant repeats the weighing and mixing cycle until the operator stops it or until it stops itself because of material shortage or some other extraordinary event. Automatic plants also normally provide a record of the amount of material incorporated into each batch.

In drum-mix plants, aggregate must be properly graded prior to its entry into the mixing drum. This is normally accomplished automatically with precision belt feeders that control the amounts of each aggregate being fed into the mixer. The aggregate is weighed before drying, and the weights of aggregate are electonically corrected to dry aggregate readings.

The asphalt being delivered to the drum mixer is controlled by a metering system that is interlocked with the aggregate weigh system to ensure the exact asphalt content in the mix.

A sensing device located at the discharge end of the drum mixer continuously monitors the temperature of the mix. The flow of asphalt mix is delivered to a surge silo, where a weigh system monitors the amount of material delivered into each truck (5).

19-17 Transportation of Mixture

The mixture is discharged from the plant into dump trucks or trailers for transportation to the job site. The vehicles that are used are required to have tight and smooth metal beds that have previously been cleaned of all foreign material. The vehicle bed may be sprayed with a light coat of limewater, soap solution, or some similar substance to prevent adherence of the mixture. Fuel oils should not be used as they have a detrimental effect on the mixture. Sometimes the vehicle is required to be insulated against excessive heat loss in the mixture during hauling and frequently is covered with a canvas to protect the mix from the weather.

19-18 Preparation of Base

Surface courses of asphaltic concrete are frequently placed on new or existing bases that require very little preparation prior to the placing of the new surface other than thorough sweeping and cleaning to remove loose dirt and other foreign materials. In other cases, the existing base or surface on which the mixture is to be placed may require extensive corrective measures. Quite frequently, when the existing surface is disintegrated, broken, or irregular in nature, specific defects may be corrected by the application of "patches" of asphaltic concrete. Excessive joint sealing compound and fatty areas should also be removed. In certain cases it may be desirable to place a leveling course of asphaltic concrete to correct irregularities in the existing surface. At points where the asphaltic concrete mixture is to be placed in contact with surfaces of manholes, curbs, gutters, and such, these surfaces are generally painted with a light coat of hot asphalt cement or liquid asphaltic material.

19-19 Placing of Mixture

Placing of the asphaltic concrete mixture is permitted only when the base is dry and under favorable weather conditions. Placing of hot mixtures is usually suspended when the air temperature becomes less than 4°C (40°F), except under unusual circumstances. Reference 6 gives specific laydown temperatures required for various conditions. It specifies minimum temperatures of the surface of the base for mixture placement, which vary with the mixture temperature measured just prior to dumping in the spreader and the lift thickness. The values range from 2° to 49°C (35° to 120°F), depending on these variables.

Base, leveling, and surface courses are placed and compacted in separate operations. In certain cases, unusually thick layers composed of the same mixture may be placed in two or more courses.

The steps of spreading and finishing are, in the large majority of cases, accomplished by the use of mechanical spreading and finishing machines (pavers) such as were described in the preceding chapter. These machines generally process one lane width at a time and must be capable of producing a smooth, even surface that has the desired elevation, grade, and cross section. This is best accomplished if the speed of laying is adjusted to the output of the plant, thus avoiding frequent long stops. Figure 19-13 shows a Blaw-Knox paver with a 5.8-m (19-ft) screed engaged in the spreading and finishing of an asphaltic concrete wearing surface. The surface is checked during rolling, and any irregularities, "fat" spots, or similar defects are corrected. In areas where the mechanical paver cannot operate satisfactorily, the mixture is placed on dump boards, spread, and raked by hand.

In placing the asphaltic concrete mixture, special attention must be given to the construction of joints between old and new surfaces or between successive days' work. It is essential that a proper bond be secured at longitudinal and transverse joints between the newly placed mixture and an existing surface, regardless of its nature, and special procedures, generally performed by hand, are used to ensure the formation of adequate joints.

The best longitudinal joint is formed when the material in the edge being laid

FIGURE 19-13 Bituminous paver with 5.8-m (19-ft) screed. (Courtesy Blaw-Knox Company.)

against is still warm enough for effective compaction. This means that two finishers working in echelon can produce the most satisactory joint. When one finisher is used, a procedure of moving from one lane to the other may be used so that the edge of the previously placed lane is still hot when the new lane is placed. Various other procedures are used when laying against a cold edge, such as cutting the cold edge back and painting with bituminous material. Some agencies allow the use of an infrared joint heater attached to the front of the finisher, which heats the cold material along the edge prior to placing a new lane.

19-20 Compaction of the Mixture

When the spreading and finishing operations just described have been completed, and while the mixture is still hot, rolling is begun.

Rolling may be carried out by steel-wheel or pneumatic-tired rollers or by a combination of the two. Steel-wheel rollers may be of three types: three-wheel rollers of 4.5 to 5.4 metric tons (10–12 tons) in mass, two-axle tandem rollers of 3.6 to 5.4 metric tons (8–12 tons), and three-axle tandem rollers of 5.4 to 8.2 metric tons (12–18 tons).

Pneumatic-tired rollers are becoming more popular as they provide a closely knit surface by kneading aggregate particles together. They will achieve, during the rolling operation, a more uniform density than usually develops after vehicular traffic has used the asphalt surface for some time. The contact pressure exerted by a pneumatic-tire compactor is dependent on tire size, ply rating, wheel load, and tire inflation pressure. Tire contact pressures generally range from 276 to 620 kPa (40–90 lb/in.²). Use of a self-propelled pneumatic-tire roller for this purpose is shown in Figure 19-14.

FIGURE 19-14 Pneumatic-tired roller on a Kansas paving project. (Courtesy *Engineering News-Record.*)

Use is now being made of vibratory rollers for compacting bituminous pavements. These vibratory rollers provide a centrifugal force of up to 209 kN (47,000 lb) and vibrations up to 3000 per min. These vibratory rollers may replace other conventional compaction equipment to some extent in the future.

Rolling of the bulk of the pavement is done in a longitudinal direction, beginning at the edges and gradually progressing toward the center, except on superelevated curves, where rolling begins on the low side and progresses toward the high side; successive trips of the roller overlap to ensure complete and uniform coverage. Rolling progresses at a uniform, slow rate of speed, with the drive roll or wheel nearest the paver.

Rolling procedures vary with the properties of the mixture, thickness of layer, and other factors. In modern practice, rolling is divided into three phases, which follow closely behind one another: initial or "break-down" rolling, intermediate rolling, and finish rolling. The breakdown and intermediate phases provide needed density and the final rolling gives the required final smoothness. A procedure used by most organizations calls for the use of three-wheel or tandem rollers for breakdown, pneumatic-tired rollers for intermediate rolling, and tandem rollers for finish rolling. Use of a steel-wheel roller in compaction of a hot-mix asphaltic pavement is illustrated in Figure 19-15.

Standards applied to the completed wearing surface are as high as is deemed practically possible. Requirements related to the finished surface frequently stipulate that the surface shall be smooth, even, and true to the desired grade and cross section, with no deviation of more than $\frac{1}{8}$ in. from a line established by a template or a 3-m (10-ft) straightedge being permitted. This seems to be a very high standard, but such surfaces are readily attainable in most cases with modern construction equipment. Deviations from the specified thickness of surface course are generally permitted to be no more than 6 mm ($\frac{1}{4}$ in.). The density to be obtained in the completed wearing surface may be stipulated in terms of a percentage of the theoretical maximum density or a percentage of the density of the laboratory-compacted mixture. A typical requirement is that the density of the completed surface course should not be less than 95 percent of the laboratory compacted density for the same mixture. Requirements related to the density of the completed pavement are very important and must be such as to guarantee,

FIGURE 19-15 Three-axle tandem roller compacting an asphaltic mixture. (Courtesy Galion Iron Works and Manufacturing Company.)

on the one hand, the attainment of a satisfactory density, and on the other, that a density is not specified that cannot be economically secured or that is impossible of attainment without crushing the aggregates.

The density of the compacted mixture is determined on samples cored or sawed from the completed mat by the procedure discussed in Section 19-9. Many agencies, in addition to determining density on compacted mixtures, determine for control purposes the percentage of voids and voids filled and obtain large enough samples to determine the gradation and asphalt content of the compacted mixture. Experimental use is being made of nuclear density gages and airflow meters (air permeability measuring devices) to measure density and control the field rolling process.

Rolling completes the construction of most asphaltic concrete pavements, and traffic may be permitted on the surface as soon as the compacted mixture has adequately cooled. In some cases, a seal coat is also applied to the newly completed pavement, as described in a previous chapter.

SHEET ASPHALT

Sheet asphalt is a type of paving mixture that is composed of carefully graded fine aggregate (sand), mineral filler, and asphalt cement. It is used almost exclusively as a surface mixture, being laid on a binder course which is generally 25 to 38 mm (1 to 1.5 in.) thick or directly on a prepared base course, such as an old concrete or brick pavement. The sheet-asphalt pavement itself may be from 13 to 76 mm (0.5 to 3 in.) thick, although the majority of these surfaces are from 25 to 38 mm (1 to 1.5 in.) thick. A binder course used in conjunction with a sheet-asphalt surface is generally composed of a mixture that closely resembles an open-graded asphaltic concrete mixture, as no mineral filler is generally used in the binder course.

At the present time, sheet asphalt is principally used in the surfacing of city streets, and many of the principal streets in the larger cities of the United States are paved with this material. Properly constructed, it is capable of withstanding the effects of very heavy wheel loads and very heavy volumes of traffic without damage and with very low maintenance costs. Surfaces of this type are frequently preferred in cities because of their added virtue of being extremely smooth, noiseless riding surfaces that are waterproof and easily cleaned. Sheet-asphalt surfaces are somewhat more difficult to design, and their construction is slightly more difficult to control than that of asphaltic concrete mixtures of comparable stability and durability. One of the principal faults of sheet asphalt is its suscep- tibility to contraction cracking in cold weather.

Sheet-asphalt mixtures are distinguished from the asphaltic concrete mixtures that have been discussed by the complete lack of coarse aggregate and very close control of the gradation of the sands used. More mineral filler is used in these mixes than in somewhat comparable asphaltic concretes, and the percentage of mineral filler frequently approaches 16 in a sheet-asphalt mix. The asphalt cement that is used may be slightly harder than that used in asphaltic concrete, asphalt cements with penetration ranges of 50 to 60 and 60 to 70 having been quite commonly used. Asphalt contents are also generally somewhat higher than for dense-graded asphaltic concrete mixtures. Construction procedures related to the preparation and placing of sheet-asphalt mixtures are practically the same as those described for asphaltic concrete pavements. Somewhat higher temperatures are sometimes used in heating the sand and asphalt cement that are to be used in a sheet-asphalt mix. As a final step, the surface of a sheet-asphalt pavement may be lightly brushed with portland cement or limestone dust immediately after completion of the rolling.

SAND ASPHALT

In some sections of the country, a type of hot plant mixture that is of importance is sand asphalt. "Sand asphalt" may be defined as a surface or base course formed from a mixture of sand and asphalt cement, with or without mineral filler. This type of mixture has been extensively used in areas where suitable sands are economically available on highways that carry moderate amounts of traffic. The construction of sand-asphalt pavements is accomplished by methods that closely resemble those used in building sheet-asphalt surfaces, the chief difference be- tween the two being that sand-asphalt mixtures are not as carefully specified or controlled and consequently are usually somewhat less dense and less stable than comparable sheet asphalts.

The usual thickness of a base course of this type is from 76 to 102 mm (3–4 in.), whereas a surface course would generally be about 50 mm (2 in.) in com- pacted thickness. Surface courses of sand asphalt usually contain from 5 to 10 percent of asphalt, and base courses may contain slightly less. Requirements related to the composition of sand-asphalt mixtures vary somewhat with different organizations, as might be expected, because it is usually desired to use local sand in a mixture of this kind. Some specifications are quite open in nature, whereas others quite closely control the grading of the sand used. A requirement related to density and stability is also frequently included as part of a specification for sand asphalt, with requirements in this respect being generally somewhat less severe than for sheet asphalt or asphaltic concrete mixtures.

TABLE 19-9 Gradation Requirements for Sand Asphalt

| Sieve Size | Percentage Passing (by weight) | |
	Type I	Type II
¼ in. (6.25 mm)	100	100
No. 8 (2.36 mm)	75–100	50–90
No. 16 (1.18 mm)	60–90	25–65
No. 30 (0.600 mm)	45–75	15–45
No. 50 (0.300 mm)	15–45	5–30
No. 100 (0.150 mm)	5–15	3–20
No. 200 (0.075 mm)	2–6	2–6

Source: Kentucky Standard Specifications for Road and Bridge Construction. Transportation Cabinet, Department of Highways, Frankfort (1994).

The variation in specifications of this sort makes it undesirable to attempt to include a "typical" specification for this type of mix. One specification has been selected, however, which is fairly representative of those in current use. Table 19-9 gives the requirements related to gradation of sand asphalt as contained in the *Kentucky Standard Specifications for Roads and Bridges.*

PROBLEMS

19-1 Given below are the requirements of a specification related to the grading of the mineral aggregates in an asphaltic concrete mixture and the sieve analyses of two aggregates (A and B) that are economically available for this use. Determine the range of blends of aggregates A and B that will produce a combined aggregate that will meet the limits of the specification, and given the gradings of the aggregate combinations selected.

| Sieve Designation | Percentage Passing (by weight) | | |
	Mix	Aggregate A	Aggregate B
1 in. (25 mm)	100	100	—
¾ in. (19 mm)	95–100	95	100
⅜ in. (9.5 mm)	60–75	72	76
No. 4 (4.75 mm)	40–55	54	52
No. 10 (2.00 mm)	30–40	43	38
No. 40 (0.425 mm)	12–22	30	18
No. 200 (0.075 mm)	3–6	8	5

19-2 A design is being prepared for an asphalt concrete paving mixture. The following ingredients are to be used in the preparation of a trial mixture:

	Percentage of Total Mix by Weight	Specific Gravity
Coarse aggregate	56.0	2.611 (bulk)
Fine aggregate	30.0	2.690 (bulk)
Mineral filler	7.0	3.100
Asphalt cement	7.0	1.030

By ASTM D2041, the maximum specific gravity of the paving mixture $G_{mm} = 2.478$, and by ASTM D2726, the bulk specific gravity of the compacted paving mixture sample

G_{mb} = 2.384. Compute the percentage of voids in the compacted mineral aggregate and the percentage of voids in the compacted mixture. What percentage of the air voids are filled with asphalt?

19-3 A sample cut from the completed pavement constructed by using the mix of Problem 19-2 weighs 3540 g in air and 1962 g in water. Compare the density obtained in the field with that of a laboratory-compacted specimen of the same mixture. Has the rolling process achieved a satisfactory degree of density in this case?

19-4 An asphaltic concrete surface course mixture is being designed by the Marshall method. The aggregate combination in a trial mixture contains 40 percent coarse aggregate, 52 percent fine aggregate, and 8 percent mineral filler; specific gravities of these materials are 2.68, 2.72, and 2.83, respectively. Asphalt cement to be used has a specific gravity of 1.02.

Results of tests on specimens prepared from this aggregate combination and various percentages of asphalt (by weight of total mixture) are as follows:

Percentage of Asphalt	Specimen Weight (g)		Stability		Flow (1/100 in.)
	In Air	In Water	kN	(lb)	
4.5	1308.7	765.0	7.659	(1722)	8
5.0	1314.2	772.3	7.851	(1765)	9
5.5	1318.9	779.7	7.864	(1788)	11
6.0	1323.3	782.1	7.259	(1632)	14
6.5	1328.6	782.9	6.254	(1406)	18

The mixture is to be designed for very heavy traffic. Determine the optimum asphalt content for this mixture. Is the mixture satisfactory?

19-5 A paving mixture is to be designed by the Hveem method. Gradation of the aggregate is as follows:

Sieve	Percentage Passing (by weight)
¾ in. (19.0 mm)	100
½ in. (12.5 mm)	94
⅜ in. (9.5 mm)	78
No. 4 (4.75 mm)	62
No. 8 (2.36 mm)	45
No. 16 (1.18 mm)	34
No. 30 (0.60 mm)	25
No. 50 (0.30 mm)	18
No. 100 (0.150 mm)	14
No. 200 (0.075 mm)	8

The CKE for the aggregate is 3.8. Specific gravity of the coarse aggregate is 2.62, fine aggregate, 2.72. The percentage oil retained by the coarse aggregate is 2.5.

Estimate the quantity of asphalt required for this mixture, assuming that the binder is to be (1) a liquid asphalt, grade MC-800; (2) an asphalt cement, penetration grade 85-100.

Suppose the stabilometer value of the mixture of (2), using the computed quantity of asphalt, is 42; is the mixture satisfactory for heavy traffic? What other requirements pertain to the mixture?

19-6 Visit a plant in your area that is producing hot-mix asphaltic concrete or a similar high-type mixture. Prepare a report describing the controls that are applied to ensure production of a mixture of uniformly high quality. Include such factors as aggregate gradings and proportions, quantity of bituminous material, temperatures, mixing times, and so on.

19-7 Describe in detail the laydown procedure used for a surface of asphaltic concrete, or a similar type of mixture, in your area. Give facts related to temperature of the mix, spreader, rolling equipment and procedure, variations in the surface of the compacted layer, density achieved by rolling, and so on.

REFERENCES

1. *Mix Design Methods for Asphalt Concrete and Other Hot-Mix Types.* Manual Series No. 2, 6th Ed., The Asphalt Institute, Lexington, KY.

2. *Kentucky Standard Specifications for Road and Bridge Construction.* Transportation Cabinet, Department of Highways, Frankfort, KY (1994).

3. *Standard Specifications for Highways and Bridges.* Alabama Highway Department, Montgomery, AL (1992).

4. *AASHTO Materials, Part II, Tests,* 16th Ed., American Association of State Highway and Transportation Officials, Washington, DC (1993).

5. *Asphalt Plant Manual.* Manual Series No. 3, 5th Ed., The Asphalt Institute, Lexington, KY (December, 1986).

6. *Standard Specifications for Construction of Roads and Bridges on Federal Highway Projects.* Federal Highway Administration, Washington, DC (1992).

CHAPTER 20

CONCRETE PAVEMENTS

The many and varied uses of portland cement concrete in the United States have made this material such an integral part of our everyday lives that it hardly seems necessary to define it. "Portland cement concrete" may be defined, however, as a plastic and workable mixture composed of mineral aggregate such as sand, gravel, crushed stone, or slag, interspersed in a binding medium of cement and water. When first combined, the materials listed form a plastic, workable mass that may be easily handled and shaped into any desired form. A short time after mixing, the concrete begins to stiffen or "set" because of chemical reaction between the cement and water in the mixture, and in a relatively short time it forms a dense, hard mass that possesses considerable compressive and flexural strength.

When properly designed and constructed, concrete roads and streets are capable of carrying almost unlimited amounts of any type of traffic with ease, comfort, and safety. Surfaces of this type are smooth, dust-free, and skid-resistant, having a high degree of visibility for both day and night driving and generally having low maintenance costs. They are economical in many locations because of their low cost of maintenance and their relative permanence. They are, of course, classed as high-type pavements. The principal use of surfaces of this type has been in the construction of heavily traveled roads and city streets, including those in residential, business, and industrial areas. It is the standard material for urban expressways, even in states where asphalt surfaces are widely used. A wearing surface of portland cement concrete usually consists of a single layer of uniform cross section that has a thickness of 152 to 279 mm (6 to 11 in.) and is placed either directly on the prepared subgrade or on a single layer of granular or stabilized material. This underlying layer is called the base course by some and the subbase by others. A new concrete base course may be constructed to serve as a support for one of the several types of bituminous wearing surfaces. Old concrete pavements have been extensively used as bases for new bituminous wearing surfaces in many areas.

A further distinction may be noted between the surfaces and bases that are discussed in this chapter and the so-called flexible pavements that have previously been described in this book. Concrete surfaces and bases are frequently classed as rigid pavements, the term "rigid" implying that pavements constructed of this material possess a certain degree of "beam strength" that permits them to span or "bridge over" some minor irregularities in the subgrade or subbase on which they rest. Thus minor defects or irregularities in the supporting foundation layer may not be reflected in the surface course, although, of course, defects of this type are certainly not desirable, as they may lead to failure of the pavement through cracking, breaking, or similar distress.

Concrete pavements can be subdivided into three types:

1. *Jointed plain concrete pavements (JPCP)*
2. *Jointed reinforced concrete pavements (JRCP)*
3. *Continuously reinforced concrete pavements (CRCP)*

Jointed plain concrete pavements are characterized by short transverse joint spacing, typically of 15 to 20 ft, with no reinforcing steel in the concrete slab, although some types of reinforcing steel may be used across the joints. Jointed reinforced concrete pavements are characterized by long joint spacing, typically of 9.1 to 30.5 m (30 to 100 ft), with reinforcing steel in the concrete slab to hold the cracks developed in the concrete by shrinkage tightly together. Continuously re-inforced concrete pavements theoretically have no joints and contain a greater percentage of steel to control cracking from shrinkage of the concrete.

The design and construction of concrete pavements is a fairly complex subject, and it seems desirable to list the major topics that will be covered in this chapter. Major subjects that will be discussed in subsequent sections include materials, proportioning of concrete mixtures, structural pavement design, and the construction of portland cement concrete pavements.

In the interests of simplificaiton, in the remainder of this text the term ''concrete'' will be taken to be synonymous with ''portland cement concrete.''

MATERIALS

The materials included in concrete, as generally used in highway construction, are coarse aggregate, fine aggregate, water, cement, and one or more admixtures. These materials will be discussed in turn in the following sections, along with a discussion of typical requirements of specifications related to each.

20-1 Coarse Aggregate

Coarse aggregates most frequently used in portland cement concrete include crushed stone, gravel, and blast-furnace slag. Other similar inert materials may also be used, and the listed materials may be used singly or in combination with one another. Specific requirements related to coarse aggregations to be used for this purpose may be divided into five groups: deleterious substances, percentage of wear, soundness, weight per cubic foot (slag), and grading.

As mentioned in Chapter 15, highway agencies specify the maximum percentages of various deleterious substances that are permitted to be present in aggregates. Specific requirements related to the cleanliness of the aggregate vary somewhat with different agencies. The requirements in Table 20-1 related to deleterious substances are indicative of those in current use (*1*).

In certain areas, considerable difficulty has been experienced with aggregates containing deleterious substances that react harmfully with the alkalies present in the cement. Such reactions generally result in abnormal expansion of the concrete. Methods have been devised (ASTM Methods C227 and C289) for detecting aggregates with these harmful characteristics, and suitable stipulations are included in typical specifications (for example, ASTM C33).

The ability of the coarse aggregate to resist abrasion is generally controlled

TABLE 20-1 Limits for Deleterious Substances and Physical Properties of Coarse Aggregate for Concrete

Typical Uses	Weathering Exposure	Maximum Allowable (percent)		
		Clay Lumps and Friable Particles	Chert (less 2.4 sp. gr. SSD)[a]	Sum of Clay Lumps, Friable Particles, and Chert[a]
Bridge decks and other uses where surface disfigurement due to pop-outs is objectionable	Severe	2.0	3.0	3.0
	Moderate	3.0	3.0	5.0
	Negligible	5.0	5.0	7.0
Pavements, base courses, and sidewalks where a moderate number of pop-outs can be tolerated	Severe	3.0	3.0	5.0
	Moderate	5.0	5.0	7.0
	Negligible	5.0	8.0	10.0
Concealed concrete not exposed to weather	All types	10.0	—	—

General requirements, all types of construction and weathering regions:
Maximum allowable material finer than No. 200 sieve (0.075 mm) = 1.0%
Maximum allowable coal and lignite = 0.5%
Maximum wear, Los Angeles abrasion test = 50%
Maximum loss, magnesium sulfate soundness test (five cycles) = 18%

[a]These limitations apply only to aggregates in which chert appears as an impurity. They are not applicable to gravels that are predominantly chert. Limitations on soundness of such aggregates must be based on service records in the environment in which they are used. (SSD = saturated, surface dry.)

Source: AASHTO Materials, Part I, *Specifications,* American Association of State Highway and Transportation Officials (1982).

by inclusion of the specifications of a maximum permissible percentage of wear in the Los Angeles abrasion test. Maximum permissible percentages of wear in this test (AASHTO Designation T96) range from as low as 30 to as high as 65 in the specifications of the various state highway departments. On the average, however, the maximum permissible loss varies from 40 to 50 percent for all three of the principal types of coarse aggregate, although many specifications do not contain a requirement related to the percentage of wear of slag.

A number of states, particularly those in the northern portion of the country, include requirements related to soundness, as measured by the use of sodium (or magnesium) sulfate, in their specifications for coarse aggregate. The soundness test used is an accelerated weathering test. The weighted loss in the sodium sulfate soundness test (AASHTO Designation T104) may generally be said to have a maximum permissible value ranging from 10 to 15 percent, as measured at the end of five cycles of alternate wetting and drying. A small number of states use a freezing and thawing test for the same purpose. In states in which slag is used as a course aggregate, specifications generally require that the unit weight of this material be not less than 1120 or 1200 kg/m³ (70 or 75 lb/ft.³).

As would be expected, requirements related to the grading of the coarse aggregates used in portland cement concrete vary considerably with different organizations. The material is generally required to be well graded from coarse to fine, with the maximum size of coarse aggregate permitted by most highway organizations being from 75 to 50 mm (3–2 in.). In this work the division between coarse and fine aggregate is made on the No. 4 sieve. That is, material which is retained on a No. 4 sieve is classed as coarse aggregate. Grading requirements are generally quite open, permitting the use of a large number of locally available aggregates which have proved to be satisfactory for use in portland cement concrete.

Typical gradings for coarse aggregate are given in AASHTO Designation M43 (*1*).

20-2 Fine Aggregate

The fine aggregate that is most generally used in concrete is sand (either natural or manufactured) composed of tough, hard, and durable grains.

The cleanliness of the sand used may be controlled by the inclusion of requirements related to the maximum amounts of various kinds of deleterious substances that may be present, in a fashion similar to that indicated above for coarse aggregate. Particular emphasis is frequently placed upon the maximum percent of "silt" (material passing the No. 200 sieve) that may be contained in the fine aggregate. This amount must generally not exceed 2 to 5 percent of the total. An additional cleanliness requirement is generally imposed on sands, in that excess amounts of organic impurities must not be present. The amount of organic material is generally controlled by subjecting the material to the colorimetric test for organic impurities (AASHTO Designation T21).

Further requirements of specifications related to sands may include stipulations as to soundness. The soundness of these materials is controlled by specifying the maximum loss that will be permitted in five cycles of alternate wetting and drying in the sodium sulfate soundness test. A typical requirement is that the loss in the sodium sulfate soundness test may not exceed 10 percent. Modern specifications also frequently include stipulations intended to control the potential alkali reactivity of fine aggregates.

The grading requirements of the various highway agencies related to fine aggregate are somewhat more uniform than similar requirements related to coarse aggregate. Many of the state highway departments use grading requirements that are the same as, or very similar to, those given in the "Standard Specifications for Fine Aggregate for Portland Cement Concrete," AASHTO Designation M6 (Table 20-2).

20-3 Water

Almost any normal source of water may be used in the manufacture of portland cement concrete. Specifications for this material frequently require simply that water to be used in the mixture is to be suitable for drinking, although other water may be used in some instances if it has been demonstrated by laboratory test or field experience that it is suitable for this purpose. Water that is used must be free from an excess of alkalies, acids, oil, or organic matter. Detergents in water will cause high entrained air content. Seawater has been successfully used in some concrete mixtures, although it is not generally used in paving operations.

20-4 Portland Cement

Portland cement is produced in five basic categories, designated type I through type V. Three of these classes of cement are commonly used in highway construc-

TABLE 20-2 Grading of Fine Aggregate (AASHTO Designation M6-81)

Sieve Designation	Percentage Passing (by weight)
³⁄₈ in. (9.5 mm)	100
No. 4 (4.75 mm)	95–100
No. 16 (1.18 mm)	45–80
No. 50 (0.30 mm)	10–30
No. 100 (0.150 mm)	2–10

tion: types I, II, and III. These types are defined in Chapter 15. Where an air-entraining agent is introduced into these compositions, they are designated types IA, IIA, and IIIA, respectively.

AASHTO has published specification (*1*) for portland cement giving the requirements for such properties as fineness, soundness, strength, time of setting, and air content of the mortar. AASHTO has also published standard methods for sampling and testing of portland cement for proper classification and quality control (*2*).

Cements conforming to the specifications for all of these types may not be readily available in some areas. The engineer should determine whether the proposed type is available before specifying any type of portland cement other than type I.

20-5 Admixtures

By an "admixture" is meant, in this case, any substance other than aggregate, water, or portland cement that may be added to a concrete mixture. A large number of admixtures may be used in conjunction with the standard ingredients of portland cement concrete for various purposes and in various ways. Only a few of these admixtures are of importance in highway construction, and certain of these will be discussed in the following paragraphs.

Perhaps the most important admixtures that may be added to concretes used in highway construction are those used to produce air-entrained concrete. Numerous materials can be used as air-entraining agents, including natural wood resins, fats, various sulfonated hydrocarbons, and oils. Some of these materials are insoluble in water and must be saponified before they can be used as admixtures. ASTM C233 gives a program of testing for the evaluation of materials proposed for use as air-entraining agents, while ASTM C260 is a tentative specification for these materials.

Another admixture that is of some consequence is calcium chloride, which is generally used as an accelerating agent. That is, calcium chloride materially decreases the time of hardening of the cement and thus leads to increased early strength of the concrete mixture. The amount of calcium chloride is generally limited to less than 2 percent by weight of standard cement. Caution should be exercised to avoid excessive use of calcium chloride, which could reduce the durability of concrete. The same properties obtained by the use of calcium chloride usually can be obtained by using type III cement or by increasing the cement factor.

Powdered materials such as diatomaceous earth, pumice, fly ash, and hydrated lime are occasionally used as admixtures in concrete, primarily as workability agents. Researchers have reported that the use of fly ash in concrete has a number of advantages including better protection against alkali-aggregate reaction and increased workability and strength from the use of this material. Widespread use is being made of various admixtures in casting concrete bridge decks to control the setting time, thus controlling dead load deflections and helping to ensure composite action between the deck and the underlying steel girders.

Admixtures may also be used to produce colored concrete. For example, red iron oxide has been used in concrete in some states for the purpose of marking traffic lanes. Similarly, the center strip in a concrete pavement is sometimes formed from concrete into which black iron oxide has been mixed.

PROPORTIONING OF CONCRETE MIXTURES

As has been indicated, a concrete mixture contains four basic ingredients, coarse aggregate, fine aggregate, cement, and water, with possible addition of admixtures to obtain specific properties. Before concreting operations may begin on any given project, the proportions of these materials must be established. The procedure of determining the amount of each of these materials that must be used to produce a concrete mixture of the desired characteristics is known as the "design of the mix." In the brief discussion that follows, no admixtures will be considered, and the design of air-entrained mixtures will be given mention in a later portion of this chapter. The objective of any procedure that may be used for the design of a given mixture is to determine the most practical and economical combination of materials that will produce a workable concrete that will have the desired properties of durability and strength after it has hardened.

Certain basic factors govern the design of satisfactory concrete mixtures, regardless of the detailed method that may be used. Detailed procedures that may be used may be quite complex and lengthy, involving considerable laboratory and field testing, or they may be very simple in nature, depending largely on the size of the project, the desired qualities of the hardened mixture, and the degree of experience which has been attained by the individual designer or organization in the use of materials suitable for concrete in a given area. The procedure outlined here is largely based on that contained in a standard of the American Concrete Institute (3). It involves the selection of a trial mixture to be used as a starting point for field concreting operations or for a detailed laboratory study. The procedure presented here is not necessarily one that is typical of the present practice of a majority of highway agencies, but it is given here because of its exposition of the fundamental factors involved and because of its universal applicability. A procedure of this type may be regarded as being especially desirable when concrete of a given flexural strength is desired.

Concrete should be placed with the minimum quantity of mixing water consistent with proper handling. Proportions should be selected to produce concrete

- of the stiffest consistency (lowest slump) that can be placed and finished efficiently to provide a homogeneous mass.
- with the maximum size of aggregate economically available and consistent with adequate placement.
- of adequate durability to withstand satisfactorily the weather and other destructive agencies to which it may be exposed.
- of the strength required to withstand the loads to be imposed without danger of failure.

A basic step to be taken in the design of a trial mixture is to select the water-cement ratio that is consistent with the desired durability and compressive strength desired in the hardened mixture. The "water-cement ratio" is a quantity of fundamental importance in the design of concrete mixture and is defined as the ratio of the amount of water, exclusive of that absorbed by the aggregate, to the amount of cement in a concrete mixture. It is preferably expressed as a decimal by weight of the water and cement. Formerly, it was commonly expressed in terms of gallons per sack of cement. A water-cement ratio of 6 gal per sack of cement is equivalent to a water-cement ratio of 0.53 by weight, calculated as follows: $w/c = 6$ gal/sack $= 6(8.34)/94 = 0.53$. Generally speaking, both the durability and compressive strength of a given concrete mixture increase with a *decrease* in the water-cement ratio; that is, the lower the water-cement ratio, the higher will be the compressive strength and the greater the durability of a given concrete mixture, all other things being the same and assuming that a workable, homogeneous mixture is secured. Maximum values of the water-cement ratio to be used in designing concrete with average materials for a desired compressive strength and to resist known conditions of exposure are well established. State highway department specifications related to concrete to be used in paving establish maximum permissible water-cement ratios varying from 0.47 to 0.55, depending on the experience of the organization concerned. The average value used is probably between 21 and 23 L (5.5 and 6.0 gal) per sack. They also require a miminum of 280 to 360 kg/m³ (470–610 lb cement per yd), feeling that the use of less cement than this will result in surface abrasion. The relationship between water-cement ratio and flexural strength is not so clearly established and should be determined by laboratory or field testing, although the flexural strength of a given mixture may generally be expected to be from 15 to 20 percent of the compressive strength for comparable conditions of curing and age. Compressive strengths of 20,700 to 27,600 kPa (3,000–4,000 lb/in²), (28-day curing) are commonly specified in highway work, whereas a typical required flexural strength is 4,500 kPa (650 lb/in²) (28-day curing, third point loading). Lower minimum flexural strengths may be specified for opening to traffic.

Another step required in the design of the mix is to select the desired consistency of the concrete to permit proper handling and placing of the mixture under the anticipated job conditions. The consistency of the fresh concrete mixture may be measured by the slump test (AASHTO Designation T119). The slump is measured by placing the fresh concrete mixture into a galvanized metal mold formed into the shape of the frustum of a cone, which is 8 in. in diameter at the base, 4 in. in diameter at the top, and 12 in. high. The concrete is placed in the mold in three layers, each of which is rodded by 25 strokes of a ⅝-in. diameter rod 24 in. long. The top layer is struck off even with the top of the mold and the mold raised vertically away from the concrete. The concrete mass subsides, and its height after subsidence is measured immediately. The difference between this height and the original height of the mass (12 in.) is the "slump." The slump test is illustrated in Figure 20-1. The desired slump for concrete to be used in pavements is from 25 to 75 mm (1–3 in.) with a lower slump for slip-form operations and higher slump for fixed form paving operations. Consistency may also be measured in the field by observing the penetration of a 153-mm (6-in.), 13.6-kg (30-lb) metal ball into the fresh concrete (ASTM C360).

Another step that must be taken in the general design procedure is the selection of the maximum size of aggregate that may be used. The maximum size of aggregate that is used in pavements that are of normal thickness and that contain

FIGURE 20-1 Equipment for slump test. (Courtesy Soil-test, Inc.)

relatively small amounts of reinforcing can be as large as 64 mm (2.5 in.); however, a maximum size of 25 to 38 mm (1–1.5 in.) is much more commonly used in paving primarily due to workability and ease of placing and finishing considerations. The maximum size of aggregate may be considerably less when the slab is to be heavily reinforced.

Additional variables that must be established in the design of a trial concrete mixture are the unit water content of the mixture and the proportion of coarse aggregate to be used in the aggregate combination. Various methods have been presented for determining these quantities, and detailed design procedures now in use by the various organizations show considerable deviation at this point in the design process. The general method being discussed defines the "unit water content" as the amount of water required per cubic yard of concrete.

The quantity of water per unit volume of concrete necessary to produce a mixture of the desired consistency is influenced by the maximum size, particle shape, and grading of the aggregate and by the amount of entrained air. It is relatively unaffected by the quantity of cement.

The approximate quantities of mixing water for different slumps and maximum sizes of aggregates are as follows:

	Recommended Water Quantities, kg/m³ (lb/yd³) for Maximum Size of Aggregate, mm (in.)			
Slump, mm (in.)	*25 (1)*	*38 (1.5)*	*50 (2)*	*76 (3)*
	Non–Air Entrained Concrete			
25–50 (1–2)	178 (300)	163 (275)	154 (260)	130 (220)
76–102 (3–4)	193 (325)	178 (300)	169 (285)	145 (245)
	Air-Entrained Concrete			
25–50 (1–2)	160 (270)	148 (250)	142 (240)	122 (205)
76–102 (3–4)	175 (295)	163 (275)	157 (265)	133 (225)

These quantities may be used with sufficient accuracy for computing cement factors for trial batches. They are maximum quantities that should be expected for fairly well shaped, angular coarse aggregates graded within the usual limits. If aggregates that are otherwise suitable lead to higher water requirements than given in the tabulation, the cement content should be increased to maintain the desired water-cement ratio.

The minimum amount of mixing water and maximum strength will result for given aggregates when the largest quantity of coarse aggregate is used, consistent with desired workability. Precise quantity of coarse aggregate for a given mixture is best determined by laboratory investigation, with later adjustment in the field; however, in the absence of laboratory data, an estimate of the proper proportions can be made for aggregates of conventional grading from empirical relationships shown in Table 20-3 (*3*).

Volumes are based on aggregates in dry-rodded condition as described in Method of Test for Unit Weight of Aggregate (ASTM Designation C29).

These volumes are selected from empirical relationships to produce concrete with a degree of workability suitable for usual reinforced construction. For less workable concrete, such as required for concrete pavement construction, they may be increased about 10 percent.

The information presented has been taken from the publication of the American Concrete Institute, which was previously cited (*3*). Additional information of this type is contained in that publication, and expanded tables of this sort are also to be found in the Design and Control of Concrete Mixtures, published by the Portland Cement Association (*4*).

Because of lack of space, no attempt will be made here to illustrate the specific use of the fundamental data that have been presented. Suffice it to say that various trial mixtures may be designed and examined in the laboratory, following the determination of certain basic "design properties" of the cement and aggregates involved. The mix may be designed for a given flexural strength, compressive strength, or both, and for the desired consistency. Some adjustment must generally be made in the selected mix on the basis of field conditions in order to secure concrete of satisfactory quality and economical proportions.

20-6 Air-Entrained Concrete Mixtures

Air-entrained concrete, as has been indicated, contains a small amount of entrapped air which is present in the form of small, disconnected air bubbles

TABLE 20-3 Volume of Coarse Aggregate per Unit Volume of Concrete

Maximum Size of Aggregate, mm (in.)	Volume of Dry-Rodded Coarse Aggregate per Unit Volume of Concrete for Different Fineness Moduli of Sand			
	2.40	*2.60*	*2.80*	*3.00*
25 (1)	0.71	0.69	0.67	0.65
38 (1½)	0.75	0.73	0.71	0.69
50 (2)	0.78	0.76	0.74	0.72
76 (3)	0.82	0.80	0.78	0.76

uniformly distributed throughout the mass. The desired amount of air is generally from 4 to 8 percent of the total mix. The chief advantage accruing from the inclusion of this amount of air in the mixture is increased durability, including resistance to calcium chloride and other salts that are widely used in northern states for ice control and that have frequently resulted in surface scaling and spalling of concrete pavements. It also increases resistance to sulfate action and to freezing and thawing. Air-entrained concrete mixes have a high degree of workability compared to similar regular mixtures, being generally somewhat sticky and plastic in nature and showing little tendency to segregation and much less tendency to "bleed," a phenomenon during which excessive water in the concrete migrates to the surface. These mixes may be placed at slightly lower slumps than regular mixtures. Because of the high degree of workability possessed by air-entrained concrete mixtures, they have been widely used in paving.

Air-entrained concrete mixtures should also be designed by the trial-mix method. It is necessary to determine and control the amount of air that is actually incorporated in the mixture. Several methods have been devised to measure the air content of air-entrained concrete mixtures, including the gravimetric method (AASHTO T121 and ASTM C138), the volumetric method (ASTM C173), and the pressure method (AASHTO T152 and ASTM C231). The last of these is most widely used and is based on the principle of Boyle's law ($P_1 V_1 = P_2 V_2$); a known pressure is applied to a known volume of concrete and the change in volume determined. A pressure device for determining the air content is shown in Figure 20-2.

20-7 Plate Bearing Test Method and Modulus of Subgrade Reaction

In the structural design of concrete pavements used by most of the design methods, the subgrade soil properties and the combined subgrade and base properties are represented by the modulus of subgrade reaction. The method for determining this property is described in the following paragraphs.

FIGURE 20-2 Determination of air content by the pressure method. (Courtesy Soiltest, Inc.)

During 1945 and 1946 the Department of Transport of Canada undertook a very comprehensive investigation of runways at a number of principal airports in Canada. This work was reported by Norman W. McLeod (5). Later, McLeod expanded the results of this study and related it to pavement design methods based on the theory of elasticity (6, 7). The plate bearing test is used to evaluate the support capability of subgrades, bases, and, in some cases, complete pavements.

The schematic arrangement of equipment used in performing the plate load test is shown in Figure 20-3. Although various sizes of steel plates are used, most tests are conducted with a circular steel plate, 1 in. thick and 30 in. in diameter. A thin bed of sand is commonly used to ensure an even bearing between the plate and the underlying material. Load is applied to the plate by means of a large-capacity hydraulic jack equipped with pressures gauges graduated in increments of 1000 or 2000 lb. Reaction to the load applied by the jack can be supplied in each test by a loaded tractor-trailer unit, scraper, or other heavy equipment.

Deflection of the bearing plate is measured to the nearest 0.001 in. by means of three dial gages set at third points around the circumference of the plate.

The results of plate bearing tests are often expressed in terms of a modulus of subgrade reaction as determined by

$$k = p/\delta \tag{20-1}$$

where k = modulus of subgrade reaction (psi/in.)
p = plate pressure (lb/in.²)
δ = plate deflection (in.)

In metric units, the modulus of subgrade reaction should be expressed in kPa per mm. To convert the modulus of subgrade reaction in psi per in. to kPa per mm, multiply it by 0.27.

STRUCTURAL DESIGN OF CONCRETE PAVEMENTS

The following several sections of this chapter will be devoted to what has been called the "structural design" of concrete pavements. Included in these sections is the design of the various elements of the pavement structure itself, including considerations related to the various types of joints and their spacing, the use of reinforcing steel, and the thickness required in the slab cross section.

JOINTS AND JOINT SPACING

Joints are installed in concrete pavements to control the stresses induced by volume changes in the concrete. These stresses may be produced in a concrete slab because of (1) its contraction due to a uniform temperature drop or a decrease in moisture, (2) its expansion due to a uniform temperature increase, and (3) the effects of the "warping" of pavements due to a vertical temperature or moisture differential in the slab.

20-8 Contraction Stresses

As do most construction materials, concrete changes in volume with a change in temperature. Thus a concrete slab, if it is free to move, contracts with a drop in

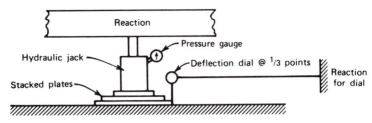

FIGURE 20-3 Equipment used in performing plate bearing tests.

temperature; however, if the contraction or "movement'" of the slab is wholly or partially prevented, tensile stresses are developed. In a pavement, resistance to movement of the slab is caused by friction between the bottom of the slab and the subgrade.

It is also known that concrete generally contracts or "shrinks" with a decrease in moisture content and expands to some extent with an increase in moisture. Because a concrete slab tends to decrease in moisture content with the passing of time, the slab might be expected to contract slightly with a resultant opening of transverse joints. This circumstance would somewhat tend to offset later expansion of the slab due to temperature change and thus further tend to substantiate the general omission of expansion joints.

20-9 Expansion Stresses

If a concrete pavement slab is subjected to a uniform increase in temperature, the slab will increase in length. Under extreme combinations of circumstances it may be imagined that a long length of concrete slab may buckle or "blow up" if this expansion is prevented. In order to prevent blow-ups, relief may be provided by the installation of transverse expansion joints. These joints, which are described in more detail later, generally provided a space an inch or so in width in which expansion may take place without damage to the slab.

Friction between the slab and the subgrade prevents much of this expansion, however, and the compressive stresses created by this restraint are generally quite small as compared with the compressive strength of the concrete. This fact, coupled with the fact that concrete shrinks somewhat during hardening, has led many designers to the conclusion that expansion joints may be spaced at very great intervals, provided only that adequate contraction joints are provided and that the concrete does not have unusual expansion qualities and was not placed at exceedingly low temperatures.

20-10 Warping Stresses

Very frequently a differential in temperature exists between the top and bottom of a concrete pavement. For example, during the day the temperature of the top portion of the slab may be considerably greater than that of the bottom. Under this circumstance the top portion of the slab tends to expand more than the bottom. This expansion and its effect in producing a slightly convex slab surface are resisted by the weight of the slab, with the consequence that fibers in the top portion of the slab are placed in compression and those in the bottom in tension. Stresses in the pavement from this case are termed "temperature warping stresses." Conversely, at night the top of the slab is frequently cooler than the bottom, with the result that the top of the slab is in tension and the bottom in compression.

Temperature warping is of importance in the design of concrete pavements in two ways. First, the flexural stresses due to this cause may, under certain conditions, be of considerable magnitude in themselves, without regard to their combination with stresses due to traffic loads. Second, the warping that does take place due to a temperature differential may partially destroy the subgrade support beneath portions of the slab, with the result that stresses due to traffic loads may be considerably increased over those that would exist if the pavement received uniform subgrade support.

If a difference in moisture content exists between the top and the bottom of the slab, the effect is somewhat similar to that produced by a temperature differential. In perhaps the majority of cases the moisture content in the bottom of the slab is greater than in the top, with the result that the bottom tends to expand, resulting in compression of the bottom fibers and somewhat offsetting stresses resulting from temperature warping during the daytime, when the temperature differential might be expected to be greatest.

Joint Design

To control the stresses resulting from the combined effects of temperature and moisture changes and wheel loadings, four types of joints are commonly provided for concrete pavements:

1. Transverse contraction joints.
2. Transverse construction joints.
3. Transverse expansion joints.
4. Longitudinal contraction and construction joints.

20-11 Transverse Contraction Joints

Transverse contraction joints are used for two purposes: to control cracking of the slab resulting from contraction and to relieve warping stresses. Two types of contraction joints that are most commonly used at the present time are the weakened-plane joint and the doweled joint, as shown in Figure 20-4. As contraction occurs, the lower portion of the slab cracks at the weakened plane created by the saw cut, and the structural integrity of the joint is maintained by the dowel

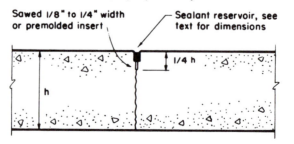

Sawed 1/8" to 1/4" width or premolded insert

Sealant reservoir, see text for dimensions

1/4 h

h

Sawed or Premolded Insert

For reinforced pavements, smooth dowel bars installed at depth h/2. See text for use of dowel bars.

FIGURE 20-4 Weakened-plane contraction joint. (Courtesy Portland Cement Association.)

bars and aggregate interlock. This type of joint is also called "dummy" contraction joint.

Details of the designs of contraction joints vary to some extent, with dowel bar diameters varying from $5/8$ to $1\frac{1}{2}$ in. being used (commonly one-eighth of the slab thickness) and with lengths varying from 12 to 20 in. Bar spacings generally are 12 to 15 in. from center to center of the bars. The groove made in the pavement is frequently either of constant $1/8$- or $1/4$-in. width, as shown in Figure 20-4. This groove is most often formed by sawing but may be formed by use of a premolded strip inserted in the fresh concrete. A double saw cut may be used to construct a wider joint sealant reservoir at the top of the pavement. Figure 20-5 illustrates the wider reservoir which may be needed to provide a better shape factor for the sealant material. The groove is later filled with one of a number of materials, including field molded or poured sealants or preformed compression seals. When field-molded sealants are used, a tape or a backer rod should be placed at the bottom of the sealant space to prevent the sealant from seeping into and bonding to the concrete surfaces fomed by the cracks.

Dowel bars used in this type of joint (and in expansion joints) are not bonded to the concrete on one side of the weakened plane, and freedom of movement is ensured by painting or lubricating one end of the dowel, by enclosing one end in a sleeve, or by other similar methods. It is essential that freedom of movement be ensured in the design and placing of the joint, since the purpose of the joint will be largely destroyed if movement is prevented. The dowels must be carefully placed so that they are horizontal and parallel to the centerline of the slab. They must also be designed to ensure their proper functioning under the stresses imposed on them in their capacity as load-transfer devices. Field evaluations have shown that use of dowel bars in transverse joints is an effective way to minimize the problem of faulting between adjacent slabs.

For plain concrete pavements, the spacing of contraction joints must be close enough to control cracking. Spacing is dependent on local experience and often is 4.6 or 6 m (15 or 20 ft) for concrete slabs of 200 to 250 mm (8–10 in.) thick. Longer joint spacing tends to have wider joint crack openings and render the aggregate interlock between the adjacent slabs less effective. Thus, load-transfer devices should be used at the joints.

In the past, engineers have discovered two ways to improve joint performance and extend the life of plain concrete pavements. One way is to construct tranverse joints that are skewed slightly with respect to a line at right angles to the center of the pavement 1.2 or 1.4 m in 7.3 m (4 or 5 ft in 24 ft). A skewed joint causes

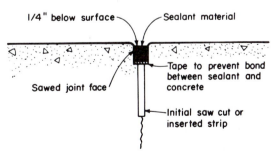

FIGURE 20-5 Detail of a weakened-plane contraction joint. (Courtesy Portland Cement Association.)

the wheel loads of each axle to cross the joint one at a time, thereby reducing the deflection and stress at the joints. Another technique for improving the performance of plain concrete pavements is to construct the joints at random spacings such as 4, 6, 5.5, and 3.6 m (13, 19, 18, and 12 ft) to avoid rhythmic effects of automobiles associated with spacings that are exact multiples of 2.3 m (7.5 ft). Use of these types of joint construction practices has been less common in recent years. Better concrete quality control and the use of dowel bars for better load transfers have diminished the cost-effectiveness of these types of joint construction practices.

Distance between transverse contraction joints may be somewhat greater when the pavement is reinforced, because the distributed steel aids in controlling cracking. Joint spacing is dependent on several factors including the quantity of steel and the thickness of slab. The Portland Cement Association (4) recommends joint spacings of not more than 12.2 m (40 ft) for reinforced pavements.

Continuously reinforced pavements, which have relatively heavy continuous steel reinforcement in the longitudinal direction, tend to develop small transverse cracks at close intervals. A high degree of load transfer is provided at these crack faces, which are restrained by the steel and by aggregate interlock. Such pavements therefore are normally built without transverse contraction joints.

20-12 Transverse Construction Joints

An additional type of joint that may be used in a concrete pavement may be called a "construction joint." Transverse construction joints may be placed at the end of a day's "run" or when work ceases due to some other interruption. If construction is stopped at the location of a transverse joint, the joint assembly may be installed as usual, concrete placed on one side of the joint, and the other side of the joint protected in an appropriate fashion until concreting operations can be resumed. If a transverse construction joint falls within the middle third of the regular joint interval, it should be a tied joint. This prevents the joint from opening and possibly causing "sympathetic" cracking of the pavement in the adjacent lane.

20-13 Transverse Expansion Joints

Expansion joints that are used at the present time are usually from 19 to 25 mm ($^3/_4$–1 in.) wide and extend the full depth of the slab. The joint space is filled with some compressible, elastic, nonextruding material. A variety of materials have been used as expansion joint fillers, including bituminous material, cork, rubber, cork-rubber compounds, bitumined fabrics, wood, and many others. Dowel bars are normally used in expansion joints, and their diameter, spacing, and length would generally be similar to that indicated for contraction joints. The dowels must be adequate to perform their load-transfer function and they must be designed and placed so as not to interfere with proper functioning of the joint. Figure 20-6 shows a somewhat typical design of a transverse expansion joint. Whenever salt corrosion is anticipated to be a problem, epoxy-coated dowel bars may be specified. Specifications for corrosion-resistant coated dowel bars are covered in AASHTO M254, and testing of coated dowel bars is covered in AASHTO T253.

After many years of laboratory and field studies, practice relative to the spacing of transverse joints has at last reached some degree of uniformity over the

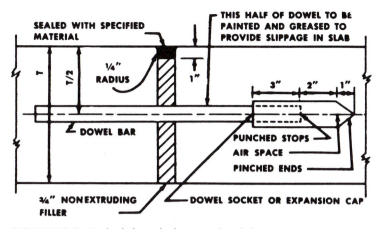

FIGURE 20-6 Typical doweled expansion joint.

country. Details vary, but the following discussion reflects current practice for both plain and reinforced concrete pavements.

The Portland-Cement Association (8) has reported that, except at intersections and structures, no expansion joints need be provided in a concrete pavement when

- the pavement is constructed of materials that have normal expansion charicteristics.
- the pavement is constructed during those periods of the year when ambient temperatures are well above freezing.
- the pavement is divided into relatively short panels by contraction joints so spaced as to prevent the formation of intermediate cracks.
- the contraction joints are properly maintained to prevent the infiltration of relatively incompressible materials.

If concrete pavements are built in cold weather, or of materials that have high coefficients of expansion, expansion joints may be necessary at intervals of several hundred feet. Otherwise, they are required only adjacent to structures and intersections.

20-14 Longitudinal Joints

A longitudinal joint in a concrete pavement is, of course, a joint running continuously the length of the pavement. The joint divides, for example, a two-lane pavement into two sections, the width of each being the width of a traffic lane. The purpose of longitudinal joints is simply to control the magnitude of temperature warping stresses in such a fashion that longitudinal cracking of the pavement will not occur. Longitudinal cracking has been almost completely eliminated in concrete pavements by the provision of adequate longitudinal joints. In two-lane pavements, the two slabs are generally tied together by means of steel tie bars extending transversely across the joint and spaced at intervals along the length of the joint. The purpose of the tie bars in a longitudinal joint is to prevent movement of one slab with respect to the other, since, in addition to serving the function of relieving temperature warping stresses, the joint functions in transmitting a portion of the load carried by one slab to the other, thus reducing the stresses in the loaded slab. Adequate "load transfer" is dependent on maintaining close contact between the two slabs (or portions of the joint) across the longitudinal joint. In multiple-lane pavements, no more than four lanes should be tied together.

Longitudinal joints used in concrete pavements are of several different types, including both patented and nonpatented varieties. Two of the most commonly used types are those designated as "deformed" or "keyed" and "weakened-plane" joints. A typical longitudinal joint of the former type is shown in Figure 20-7. This type of joint often is used as a construction joint when the pavement is built one lane at a time. A weakened-plane joint generally is used as a contraction joint when two or more lanes are paved at one time. The weakened plane type of joint is formed mostly by sawing and is described in detail in the portion of the discussion related to contraction joints.

Steel tie bars are used to tie the two slabs together and are firmly "bonded" to the concrete; the behavior of these bars should not be confused with that of the dowel bars mentioned in later sections. Tie bars that are used in longitudinal joints are generally designed to withstand the entire stress created by contraction of the slab in a transverse direction.

Reference 8 gives equations for the calculation of the cross-sectional area and length of steel tie bars at joints in concrete pavements.

20-15 Pumping of Joints

Special attention is given in this section to a problem that may be encountered in the design (and maintainance) of joints in concrete pavements. This problem is "pumping." The phenomenon of pumping may be defined as follows. With a certain combination of factors present, the deflection of slab ends under traffic loads causes the extrusion or "pumping" of a portion of the subgrade material at joints, in cracks, and along the edges of the pavement. Pumping through joints will only occur under the following circumstances: (1) frequent occurrence of heavy wheel loads, (2) the existence of a surplus of water in the subgrade soil, and (3) the presence of a subgrade soil that is susceptible to pumping. All three of these elements must be present in order for pumping to occur. The amount of soil removed by pumping may be sufficient to cause a sizable reduction in subgrade support for the slab and may result in eventual failure of the pavement. Pumping is also the main contributing factor for the development of faulting

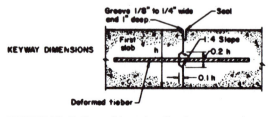

FIGURE 20-7 Keyed longitudinal construction joint with tie bars. (Courtesy Portland Cement Association.)

across the joints. Design procedures intended to prevent pumping will be briefly discussed here, whereas maintenance operations intended to correct pumping that occurs in existing pavements will be discussed in the next chapter.

Because much of the difficulty with pumping has been encountered at expansion joints, an obvious solution is to minimize the number of these joints or eliminate them altogether; this consideration has doubtless been a contributing factor in the present trend to minimize the use of expansion joints. Also, because pumping is associated with fine-grained soils, another obvious solution is to replace or improve these so-called pumping soils. This may be done, for example, by the use of a nominal thickness of granular material or a selected sandy soil, or by stabilization of the existing subgrade soil through the use of bituminous material or portland cement.

In modern practice, a granular subbase from 75 to 150 mm (3 to 6 in.) thick is employed by many organizations in locations where the subgrade soil is such as to be susceptible to pumping. Much thicker subbases may be used in areas subject to frost action or on very weak soils. Requirements for subbase materials vary; a suitable specification is AASHTO M147 (see Table 18-1). A desire to prevent pumping is one factor in the increasing use of cement-treated base courses beneath concrete pavements. Some states now use asphaltic plant mixtures under concrete pavements for the same purpose.

USE OF DISTRIBUTED STEEL REINFORCING

Distributed steel reinforcing is used primarily to control cracking of a concrete pavement and to maintain the structural integrity of the slab between transverse joints. In accordance with principles previously stated, some organizations believe that reinforcing steel is unnecessary where contraction joints are spaced at intervals of 4.5 to 6 m (15 to 20 ft). If the distance between contraction joints is appreciably greater than this, it may generally be regarded as desirable, with the amount of steel required being a direct function of the spacing between joints. Cracking of the slab will occur even though the steel is present, but the steel serves to hold the edges of the cracks close together, thus preventing the progressive opening of these cracks and accompanying detrimental effects. Distributed steel, as commonly used in concrete pavements, is not counted on to contribute to the flexural strength of the slab. By holding cracks tightly closed, however, it maintains the shearing resistance of the slab and, consequently, its load-carrying capacity. These properties are greatly reduced if cracks are permitted to open.

Distributed reinforcing is used in the form of welded wire "fabric" or bar mats, both of which contain both longitudinal and transverse elements. Different sizes and spacing of the elements of the reinforcing "mat" are used by the various highway agencies in order to conform with requirements related to the area of steel required across the width and length of the slab. The amount of steel required in a given case is based on the assumption that it must be strong enough to drag both ends of the slab over the subgrade toward its center.

Generally, the cross-sectional area of longitudinal steel ranges from about 0.1 to about 0.2 percent of the cross-sectional area of the slab. Weights required typically range from about 2.7 to 4.3 kg/m² (0.55 to 0.8 lb/ft²) of pavement surface.

20-16 Continuously Reinforced Pavement

Following World War II, interest was focused on the possibility of building continuously reinforced pavements without joints with the construction of pavement of this type in Illinois and New Jersey. Longitudinal steel was used continuously over the length of 1 mile in the New Jersey experiment and 3500 ft in Illinois. The amount of steel was varied in different sections of these test roads, ranging from 0.3 to 1.0 percent. Subsequently, test sections of continuously reinforced concrete pavement were built in Texas, California, Pennsylvania, and several other states, and tens of thousands of miles of these pavements have been built. Nevertheless, continuously reinforced pavements have not enjoyed widespread popularity in the U.S.

The required amount of longitudinal reinforcing steel in this kind of pavement is not in direct proportion to the length of the slab. Observations of long (1 km or more) continuously reinforced slabs indicate that less than 10 percent of the pavement at each end is subject to longitudinal movement; for all practical purposes, the long central section is fully restrained. Seasonal movements at the slab ends total less than 50 mm (2 in.) regardless of the slab length.

It is believed that the optimum quantity of reinforcing steel is that which results in a crack spacing of about 1 to 3 m (3 to 10 ft). Crack intervals in this range typically occur with steel ranging from 0.5 to 0.7 percent of the cross-sectional area of the slab. The *1986 AASHTO Guide for Design of Pavement Structures* (9) has provided a detailed procedure for the design of the reinforcing steel.

Recommended slab thickness varies from 150 to 228 mm (6 to 9 in.), depending on traffic conditions. The design of the thickness of continuously reinforced pavements is often based on that of conventional concrete pavements.

Both welded wire fabric and bar mats are used for continuous reinforcement. For wire fabric, spacing of the longitudinal wires should be not less than 4 in. nor more than 12 in.; spacing of transverse wires should not exceed 24 in.; and the clearance between the reinforcement and the slab edges should not be less than 2 in. nor more than 6 in.

DIMENSIONS OF THE CROSS SECTION (THICKNESS DESIGN)

It is the purpose of this section to discuss the determination of the design thickness of a concrete pavement. Several different factors enter into the thickness determination, as will be outlined in following sections. Basically, however, the pavement must have a thickness that is adequate to support the loads which will be applied to it during its service life, and the design must be economical.

Many organizations and individuals have given their attention to the problem of thickness design of concrete pavements during the last 50 years or so. Space will not permit the chronicling of the many contributions that have been made by these investigators and others. Rather, it has been decided to present the essentials of only one design method in the belief that this method is typical of modern approaches to this subject. The method presented is that contained in the publication, "Thickness Design for Concrete Pavements," published by PCA in 1984 (10). The other commonly used design method is the *1986 AASHTO Guide for the Design of Pavement Structures* (9).

20-17 Portland Cement Association Method

Design Criteria

The PCA design method is based on the following two design criteria:

1. Fatigue. It attempts to keep pavement stresses from repeated loads within acceptable limits to prevent fatigue cracking.
2. Erosion. It attempts to limit the effects of pavement deflections at joints and corners of slabs in order to control the erosion of subgrade materials and thus minimize joint faulting.

Fatigue analysis is based on the edge stress midway between the transverse joints. Because the load is near the midslab far away from the joints, the presence and types of joint (weakened plane or doweled) have practically no effect on the edge stress. On the other hand, when a concrete shoulder is used and is tied onto the mainline pavement, the magnitude of the critical stress is substantially reduced. The cumulative damage concept shown in Equation (20-2) is used to take into account the cumulative effects of the repeated load–induced stresses due to all traffic load groups within the pavement design period. Stresses induced in the concrete slab due to warping and curling are excluded from the fatigue analysis because the moisture content and temperature are usually higher at the bottom of the slab than at the top, and thus the combined effects of warping and curling stresses are substractive from the loading stresses.

$$D_r = \sum_{i=1}^{m} n_i / N_i \tag{20-2}$$

where D_r is the damage ratio accumulated over the design period resulting from all load groups (m groups). When this cumulative damage ratio reaches 1, transverse cracking most likely will occur on the concrete pavement.

n_i is the estimated, or projected, number of repetitions of the i-th load group.

N_i is the allowable number of repetitions for the i-th load group.

The allowable number of repetitions is dependent on the ratio of the flexural stress developed in the slab as a result of the particular group to the flexural strength of the concrete. This number increases as the stress ratio decreases. Studies show that, if the stress ratio is less than 0.55, concrete will withstand virtually unlimited stress repetitions without causing the concrete to fail. To be conservative, PCA reduced this ratio to 0.5.

The other design criteria used in the PCA design method is erosion analysis, which take into consideration the effect of excessive deflection at the corner on pavement distresses, such as pumping, erosion of foundation, and joint faulting. The type of joint (weakened plane or dowelled), and whether there is tied concrete shoulder, will affect the pavement corner deflection and thus affect the erosion analysis.

It is known that only a small percentage of trucks travel with their outside wheels placed at the edge, whereas most truck drivers drive with their outside wheels placed about 0.6 m (2 ft) from the pavement edge. The flexural stress and the corner deflection induced in the slab are substantially higher if the wheel loads are applied at the edge of the pavement than 0.6 m (2 ft) from the pavement edge. The PCA design procedure is based on the assumption that 6 percent of the trucks travel with the outside wheels at or beyond the pavement edge.

Design Factors

The design of concrete pavement by the PCA method is governed by four design factors: concrete modulus of rupture, subgrade and subbase support, design period, and traffic.

Concrete Modulus of Rupture Flexural strength or modulus of rupture of concrete is determined by a 15- × 15- × 52.5-cm concrete beam simply supported at 45 cm span and subjected to third-point loading. PCA recommends the use of 28-day strength tests for roads and streets.

Subgrade and Subbase Support Subgrade and subbase support is defined by the modulus of subgrade reaction k, described in Section 20-7. Because the plate-

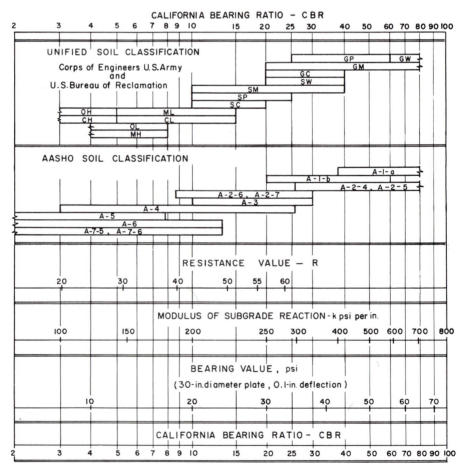

FIGURE 20-8 Approximate interrelationships of soil classifications and bearing values. (Courtesy Portland Cement Association.)

loading test to determine the *k* value is time-consuming and expensive, the *k* value is usually estimated by correlation to simpler tests, such as the California Bearing Ratio test, or directly from the soil classifications as shown in Figure 20-8. It is usually not economical to use subbase for the purpose of increasing *k* values and thus reducing the concrete slab thickness. Where a subbase is used for the purpose of reducing such things as pumping and erosion, there will be an increase of *k* value that should be used in the thickness design. Table 20-4 and Table 20-5 can be used to estimate the combined *k* values.

Design Period The design period of 20 years has been commonly used in pavement design; however, there are circumstances where the use of a shorter or longer design period may be warranted.

Traffic The PCA design method details two methods for determining traffic volumes and the number and magnitude of heavy wheel load repetitions. One approach is to use figures based on traffic volume and classification counts plus special loadometer surveys. Such information is generally available from the state highway departments or other traffic planning agencies for the design of any major route. Another way is to base the design on traffic capacity considerations for the route in question. Capacity is discussed in Section 6-6.

In the design procedure, the axle loads are multiplied by a load safety factor to account for the allowance of overload. The following load safety factors are recommended: interstate and other heavy truck traffic, 1.2; highways and streets with moderate volume of truck traffic, 1.1; and street and highways that carry small volumes of truck traffic, 1.0.

Design Procedure

PCA engineers worked out a number of design examples, one of which is summarized below and is used to illustrate the design procedure.

TABLE 20-4 Effect of Untreated Subbase on *k* Values

Subgrade k Value, kPa/mm (psi/in.)	Subbase k Value, kPa/mm				Subbase k Value, psi/in.			
	102 mm	152 mm	228 mm	305 mm	4 in.	6 in.	9 in.	12 in.
14 (50)	18	20	23	30	(65)	(75)	(85)	(110)
27 (100)	35	38	43	51	(130)	(140)	(160)	(190)
54 (200)	59	62	73	86	(220)	(230)	(270)	(320)
81 (300)	86	89	100	226	(320)	(330)	(370)	(430)

Source: Thickness Design for Concrete Highway and Street Pavements, Portland Cement Association (1984).

TABLE 20-5 Design *k* Values for Cement-Treated Subbases

Subgrade k Value, kPa/mm (psi/in.)	Subbase k Value, kPa/mm				Subbase k Value, psi/in.			
	102 mm	152 mm	228 mm	305 mm	4 in.	6 in.	9 in.	12 in.
14 (50)	46	62	84	105	(170)	(230)	(310)	(390)
27 (100)	76	108	140	173	(280)	(400)	(520)	(640)
54 (200)	127	173	224	—	(470)	(640)	(830)	—

Source: Thickness Design for Concrete Highway and Street Pavements, Portland Cement Association (1984).

EXAMPLE 20-1* **Thickness Design, Rigid Pavements** A four-lane rural interstate project is to be built in rolling terrain. The design period for the concrete pavement is 20 years. The current traffic volume (ADT) is 12,900 and the projection factor for 20 years is 1.5. Designers estimate that truck traffic volume ($ADTT$) is 19 percent of the total.

The design ADT is $12,900 \times 1.5 = 19,350$ vehicles per day, and the $ADTT$ is $19,350(0.19) = 3680$. Truck traffic each way is $3680/2 = 1840$. Data not shown here reveal that 81 percent of the trucks will be in the right lane. Thus, for this lane and a design period of 20 years, there will be $1840 \times 0.81 \times 365 \times 20 = 10.88$ million trucks.

PCA engineers evaluated five designs (1A through 1E). For design 1A, the k of the subgrade was taken to be 100 lb/in.³ with a 4-in. untreated granular subbase. Combined k for the subgrade and subbase was taken to be 130 lb/in.³ (Table 20-4). The load safety factor is 1.2, with a modulus of rupture of the concrete of 650 lb/in.²

The engineers assumed a trial depth of 9.5 in. Key calculations are shown on the design worksheet (Fig. 20-9).

For the fatigue analysis, equivalent stresses are determined from Table 20-6 for single and tandem axles and entered on the design worksheet as items 8 and 11, respectively. The equivalent stresses are then divided by the concrete modulus of rupture and entered as items 9 and 12 (stress ratio factors). Next, the allowable repetitions are determined for each loading category from Figure 20-10. The percentage of fatigue is then calculated by dividing the expected repetitions (column 3) by the allowable repetitions (column 4). The sum of the percentages of fatigue used is totaled for all loading categories, and for this example is 62.8 percent.

In a similar way, the erosion analysis involves determining the allowable repetitions to ensure that harmful erosion of foundation and shoulder materials does not occur from pavement deflections at slab edges, joints, and corners.

To perform this analysis, erosion factors are determined from Table 20-7 and recorded on the design worksheet as items 10 and 13 for single- and tandem-axle loadings, respectively. Allowable repetitions are then determined from Figure 20-11 and recorded in column 6 for each loading category.

The damage due to erosion is then compiled for each loading condition by dividing the expected repetitions (column 3) by the allowable repetitions (column 6). The total erosion damage (percent) is determined by summing the incremental damages (column 7). For this example, the erosion damage is 38.9 percent.

The totals of fatigue use and erosion damage use of 62.8 and 38.9 percent, respectively, show that the 9.5 in. thickness is satisfactory for the given conditions.

In the preceding example, we have illustrated the PCA design procedure for one set of conditions: doweled joints, untreated subbase, and no concrete shoulder. Reference 6 provides similar illustrations and examples for designs of other combinations of subbase support, shoulder treatment, and joint design.

---

*Conventional U.S. units are used in this example and in the assigned problems to be consistent with available nomographs.

Calculation of Pavement Thickness

Project __Design 1A, four-lane Interstate, rural__

Trial thickness __9.5__ in. Doweled joints: yes ✓ no ____

Subbase-subgrade k __130__ pci Concrete shoulder: yes ____ no ✓

Modulus of rupture, MR __650__ psi

Load safety factor, LSF __1.2__ Design period __20__ years

__4-in. untreated subbase__

Axle load, kips	Multiplied by LSF 1.2	Expected repetitions	Fatigue analysis		Erosion analysis	
			Allowable repetitions	Fatigue, percent	Allowable repetitions	Damage, percent
1	2	3	4	5	6	7

8. Equivalent stress __206__   10. Erosion factor __2.59__

9. Stress ratio factor __0.317__

Single Axles

30	36.0	6,310	27,000	23.3	1,500,000	0.4
28	33.6	14,690	77,000	19.1	2,200,000	0.7
26	31.2	30,140	230,000	13.1	3,500,000	0.9
24	28.8	64,410	1,200,000	5.4	5,900,000	1.1
22	26.4	106,900	Unlimited	0	11,000,000	1.0
20	24.0	235,800	"	0	23,000,000	1.0
18	21.6	307,200	"	0	64,000,000	0.5
16	19.2	422,500			Unlimited	0
14	16.8	586,900			"	0
12	14.4	1,837,000			"	0

11. Equivalent stress __192__   13. Erosion factor __2.79__

12. Stress ratio factor __0.295__

Tandem Axles

52	62.4	21,320	1,100,000	1.9	920,000	2.3
48	57.6	42,870	Unlimited	0	1,500,000	2.9
44	52.8	124,900	"	0	2,500,000	5.0
40	48.0	372,900	"	0	4,600,000	8.1
36	43.2	885,800			9,500,000	9.3
32	38.4	930,700			24,000,000	3.9
28	33.6	1,656,000			92,000,000	1.8
24	28.8	984,900			Unlimited	0
20	24.0	1,227,000			"	0
16	19.2	1,356,000				
				Total 62.8		Total 38.9

FIGURE 20-9 Design worksheet for example problem using PCA procedure. (Courtesy Portland Cement Association.)

TABLE 20-6 Equivalent Stress—No Concrete Shoulder (Single Axle/Tandem Axle)

Slab Thickness (in.)	k of Subgrade—Subbase (lb/in.³)						
	50	100	150	200	300	500	700
4	825/679	726/585	671/542	634/516	584/486	523/457	484/443
4.5	699/586	616/500	571/460	540/435	498/406	448/378	417/363
5	602/516	531/436	493/399	467/376	432/349	390/321	363/307
5.5	526/461	464/387	431/353	409/331	379/305	343/278	320/264
6	465/416	411/348	382/316	362/296	336/271	304/246	285/232
6.5	417/380	367/317	341/286	324/267	300/244	273/220	256/207
7	375/349	331/290	307/262	292/244	271/222	246/199	231/186
7.5	340/323	300/268	279/241	265/224	246/203	224/181	210/169
8	311/300	274/249	255/223	242/208	225/188	205/167	192/155
8.5	285/281	252/232	234/208	222/193	206/174	188/154	177/143
9	264/264	232/218	216/195	205/181	190/163	174/144	163/133
9.5	245/248	215/205	200/183	190/170	176/153	161/134	151/124
10	228/235	200/193	186/173	177/160	164/144	150/126	141/117
10.5	213/222	187/183	174/164	165/151	153/136	140/119	132/110
11	200/211	175/174	163/155	154/143	144/129	131/113	123/104
11.5	188/201	165/165	153/148	145/136	135/122	123/107	116/98
12	177/192	155/158	144/141	137/130	127/116	116/102	109/93
12.5	168/183	147/151	136/135	129/124	120/111	109/97	103/89
13	159/176	139/144	129/129	122/119	113/106	103/93	97/85
13.5	152/168	132/138	122/123	116/114	107/102	98/89	92/81
14	144/162	125/133	116/118	110/109	102/98	93/85	88/78

Source: Thickness Design for Concrete Highway and Street Pavements. Portland Cement Association (1984).

20-18 Present Status of Design

It should be noted at this stage that, although design principles and methods such as those illustrated in the preceding section are in existence at the present time, these methods are not commonly applied as a part of the routine operations of a majority of highway agencies, including many state highway agencies. Thicknesses of concrete pavements as used in this country commonly vary from 7 to 10 in., and each organization normally has developed "standard designs" that are satisfactory for its purposes. Usually, a concession is made to the principles outlined just by the use of greater thicknesses for heavily traveled routes and areas of poor subgrade support.

20-19 Composite Pavements

In broad terms, a "composite" pavement is one that combines dissimilar pavement types, that is, "flexible" and "rigid" pavements. Usually, a composite pavement comprises a concrete- or cement-treated base course and a wearing surface of asphaltic concrete.

Comparatively few composite pavements have been designed and built as new pavements, although New York City has for many years built city streets with a portland cement concrete base, an asphaltic concrete binder course, and a sheet-asphalt wearing surface. Other cities build similar pavement structures.

Many composite pavements have, of course, been produced by stage construction or by resurfacing operations. Overlays of existing pavements are discussed in Section 21-5.

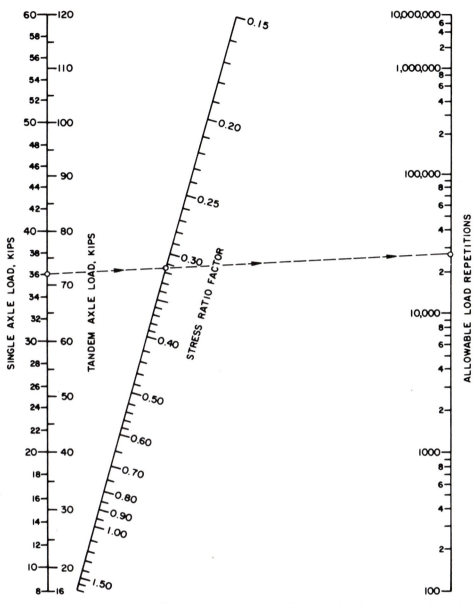

FIGURE 20-10 Nomograph for fatigue analysis—allowable load repetitions based on stress ratio factor (with and without concrete shoulder). (Courtesy Portland Cement Association.)

Relatively little design information is available on composite pavements, although some experimental installations have been built, and the Corps of Engineers has extended methods used to design composite pavements for airfields to roads and streets.

CONSTRUCTION OF CONCRETE PAVEMENTS

The construction of portland cement concrete pavements is marked by the use of a large number of specially designed machines, each of which performs a

TABLE 20-7 Erosion Factors—Doweled Joints, No Concrete Shoulder (Single Axle/Tandem Axle)

Slab Thickness (in.)	k of Subgrade—Subbase (lb/in.³)					
	50	100	200	300	500	700
4	3.74/3.83	3.73/3.79	3.72/3.75	3.71/3.73	3.70/3.70	3.68/3.67
4.5	3.59/3.70	3.57/3.65	3.56/3.61	3.55/3.58	3.54/3.55	3.52/3.53
5	3.45/3.58	3.43/3.52	3.42/3.48	3.41/3.45	3.40/3.42	3.38/3.40
5.5	3.33/3.47	3.31/3.41	3.29/3.36	3.28/3.33	3.27/3.30	3.26/3.28
6	3.22/3.38	3.19/3.31	3.18/3.26	3.17/3.23	3.15/3.20	3.14/3.17
6.5	3.11/3.29	3.09/3.22	3.07/3.16	3.06/3.13	3.05/3.10	3.03/3.07
7	3.02/3.21	2.99/3.14	2.97/3.08	2.96/3.05	2.95/3.01	2.94/2.98
7.5	2.93/3.14	2.91/3.06	2.88/3.00	2.87/2.97	2.86/2.93	2.84/2.90
8	2.85/3.07	2.82/2.99	2.80/2.93	2.79/2.89	2.77/2.85	2.76/2.82
8.5	2.77/3.01	2.74/2.93	2.72/2.86	2.71/2.82	2.69/2.78	2.68/2.75
9	2.70/2.96	2.67/2.87	2.65/2.80	2.63/2.76	2.62/2.71	2.61/2.68
9.5	2.63/2.90	2.60/2.81	2.58/2.74	2.56/2.70	2.55/2.65	2.54/2.62
10	2.56/2.85	2.54/2.76	2.51/2.68	2.50/2.64	2.48/2.59	2.47/2.56
10.5	2.50/2.81	2.47/2.71	2.45/2.63	2.44/2.59	2.42/2.54	2.41/2.51
11	2.44/2.76	2.42/2.67	2.39/2.58	2.38/2.54	2.36/2.49	2.35/2.45
11.5	2.38/2.72	2.36/2.62	2.33/2.54	2.32/2.49	2.30/2.44	2.29/2.40
12	2.33/2.68	2.30/2.58	2.28/2.49	2.26/2.44	2.25/2.39	2.23/2.36
12.5	2.28/2.64	2.25/2.54	2.23/2.45	2.21/2.40	2.19/2.35	2.18/2.31
13	2.23/2.61	2.20/2.50	2.18/2.41	2.16/2.36	2.14/2.30	2.13/2.27
13.5	2.18/2.57	2.15/2.47	2.13/2.37	2.11/2.32	2.09/2.26	2.08/2.23
14	2.13/2.54	2.11/2.43	2.08/2.34	2.07/2.29	2.05/2.23	2.03/2.19

Source: Thickness Design for Concrete Highway and Street Pavements. Portland Cement Association (1984).

specific function in the construction process. Once paving operations are begun, the various steps in the construction procedure are arranged in the form of a continuing series of separate operations that are planned and coordinated so that the construction proceeds with a minimum loss of time and effort. Each of the separate steps must be done carefully and precisely so that the completed pavement will meet the exacting standards for structural strength and smoothness that are applied to it. The exact methods and machines used in the construction process vary somewhat from job to job, and no attempt will be made here to discuss all the possible variations in job methods and procedures. The following, however, is the sequence of separate steps on typical projects:

1. Preparation and preliminary finishing of the subgrade.
2. Placing of forms (where used).
3. Final finishing of the subgrade.
4. Installation of joints.
5. Batching of aggregates and cement.
6. Mixing and placing concrete.
7. Placing and finishing concrete.
8. Slipform paving.
9. Curing.

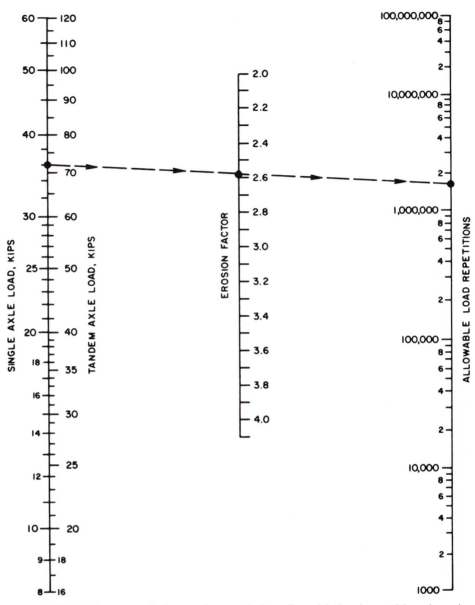

FIGURE 20-11 Nomograph for erosion analysis—allowable load repetitions based on erosion factor (without concrete shoulder). (Courtesy Portland Cement Association.)

20-20 Preparation and Preliminary Finishing of the Subgrade

The preparation of the subgrade on which a concrete pavement is to rest is, of course, a very important step in the overall construction process. It is essential that uniform subgrade support be provided for the completed pavement during its service life and that the pavement be free from other detrimental effects associated with unsatisfactory subgrade soils. Included among the most commonly encountered problems that are related to the character and condition of the subgrade soil are pumping and frost action, whereas other difficulties may become evident when soils of inadequate shearing strength, high-volume-change soils, organic soils, alkali soils, soils that are difficult to drain, and others are encountered. Provision must be made for elimination of these defects in the subgrade soil before the pavement is placed.

In some cases the existing subgrade soil may be entirely suitable as a foundation and thus require only compaction and adequate drainage prior to the placing of a concrete slab. In other instances a layer of imported soil may be used between the existing soil and the slab, or the existing soil may be stabilized, as with portland cement, bituminous material, or granular materials. It is common practice to provide a blanket or "insulation" course of suitable granular material between the existing soil and the pavement itself. Various materials are used to form this layer or "subbase," including gravel, sand, sand and gravel, crushed rock, and so on. The depth of treatment varies from a minimum of about 75 to 150 mm (3 to 6 in.) or so if the primary purpose is to prevent pumping; very much thicker layers of granular material may be used to prevent frost action in northern states.

In recent years many states, particularly those in the Far West, Southeast, and Southwest, have gone to the use of treated materials for subbases beneath concrete pavements. Most frequently used is a cement-treated base course, described more fully in Section 20-31, but some states now use asphaltic mixtures for this purpose.

Whatever the solution used in a specific case, the layer immediately beneath the slab must be brought to a high degree of density and stability and must be adequately drained. Preliminary finishing of this layer will generally be accomplished by the use of motor graders or similar equipment that shape the layer to the desired cross section, elevation, and line in what have previously been termed "fine-grading operations." In most cases the subgrade layer will be brought very close to final shape before construction associated with the slab itself actually begins.

20-21 Placing of Forms

Most concrete highway pavements are constructed with the slipform paver (see Section 20-26); however, steel forms are used for many city streets and some small highway projects. Where steel forms are used, they must be very carefully placed and secured in position so that the desired position, width, elevation, and grade may be secured in the final slab. Forms commonly used in highway work are straight 10-ft sections that are aligned both vertically and horizontally by slip joints and are held in position by three or more steel stakes driven at intervals at the back of the form. Forms of this type vary somewhat in size and weight, being available in heights of 6 to 12 in., with corresponding base widths over a similar range. One frequently used size, for example, has a height of 9 in. and a base width of 8 in. Special forms are available for curbs, curb and gutter sections, and for the construction of sharp curves, including both rigid and flexible-type curved forms.

Several machines are frequently used in connection with the placing of the forms. One of these is the so-called "form grader" that cuts a trench of exact size in the correct position to receive the forms, as shown in Figure 20-12. With this machine the trench is cut with a hydraulic cutter, which is adjustable hy-

FIGURE 20-12 Cutting of trench to exact line and grade, prior to placing of forms. (Courtesy Portland Cement Association.)

draulically by the operator. Forms are then generally aligned and placed by hand, with the steel pins being driven either by hand or with the aid of an air hammer. A second machine may then be used to tamp the forms securely in place. The placing of the forms on most jobs is characterized by extreme care, as many of the machines used in later operations ride on the forms and it is imperative that the forms be securely and carefully aligned and placed. Their elevation and alignment are, of course, carefully checked prior to continuance of construction operations. Concrete pavements generally are constructed in two-lane widths; forms are used on both sides. Sufficient forms are generally required to be available on the job to permit the forms to remain in place during preliminary curing of the slab and to provide several hundred feet of forms in place ahead of the paver.

20-22 Final Finishing of the Subgrade

The next step in the indicated sequence of operations at the job site is the final shaping of the subgrade to the exact dimensions required by the plans and specifications. This operation is generally accomplished by a machine called a "subgrader" or "fine-grader," which rides on the forms and cuts off the subgrade to the exact shape desired. The subgrade is generally left a little high prior to this step, and the earth (or subbase material) excavated by this machine is thrown outside the forms. Extensive use also is now being made of electronically controlled fine-graders without forms, particularly when slipform pavers are used. This machine operates from a taut wire. Trimming is done by cutting blades on a rotating drum. In some instances, trimming of the subgrade is accompanied by final rolling with steel rollers, particularly where a granular subbase is being built.

A scratch template is generally employed at this stage to check the final finish of the subgrade.

20-23 Installation of Joints

The installation of the various types of joints that may be used in a concrete pavement is also a very important step in the construction process. A portion of the process of construction of the required joints, which we may call the "installation of joint assemblies," normally takes place between the final finishing of the subgrade and the beginning of actual concreting operations. Such assemblies may not be required, as for a plain concrete pavement in which transverse joints are sawed and dowel bars are not used. Sawing of joints is covered in Section 20-28. Extreme care must be used in all operations accompanying the construction of joints if they are to function properly. Dowel bars, as commonly used in transverse joints, must be carefully placed and aligned parallel to the centerline and subgrade so they will not inhibit free movement of the slab ends in longitudinal direction and will properly perform their load-transfer functions.

Joint assemblies which are used to facilitate the construction of joints are quite varied in nature, depending on the exact type of joint that is to be used. Typical joint assemblies are shown in Figure 20-13, which illustrates the use of the dowel unit manufactured by the Bethlehem Steel Company. The removable cap shown in Figure 20-13a is removed after the concrete has been placed and subjected to preliminary finishing. The slot shown in Figure 20-13b may be made just after the concrete has been finished by pressing a thin steel plate of appropriate dimensions into the concrete at the desired location or by sawing the hardened concrete. The spaces above the joint filler or the surface slot in a weakened plane joint, Figure 20-13b, are filled with one of several sealers previously discussed

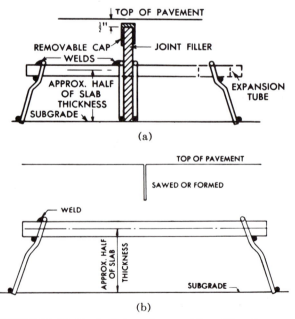

(a)

(b)

FIGURE 20-13 Typical dowel assemblies. (Courtesy Bethlehem Steel Co.)

after the concrete has hardened. The sealer is forced into the joint spaces by special machines. Edges of the surface groove except for sawed joints are usually finished to a curve that has a $\frac{1}{4}$-in. radius. It should be noted that dowel bars used in transverse joints are commonly treated with paint, grease, or other materials on one side of the joint to prevent bonding of the bar to the concrete; the other end of the bar will, of course, be bonded. In transverse joints in which dowel bars are not used, the only steps involved in the construction are the formation of the surface groove, subsequent edging of the joint, and installation of joint sealer. Longitudinal joints are installed in a similar fashion, except that the tie bars are bonded to the concrete on both sides of the joint.

20-24 Mixing of Concrete

At this point in our discussion of the construction of concrete pavements, all preparations have been made at the job site for the placing of concrete between the forms. Let us now examine the steps of preparing and mixing the concrete.

Formerly, aggregates and cement were batched "dry" at a convenient location some distance away from the job site, brought to the site in trucks, and mixed with water on the job, usually in a boom-and-bucket paver. This method is rarely used today and is not described here.

Today, on most highway projects, all ingredients of the concrete, including the water, are batched and mixed at a central location. The fresh concrete is then transported to the paving site in transit mixers or in trucks with agitating or nonagitating bodies.

Central mixing plants vary widely. In urban areas, the plant often is a highly sophisticated, stationary "ready-mix" plant capable of producing automatically many mixtures of varying composition for many different purposes. Other plants are portable or semiportable; they are automatically controlled, can be moved from one location to another quite easily, and have production rates up to about 418 m³/hr (500 cu yd/hr) with a single mixer.

Nearly all modern concrete plants are fully automatic and equipped with recording systems. A single operator at a control panel performs simple steps that control all the functions of charging, mixing, and discharging. Ingredient proportions can be controlled by preset dials, tapes, or batch plugs. The control system incorporates an automatic moisture-sensing device in the fine aggregate; adjustments are made automatically in batch quantities to correct for changes in moisture content. A plant also may be equipped with a "slump meter," which measures the current required to turn the mixer; the meter is calibrated with slump tests and gives an immediate check on the consistency of the concrete.

Permanent plants have more elaborate control systems that sense and govern the flow the all materials from the aggregate stockpiles through the plant to discharge of the concrete; ready-mix trucks often are dispatched from a central location by two-way radio.

Figure 20-14 shows a close-up view of a tilting-drum mixer discharging concrete into a truck with an elevating discharge, agitating truck body.

20-25 Placing and Finishing

Placing of the fresh concrete at the job site and finishing the slab to required smoothness are important operations that are done in a variety of ways. The

FIGURE 20-14 Tilting-drum mixer discharging concrete into agitating body truck. (Courtesy Wire Reinforcement Institute.)

number and kind of machines used vary from state to state and among contractors.

If a boom-and-bucket paver is used, the bucket dumps the concrete on the subbase ahead of a spreader. The most commonly used spreader is one that incorporates a long screw or auger across the front of the machine; the screw distributes the concrete uniformly across the subbase between the forms. The spreader may carry vibrators and often has a transverse screed at the rear to give the slab a preliminary strike-off.

In a conventional paving train, the spreader is followed by a finishing machine that incorporates two transverse, oscillating screeds. Vibrators may be mounted on the front of this machine. This unit often pulls along a long-wheelbase, float-type finisher. The float, which is suspended from a frame, operates independently of the side forms to iron out any irregularities left by the transverse screeds.

Also available are "one-pass" finishing machines that ride on the side forms, combine in one unit a spreader, vibrators, transverse screeds, and longitudinal float. These units have been described as "slipform pavers riding on side forms." (See Section 20-26.)

Contractors often use special machines to place central mix concrete between

FIGURE 20-15 Agitating-body trucks discharging concrete into hopper-type spreader. (Courtesy *Engineering News-Record.*)

forms. These machines use hoppers to receive the concrete. In one machine of this type, the hopper travels across the grade to deposit the fresh concrete between the forms. Another machine of this type is shown in Figure 20-15; on this project, two agitator-body trucks dumped concrete simultaneously into hoppers on either side of the grade. A transverse screw moved the concrete across the grade.

Simple box-type spreaders often are used with slipform pavers.

The placing of distributed steel reinforcing in the form of either bar mats or wire mesh, complicates the spreading operation but generally leaves the finishing operations unchanged. One way of placing the steel is to lay one course of concrete with a spreader; workmen then place the sheets or mats of steel on top of the fresh concrete. The spreader then makes a second pass over the steel (or a second spreader may be used).

A second method is to spread the concrete full depth, lay the steel on top, and then use another machine (a mesh installer) to force the steel into the concrete to the required depth. This operation is shown in Figure 20-16.

Special methods are used to place continuous reinforcement.

The pavement is generally checked during final finishing operations by means of long wooden, steel, or aluminum straightedges and any irregularities removed.

The final finish is applied to the surface by belting, brooming, or use of a burlap drag. Belting is done by the use of a narrow canvas or rubber belt that is moved longitudinally along the surface with a slight transverse motion; two people handle the belt, one on either side of the slab. To secure a broomed finish, which is considered to have slightly greater skid resistance than a belted surface, long-handled fiber brooms are used. The finish is obtained by placing the broom at one edge of the pavement and drawing it transversely across the surface in such a way that corrugations are produced that are not more than $\frac{1}{8}$ in. in depth. Mechanical brooms are now common. A burlap drag is pulled longitudinally along the surface of the pavement, thus producing shallow surface corrugations; this is the method most commonly used in modern practice. The joints and edges of the pavement may be given a final finish, generally by the use of hand tools, with workmen operating from bridges as required.

FIGURE 20-16 Blades of mesh placer push and vibrate reinforcing steel into fresh concrete. (Courtesy Portland Cement Association.)

The surface of the pavement is generally checked again after the concrete has hardened, with surface irregularities up to $\frac{1}{8}$ in. (as measured by a 10-ft straight-edge) being permitted. High spots detected at this stage may be removed by grinding. Use is also being made of "profilographs" or "road roughness indicators," which measure surface variations over a length of pavement by an accumulative measurement of deflections of a test tire. Roughness is usually expressed in terms of inches per mile.

20-26 Slipform Paving

Introduction of the first production models of slipform pavers in 1954 revolutionized concrete paving in the United States (*11*). By 1985, slipform paving was being used almost exclusively for highway concrete paving projects.

Principal advantage of the slipform paver is the fact that one machine, under the control of a single operator, replaces the several machines in a conventional paving train. Hand finishing is held to a minimum. Because there are no side forms used, the labor of setting and handling the forms is eliminated. Pavements of outstanding smoothness have been built by the slipform method.

Several manufacturers produce slipform pavers, which vary in details of design and operation. All operate on the same general principle, however, to combine several operations in the one machine. Each of the machines has an electronic guidance system that operates from a taut wire to maintain line and grade; the taut wire often is the same one used to guide the subgrader (Section 20-20) during final finishing of the subbase.

Concrete is dumped on the subbase ahead of the paver and spread to a reasonably uniform depth, then struck off by an oscillating screed. The concrete is then vibrated heavily and forced through the space formed by the traveling side forms that are part of the paver, the subbase, and a heavy transverse pan or beam (or beams) at the rear of the machine. The beam forms the surface of the slab and can be adjusted to give the desired cross section. Some machines of this type have trailing side forms of variable length behind the machine to restrain the fresh concrete while the machine moves slowly forward; others carry no trailing forms. Careful control of the consistency (usual slump is 1 to $1\frac{1}{2}$ in.) and entrained air content is essential for successful results.

Figure 20-17 shows a slipform paver in use on a highway project; conventional dump trucks supply concrete to the paver, which has no trailing forms. The edge of the slab behind the end of the trailing forms (on another make of paver) is shown in Figure 20-18; a mason is touching up the slab edge.

Finishing operations vary, but handwork is held to a minimum. Figure 20-19 shows a long aluminum tube to float the surface of the slab on some projects; the keyed joint is to provide vertical load transfer for an additional lane to be placed in a subsequent operation.

20-27 Curing

Curing of a concrete paving slab is necessary in order that the concrete harden properly. It should be noted that water is absolutely necessary for the proper hydration of the cement and that the hardening of the concrete is not a drying-out process. Steps then must be taken to prevent loss of moisture from the concrete during the curing period. A large number of different curing methods are available, and the specifications of highway agencies related to this phase of concrete construction may permit several different alternative procedures to be used.

Earth or straw may be spread over the surface of the pavement and kept

FIGURE 20-17 Slipform paver at work on an interstate project. (Courtesy Portland Cement Association.)

FIGURE 20-18 Pavement edge left by trailing form of slipform paver. (Courtesy Wire Reinforcement Institute.)

constantly wet during the curing period. Another "wet-curing" process involves the spreading of burlap, felt, or cotton mats over the surface; the mats are sprinkled and kept constantly wet. Waterproof paper may be placed over the slab to retain the moisture and is widely used in some areas. Burlap, cotton mats, and paper are commonly furnished in rolls that cover the entire width of the pavement.

FIGURE 20-19 Aluminum tubular float finishing a concrete pavement. (Courtesy Wire Reinforcement Institute.)

By far the most popular method in current use involves the spray application of light-colored fluid curing compound to the entire area of the wet concrete. This is the commonly used "membrane" method. The fluid forms a film over the pavement that prevents moisture loss. The color in the fluid disappears after the passage of time. Membrane fluid is commonly applied by special spray machines that ride on the forms, or on rubber tires if a slipform paver is used, and ensure a uniform application of material over the entire area. The pavement edges as well as the riding surface must be cured.

Preliminary wet curing is required by some highway departments, and the period of preliminary curing generally varies from 1 to 3 days. Total curing time, as evidenced by the period which must elapse before the pavement may be opened to traffic, commonly varies from 7 to 14 days. Opening of the pavement to traffic is very frequently based on the attainment of a certain minimum flexural strength of the hardened concrete rather than on an arbitrary time period.

20-28 Joint Sawing

Most states in which a substantial mileage of concrete pavements is being built either permit or require the sawing of transverse joints; longitudinal joints are also being sawed in some cases. Sawing is done by the use of self-propelled, manually guided, single-blade concrete saws, as illustrated in Figure 20-20, or by multiblade saws that ride directly on the pavement or the side forms. To make joint sealing more effective, some states now require a step-down joint with a wider groove at the surface; such joints are cut by two or more saws operating in tandem. The geometry of the joint varies with the type of sealant.

Joints are sawed a short time after the concrete has been given its final finish. The time at which sawing is done is critical. On the one hand, the sawing must be done before random cracking occurs; however, premature sawing causes excessive spalling, water erosion, and excessive blade wear. The best way to determine the best time for sawing probably is to make a short trial cut a few hours

FIGURE 20-20 Concrete saw with a diamond blade cutting a pavement joint. (Courtesy Portland Cement Association.)

after final finishing operations; appearance of the cut is then evaluated. Time of sawing varies from as little as 4 to as much as 24 hours after placing.

20-29 Other Considerations

Forms may generally be removed from the hardened concrete after 12 to 24 hr. Special devices are usually used to pull the forms so that the pavement will not be damaged in this process. Edges of the slab may be given their final finish at this time. At the end of the curing period and before opening the road to traffic, the surface grooves in longitudinal and transverse joints are cleaned and filled with joint sealing compound or some other material, like the preformed compression seal described in Section 20-11.

Some mention may be made of concreting in cold weather. Most highway organizations suspend concreting operations when the air temperature is 4°C (40° F) (or less) and falling and resume operations when the temperature is 2°C (35° F) and rising. In many sections of the country concrete construction is completely suspended during the winter months. When concrete must be placed in cold weather, the heating of aggregates and water to moderate temperature is quite common practice. In such circumstances the temperature of the concrete prior to placing may generally be required to be from 50° to 70° F. Calcium chloride frequently is used as an accelerator to hasten the initial set of concrete placed in cold weather.

20-30 Concrete Base Courses

Concrete base courses are used occasionally to provide a support for a high-type bituminous pavement (as in a vehicular tunnel). Their principal use at the present time is in urban areas. They are generally built of uniform thickness, with thickness of 6 to 9 in. being common.

Concrete bases have been built of both plain and reinforced concrete, with and without joints to control cracking of the base. Authorities differ as to the exact structural behavior of concrete bases that support a bituminous wearing surface. The combined thickness of the base and wearing surface must, of course, be adequate to support the loads placed on the pavement structure. Many designers believe that a bituminous wearing surface of nominal thickness contributes but little to the structural strength of the pavement. It is customary in some organizations, however, to use concrete in the base course that has slightly less strength than that required of concrete for paving. A requirement of a flexural strength of 450 lb/in.^{2} is perhaps typical of such practice. This decrease in strength is generally justified by the fact that temperature warping stresses should be somewhat less in concrete bases than in pavements.

Another consideration related to structural behavior of the base is the fact that extensive cracking that occurs in the base will generally be reflected in the wearing surface over a period of time. With this fact in mind, many designers feel that enough joints and reinforcing steel should be used in the base to effectively control cracking. Again, because of lessened temperature warping stresses, longitudinal joints may not be necessary in a concrete base which has a width of 20 ft or so. Contraction joints of the weakened-plane type are generally provided in concrete bases to control cracking, while transverse expansion joints are less commonly used. It is recommended that transverse joints in concrete base courses be placed at short intervals (4.5 to 6 m or 15 to 20 ft) in order to prevent excessive opening of the joints and resultant cracking of the surface course.

Methods used in the construction of concrete base courses are very similar to those used in the construction of concrete pavements. Requirements relative to the surface finish of the completed base may be somewhat less severe than similar requirements relative to pavements. The surface of the concrete is usually given a roughened, broomed finish where a bituminous wearing surface is to be placed. The base course must be cured as indicated previously for concrete pavements.

20-31 Cement-Treated Bases

Cement-treated bases are used widely beneath concrete pavements and to a lesser extent beneath asphaltic wearing surfaces. Principal purposes of a cement-treated base are to provide a stable base for the construction of the concrete slab, particularly when a slipform paver is used, and to prevent pumping.

A cement-treated base course mixture is very similar to the soil-cement mixtures described in Chapter 18. In many cases, however, there are two major differences: The material with which the cement is mixed is a crushed or un-crushed granular material of moderate to low plasticity, and cement contents are relatively low (from $2\frac{1}{2}$ to 6 percent by dry weight of aggregate in California, where bases of this type have been built for many years).

Specifications for materials vary, but typically a clean graded aggregate with a maximum size of 25 mm (1 in.) and from 3 to 15 percent passing a No. 200 sieve would be specified for a class A cement-treated base. Cement-treated bases vary from about 100 to 175 mm (4 to 7 in.) in thickness. State highway agencies usually specify a minimum compressive strength at 7 days. PCA (*12*) recommends that enough cement be mixed with the granular materials to develop an unconfined compressive strength of not less than 2068 kPa (300 lb/in.²) at 7 days.

Construction of a cement-treated base course is by either road mixing or plant mixing, using the general methods described in Chapter 18. The mixture is compacted at or near optimum moisture to maximum density. It is cured as is soil-cement, often by applying a bituminous material to the top surface.

PROBLEMS

20.1 A concrete pavement is to be built on a subgrade with a k value of 150 lb/in.³ The combined k of the subgrade and a 4-in. thick untreated granular base is 200 lb/in.³ The design period is 20 years. Over this period, the expected loadings are tabulated below:

Single Axle Load, kips	Expected Repetitions
30	20,000
26	75,000
22	300,000
18	700,000
14	900,000

Tandem Axle Load, kips	Expected Repetitions
48	50,000
40	480,000
32	1,700,000
24	2,000,000

1. Determine the minimum thickness of slab that can be used safely for this rural interstate route if the modulus of rupture of the concrete is 650 lb/in.^{2} (28 days). Assume doweled joints and no concrete shoulder.
2. Can a thinner slab be used for this pavement if the modulus of rupture of the concrete is increased to 700 lb/in.^{2} at 28 days?

20-2 Solve Problem 20-2 if a cement-treated base with a design k of 500 lb/in.^{3} is substituted for the untreated granular base.

REFERENCES

1. *Standard Specifications for Transportation Materials and Methods of Testing, Part I, Specifications.* American Association of State Highway and Transportation Officials, Washington, DC (1982).

2. *Standard Specifications for Transportation Materials and Methods of Testing, Part II, Tests.* American Association of State Highway and Transportation Officials, Washington, DC (1982).

3. *ACI Manual of Concrete Practice 1984, Part I, Materials and General Properties of Concrete.* American Concrete Institute, Detroit, MI (1987).

4. *Design and Control of Concrete Mixtures.* Portland Cement Association, Skokie, IL (Revised 1990).

5. McLeod, N. W. *Airport Runway Evaluation in Canada.* Highway Research Board Research Report No. 4B (1947) and Supplement (1948).

6. *Flexible Pavement Design. Highway Research Record 13,* Transportation Research Board, Washington, DC (1965).

7. *Flexible Pavement Design—1963 and 1964. Highway Research Record 71,* Transportation Research Board, Washington, DC (1965).

8. *Joint Design for Concrete Highway and Street Pavements.* Portland Cement Association, Skokie, IL (1980).

9. *1986 Guide for the Design of Pavement Structures.* American Association of State Highway and Transportation Officials, Washington, DC (1986).

10. *Thickness Design for Concrete Highway and Street Pavements.* Portland Cement Association, Skokie, IL (1984).

11. *Symposium on Slip Form Paving. Highway Research Record No. 98,* Transportation Research Board, Washington, DC (1965).

12. *Subgrades and Subbases for Concrete Pavements.* Portland Cement Association, Skokie, IL (1971).

CHAPTER 21

HIGHWAY MAINTENANCE AND REHABILITATION

When a highway construction project has been completed, accepted from the contractor, and final payment has been made to the contractor, a new facility is available for use by the traveling public. By the same token, a new responsibility is created to preserve the new investment and to serve and protect the interests of the traveling public. This is true not only for newly constructed highways but also for all highways of our road system. Sudden failures, damage by storms, gradual deterioration, and unexpected obstructions can cause personal injury, death, or delay.

"Highway maintenance" is defined as the function of preserving, repairing, and restoring a highway and keeping it in condition for safe, convenient, and economical use. "Maintenance" includes both physical maintenance activities, such as patching, filling joints, mowing, and so forth and traffic service activities, including painting pavement markings, erecting snow fences, and removing snow, ice, and litter. It does not include major rehabilitation or reconstruction activities, such as widening the roadbed, or extensive resurfacing projects.

Highway maintenance programs are designed to offset the effects of weather, vandalism, vegetation growth, and traffic wear and damage, as well as deterioration due to the effects of aging, material failures, and design and construction faults (1). As Figure 21-1 illustrates, in spite of vigilant and determined efforts to maintain the serviceability of a highway, there comes a time when a major rehabilitation of the facility is required. The rehabilitation work, termed "betterment," is not considered maintenance, but the periodic routine activities before and after rehabilitation are. The AASHTO Maintenance Manual (1) includes a table that clearly distinguishes between the four general classes of roadway maintenance and construction activities:

1. Maintenance
     Traffic services
     Physical maintenance
2. Construction
     Betterment
     Construction and reconstruction

For example, resurfacing a pavement with bituminous material $3/4$ in. thick or more for a length of 500 continuous feet or more is considered to be "betterment." Resurfacing a pavement with material less than $3/4$ in. thick or replacement of pavement for a length less than 500 continuous feet is defined as "physical maintenance."

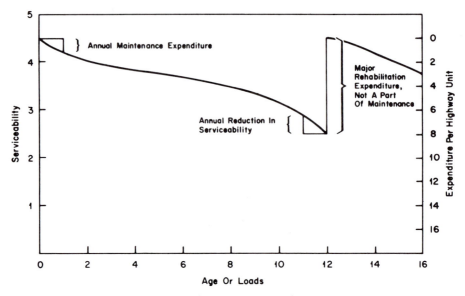

FIGURE 21-1 Typical highway performance curve. (Courtesy Transportation Research Board.)

21-1 Importance and Challenge of the Highway Maintenance Function

The importance of highway maintenance lies mainly in the need to protect the substantial investment that has been made in the highway plant. Furthermore, the significance of the maintenance function is seen in the vital need to continue the economic and safety benefits of the 3.9 million mile public road system. The traveling public is acutely aware of the condition of public streets and highways and is inclined to be quite vocal in its displeasure about potholes, ragged shoulders, illegible signs, and highway drainage structures that do not function properly.

In recent years, the highway maintenance function has increased in both complexity and scope. These changes have been brought about by the addition of significant mileage to the system, notably the Interstate System, by increases in vehicle mileage of travel, and, in many states, increases in the maximum size of allowable wheel loads.* In addition, the traveling public now expects a higher level of maintenance than it tolerated previously. Systematic mowing, large-scale landscaping, and litter control are examples of the escalation of maintenance standards that has occurred in recent years. Furthermore, the maintenance must frequently be performed under conditions of heavy traffic, entailing much greater difficulties as well as potential hazards to both workers and the traveling public.

Maintenance personnel face an enormous challenge in adequately maintaining the vast system of public roadways. The magnitude of the problem is seen in the size of the maintenance budgets for the various state highway and transportation

---

*Between 1970 and 1992, annual average daily traffic on the rural Interstate System increased almost 100 percent, and the total daily load based on truck weight study data increased nearly 400 percent (*2*).

agencies. In 1993, more than $9.5 billion was spent by states for maintenance and traffic services (2), representing about 15.3 percent of the total highway dollars disbursed. It is estimated that an additional $13.3 billion was spent for maintenance by counties, townships, and municipalities.

During the past decade, annual maintenance costs have increased at an average rate of about 5.5 percent, aggravating the maintenance engineer's difficult task of coping with rising traffic volumes and public demands for increased services.

Figure 21-2 shows the approximate services and cost distribution for highway maintenance in a typical year, on a national basis. There is, of course, considerable variation among the various states in how the maintenance dollar is spent, reflecting differences in climate, topography, and maintenance practices.

21-2 Maintenance Management

To increase the productivity of labor and equipment used in maintenance operations, highway agencies are increasingly applying modern management methods to the maintenance functions. By this technique, maintenance budgets are based on the estimated costs of accomplishing specific work programs to stated levels of service and performance standards. Maintenance management includes control procedures that are applied to ensure that the work accomplished by field personnel is in line with the objectives of the program and the budgeted funds available.

The basic purpose of maintenance management systems is to capture information about maintenance activities performed and resources expanded. Maintenance management systems do not manage programs, reduce costs, or improve performance; rather, they provide maintenance engineers and managers with the information and analytical tools needed to allow them to do so (3).

Although the details of maintenance management systems vary a great deal among the various highway agencies, there is considerable agreement as to the basic concepts and objectives of such programs. In the following paragraphs,

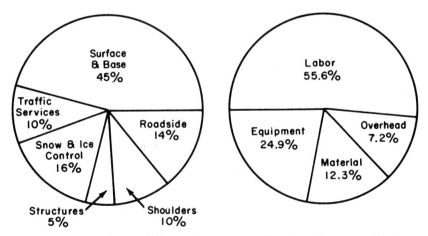

FIGURE 21-2 Pie chart of the maintenance dollar showing approximate services and cost distribution for a typical year. (Courtesy Highway Users Federation for Safety and Mobility.)

some of the essential elements of a maintenance management program are described. The subject is covered in more detail in Refs. 3, 4, and 5.

Typical features of a maintenance management system include

- development of an annual work program.
- budgeting and allocating resources.
- work authorization and control.
- scheduling.
- performance evaluation.
- fiscal control.

Work Program Development

This phase of the program normally involves the following types of activities or steps:

1. Defining work activities.
2. Establishing work quantity planning values.
3. Establishing maintenance performance standards.
4. Conducting a road inventory and inspection.
5. Estimating the size of the work program.

It is first necessary to define the various work activities in terms that are specific and measurable. Studies have shown that in a typical maintenance operation only 20 to 30 activities will account for 75 to 80 percent of the workload. Thus, the number of activities defined is relatively small and manageable. Typical work activity definitions are shown in Figure 21-3.

For each work activity, work quantity planning values must be established. These are the number of units of work required per year to provide a desired level of service. Such quantities are usually based on appearance, safety, and protection of the highway investment. Examples of work quantity planning values include 0.25 metric tons of asphaltic mix per lane kilometer per year, three mowings per acre per year, and so forth.

It is then necessary to establish, for each activity, performance standards indicating productivity of a typical crew. Because productivity depends on the size and makeup of the crew and the work method and equipment employed, such details constitute an essential part of the performance standard. The performance standard will normally show the productivity in terms of work units such as those indicated in Figure 21-3 (e.g., tons of premix placed per day, miles of shoulders bladed per day, etc., Fig. 21-4).

A road inventory and inspection may be required to identify and describe field conditions in need of maintenance. Many agencies maintain roadway inventory records, and it may only be necessary to supplement existing data. The inventory data provide up-to-date information on the types, amounts, and locations of roadway maintenance work.

Budgeting and Allocating Resources

The next step involves estimating the amount of labor, equipment, and materials required to carry out the program and computing the annual costs of the work.

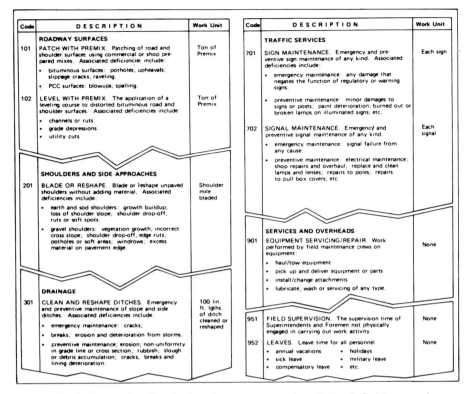

FIGURE 21-3 Example of typical maintenance work activity definitions and work measurement units. (Courtesy National Association of County Engineers.)

In most highway agencies, average costs of labor, equipment, and materials can normally be provided by the accounting division. It is then a simple matter to convert resources requirements of each work activity to the overall estimated activity costs. For example, the procedure for computing the cost of premix patching might be as follows:

Type	Resource Requirements		Average Unit Cost	Annual Cost
Labor	960 person-hr	×	$ 6.00	$5760
Equipment	320 dump-truck-hr	×	$ 3.00	960
Materials	250 tons	×	$12.00	1800
Total (activity code 101)				$8520

Similar calculations are performed for each work activity and the results are summarized for each class of work activities, for example, roadway surface, shoulders and approaches, and so forth.

Typically, cost data of this type are used by the highway agency's accounting staff to prepare a formal budget in accordance with established policy. After appropriate review and modification, the budget is approved. The approved budget then becomes the basis for legislative appropriation and authorization.

PERFORMANCE STANDARD	ACTIVITY: PATCH WITH PREMIX ACTIVITY CODE: 101 EFFECTIVE DATE:

DESCRIPTION AND PURPOSE:

Patching of road and shoulder surfaces using commercial or shop prepared mixes. Associated deficiencies include:

- bituminous surfaces: potholes, upheavals, slippage cracks, raveling.
- PCC surfaces: blowups, spalling.

PERFORMANCE CRITERIA:

Work program category is: Routine-Unlimited
Authority for performing work activity is: Foreman
Quantity control category is: Unlimited
Major attention should be placed during: Spring and Fall

CREW, EQUIPMENT, MATERIAL	WORK METHOD
Crew Size (incl. flagmen): 1 driver/workman 1 workman 1* flagman * Adjust to traffic conditions.	1. Place safety devices and signs. 2. Square up and clean potholes, depressions. 3. Tack sides and bottom of hole. 4. Shovel mix into hole and level. 5. Compact with hand tamper or truck wheels. 6. Place top layer. 7. Screed and compact again.

Equipment:

Description	No.	Class
3T truck	1	678

Material:
01 — hot or cold premix
misc. tack

PRODUCTIVITY	WORK UNIT
Daily Production: 3.0 to 4.5 tons	ton of premix

FIGURE 21-4 Sample performance standard. (Courtesy National Association of County Engineers.)

Work Authorization and Control

"Historically, there has been reasonable control of budgeted expenditures, but little control of operations. There is abundant and consistent evidence of great and unwarranted variations in the quantities of work performed for comparable situations. Also, the ways in which work is performed and the productivity of work forces have varied tremendously" (*4*). Operational control of the work is required if performance budgeting is to be meaningful.

In a state agency, the authorization to proceed with the maintenance work must be passed along the various levels of administration: from top management, to the district engineer, to the subdistricts, to the areas where the work is to be performed.

The recommended system of authorizing work employs a work-order approach in which crew schedule cards are distributed through upper management personnel to first line supervisors (foremen) of work crews. An example crew

CREW SCHEDULE CARD		CREW SIZE 3	ACTIVITY 101

RD. CLASS RRI	AREA 01	CARD NO. 26-1	

ACTIVITY *Patch with Premix*

LOCATION *Route # 68 & 57*

FOREMAN *Smith*   DATE *1-16-*

LABOR		EQUIPMENT	
HRS.	EMPLOYEE	NUMBER	HRS.
8	Al Ray	678-0609	7
8	Joe Dobb		
8	Pete Jones		
24	TOTAL		

MATERIAL			
DESCRIPTION		UNIT	AMOUNT
Bituminous Material		Tons	3

ACCOMPLISHMENT _____3_____ UNIT *Tons of Bit*

(front)

PATCH WITH PREMIX

DESCRIPTION AND PURPOSE:

Patching on road and shoulder surfaces using commercial or shop prepared mixes.

Associated deficiencies include:

- bituminous surfaces: potholes, upheavals, slippage cracks, edge raveling.

- PCC surfaces: blowups, spalling.

WORK METHOD:

1. Place safety devices and signs.

2. Square up and clean potholes, depressions.

3. Tack sides and bottom of hole.

4. Shovel mix into hole and level.

5. Compact with hand tamper or truck wheels.

6. Place top layer.

7. Screed and compact again.

(Back)

FIGURE 21-5 Sample crew day card. (Courtesy National Association of County Engineers.)

schedule card is shown in Figure 21-5. Each card should represent a single day's work for a standard size crew performing the specified work. Crew schedule cards also provide a convenient means for supervisors to report on the work performed. Typically, the area foreman records the actual times required, the equipment used, the travel time, materials used, and so forth. Information from the crew schedule cards are periodically summarized and evaluated by the maintenance engineer and other maintenance managers. By such means, it is possible to determine the progress of work in each area and how well foremen are adhering to recommended crew sizes and levels of performance.

Scheduling

In order to provide a reasonable balance of workload throughout the year, a seasonal schedule of maintenance operations should be developed to facilitate the schedule of appropriate work activities during specific periods of the year (Fig. 21-6). Normally, each maintenance foreman schedules a designated number of crew-days for each work activity under his supervision. He will also usually be required to make appropraite arrangements to ensure availability of needed materials and equipment.

Typically, each road maintenance division has equipment scheduling boards that indicate reservations for and locations of available equipment at all times. Some agencies assign a priority rating to each project to ensure proper scheduling of the maintenance equipment.

OPERATION   NORMAL TIMELINESS

JAN. FEB. MAR. APR. MAY JUNE JULY AUG. SEPT. OCT. NOV. DEC.

1. Snow & Ice Removal
2. Tree Removal
3. Roadside Cleanup
4. Markers & Signs
5. Tree Trimming
6. Repair Picnic Tables
7. Snow Fence–Dismantling & Erection
8. Blading Of Shoulders
9. Repair Drainage Structures & Ditches
10. Patching
11. Seeding & Sodding
12. Painting Guard Rail & Posts
13. Regalvanize Guard Rails & Bridge Railings
14. Stabilization Of Shoulders
15. Crack Filling
16. Joint Sealing
17. Mudjacking
18. Pavement Marking
19. Bituminous Surface Treatment
20. Erosion Control & Repair
21. Maintenance Of Special Roadside Areas
22. Roadside Mowing
23. Weed Spraying
24. Soil Sterilants
25. Preparation For Winter Maintenance
26. Structure Painting
27. Bridge Deck Cleaning

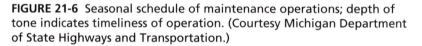

FIGURE 21-6 Seasonal schedule of maintenance operations; depth of tone indicates timeliness of operation. (Courtesy Michigan Department of State Highways and Transportation.)

Performance Evaluation

The work performance of maintenance crews, as reported on the crew schedule cards, should be periodically summarized and reviewed by management personnel. Such summaries typically include, for each maintenance activity, (1) the number of crew schedule cards used to date, (2) the number of person-days or person-hours reported, and (3) the work units accomplished. This permits the maintenance engineer and other managers to know the progress of work, the productivity of the various work crews, and how well the foremen are adhering to established practices.

Fiscal Control

Most highway agencies also have fiscal control procedures to ensure that budgetary limitations are observed. Control is exercised by the issuance of timely periodic status reports to persons with the authority for expending money or obligating expenditures. Maintenance managers should carefully monitor the status of expenditures and costs and their relationship to the performance budget. Careful fiscal review of maintenance operations will make it possible to avoid overexpenditures of budgetary allotments and to recommend adjustments in such allotments when unexpected contingencies arise.

In this section, we have assumed that the maintenance activities are performed by agency employees. A number of agencies have considered performing highway maintenance by private contractors. Reference 6 gives recommended methods and procedures in contract maintenance and guidelines for preparing contract bids, along with sample forms.

MAINTENANCE OPERATIONS

In the following paragraphs, we will briefly describe typical maintenance operations in highway agencies in the United States. It is convenient to categorize our discussion of these practices in terms of the following types or classes of highway maintenance, the maintenance of (1) roadway surfaces; (2) shoulders and approaches; (3) roadsides; (4) bridges, tunnels, and drainage structures; (5) traffic control and safety devices; and (6) the control of snow and ice. For further guidance on these matters, the reader should consult Ref. 1.

MAINTENANCE OF ROAD SURFACES

21-3 Earth Aggregate Road Surfaces

This classification of roads includes those constructed from natural earth, sand-clay mixes, and various coarse-graded aggregate materials such as gravel, crushed stone, crushed slag, and chert. Water-bound and traffic-bound macadam surfaces are included in this classification.

To maintain this type of road properly, it is necessary that the surface be kept smooth, firm, and free from excess loose material, with the proper crown for adequate drainage. A crown of not less than 1:24 nor more than 1:12 sloping outward from the centerline is recommended (*1*). The use of a dust palliative is frequently necessary in order to minimize the loss of material and to eliminate as much as possible the hazards due to dust. Maintenance of this type of road can include (1) patching, (2) blading, (3) scarifying and resurfacing, (4) stabilization and dust control, and sometimes (5) application of a bituminous surface treatment.

Patching

Failures of soil-aggregate roads are principally due to improper drainage, poorly mixed materials, or an inadequate foundation. A careful investigation should be made to determine which of these may be the cause of failure. Surface drainage can be ensured through the use of the proper amount of crown. Excessive moisture in the subgrade may be eliminated by lateral or side drains to intercept free water or to lower the water table. Perforated pipe underdrains may also be installed.

When road failure is due to a poor mixture or gradation of materials, it is usually necessary to remove the unsuitable material and replace it with suitable materials of the proper gradation. The edges of the area removed from the traveled roadway should be squared and the patching material placed and thoroughly tamped, adding water when necessary to obtain the proper compaction. When the area to be repaired is large, or where the surface is worn away, it may be better to scarify, reshape the surface, and add the necessary additional material.

Blading

Blading is a general operation carried out on all soil and stabilized surfaces. Its purpose is to fill in the ruts and smooth out any irregularities in the surface.

Blading is done by a motor grader. The versatile motor grader also can do many other maintenance operations such as shaping of shoulders, cleaning and shaping of ditches, and similar work. Blading also may be done by trucks with underbody blades. This equipment is usually attached to four-wheel-drive trucks and hydraulically operated.

Blading should be done as soon as practicable after a rain, when the surface materials are moist. The work is accomplished by dragging or blading the surface materials from the edges toward the center and then drawing the loosened material back again, depositing it in depressions in the surface. Some blading may be required during dry weather in order to remove surplus loose materials from the traveled roadway. This material is bladed to the side of the road and bladed back during wet weather or when the material is moist.

Scarifying and Resurfacing

When the areas to be patched are numerous, or when they extend over a considerable area, it is often more efficient to recondition the entire surface. The usual procedure is to scarify the surface of the roadway to the full depth of the surface material, proper care being exercised so that any undesirable material from the shoulder or the subgrade is not mixed with the surface-course materials. When additional material must be added, it should be selected with care. A visual inspection and a knowledge of local materials may aid in the selection of the necessary materials; however, if the section being repaired is of considerable extent, laboratory and field testing of the separate materials and their combinations may be necessary.

When the scarifying of the old road surface has been completed and any necessary material added and mixed, a blade grader is used to shape the road surface to proper crown and thickness. The necessary compaction may be accomplished by permitting traffic to use the surface while continuing blading or dragging operations, or by rolling and watering if necessary until the surface is thoroughly compacted.

Stabilization and Dust Treatment with Palliatives

Stabilization of roads with the use of binder soil such as clay, the chlorides, or other additives is often done by maintenance forces. The methods of stabilization are similar to those discussed in Chapter 18. The use of dust palliatives serves two purposes: (1) elimination of dust and (2) preservation of the soil-aggregate surface. Tests have indicated that the annual loss of material on gravel roads is equivalent to 13 to 25 mm ($\frac{1}{2}$ to 1 in.) of surface. The use of palliatives reduces this loss and at the same time aids in the stabilization of the surface.

The materials most commonly used as dust layers are calcium chloride, sodium chloride, and bituminous substances. Calcium chloride is most generally used because of its "hygroscopic" nature, that is, it attracts moisture and it is "deliquescent"—it dissolves itself in the moisture that it attracts. When calcium chloride is scattered over the road, it absorbs and holds two to five times its weight in water and then serves to keep the surface damp and dustless for prolonged periods.

In dry weather, calcium chloride will absorb moisture from the air during periods of high humidity, such as at night, in sufficient amounts to maintain a dustless surface during the day. Calcium chloride is susceptible to dilution by rain and gradually leaches into the soil, where, if present in sufficient concentration, it tends to reduce the severity of frost heaves.

Calcium chloride is available in flake and granular form, and best results are obtained when mechanical spreaders are used. Either a lime-drill type or a centrifugal spreader may be used. Overlapping of the spreads in the middle of the road will safeguard against leaving an untreated portion in the middle of the roadway.

The best time to apply the material is directly after a rain. The first application is usually made in the spring when dust begins to appear, followed by periodic applications during the summer months when needed. The initial application is made at a rate of 0.4 to 0.8 kg/m^2 ($\frac{3}{4}$–$1\frac{1}{2}$ lb/yd^2). Later application requirements will vary, depending on the texture of the surface, previous amount of stabilization, dryness of the season, and amount of traffic. Shaded areas require smaller amounts of palliatives than do areas exposed to the sun. The same is true for low-lying sections of road. Frequent applications may be desirable on curves, grades, intersections, and other places where excessive wear occurs.

Use is also made of various bituminous materials for laying dust on soil-aggregate roads. The surface to be treated must be bladed to proper crown before the material is added, and any loose material must be removed. Materials used may be tars RT-1 and RT-2; the slow-curing road oils SC-70 and SC-250; or the medium-curing cut-back MC-70. Asphalt emulsions are also used.

The roadway should be fairly dry and free from dust when the bituminous materials are applied, except when an emulsion is used, when the presence of moisture is desirable. Emulsions require a certain period for breaking and curing. If traffic must follow soon after the use of the emulsion, the surface is covered with a light coating of clean coarse sand.

The amount of bituminous material for dust-laying ranges from 0.5 to 1.1 L/m^2 (0.10 to 0.25 gal/yd^2). Emulsions often are diluted, one part emulsion to three parts water. Application should be made with a pressure distributor.

Bituminous Surface Treatment

Bituminous surface treatment operations, which are sometimes done by maintenance forces, have been described in Chapter 18. As traffic increased, the road surface may be improved in various ways. Care should be exercised to segregate maintenance costs from those of betterments as much as possible.

21-4 Maintenance of Bituminous Surfaces

The various types of bituminous surfaces and their construction have been described in Chapters 18 and 19. Weaknesses develop in these surfaces due to weathering, the action of alternate freezing and thawing, or failure of the base or subgrade. The maintenance of these surfaces may be divided into five distinct operations: (1) patching, (2) paint patching, (3) scarifying, (4) resealing, and (5) nonskid surface treatment. Reference 7 is recommended for more detailed information on the maintenance of asphalt pavements.

Before maintenance repairs are made to the surface, it is important to investigate the cause of failure. If failure indicates a poor subgrade or faulty drainage, these items should be corrected before any repairs are made on the surface. Some states are now using road roughness measurements and the pavement serviceability concept as an aid in setting up maintenance programs.

Patching may consist simply of an application of a thin coat of bituminous material to an area where cracking or raveling appears and spreading a coarse mineral aggregate over the affected area. Care must be exercised in the application of the proper amount of bituminous material in order to prevent bleeding, with resulting slipperiness of the repaired surface.

In areas where the surface shows raveling and disintegration, a bituminous mixture may be placed over the weakened area and then compacted. Both hot and cold bituminous mixtures are used, with asphaltic concrete and sheet asphalt being the most common. Where the surface has disintegrated to a depth of 50 to 75 mm (2 or 3 in.), the failed area should be cut out, all loose material removed, the base course made true and firm, and the bituminous patching material placed in the hole. The edge of the hole should be cut vertically and should receive a coat of bituminous material to seal out any moisture that may enter the subgrade. The final patch should conform to the grade and contour of the existing surface. The steps involved in making a permanent repair of an asphalt pavement are illustrated by Figure 21-7.

Patches may also be made by a penetration method that consists of filling the hole with aggregate, compacting the aggregate, and then applying the bituminous material. Crushed stone is then applied, and the patch is rolled. Many different patching methods are used with good results. The method used will depend on the type of bituminous surface and current practice prevailing in the particular locality.

Paint patching is essentially a preventive maintenance operation. Its purpose is to seal the surface to prevent penetration of water and to strengthen the wearing surface. The work consists of applying a thin coating of bitumen over the surface and placing fine stone chips, pea stone, or coarse sand over the patched area. The patch is then broomed and rolled with a light roller.

On low-type bituminous surfaces, such as bituminous treated gravel, where large areas are in poor condition, the entire surface should be scarified and reworked. This can be accomplished by the use of a motor grader with a scarifier attachment or with a scarifier and a mechanical pulverizer or mixer. A combination of a ripper (*not shown*) and the cutter-crusher-compactor shown in Figure 21-8 also may be used to salvage old bituminous roads and streets. The material is reworked, using additional bituminous material as required. It is then bladed to proper crown and grade and compacted. The compacted surface is then sealed with a bituminous material.

The seal coats and surface treatments described in Chapter 18 are widely used to rejuvenate old bituminous surfaces or to correct a slippery condition. Slipperiness is usually caused by an excess of bitumen on the road surface. This may be due to the application of too much bituminous material or an improper gradation of aggregate, or it may be caused by the kneading action of heavy traffic pushing the asphalt binder to the surface. Skidproofing may be achieved by application of a surface treatment using sharp, angular aggregate, such as blast-furnace slag, or by disking the surface. Slurry seal coats (Chapter 18) are widely used for this purpose. In severe cases the complete surface may have to be reworked in order to correct slipperiness.

Resurfacing of Old Bituminous Pavements

When maintenance operations become excessively expensive, it is difficult to maintain an adequately smooth surface, the pavement has to be widened, or for other reasons, the old pavement may be resurfaced. Before undertaking a resurfacing project, a careful study of its economic justification should be made. A general rule of thumb is that resurfacing will probably be economical when from 2 to 3 percent of the total pavement area requires patching each year. Both the availability of money and the priority of the project will influence the decision. A comprehensive study of the condition of the existing surface and related factors is desirable before undertaking a resurfacing project. Included among the factors that must be given detailed consideration are the following:

1. History of pavement.
2. Existing physical condition.
3. Character of existing and anticipated traffic.
4. Relation of geometric design to safe and efficient operation of anticipated traffic.

Corrective measures that may be required in the preparation of an old surface for resurfacing may be very considerable in extent or somewhat minor, depending on the type and condition of the existing surface and its adequacy to serve its new function as a base course. For example, major reconstruction measures may be required when substantial changes in grade, profile, or crown are called for in the new surface. Recommended reductions in grade and crown may be accomplished by scarifying and recompacting some old surfaces such as water-bound macadam or stabilized soils; higher types of surfaces may require removal and reconstruction to the desired grade and section. Changes in width are generally accomplished by duplicating the original construction as closely as possible, with particular attention being paid to the development of adequate supporting power in the widened section so that this section will not subside under service to a lower elevation than exists in the original pavement.

The old surface, regardless of its type, may be rough, with waves and bumps that are too considerable in magnitude to be compensated for by a relatively thin wearing surface. These bumps and waves may be removed by scarifying and recompacting certain surface types. When old bituminous wearing surfaces have these defects and are to be resurfaced, the bumps and waves may be removed by cutting or burning away the old surface with a heater planer. Burners in a big, insulated furnace heat and soften the surface, which is then scarified and remixed if desired, with additional bituminous material added as needed. Sometimes it may be more desirable to correct existing surface irregularities by the construction of a special leveling course, which may be the same bituminous mixture used in the original construction or a new mix. The construction of a leveling course is a major operation that usually is quite similar to the placing of a bituminous wearing surface. The course will have a variable thickness and will be constructed to the desired section and grade. The leveling course will have a lower bitumen content than the wearing course.

(a) Removing surface and base

(b) Applying tack coat to vertical surfaces

(c) Backfilling hole with plant—mix

FIGURE 21-7 Patching an asphalt pavement. (Courtesy The Asphalt Institute.)

(*d*) Spreading the mix

(*e*) Compacting the mix

(*f*) Straightedging the patch

FIGURE 21-7 *Continued*

FIGURE 21-8 Compaction-cutter-crusher attachment processes an old bituminous surface. (Courtesy American Tractor Equipment Company.)

Holes and depressions that occur in the existing surface must also be corrected before resurfacing (or the placing of a leveling course) may begin. The holes are filled with the same mixture used in the original construction or with a special patching mixture.

In many cases the preparation of the surface may consist simply of the removal of loose material on the surface by the use of rotary brooms or power blowers. If excessively dusty or unconsolidated areas remain after this treatment, they may have to be primed before construction may proceed.

When preparations are complete, the new surface is constructed by methods that are very similar to those described previously for new surfaces. Multiple-lift surface treatments and hot asphaltic paving mixtures, particularly asphaltic concrete and sheet asphalt, are used most frequently. Road-mix surfaces of various types are used to some extent.

21-5 Maintenance of Portland Cement Concrete Surfaces

The maintenance of concrete pavements consists for the most part of (1) filling and sealing joints and cracks in the pavement surface; (2) repairing spalled, scaled, and map-cracked areas; (3) patching areas where failure has occurred; (4) repairing areas damaged by settlement or pumping; and (5) treating buckled pavements.

The purpose of filling and sealing joints and cracks is to prevent the seepage of moisture to the subgrade and to preserve the original joint space. At the time the joint is sealed, the concrete must be dry and the joint space thoroughly cleaned of all scale, dirt, dust, and other foreign matter, including old joint sealer.

Power cutters are used to cut and groove the joints prior to resealing; a mechanically driven wire brush is used to thoroughly clean the joint. Immediately before sealing, the joint is blown out by the use of a jet of compressed air. Hot-applied rubber-asphalt sealing compounds are often used; they are generally heated and melted in a portable (rubber-tired) melting kettle of the double-boiler, indirect heating type which uses oil or other suitable material as a heat transfer medium. The recommended type of equipment for applying the sealer is a mechanical pressure-type applicator. Mechanical applicators of the gravity type are also used, as are hand-pouring pots. It is recommended that joints be filled from the bottom up to within 3 mm ($\frac{1}{8}$ in.) of the top (*1*). Cold-applied, two-component elastomeric polymers also are used as joint sealers. This maintenance operation is carried on throughout the year. The most favorable time, however, is during the warm summer months, when working temperatures are higher and better results can be obtained.

The repair of spalled, scaled, and map-cracked surfaces is a common maintenance problem for this type of pavement. "Map-cracking" is distinguished by irregular cracking over the pavement surface. "Spalling" is a chipping or splintering of sound pavement and usually occurs along the joints or cracks in the pavement. Scaling is caused by the deterioration or disintegration of the concrete and may occur anywhere on the pavement surface.

Minor cracks in a portland cement concrete surface are usually repaired by filling with a joint sealing compound. It is recommended that cracks more than $\frac{1}{8}$ in. wide be filled with a material that will allow for expansion and contraction of the concrete. A concrete or epoxy repair should be made when cracked, chipped, or splintered areas are too large to control by routine joint sealing methods.

Failed areas in concrete pavement should be repaired with cement concrete. The failed areas are marked out by a person qualified for this type of work. A crew then follows with jackhammers or other mechanical equipment, breaks and removes the broken concrete, and prepares the area for the new surface. At the edges of the patch the old slab of concrete should be undercut and the patch placed as shown in Figure 21-9. High early strength cement should be used in the new concrete. The thickness of the patch should never be less than that of the existing slab, and the use of reinforcing steel is very often desirable. When patches are made adjacent to expansion joints, the expansion joint should be replaced. Contraction or construction joints may be omitted, however, if the patch extends across the entire width of pavement.

Surfacing irregularities can be removed from concrete pavemens by grinding or cutting the surface with a special machine known as a "bump-cutter." Such machines have a cutting head with diamond blades that produce a series of shallow, parallel grooves in the surface, giving it a nonskid texture.

Pumping of Concrete Pavements

One of the major problems in the maintenance of concrete surfaces is correcting settlement of the slab caused by pumping. Pumping is usually indicated by (1) spalling of the pavement near the centerline and a transverse joint or crack, (2) ejection of water from joints and cracks, (3) discoloration of pavement surface by subgrade soil, (4) the presence of mud boils at the pavement edge, and (5) breakage of the pavement.

FIGURE 21-9 Concrete patching. (Courtesy Federal Highway Administration.)

Pumping of concrete pavements may be prevented by maintaining adequate drainage, correcting faulty drainage, and sealing joints and cracks. Where pumping has progressed to an appreciable degree, this condition is corrected by mudjacking or subsealing. This procedure consists of drilling holes into the slab and forcing in a suitable slurry to fill the voids between the subgrade and the slab.

The usual method involves the drilling of one hole in the lowest portion of the settled pavement. This hole is about 38 mm (1.5 in.) in diameter and usually placed about 300 mm (12 in.) from the point. If needed, an additional hole is drilled about 750 mm (30 in.) from the edge. Water and sand are blown out of the area through the drilled hole by use of compressed air. This is necessary to avoid dilution of the slurry to be forced under the pavement. As the slurry is forced into the drilled hole, pumping is continued until all voids are filled and the settled slab is raised to its proper position. The holes are then plugged.

The slurry mixtures used vary a great deal, but for the most part they consist of a combination of loam, cement, and water or of asphaltic materials. Two widely used mixtures include

- soil, 60 to 84 percent, cement, 16 to 40 percent.
- soil, 77 percent; cement, 16 percent; and cut-back asphalt (SC-250), (MC-70), (RC-250), 7 percent.

The mix used should be fluid enough to spread rapidly and flow into small cavities.

The buckling or blow-up of old concrete pavements is a continuing problem; it is usually caused by longitudinal expansion and the failure of transverse expansion joints to function properly. When this occurs, it is often necessary to remove the damaged portion of the pavement and replace it with a patch of concrete or bituminous material. Buckling or blow-ups may be prevented, when evidence of extreme compression is noted in the pavement, by cutting a wider expansion joint if one already exists or providing a new joint across the pavement.

Resurfacing of Old Concrete Pavements

When justified, old concrete pavements may be completely resurfaced. Two materials are in common use: asphaltic concrete and portland cement concrete.

In preparation for resurfacing with asphaltic concrete, joints and cracks may be cleaned and filled with asphaltic concrete. Bituminous patches that show an excess of bituminous material must be removed. Major defects must be corrected, and the cause of the failure must be corrected before the new surface is placed. A common problem with this type of resurfacing is that, over a period of time, joints and cracks in the old concrete pavement show up in the new wearing surface; this phenomenon is called "reflection cracking." In attempts to prevent reflection cracking, several different approaches have been used, many with very limited success.

One approach that has met with some success involves covering the crack in the old pavement with a stress-relief fabric mastic system (*8*). One such system, which goes by the commercial name of PavePrep, consists of a layer of heavy duty mastic material sandwiched between two synthetic fabrics, one woven, the other non-woven. This product comes in rolls in widths of 30, 50, 90, and 100 cm (12, 20, 36, and 40 in.). It can be applied directly to cracks up to 32 mm (1.25 in.) in width; larger cracks should be filled with a polymeric crack filler or other suitable material. After the crack has been cleaned of any dirt, water, or vegetation, it is covered with a bituminous tack coat. The stress-relief interlayer is then rolled over the crack and into the tack coat. It is then usually covered with a hot mix overlay with a minimum thickness of 38 mm (1.5 in.).

Old concrete pavements may be resurfaced with portland cement concrete. Considerable success has been achieved in recent years with thin overlays, from 25 to 75 mm (1 to 3 in.) thick. The old pavement must be thoroughly cleaned and all unsound material removed; the bumpcutter described earlier may be used to scarify the surface. The surface then may be washed down with detergent and acid baths (muriatic acid is often used). Best results are achieved if a thin grout layer is used to provide bond between the old pavement and the new surface. A 1:1 sand-cement grout is used; it is brushed onto the pavement just before the new concrete is applied. In some cases the grout has been placed pneumatically. Concrete must be of high quality and resistant to freezing and thawing, particularly where it will be subjected to salt applications for ice control. In the northern states, air-entrained concrete is normally used. To obtain adequate bond, thorough compaction of the fresh concrete is essential. This is usually done by the use of special vibratory screeds. Whenever the new surface extends over a joint, or butts against an expansion joint in the old pavement, a joint will be necessary in the new concrete. Joints may be made by tooling the plastic concrete or by sawing. Adequate curing of the new surface is essential. In some cases, particularly where the old pavement is excessively cracked, wire mesh reinforcing may be used in the new surface.

MAINTENANCE OF SHOULDERS AND APPROACHES

21-6 Shoulder Maintenance

The importance of properly maintaining road shoulders becomes apparent when they are considered as a continuation of the traveled surface that must be used in cases of emergency. Adquate shoulders are also necessary if the full capacity of the roadway surface is to be used.

Shoulders may be made of many types of bases and surfaces including soil aggregate, soil surfaces that are capable of supporting vegetation and may be seeded, and various bituminous surfaces. It is now almost standard practice to pave shoulders on heavily traveled roads; often the shoulders differ in color and texture from the travel lanes. Each of these approaches presents a slightly different type of maintenance problem.

A well-graded gravel shoulder will provide a satisfactory all-weather shoulder for light traffic; the maintenance of shoulders of this type will follow a pattern similar to that of soil-aggregate surfaces. This will include the dragging and blading of the shoulder surface for the proper slope and the filling of any ruts. It may be necessary to replace worn-out and lost materials at periodic intervals. This work can be done with a light motor grader.

Turf shoulders require the filling of holes and ruts, rolling or blading down high spots, and seeding, sodding, and fertilizing to maintain a satisfactory turf. It is also necessary to mow and clean the shoulders and provide some means of weed control or weed eradication. Mowing is usually accomplished with light equipment manufactured for this purpose. Treated bituminous and paved shoulders require maintenance that is similar to that for surfaces made of these same materials, which were previously described, including patching, surface, and so on.

21-7 Maintenance of Approaches

Closely related to shoulder maintenance is that of road approaches. Approaches include public side roads, private driveways, ramps, speed-change lanes, and turnouts for mailboxes and bus stops. The methods of maintaining approach areas do not differ much from those used for regular shoulder maintenance; however, because of the traffic in these areas, special maintenance efforts may be required to keep approaches free of potholes, ruts, and other types of deterioration.

MAINTENANCE OF ROADSIDES

Roadside maintenance in general may include the care of the area between the traveled surface and the limits of the right-of-way. This will include any median strip or landscaped areas on divided or dual-lane highways; roadside parks and picnic and recreational areas; and various appurtenances such as right-of-way fences, picnic tables, and the like.

In the following paragraphs, we group roadside maintenance activities into three classes: (1) vegetation management and control, (2) maintenance of rest areas, and (3) litter control.

21-8 Vegetation Management and Control

Vegetation management and control includes mowing; weed eradication and control; seeding, sodding, and the planting of vegetation; and care of trees and shrubs.

The mowing of roadsides is done to provide better sight distances, improve drainage, reduce fire hazards, and improve the appearance of the roadway. Mowing should be started as soon as the grass or weeds are high enough to be cut, and it should be continued at periodic intervals, when necessary, throughout the growing season. Mowing can be accomplished with machines or by hand. Some heavy-duty mowing machines are available that are capable of mowing roadside brush up to 60 mm (2.5 in.) in diameter. Cleaning of roadsides after mowing can be accomplished by the use of hand or power rakes. Grass cuttings can be left as a mulch, but brush cuttings should be removed. The extent of mowing recommended by AASHTO (*1*) is given in Table 21-1.

Chemicals may be used on roadside areas to retard vegetation growth and lessen the frequency of mowing, especially in areas that are difficult to mow. Chemicals are also used to control weeds (herbicides), insects (insecticides), and fungi (fungicides). The spraying of chemicals along roadsides should be done by qualified personnel following the manufacturer's recommendations and guidelines of the Environmental Protection Agency. Precautions are required to avoid harm to native plants and wildlife and to ensure the safety of the traveling public, nearby residents, and maintenance personnel.

Seeding, sodding, and the planting of vegetation are important maintenance operations for the prevention of erosion. Seeding may be done on relatively flat areas, while sodding is necessary on steeper slopes. On steep slopes where seeding or sodding is not practical, the ground surface may be protected by the planting of vines or similar ground cover. The planting of dune grasses on sandy slopes to prevent wind erosion has been quite successful.

The care of trees and shrubs along the roadsides may include planting, trimming, fertilizing, spraying, and the construction of tree wells. Major tree surgery and the removal of broken limbs caused by storms are often necessary. A large number of highway departments have forestry departments that take care of all trees and shrubs along major highways.

21-9 Maintenance of Rest Areas

Special roadside areas that have been developed for picnic and recreational areas, such as roadside parks and turnouts, should be maintained in a manner necessary to meet minimum health and aesthetic standards. Tables, benches, and refuse containers should be placed at convenient locations. The grounds should be kept cleaned, trees and shrubs trimmed, and stoves kept in good condition. Toilet facilities should be kept in a sanitary condition, and any drinking water should be frequently tested for purity. The parking areas and access roads must also be properly maintained.

21-10 Litter Control

Removal of rubbish and debris (litter) is one of the vexing and expensive problems of roadside maintenance. Debris such as fallen branches, rocks, landslides, and articles that have fallen from trucks should be removed immediately in order to protect the traveling public. The remains of animals killed by motor vehicles should also be removed promptly and buried at a convenient location. Garbage and trash dumped within the highway right-of-way must be removed because of its unsightliness and for reasons of health. Generally speaking, roadsides should be thoroughly cleared of litter in the spring and periodically as needed thereafter (*1*).

TABLE 21-1 Mowing Recommendations of AASHTO

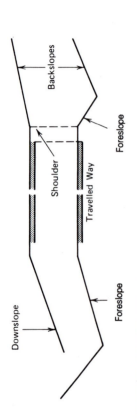

SUGGESTED MOWING LIMITS (Measured Along Surface of Slopes)

	On slopes 2½:1 or less	On slopes more than 2½:1, up to 1:1 within 6 m outside of pavement edge	More than 1:1 slopes within 3 m outside of pavement edge
Backslopes	Mow 4.5 m outside of pavement edge.	With side-mounted units mow one swath width past slope that does not exceed 2½:1.	Mow brush and as required for sight distance.
Foreslopes	Mow 4.5 m or entire width up to 7.5 m outside of pavement edge or one pass on the backslope to maintain drainage.	With side-mounted units mow one swath width past slope that does not exceed 2½:1.	Mow brush up to 4.5 m or entire width and as required for sight distance.
Downslopes	Mow 4.5 m outside of pavement edge.	With side-mounted units mow one swath width past slope that does not exceed 2½:1.	Mow brush and as required for sight distance.
Median areas	Mow 4.5 m outside of pavement edge or entire width if less than 15 m.	With side-mounted units mow one swath width past slope that does not exceed 2½:1.	Mow brush and as required for sight distance.
Interchange areas	Mow 4.5 m from pavement edge and any areas that obstruct sight distances or areas used for snow storage or entire area in urban areas.	With side-mounted units mow one swath width past slope that does not exceed 2½:1.	Mow brush and as required for sight distance.
Guard rail	Mow 1 m behind guard rail.	Mow brush for 1 m behind and as required for sight distance.	Mow brush for 1 m behind and as required for sight distance.

Source: Adapted from *AASHTO MAINTENANCE MANUAL 1987.* American Association of State Highway and Transportation Officials (1987).

MAINTENANCE OF BRIDGES, TUNNELS, AND DRAINAGE STRUCTURES

21-11 Bridge Maintenance

Highway agencies continually inspect bridges and elevated structures, then take any necessary steps to minimize deterioration or repair damage caused by accidents, floods, or other unforeseen events.

Bridge superstructures require periodic maintenance to preserve and protect the investment made in them. For instance, steel bridges (unless they are made of special alloys or have been galvanized) are cleaned and painted regularly to prevent corrosion.

A serious problem for the maintenance engineer is the deterioration of concrete bridge decks, particularly on heavily traveled routes in northern states, where chlorides are used extensively for snow and ice control. Concrete normally protects the reinforcement bars from corrosion, but, the penetration of water and deicing chemicals results in the corrosion of the reinforcement bars and causes the concrete to crack.

Bridge decks that have experienced only minor deterioration may be patched with specialized concretes. Advantages and disadvantages of some of the products used for bridge deck patching are given in Table 21-2. Spalled areas may be temporarily repaired with asphaltic patching materials.

When the deterioration is severe, it may be necessary to cover the bridge deck with an overlay or to remove and reconstruct the entire deck. Two types of overlays have been used successfuly (9): (1) a bituminous concrete wearing surface placed over a waterproofing membrane and (2) a portland cement mortar or concrete (possibly including a liquid latex admixture). In such work, it is recommended that all bad concrete be removed and a grout carefully applied to the surface before the new surface is constructed (10).

Bridge inspections should be complete enough to discover any undermining of the bridge footings or any damage to the substructure. Timber bulkheads should be repaired as soon as evidence of failure is shown. Piling that shows deterioration caused by erosion, corrosion, or attack by organisms should be replaced when necessary.

The stream bed should be kept clean and free from debris in order that the free flow of water may be maintained. Provision for the installation of some sort of barrier to prevent floating debris from damaging the bridge floor during periods of high water must sometimes be made.

TABLE 21-2 Comparison of Products Used for Bridge Deck Patching

Patch Type	Advantages	Disadvantages
Regular portland cement	Ease of handling; low cost	Slow cure; edge cracks
High early strength cement	Reduced cure; low cost	Shrinkage
Metallic aggregates	Reduce shrinkage	High cost; may contain salt
Latex additives	Increase bond	High cost
Epoxy resin	Fast cure; good bond	High cost; edge cracks

Source: Concrete Bridge Deck Durability. National Cooperative Highway Research Program Synthesis of Highway Practice 4 (1970).

The maintenance of pumping facilities at highway underpasses is also a very important operation.

The maintenance of bridges is usually carried on by a special bridge crew trained and skilled in this type of work. Some state highway departments have heavy equipment that is necessary for all phases of bridge maintenance, including mobile shops.

The maintenance of bridges is an extremely important and complex phase of highway maintenance. Additional information on this important topic is given in Ref. 9.

21-12 Maintenance of Tunnels

It is important that tunnels be inspected by qualified maintenance personnel on a regular basis. Areas requiring special concern by inspectors are (1) supports, (2) tunnel walls and ceilings, (3) portals, (4) lighting, and (5) ventilation. AASHTO (*1*) recommends that timber supports be inspected annually for evidence of decay, insect infestation, and other deterioration. Tunnel walls and ceilings should be examined from time to time for evidence of water leaks. When used, ceramic tiles should be washed annually and kept in proper repair. Tunnel portals should be kept free of loose rocks, snow deposits, or other materials that could pose a hazard to users of the facility.

Special attention should be given to tunnel lighting to ensure that an adequate level of visibility is maintained. Generally, burned-out lamps should be replaced immediately, and group replacement of lamps should be scheduled before 90 percent of the rated life elapses (*1*).

Ventilation systems for tunnels should be inspected at least monthly, and airflow meters should be employed to ensure that the ventilation system is functioning as designed.

21-13 Drainage and Drainage Structures

All drainage structures and appurtenances have to be kept in good working condition so as to provide free and unobstructed flow. Generally, maintenance operations include those related to (1) surface drainage, (2) ditches, and (3) culverts. Modern roads are generally well developed for drainage, but many of the older ones lack adequate drainage. Observations of conditions at times of heavy rains and high water will give the best indications of the ability of the drainage system to perform its function properly without erosion or damage. The entire system must be so maintained as to handle heavy rainfalls and to function properly under all conditions, even under winter conditions of ice and snow.

The first objective of surface drainage is to remove the water quickly so that it will not interfere with the use of the road. To accomplish this, the roads are crowned and elevated. On the high-type road surfaces, these crowns are more permanent than on the lower type of road surfaces. Thus, on gravel roads and similar types, it is necessary to maintain a proper crown by blading.

Wide shallow ditches are preferable for maintenance and are less dangerous when vehicles have to leave the road. Certain types of roadside ditches may be kept clean and the proper slope maintained by the use of a power grader. Care should be exercised to preserve seeded and sodded areas and shrubs during maintenance operations. The original line or grade of the ditch should also be maintained. When ditches become eroded owing to excessive grade, handling of large volumes of water, or a combination of both, it may be necessary to retard the flow or to line them, as described in Chapter 11.

All culverts should be checked regularly to see that they are free from obstructions. Inlets and outlets should be kept open and free from refuse, and culverts should be cleaned and repaired when necessary.

MAINTENANCE OF TRAFFIC CONTROL AND SAFETY DEVICES

21-14 Maintenance of Traffic Control Devices

Signs, signals, and markings for the direction, warning, and regulation of traffic are essential to traffic safety and must be properly maintained. This work includes the installation, repair, and painting of signs and the care of pavement markings. Because of their location, many pavement markings must necessarily be renewed practically every year. Many departments have mechanized marking equipment for the application of the painted stripe. Large machines of this type can apply simultaneously as many as three stripes, either solid or broken, and use two colors of paint, if desired. Reflectorized glass beads may be added to the painted surface during the striping process. The beads have proved beneficial for night driving. Thermoplastic compounds are used for pavement markings, also, as are raised plastic disks cemented to the road surface. Many highway departments maintain shops for the fabrication and repair of signs and other traffic control devices. Automatic traffic signals have to be checked, repaired, and often readjusted to traffic conditions. Highway departments are continually standardizing and modernizing traffic signs and signals, and much of this work is done by maintenance forces. During the winter season in the northern states, special attention is given to warning signs to keep them free from snow and visible as much as possible.

21-15 Maintenance of Safety Devices

A variety of appurtenances within highway rights-of-way are provided for the safety of the users. Such features include guardrails, barriers, impact attenuators, pedestrian overpasses and underpasses, and fences to restrict access of pedestrians and animals. Highway safety devices should be frequently and systematically inspected and repaired on a high-priority basis. Because of variations in the design and the materials used in the construction of such devices, each type of appurtenance has its special maintenance procedures. Guidance on the maintenance of various types of highway safety devices is given in Ref. 1.

THE CONTROL OF SNOW AND ICE

Considering the public's complete dependence on the uninterrupted use of highways throughout the year, winter maintenance has become one of the most essential activities of many state highway departments. This is particularly true in the northern states, where a program for snow removal and ice control has to be considered when the annual maintenance budget is prepared. The amount of money expended for these services will vary a great deal in the many states, and it is dependent in large measure on the length and intensity of storms and the amount of snowfall in a particular area.

Measures preparing for snow removal and ice control are taken before the start of the winter season. The most important of these are (1) the preparation of the necessary equipment; (2) stockpiling of ice-control chemicals and abrasives, if used; (3) placing of snow fences, snow guides, and containers for ice-control abrasives; and (4) organization arrangements.

The necessary equipment needed for snow-removal operations consists of large trucks, various types of snowplows, power graders, bank-slicers, and sanders. Some trucks are equipped with underbody blades, others with "Vee" plows or with bank-slicers. All equipment should be put into working order before the winter season starts. An ample supply of snowplow shoes and cutting edges should be on hand for each type of plow. Tire chains, flags, blinker lights, tow chains, and other safety devices should also be checked and placed, ready for use.

Ice-control chemicals and abrasives should be stockpiled previous to the first storm in the fall; the quantities to be stockpiled may be estimated on the basis of past experience in a particular district or division. Stockpiles should be located with care at convenient places along the road or at central locations, such as main or branch garage yards, gravel or sand pits, or other places where overhead bins or mechanical loaders are available. Generally, a coarse heavy sand is the best abrasive. Cinders or other abrasives, however, may be desirable, depending on availability and cost. Calcium chloride is frequently incorporated and thoroughly mixed with the abrasive at the time the stockpiles are made. From 50 to 100 lb of calcium chloride per cubic yard are used, and a light application of calcium chloride is usually spread over the surface of the pile to prevent freezing.

Snow fences are generally placed parallel to the road at a distance of 20 to 40 m (75 to 125 ft) from the centerline of the road in areas where observations show that drifts constantly form. A snow fence usually consists of fabricated wooden slats mouned on steel posts, which can be removed, rolled in bundles, and stored when not in use. The purpose of a snow fence is to change the velocity of the snow-laden wind so that the snow is deposited before it reaches the roadway. As the snow accumulates near the fence, the fence is raised in order that it may effectively gather snow. Extension posts are provided when this is necessary. In certain areas where right-of-way is available, natural windbreaks or snow fences may be formed by planting evergreens parallel to the road.

Snow guides may be made from peeled saplings approximately 3.6 m (12 ft) long with a maximum top diameter of 50 mm (2 in.). They are sometimes painted with alternate bands of black and white or some other suitable color, and they are often fitted with reflectorized buttons. They are used for making the roadway or road hazards in areas of heavy snowfall. Guides marking the locations of guardrails, culverts, headwalls, and curbs are essential during plowing operations when these structures are covered with snow.

The placing of containers for abrasives at strategic locations has proved very beneficial to the traveling public, who can avail themselves of these materials if needed before maintenance crews arrive. The containers are usually barrels equipped with a utensil, such as a gallon can, for removing the abrasive.

To properly carry on the work of snow removal and ice control, good organization is necessary. Many states maintain a weather reporting service, with all districts of a division reporting adverse weather or road conditions to a central office. These give such information as temperature, snowfall, and other adverse weather conditions, such as rain, fog, mist, and smog. Road conditions are then rebroadcast to serve as a source of information for the traveling public. Maintenance crews are generally subject to call at all hours, and provisions are usually made for dual crews to operate for alternate 12-hr periods during emergencies. The use of radio communications has facilitated work during snow-removal and ice-control operations. Mobile and fixed station units assist in the rapid dispatching of equipment, the reporting of adverse conditions, and the general control of operations. Use is made of both one-way and two-way radio communication systems for other maintenance operations as well.

The procedure followed to keep trunk highways clear of snow varies. One method is to start plowing with trucks equipped with underbody blades after $\frac{1}{2}$ to 1 in. of snow has fallen, using a conventional curved cutting edge to remove the snow before it becomes packed by traffic. These units are generally kept working during a continuous snowfall up to the time when the snow becomes too deep for their effective operation. When this occurs, light trucks equipped with side delivery plows start operating and stay in operation until the storm has subsided or until sufficient snow has accumulated to require heavier trucks equipped with side delivery plows or light trucks with "Vee" plows. When this equipment proves inadequate, heavy "Vee" plows or "snogos" are brought into operation. When the storm has abated, snow removal operations are continued until the snow is removed to a point beyond the outer edge of the shoulder. On trunk lines in urban areas, use is often made of mechanical loads to speed snow removal. Use is sometimes made of heavy rotary plows for opening up roads after storms of such intensity that snow-removal operations have to be temporarily suspended. Care must be exercised at railroad crossings to remove snow between the rails so that ice does not form.

Ice control means the application of quick-action emergency measures to counteract slippery conditions as early as possible and remove all ice from the pavement after skid control has been established. These operations may consist of removing sleet or thin ice, thick ice or frozen slush, and hard-packed snow from the surface of the roadway. Salt, calcium chloride, a mixture of the two, and abrasives mixed with chlorides are widely used for ice control in urban and rural areas. Many organizations use a mixture of 1 part calcicum chloride and 2 parts sodium chloride, by volume. Application rates for the mixture vary from 56 to 140 kg/km (200 to 500 lb mile) of two-lane roadways. They depend largely on straight chemicals in urban areas, because abrasives may clog storm sewers. Guidelines for chemical application rates are shown as Table 21-3.

To remove sleet or thin ice from the roadway, treated abrasive may be placed on the surface at the rate of 1 to 1.4 m³/km (2 to 3 cu yd/mile) as the ice forms. Chlorides may also be used alone. First consideration in ice-control treatment is given to approaches to railroad crossings, road intersections, and dangerous horizontal and vertical curves. This kind of work is usually done by light trucks in advance of other ice-control measures.

Sleet or thin ice on gravel roads should first be serrated. This process forms grooves, and in many cases will bring up enough gravel to control skidding. If this does not occur, chloride may be added. Thick ice or frozen slush is also scored with a serrated blade in order to concentrate the chemical on the road. When the ice or slush has been loosened, it is bladed off the road.

Abrasives treated with calcium or sodium chloride, or chlorides alone, should be used as sparingly as possible, on portland cement concrete pavements. This is because repeated freezings and thawings of concrete in contact with these salts are conductive to scaling.

On urban expressways, the overwhelming need for getting a bare pavement

TABLE 21-3 Guidelines for Chemical Application Rates

Weather conditions			Application rate (kilograms of material per kilometer of two-lane road or two lanes of divided)			
Between Temperatures	Pavement Conditions	Precipitation	Low- and High-Speed Multilane Divided	Two- and Three-Lane Primary	Two-Lane Secondary	Instructions
−1 °C and above	Wet	Snow	85 salt	85 salt	85 salt	Wait at least 0.5 hr before plowing
		Sleet or freezing rain	56 salt	56 salt	56 salt	Reapply as necessary
−1°–4°C	Wet	Snow or sleet	Initial at 113 salt Repeat at 56 salt	Initial at 113 salt Repeat at 56 salt	Initial at 113 salt Repeat at 56 salt	Wait at least 0.5 hr before plowing
		Freezing rate	Initial at 85 salt Repeat at 56 salt	Initial at 85 salt Repeat at 56 salt	Initial at 85 salt Repeat at 56 salt	Repeat as necessary
−4°–7°C	Wet	Snow or sleet	Initial at 140 salt Repeat at 70 salt	Initial at 140 salt Repeat at 70 salt	338 of 5:1 sand/ salt; repeat same	Wait about 0.75 hr before plowing; repeat
		Freezing rain	Initial at 113 salt Repeat at 85 salt	Initial at 113 salt Repeat at 85 salt		Repeat as necessary
−7°–9°C	Dry	Dry snow	Plow	Plow	Plow	Treat hazardous areas with mix of 20:1 sand/salt
	Wet	Wet snow or sleet	140 of 3:1 salt/calcium chloride	140 of 3:1 salt/calcium chloride	338 of 5:1 sand	Wait about 1 hr before plowing; continue plowing until storm ends; then repeat application
Below −9°C	Dry	Dry snow	Plow	Plow	Plow	Treat hazardous area with 338 of 20:1 sand/salt

Source: Adapted from *AASHTO Maintenance Manual.* American Association of State Highway and Transportation Officials (1987).

quickly leads to excessive use of salts, with attention being focused on building pavements that can withstand this treatment.

PAVEMENT REHABILITATION

By timely and proper maintenance, highway engineers can extend a pavement's usefulness. Eventually, however, even the best-maintained pavement will begin to disintegrate and will need to be rehabilitated.

The traditional approach to pavement rehabilitation has been to either reconstruct it with all new materials or patch and overlay it with a new wearing surface. Both asphaltic and portland cement concrete mixes have been used for pavement resurfacing.

Increasing construction costs have caused highway engineers to search for new methods of rehabilitating highway pavements. Two such methods, milling and pavement recycling, are briefly discussed in the following paragraphs.

21-16 Milling of Pavements

In recent years, equipment manufacturers have developed pavement milling machines that are capable of correcting pavement and surface distresses such as rutting, raveling, bleeding, and aggregate polishing (Fig. 21-10). Such machines are equipped with rotating drums fitted with special cutting teeth for milling asphalt and concrete pavements to predetermined depths. (*11*). Figure 21-11 illustrates a milling drum from one of these machines. Pavement milling machines can be used in a number of ways to rehabilitate worn and distressed pavements. Typical uses of these machines are listed in Table 21-4.

21-17 Pavement Recycling

Hot-mix recycling is a proven approach for the rehabilitation of worn-out asphaltic pavements. It involves (1) removing the existing pavement to full or

FIGURE 21-10 Pavement milling machine. (Courtesy Cedarapids, Inc.)

FIGURE 21-11 Rotating drum with cutting teeth in a pavement milling machine. (Courtesy Cedarapids, Inc.)

partial depth, (2) reducing the reclaimed materials to a suitable size for reprocessing, (3) blending the reclaimed material with virgin aggregates and liquid asphalt, and (4) relaying the materials as a base, binder, or surface course.

Advantages of hot-mix recycling include

- economy.
- conservation of natural resources.
- improvement of the structural strength of the pavement with little or no change in thickness.
- correction of existing deficiences in the pavement mix.

Before hot-mix recycling can proceed, all of the materials must be tested and evaluated to determine the optimum blend. Samples of the reclaimed material are randomly selected for testing, preferably from the cold feedstock piles at the plant site. Laboratory tests of the pavement samples are made to determine the aggregate gradation, asphalt content, and asphalt viscosity at 60°C (140°F). The old asphalt is extracted from the sample by a standard procedure such as ASTM Designation D2172, providing a measure of the quantity of asphalt in the old mix. The extracted asphalt is reclaimed and tested by ASTM Method D2171 to determine its viscosity at 60°C (140°F). The reclaimed aggregate is subjected to a sieve analysis to determine the aggregate gradation and to reveal any deficiencies in aggregate gradation. It will normally be necessary to add virgin aggregate to the reclaimed asphalt pavement to produce a mix with the desired gradation.

New low-viscosity asphalt must normally be added to the mixture of reclaimed pavement and new aggregate to increase the total asphalt content to the desired

TABLE 21-4 Typical Uses and Advantages of Cold Milling Machines

Problem	Solution	Advantages
Bleeding or flushing of asphalt pavement and polishing of aggregates in flexible and rigid pavements cause skid hazards.	Restore surface texture to slick pavements by skimming off the surface to a depth of $\frac{1}{2}$ in. in flexible pavements and $\frac{1}{4}$ in. in rigid pavements.	Skid-resistant surface can be produced at a low cost.
Portland cement concrete pavement faulting with differing elevations of slabs at expansion joints creates poor rideability and danger to road users.	At the expansion joint, mill the high slabs flush to the adjacent slabs, tapering off to zero, leaving a textured drivable surface.	This provides a more economical solution than the traditional approach of replacement and patching.
Light rutting and corrugating in the upper region of a wearing course cause poor drainage and poor rideability.	Mill wearing course down to the deepest wheel rut and corrugation maintaining grade and slope and leaving a new textured riding surface.	A better drained and smoother surface can be produced without the need for new materials.
Successive overlays have filled the curbs of an urban street, causing poor drainage and rideability.	Mill the pavement to predetermined grade and slope, re-exposing curb and restoring drainage. Resurface with new or recycled material.	Solution provides better drainage and rideability.

Source: Adapted from *Cold Planing.* Bulletin CP-1, Iowa Manufacturing Company (1985).

quantity and to provide an asphalt with a desired viscosity. In addition, organic materials called recycling agents are usually added to restore the old asphalt to desired specifications. The detailed procedures for determining the mix design for a recycled pavement are not given here but may be found in Ref. 12.

Old pavements may be removed (1) by cold milling or planing, as described in Section 21-16, or (2) by ripping and crushing. Where the ripping and crushing procedure is used, scarifiers, grid rollers, or rippers are used to break up the asphalt pavement. It is then loaded and hauled to a crushing and screening plant, where it is pulverized. After pulverization by one of these methods, the material is stockpiled at a mixing plant for subsequent processing.

Recycled hot mixes may be prepared either by batch plants or by drum-mix plants. The plants must be modified to allow the reclaimed asphalt pavement material to be heated and dried without exposing it directly to the high-temperature flame and combustion gases in the drier. Without these modifications, the asphalt in the reclaimed material will harden excessively, and it will not be possible to comply with regulations governing harmful exhaust stack emissions.

In batch plants, the aggregate is proportioned from cold-feed bins, heated in a conventional aggregate dryer, and then conveyed to the hot storage bins in the usual manner. The reclaimed asphalt pavement material is not heated or dried but is carried directly from the stockpile to a special cold-feed bin. From there, the material is conveyed to the weigh hopper, where it is weighed as a fifth material in a normal four-bin batch plant. The cold reclaimed material is mixed in the pugmill with the superheated untreated aggregate. After mixing, the blended material goes to a storage silo and is subsequently loaded into trucks.

Contractors and plant manufacturers have developed several schemes to modify drum-mix plants to ameliorate problems with air pollution. One approach has been to use a split feed in which the untreated aggregate enters the drum at the burner end and the reclaimed material is brought in some distance downstream from the burner in order to avoid exposure to the flame and extremely hot gases. This arrangement can be seen in Figure 18-10.

Another type of hot-mix recycling plant is a "drum-in-a-drum" design. In this design, the burner fires into a smaller drum that is partially inserted into the upper end of the main drum. The new aggregate enters the inner drum and is exposed to the flame, but the reclaimed material is fed into the main drum, where it is protected from the flame.

Still another type of hot-mix recycling plant uses a combustion chamber with a conical heat shield inserted between the burner and the drum to protect the reclaimed material from the flame (*12*).

Hot-mix recycled mixtures are spread and compacted by using conventional procedures such as those described in Chapter 19.

Looking to the future, it is expected that greater emphasis will be placed on pavement recycling as engineers continue to search for cost-effective ways to provide safe and efficient transportation and protect the vast investment in the nation's highway system.

REFERENCES

1. *AASHTO Maintenance Manual.* American Association of State Highway and Transportation Officials, Washington, DC (1987).

2. *Highway Statistics.* Federal Highway Administration, Washington, DC (1993).

3. *Street and Highway Maintenance Manual.* American Public Works Association, Chicago, IL (1985).

4. *Performance Budgeting System for Highway Maintenance Management.* National Cooperative Highway Research Program Report 131, Roy Jorgensen Associates, Transportation Research Board, Washington, DC (1984).

5. *Maintenance Management Systems.* National Cooperative Highway Research Program Synthesis of Highway Practice 110, Transportation Research Board, Washington, DC (1984).

6. *A Guide for Methods and Procedures in Contract Maintenance.* American Association of Highway and Transportation Officials, Washington, DC (1987).

7. *Asphalt in Pavement Maintenance.* Manual Series No. 16, 2nd Ed., The Asphalt Institute, Lexington, KY (1983).

8. Interlayer Solves Cracking Problem. *Better Roads,* February 1989.

9. *AASHTO Manual for Bridge Maintenance.* American Association of State Highway and Transportation Officials, Washington, DC (1987).

10. *Concrete Bridge Deck Durability.* National Cooperative Highway Research Program Synthesis of Highway Practice 4, Transportation Research Board, Washington, DC (1970).

11. *Cold Planing.* Bulletin CP-1, Iowa Manufacturing Company, Cedar Rapids, IA (1985).

12. *Asphalt Hot Mix Recycling.* Manual Series No. 20, 2nd Ed., The Asphalt Institute, Lexington, KY (1986).

SELECTED TABLES IN CONVENTIONAL U.S. UNITS

TABLE 4-2A Accident Rates by Road Type

	Rate (Number per MVM)			
Location and Road Type	Fatal Accidents	Injury Accidents	Property Damage Only Accidents	Total Accidents
Rural				
No access control				
2 lanes	0.070	0.94	1.39	2.39
4 or more lanes, undivided	0.047	0.89	1.95	2.89
4 or more lanes, divided	0.063	0.77	1.25	2.09
Partial access control				
2-lane expressway	0.051	0.52	0.76	1.33
Divided expressway	0.038	0.44	0.76	1.24
Freeway	0.025	0.27	0.49	0.79
Urban				
No access control				
2 lanes	0.045	1.51	3.38	4.94
4 or more lanes, undivided	0.040	2.12	4.49	6.65
4 or more lanes, divided	0.027	1.65	3.19	4.86
Partial access control				
2-lane expressway	0.033	6.65	1.05	1.73
Divided expressway	0.022	1.08	2.04	3.14
Freeway	0.012	0.40	1.01	1.43
Suburban				
No access control				
2 lanes	0.048	1.26	2.56	3.88
4 or more lanes, undivided	0.037	1.58	3.31	4.93
4 or more lanes, divided	0.030	1.10	2.24	3.37
Partial access control				
2-lane expressway	0.096	0.82	1.42	2.34
Divided expressway	0.060	0.82	1.29	2.16
Freeway	0.015	0.32	0.74	1.07
Statewide				
No access control				
2 lanes	0.066	1.01	1.66	2.74
4 or more lanes, undivided	0.041	1.88	3.97	5.88
4 or more lanes, divided	0.031	1.50	2.89	4.43
Partial access control				
2-lane expressway	0.053	0.54	0.81	1.40
Divided expressway	0.037	0.61	1.08	1.73
Freeway	0.015	0.36	0.86	1.23

Source: A Manual on User Benefit Analysis of Highway and Bus-Transit Improvements. American Association of State Highway and Transportation Officials (1977).

TABLE 5-4A Maximum and Minimum Dimensions of Standard American Passenger Cars Manufactured in 1994

Dimension	Minimum (inches)	Maximum (inches)
Overall length	142.5	225.1
Overall width	64.2	79.8
Overall width, doors open	127.8	173.6
Overall height	44.0	65.9
Front overhang distance	30.9	47.6
Rear overhang distance	27.4	59.6
Wheelbase length	90.7	121.5
Front tread width	53.7	62.8
Rear tread width	52.8	64.1
Bottom of front bumper to ground	5.1	17.6
Bottom of rear bumper to ground	7.3	18.5
	(feet)	*(feet)*
Turning diameter, outside front		
Wall to wall	17	46
Curb to curb	16	45
Turning diameter, inside rear		
Wall to wall	9	28
Ramp breakover angle, deg.	8.0	26.0

Source: Vehicle Dimensions, 1994 Model Year, American Automobile Manufacturers Association, Detroit, Michigan (1994).

TABLE 5-5A Summary of State Limitations on Vehicle Weights and Dimensions

	Minimum in Any State	Maximum in Any State
Width (in.)	102	108
Height (ft)	13	14.5
Length (ft)		
Tractor-Semitrailer Combinations		
Single unit truck	40	60
Semitrailer on Interstate and National Network	48	60
Semitrailer off National Network	45	60
Overall combination length, other roads	55	88
Twin Combinations		
Semitrailer or trailer on Interstate and National Network	28	95
Twin combination length on other roads	28	81
Straight Truck + Trailer	50	85
Weight (lbs)		
Axle limits		
Single	20,000	22,500
Tandem	34,000	44,000
Tridem	34,000	66,000
Maximum gross weight, Interstate	80,000	117,000
Maximum gross weight, other roads	73,280	117,000

Source: Summary of Size and Weight Limits, American Trucking Associations, Inc., Alexandria, Virginia, January, 1994.

TABLE 5-6A Rolling Resistances of Passenger Cars on Low-Grade Road Surfaces (lb/ton)

	Uniform Speed (mph)			
Type of Surface	*20*	*30*	*40*	*50*
Badly broken and patched asphalt	29	34	40	51
Dry, well-packed gravel	31	35	50	62
Loose sand	35	40	57	76

Source: Claffey, Paul J. Vehicle Operating Characteristics. *Transportation and Traffic Engineering Handbook,* Institute of Transportation Engineers, Arlington, Va. (1975).

TABLE 5-7A Typical Curve Resistances of Passenger Cars on High-Type Road Surfaces (lb)

Degree of Curve	*Speed (mph)*	*Resistance (b)*
5	50	18
5	60	36
10	30	18
10	40	54
10	50	108

Source: Claffey, Paul J. Vehicle Operating Characteristics. *Transportation and Traffic Engineering Handbook,* Institute of Transportation Engineers, Arlington, Va. (1975).

TABLE 7-2A Design Vehicle Dimensions

Design Vehicle Type	Symbol	Dimension (ft)										
		Overall			Overhang		WB_1	WB_2	S	T	WB_3	WB_4
		Height	Width	Length	Front	Rear						
Passenger car	P	4.25	7	19	3	5	11					
Single unit truck	SU	13.5	8.5	30	4	6	20					
Single unit bus	BUS	13.5	8.5	40	7	8	25					
Articulated bus	A-BUS	10.5	8.5	60	8.5	9.5	18		4[a]	20[a]		
Combination trucks												
Intermediate semitrailer	WB-40	13.5	8.5	50	4	6	13	27				
Large semitrailer	WB-50	13.5	8.5	55	3	2	20	30				
"Double Bottom" semi-trailer—full-trailer	WB-60	13.5	8.5	65	2	3	9.7	20	4[b]	5.4[b]	20.9	
Interstate Semitrailer	WB-62*	13.5	8.5	69	3	3	20	40–42				
Interstate Semitrailer	WB-67**	13.5	8.5	74	3	3	20	45–47				
Triple Semitrailer	WB-96	13.5	8.5	102	2.5	3.3	13.5	20.7	3.3[d]	6[d]	21.7	21.7
Turnpike Double Semitrailer	WB-114	13.5	8.5	118	2	2	22	40	2[c]	6[c]	44	
Recreation vehicle												
Motor home	MH		8	30	4	6	20					
Car and camper trailer	P/T		8	49	3	10	11	18	5			
Car and boat trailer	P/B		8	42	3	8	11	15	5			
Motor home and boat trailer	MH/B		8	53	4	8	20	21	6			

* = Design vehicle with 48' trailer as adopted in 1982 STAA (Surface Transportation Assistance Act)

** = Design vehicle with 53' trailer as grandfathered in 1982 STAA (Surface Transportation Assistance Act)

[a] = Combined dimension 24, split is estimated.

[b] = Combined dimension 9.4, split is estimated.

[c] = Combined dimension 8, split is estimated.

[d] = Combined dimension 9.3, split is estimated.

WB_1, WB_2, WB_3, WB_4 are effective vehicle wheelbases.

S is the distance from the rear effective axle to the hitch point.

T is the distance from the hitch point to the lead effective axle of the following unit.

Source: A Policy on Geometric Design of Highways and Streets. American Association of State Highway and Transportation Officials, Washington, D.C. (1990).

TABLE 7-4A Maximum Degree of Curve and Minimum Radius for Limiting Values of e and f, Rural Highways and High-Speed Urban Streets

Design Speed (mph)	Maximum e	Maximum f	Total (e + f)	Maximum Degree of Curve	Rounded Maximum Degree of Curve	Minimum Radius (ft)
20	0.04[a]	0.17	0.21	44.97	45.0	127
30	0.04	0.16	0.20	19.04	19.0	302
40	0.04	0.15	0.19	10.17	10.0	573
50	0.04	0.14	0.18	6.17	6.0	955
60	0.04	0.12	0.16	3.81	3.75	1528
20	0.06	0.17	0.23	49.25	49.25	116
30	0.06	0.16	0.22	20.94	21.0	273
40	0.06	0.15	0.21	11.24	11.25	509
50	0.06	0.14	0.20	6.85	6.75	849
60	0.06	0.12	0.18	4.28	4.25	1348
65	0.06	0.11	0.17	3.45	3.5	1637
70	0.06	0.10	0.16	2.80	2.75	2083
20	0.08	0.17	0.25	53.54	53.5	107
30	0.08	0.16	0.24	22.84	22.75	252
40	0.08	0.15	0.23	12.31	12.25	468
50	0.08	0.14	0.22	7.54	7.5	764
60	0.08	0.12	0.20	4.76	4.75	1206
65	0.08	0.11	0.19	3.85	3.75	1528
70	0.08	0.10	0.18	3.15	3.0	1910
20	0.10	0.17	0.27	57.82	58.0	99
30	0.10	0.16	0.26	24.75	24.75	231
40	0.10	0.15	0.25	13.38	13.25	432
50	0.10	0.14	0.24	8.22	8.25	694
60	0.10	0.12	0.22	5.23	5.25	1091
65	0.10	0.11	0.21	4.26	4.25	1348
70	0.10	0.10	0.20	3.50	3.5	1637

[a] In recognition of safety considerations, use of $e_{max} = 0.04$ should be limited to urban conditions.

Source: A Policy on Geometric Design of Highways and Streets. American Association of State Highway and Transportation Officials, Washington, D.C. (1990).

TABLE 7-5A Length Required for Superelevation Runoff, Two-Lane Pavements

Super-elevation Rate	L = Length of Runoff (ft) for Design Speed (mph) of:							
	20	30	40	50	55	60	65	70
12-ft lanes								
.02	50	100	125	150	160	175	190	200
.04	60	100	125	150	160	175	190	200
.06	95	110	125	150	160	175	190	200
.08	125	145	170	190	205	215	230	240
.10	160	180	210	240	255	270	290	300
.12	195	215	250	290	305	320	350	360
10-ft lanes								
.02	50	100	125	150	160	175	190	200
.04	50	100	125	150	160	175	190	200
.06	80	100	125	150	160	175	190	200
.08	105	120	140	160	170	180	190	200
.10	130	150	175	200	215	225	240	250
.12	160	180	210	240	255	270	290	300

Source: A Policy on Geometric Design of Highways and Streets, American Association of State Highway and Transportation Officials, Washington, D.C. (1990).

TABLE 7-6A Stopping Sight Distance on Wet Pavements

Design Speed (mph)	Assumed Speed for Condition (mph)	Brake Reaction		Coefficient of Friction f	Braking Distance on Level[a] (ft)	Stopping Sight Distance	
		Time (sec)	Distance (ft)			Computed[a] (ft)	Rounded for Design (ft)
20	20–20	2.5	73.3–73.3	0.40	33.3–33.3	106.7–106.7	125–125
25	24–25	2.5	88.0–91.7	0.38	50.5–54.8	138.5–146.5	150–150
30	28–30	2.5	102.7–110.0	0.35	74.7–85.7	177.3–195.7	200–200
35	32–35	2.5	117.3–128.3	0.34	100.4–120.1	217.7–248.4	225–250
40	36–40	2.5	132.0–146.7	0.32	135.0–166.7	267.0–313.3	275–325
45	40–45	2.5	146.7–165.0	0.31	172.0–217.7	318.7–382.7	325–400
50	44–50	2.5	161.3–183.3	0.30	215.1–277.8	376.4–461.1	400–475
55	48–55	2.5	176.0–201.7	0.30	256.0–336.1	432.0–537.8	450–550
60	52–60	2.5	190.7–220.0	0.29	310.8–413.8	501.5–633.8	525–650
65	55–65	2.5	201.7–238.3	0.29	347.7–485.6	549.4–724.0	550–725
70	58–70	2.5	212.7–256.7	0.28	400.5–583.3	613.1–840.0	625–850

[a] Different values for the same speed result from using unequal coefficients of friction.

Source: A Policy on Geometric Design of Highways and Streets. American Association of State Highway and Transportation Officials, Washington, D.C. (1990).

TABLE 7-7A Design Controls for Crest Vertical Curves Based on Stopping Sight Distance

Design Speed (mph)	Assumed Speed for Condition (mph)	Coefficient of Friction f	Stopping Sight Distance Rounded for Design (ft)	Rate of Vertical Curvature, K (length (ft) per percent of A)	
				Computed[a]	Rounded for Design
20	20–20	0.40	125–125	8.6–8.6	10–10
25	24–25	0.38	150–150	14.4–16.1	20–20
30	28–30	0.35	200–200	23.7–28.8	30–30
35	32–35	0.34	225–250	35.7–46.4	40–50
40	36–40	0.32	275–325	53.6–73.9	60–80
45	40–45	0.31	325–400	76.4–110.2	80–120
50	44–50	0.30	400–475	106.6–160.0	110–160
55	48–55	0.30	450–550	140.4–217.6	150–220
60	52–60	0.29	525–650	189.2–302.2	190–310
65	55–65	0.29	550–725	227.1–394.3	230–400
70	58–70	0.28	625–850	282.8–530.9	290–540

[a] Using computed values of stopping sight distance.

Source: A Policy on Geometric Design of Highways and Streets. American Association of State Highway and Transportation Officials, Washington, D.C. (1990).

TABLE 7-8A Design Controls for Sag Vertical Curves Based on Stopping Sight Distance

Design Speed (mph)	Assumed Speed for Condition (mph)	Coefficient of Friction f	Stopping Sight Distance Rounded for Design (ft)	Rate of Vertical Curvature, K (length (ft) per percent of A)	
				Computed[a]	Rounded for Design
20	20–20	0.40	125–125	14.7–14.7	20–20
25	24–25	0.38	150–150	21.7–23.5	30–30
30	28–30	0.35	200–200	30.8–35.3	40–40
35	32–35	0.34	225–250	40.8–48.6	50–50
40	36–40	0.32	275–325	53.4–65.6	60–70
45	40–45	0.31	325–400	67.0–84.2	70–90
50	44–50	0.30	400–475	82.5–105.6	90–110
55	48–55	0.30	450–550	97.6–126.7	100–130
60	52–60	0.29	525–650	116.7–153.4	120–160
65	55–65	0.29	550–725	129.9–178.6	130–180
70	58–70	0.28	625–850	147.7–211.3	150–220

[a] Using computed values of stopping sight distance.

Source: A Policy on Geometric Design of Highways and Streets. American Association of State Highway and Transportation Officials, Washington, D.C. (1990).

TABLE 7-9A Design Controls for Crest Vertical Curves Based on Passing Sight Distance

Design Speed (mph)	Minimum Passing Sight Distance, Rounded for Design (ft)	Rate of Vertical Curvature, K,[a] Rounded for Design (length (ft) per percent of A)
20	800	210
25	950	300
30	1,100	400
35	1,300	550
40	1,500	730
45	1,650	890
50	1,800	1,050
55	1,950	1,230
60	2,100	1,430
65	2,300	1,720
70	2,500	2,030

[a] Computed from rounded values of passing sight distance.

Source: *A Policy on Geometric Design of Highways and Streets;* American Association of State Highway and Transportation Officials, Washington, D.C. (1990).

TABLE 8-1A Clear Zone Distances in Feet from Edge of Driving Lane

Design Speed	Design ADT	Fill Slopes			Cut Slopes		
		6:1 or Flatter	5:1 to 4:1	3:1	3:1	4:1 to 5:1	6:1 or Flatter
40 mph or less	Under 750	7–10	7–10	**	7–10	7–10	7–10
	750–1500	10–12	12–14	**	10–12	10–12	10–12
	1500–6000	12–14	14–16	**	12–14	12–14	12–14
	Over 6000	14–16	16–18	**	14–16	14–16	14–16
45–50 mph	Under 750	10–12	12–14	**	8–10	8–10	10–12
	750–1500	12–14	16–20	**	10–12	12–14	14–16
	1500–6000	16–18	20–26	**	12–14	14–16	16–18
	Over 6000	18–20	24–28	**	14–16	18–20	20–22
55 mph	Under 750	12–14	14–18	**	8–10	10–12	10–12
	750–1500	16–18	20–24	**	10–12	14–16	16–18
	1500–6000	20–22	24–30	**	14–16	16–18	20–22
	Over 6000	22–24	26–32*	**	16–18	20–22	22–24
60 mph	Under 750	16–18	20–24	**	10–12	12–14	14–16
	750–1500	20–24	26–32*	**	12–14	16–18	20–22
	1500–6000	26–30	32–40*	**	14–18	18–22	24–26
	Over 6000	30–32*	36–44*	**	20–22	24–26	26–28
65–70 mph	Under 750	18–20	20–26	**	10–12	14–16	14–16
	750–1500	24–26	28–36*	**	12–16	18–20	20–22
	1500–6000	28–32*	34–42*	**	16–20	22–24	26–28
	Over 6000	30–34*	38–46*	**	22–24	26–30	28–30

* Where a site specific investigation indicates a high probability of continuing accidents, or such occurrences are indicated by accident history, the designer may provide clear zone distances greater than 30 feet as indicated. Clear zones may be limited to 30 feet for practicality and to provide a consistent roadway template if previous experience with similar projects or designs indicates satisfactory performance.

** Since recovery is less likely on the unshielded, traversable 3:1 slopes, fixed objects should not be present in the vicinity of the toe of these slopes. Recovery of high speed vehicles that encroach beyond the edge of shoulder may be expected to occur beyond the toe of slope. Determination of the width of the recovery area at the toe of slope should take into consideration right of way availability, environmental concerns, economic factors, safety needs, and accident histories. Also, the distance between the edge of the travel lane and the beginning of the 3:1 slope should influence the recovery area provided at the toe of slope. While the application may be limited by several factors, the fill slope parameters which may enter into determining a maximum desirable recovery area are illustrated in Figure 8.2.

Source: Roadside Design Guide, American Association of State Highway and Transportation Officials, Washington, D.C. (1989).

TABLE 9-1A Minimum Edge of Pavement Designs for Turns at Intersections—Simple Curves and Tapers

Angle of Turn (degrees)	Design Vehicle	Simple Curve Radius	Simple Curve Radius with Taper		
			Radius (ft)	Offset (ft)	Taper (ft:ft)
45	P	50	—	—	—
	SU	75	—	—	—
	WB-40	120	—	—	—
	WB-50	—	120	2.0	15:1
	WB-62	—	140	4.0	15:1
60	P	40	—	—	—
	SU	60	—	—	—
	WB-40	90	—	—	—
	WB-50	—	95	3.0	15:1
	WB-62	—	140	4.0	15:1
75	P	35	25	2.0	10:1
	SU	55	45	2.0	10:1
	WB-40	—	60	2.0	15:1
	WB-50	—	65	3.0	15:1
	WB-62	—	140	4.0	20:1
90	P	30	20	2.5	10:1
	SU	50	40	2.0	10:1
	WB-40	—	45	4.0	10:1
	WB-50	—	60	4.0	15:1
	WB-62	—	120	4.0	30:1
105	P	—	20	2.5	8:1
	SU	—	35	3.0	10:1
	WB-40	—	40	4.0	10:1
	WB-50	—	55	4.0	15:1
	WB-62	—	115	3.0	30:1
120	P	—	20	2.0	10:1
	SU	—	30	3.0	10:1
	WB-40	—	35	5.0	8:1
	WB-50	—	45	4.0	15:1
	WB-62	—	100	5.0	25:1
135	P	—	20	1.5	15:1
	SU	—	30	4.0	8:1
	WB-40	—	30	8.0	6:1
	WB-50	—	40	6.0	10:1
	WB-62	—	80	5.0	20:1

Source: A Policy on Geometric Design of Highways and Streets, American Association of State Highway and Transportation Officials, Washington, DC (1990).

TABLE 9-2A Minimum Edge of Pavement Designs for Turns at Intersections—Three Centered Compound Curves

Angle of Turn (degrees)	Design Vehicle	3-Centered Compound		3-Centered Compound	
		Curve Radii (ft)	Symmetric Offset (ft)	Curve Radii (ft)	Asymmetric Offset (ft)
45	P	—	—	—	—
	SU	—	—	—	—
	WB-40	—	—	—	—
	WB-50	200–100–200	3.0	—	—
	WB-62	460–240–460	2.0	120–140–500	3.0–8.5
60	P	—	—	—	—
	SU	—	—	—	—
	WB-40	—	—	—	—
	WB-50	200–75–200	5.5	200–75–275	2.0–6.0
	WB-62	400–100–400	15.0	110–100–220	10.0–12.0
75	P	100–75–200	2.0	—	—
	SU	120–45–120	2.0	—	—
	WB-40	120–45–120	5.0	120–45–200	2.0–6.5
	WB-50	150–50–150	6.0	150–50–225	2.0–10.0
	WB-62	440–75–440	15.0	140–100–540	5.0–12.0
90	P	100–20–100	2.5	—	—
	SU	120–40–120	2.0	—	—
	WB-40	120–40–120	5.0	120–40–200	2.0–6.0
	WB-50	180–60–180	6.0	120–40–200	2.0–10.0
	WB-62	400–70–400	10.0	160–70–360	6'–10'0"
105	P	100–20–100	2.5	—	—
	SU	100–35–100	3.0	—	—
	WB-40	100–35–100	5.0	100–55–200	2.0–8.0
	WB-50	180–45–180	8.0	150–40–210	2.0–10.0
	WB-62	520–50–520	15.0	360–75–600	4.0–10.5
120	P	100–20–100	2.0	—	—
	SU	100–30–100	3.0	—	—
	WB-40	120–30–120	6.0	100–30–180	2.0–9.0
	WB-50	180–40–180	8.5	150–35–220	2.0–12.0
	WB-62	520–70–520	10.0	80–55–520	24.0–17.0
135	P	100–20–100	1.5	—	—
	SU	100–30–100	4.0	—	—
	WB-40	120–30–120	6.5	100–25–180	3.0–13.0
	WB-50	160–35–160	9.0	130–30–185	3.0–14.0
	WB-62	600–60–600	12.0	100–60–640	14.0–7.0

Source: A Policy on Geometric Design of Highways and Streets, American Association of State Highway and Transportation Officials, Washington, DC (1990).

SELECTED EXAMPLES IN CONVENTIONAL U.S. UNITS

EXAMPLE 8-1B **Crash Cushion Design by Kinetic Energy Principle (Conventional Units)** A crash cushion device is to be placed at an elevated expressway gore to safely decelerate a 4500-lb vehicle traveling at a speed of 60 mph. Fifty-five gallon steel barrels with a 7-in.-diameter hole in the center of each end will serve as the basic element. Laboratory studies have indicated that a dynamic force of 9000 lb is required to crush one barrel from its original 2-ft diameter to approximately 0.5 ft. The dynamic energy consumption $e_d = 1.5 \times 9000 = 13,500$ ft-lb.

Determine the number and arrangement of barrels that will stop the car and the average deceleration level.

Solution The kinetic energy of the vehicle

$$\text{K.E.} = \frac{Wv^2}{2g} = \frac{4500(88)^2}{2(32.2)} = 541,000 \text{ ft-lb}$$

The minimum number of barrels needed

$$N_b = \frac{\text{K.E.}}{13,500 \text{ ft-lb/barrel}} = \frac{541,000}{13,500} = 40$$

Suppose the barrels are arranged as shown in Figure 8-13. The barrier stopping force is a stepped function corresponding to the number of barrels in a row, as the figure illustrates.

Vehicle penetration into the crash cushion is determined by equating the kinetic energy to the area under the force–penetration curve. The penetration of the single, double, and triple rows of barrels is $1.5 + 3 \times 1.5 + 6 \times 1.5 = 15$ ft. Let X_1 be the penetration of the barrels arranged in four rows and X_2 the total penetration. The kinetic energy

$$\text{K.E.} = 541 \text{ ft-kip} = (9 \times 1.5) + (18 \times 4.5) + (27 \times 9.0) + (36X_1)$$

$$X_1 = 5.6 \text{ ft}$$

$$X_2 = 5.6 + 15 = 20.6 \text{ ft}$$

The average deceleration level

$$G_a = \frac{v^2}{2gX_2} = \frac{(88)^2}{2g(20.6)} = 5.84g$$

EXAMPLE 8-2B **Crash Cushion Design by Conservation of Momentum** A 4500-lb impacts a sand-filled inertial barrier head on at a speed of 45 mph (66 ft/sec). The system, which is shown below, designed for a speed of 50 mph. Determine the speed after impact with each of the rows of barrels and the average rate of deceleration. The barrels are 0.91 m (3 ft) in diameter, and the length of the system is 6.4 m (21 ft). The numbers shown on the sketch show the masses of each barrel of sand in hundreds of pounds.

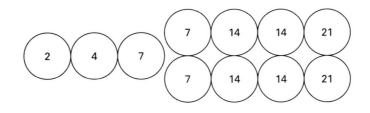

Solution By equation (8-4),

$$V_1 = (45)(4500)/(4500 + 200) = 43.1 \text{ mph}$$
$$V_2 = (43.1)(4500)/(4500 + 400) = 39.5 \text{ mph}$$
$$V_2 = (39.5)(4500)/(4500 + 700) = 34.2 \text{ mph}$$
$$V_4 = (34.2)(4500)/(4500 + 1400) = 26.1 \text{ mph}$$
$$V_5 = (26.1)(4500)/(4500 + 2800) = 16.1 \text{ mph}$$
$$V_6 = (16.1)(4500)/(4500 + 2800) = 9.9 \text{ mph}$$
$$V_7 = (9.9)(4500)/(4500 + 4200) = 5.1 \text{ mph}$$

Note that theoretically, the vehicle cannot be stopped completely by this concept; however, it is usually adequate to show that the computed vehicle velocity decreases to about 16 km/hr (10 mph). The remaining energy is dissipated as the vehicle "bulldozes" through the modules (*1*).

The average deceleration

$$G_a = \frac{V_0^2 - V_7^2}{2D} = \frac{(66)^2 - (7.5)^2}{2(21)} = 102 \text{ ft/sec}^2$$

$$G_a = 102/32.2 = 3.1g$$

INDEX